DNA Tumor Viruses

DNA Tumor Viruses

Edited by August Hudson

STATES
ACADEMIC PRESS
www.statesacademicpress.com

States Academic Press,
109 South 5th Street,
Brooklyn, NY 11249, USA

Visit us on the World Wide Web at:
www.statesacademicpress.com

ISBN: 978-1-63989-773-5

Cataloging-in-Publication Data

DNA tumor viruses / edited by August Hudson.
p. cm.
Includes bibliographical references and index.
ISBN 978-1-63989-773-5
1. Oncogenic DNA viruses. 2. DNA viruses. 3. Oncogenic viruses. I. Hudson, August.
QR372.O58 D53 2023
616.994 071--dc23

Table of Contents

Preface

The viruses which have DNA as their genetic material and replicate using a DNA-dependent DNA polymerase are known as DNA viruses. The DNA viruses which cause cancers among humans are known as DNA tumor viruses. These include the hepatitis B virus, Kaposi sarcoma herpes virus, human papillomavirus, Merkel cell polyomavirus and Epstein-Barr virus. DNA tumor viruses cause tumors in human beings and animals by transforming the infected host cells. These oncogenic viruses encode viral proteins that prevent cell death by deregulating the integrated framework of host cellular processes. These viruses typically impair two families of tumor suppressor proteins, namely, retinoblastoma proteins and tumor proteins p53. This book aims to shed light on some of the unexplored aspects of DNA tumor viruses. It will serve as a reference to a broad spectrum of readers. Experts and students actively engaged in the research on these viruses will find this book full of crucial and unexplored concepts.

This book has been the outcome of endless efforts put in by authors and researchers on various issues and topics within the field. The book is a comprehensive collection of significant researches that are addressed in a variety of chapters. It will surely enhance the knowledge of the field among readers across the globe.

It gives us an immense pleasure to thank our researchers and authors for their efforts to submit their piece of writing before the deadlines. Finally in the end, I would like to thank my family and colleagues who have been a great source of inspiration and support.

Editor

1

Regulation of Cellular Metabolism by High-Risk Human Papillomaviruses

Imelda Martínez-Ramírez [1,2], Adela Carrillo-García [2], Adriana Contreras-Paredes [2], Elizabeth Ortiz-Sánchez [2], Alfredo Cruz-Gregorio [1] and Marcela Lizano [2,3,*]

1. Programa de Maestría y Doctorado en Ciencias Bioquímicas, Facultad de Química, Universidad Nacional Autónoma de México, Ciudad Universitaria, Mexico City 04510, Mexico; immara02@yahoo.com.mx (I.M.-R.); cruzgalfredo@gmail.com (A.C.-G.)
2. Unidad de Investigación Biomédica en Cáncer, Instituto Nacional de Cancerología (INCan)/Instituto de Investigaciones Biomédicas, Universidad Nacional Autónoma de México (UNAM), Mexico City 14080, Mexico; adcarrillo2004@yahoo.com.mx (A.C.-G.); adrycont@yahoo.com.mx (A.C.-P.); elinfkb@yahoo.com.mx (E.O.-S.)
3. Departamento de Medicina Genómica y Toxicología Ambiental, Instituto de Investigaciones Biomédicas, Universidad Nacional Autónoma de México (UNAM), Mexico City 04510, Mexico

* Correspondence: lizanosoberon@gmail.com

Abstract: The alteration of glucose metabolism is one of the first biochemical characteristics associated with cancer cells since most of these cells increase glucose consumption and glycolytic rates even in the presence of oxygen, which has been called "aerobic glycolysis" or the Warburg effect. Human papillomavirus (HPV) is associated with approximately 5% of all human cancers worldwide, principally to cervical cancer. E6 and E7 are the main viral oncoproteins which are required to preserve the malignant phenotype. These viral proteins regulate the cell cycle through their interaction with tumor suppressor proteins p53 and pRB, respectively. Together with the viral proteins E5 and E2, E6 and E7 can favor the Warburg effect and contribute to radio- and chemoresistance through the increase in the activity of glycolytic enzymes, as well as the inhibition of the Krebs cycle and the respiratory chain. These processes lead to a fast production of ATP obtained by Warburg, which could help satisfy the high energy demands of cancer cells during proliferation. In this way HPV proteins could promote cancer hallmarks. However, it is also possible that during an early HPV infection, the Warburg effect could help in the achievement of an efficient viral replication.

Keywords: Warburg effect; human papillomavirus; metabolism; oxidative phosphorylation

---

1. Introduction

Cancer is a multifactorial disease involving several hallmarks. Reprogramming of energy metabolism by cancer cells causes the increase of energy intermediaries providing the cell a growth advantage associated with cell survival, migration, metastasis, and chemo/radiotherapy resistance [1–3]. Under physiological conditions the cells process glucose through glycolysis that generates pyruvate, which continues its metabolism in the mitochondria. On the other hand, in anaerobic conditions, glycolysis is favored with the production of lactate and consequently low concentrations of pyruvate are processed in the mitochondria [4]. Otto Warburg was the first scientist to observe that cancer cells can reprogram their glucose metabolism since, even in the presence of oxygen, the energy metabolism is favored essentially to glycolysis, a state that is called "aerobic glycolysis", or the Warburg effect [5]. "Aerobic glycolysis" commonly increases in cancer cells, which allows them to obtain energy and metabolic intermediaries to supply the fuel needed for tumor growth [6].

Oncogenic viruses play an important role in carcinogenesis and contribute to approximately 15–20% of incident cancer cases per year [7,8]. A common feature among them is their ability to cause persistent infections that disrupt key cellular pathways. Approximately 5% of all human cancers worldwide are caused by Human papillomavirus (HPV) [9]. HPV has mainly been associated with cervical cancer, which is the fourth cause of cancer death in women worldwide [10]. HPV uses the host cell replication machinery along the viral replicative cycle to produce infectious particles; nevertheless, in some cases, the interactions of viral proteins with cell proteins induce a switch to malignant transformation, where the cells suffer a metabolic change to face their high-energy demand due to the increase in cell proliferation. This review is focused on the role of HPV proteins in the modulation of the cellular metabolism and its participation in carcinogenesis.

2. Cellular Energy

Normal cells obtain their energy as adenosine triphosphate (ATP) through glucose, protein, carbohydrate, and fat metabolism, obtained from food. This process is regulated by hormones as glucagon, insulin, glucocorticoids, and growth hormone [11]. As shown in Figure 1 the metabolic intermediaries such as pyruvate and acetyl-coenzyme A (acetyl-CoA) are generated from the metabolism of amino acids, glucose, and metabolic intermediaries, such as fatty acids and glycerol. Subsequently, the production of reduced nicotinamide adenine dinucleotide (NADH), reduced flavin adenine dinucleotide ($FADH_2$), as well as the high production of ATP are a result of the metabolism of acetyl-CoA in the Krebs cycle (KC), the respiratory chain, and oxidative phosphorylation (OXPHOS) [3] (Figure 1). To achieve their energy requirements, cancer cells undergo metabolic reprogramming since their metabolic needs differ from those of normal cells [12].

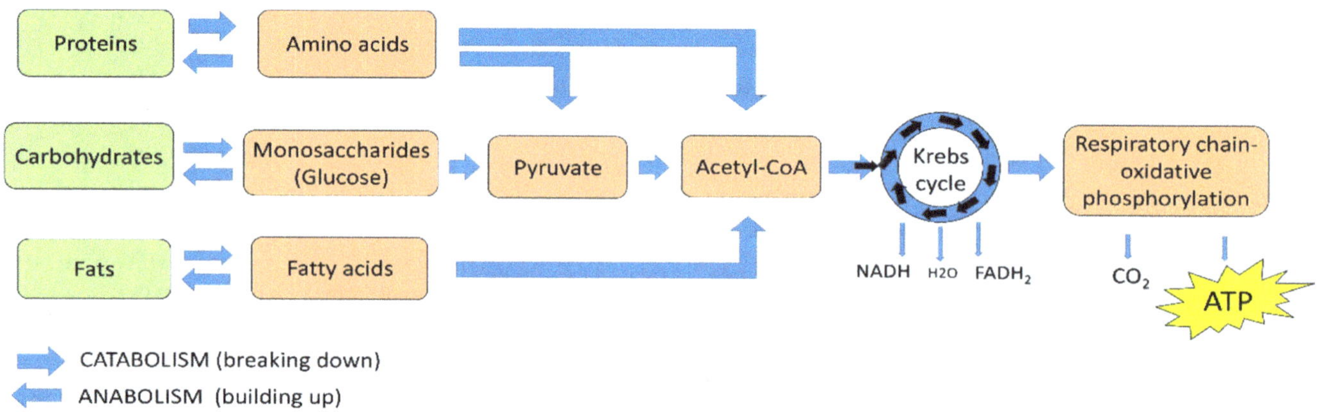

Figure 1. ATP production from food metabolism. Amino acids, monosaccharides, and fatty acids are produced from the metabolism of proteins, carbohydrates, and fats, respectively, from which pyruvate and/or acetyl-CoA are obtained, which, in turn, are generally metabolized in the Krebs cycle and the oxidative phosphorylation system.

3. The Warburg Effect

Altered glucose metabolism is one of the first biochemical features identified in cancer. Glucose metabolism provides ATP and metabolites used in a variety of anabolic pathways [13]. Most types of cancer cells increase glucose consumption and glycolytic rates. Instead of continuing its metabolism through the Krebs cycle, most of the pyruvate produced in cancer cells is converted to lactate by lactate dehydrogenase (LDH); subsequently, the lactate is secreted by the muscle into the blood to be transformed back into glucose in the liver by the Cori cycle; otherwise, lactate can immediately replace glucose as a fuel for almost all cells in the body that contain mitochondria [14]. This metabolic modification occurs even when there are enough oxygen molecules for mitochondrial respiration and the integrity of the mitochondria is optimal [15]. It is not clear why cancer cells process pyruvate into lactate instead of entering it into the Krebs cycle. Apparently, cancer cells require glycolytic intermediaries, generated by the Warburg effect, to meet anabolic needs and to promote their growth and long-term maintenance [16]. Nevertheless, cancer cells can also obtain energy from a smaller fraction of glucose that is metabolized through the Krebs cycle. Moreover, cancer cells use anaplerotic pathways to obtain energy, such as the glutaminolysis, which also feeds the Krebs cycle [17] (Figure 2).

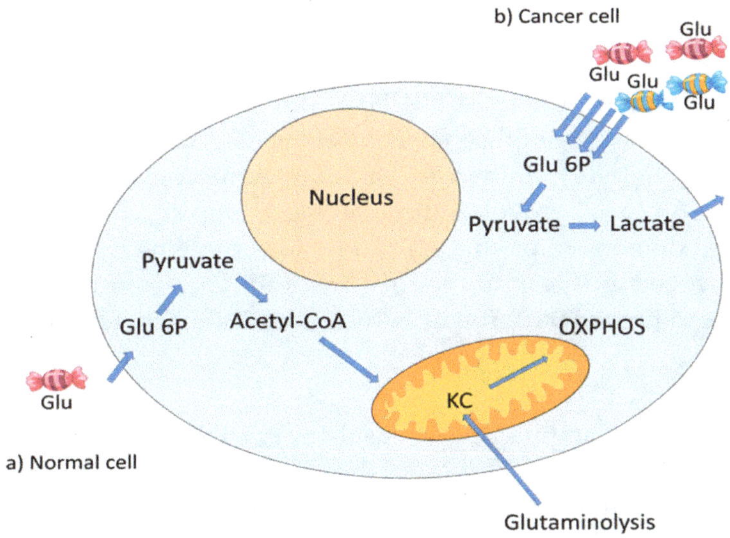

Figure 2. Metabolism of glucose in normal and cancer cells. The glucose (Glu) that enters the cell is phosphorylated to glucose 6 phosphate (Glu 6P) and, subsequently, is metabolized to pyruvate (**a**,**b**). In normal cells, pyruvate is metabolized to acetyl-coenzyme A (Acetyl-CoA) and continues its metabolism in the Krebs cycle (KC) and the oxidative phosphorylation (OXPHOS) system (**a**). In cancer cells, glucose entry is increased, and pyruvate is metabolized to lactate, which is then expelled from the cells (**b**). The Krebs cycle can be fed by intermediaries from glutaminolysis.

4. Human Papillomavirus

A persistent infection with HPV is the main risk factor for the development of cervical cancer [18]. More than 200 HPV types have been identified [19] and 40 of them are associated to anogenital cancer. HPV infections are sexually transmitted and according to their oncogenic potential and prevalence in cancer are classified in low-risk HPVs (LR-HPV) and high-risk HPVs (HR-HPV) [20]. HPV types 16 and 18 are the most frequently found in cervical cancer, in 60% and 15% of the cases, respectively [21]. Moreover, HR-HPV are related to other cancers as vulvar, vaginal, anal, penile [22], and oropharyngeal squamous cell carcinoma [23].

4.1. HPV Genome

HPVs are small non-enveloped viruses with a double-stranded circular DNA with a length of about 8 Kb. A fragment of the HPV genome of approximately 900 bases, named the long control region (LCR), regulates viral replication and transcription. The HPV genome encodes the E1, E2, E4, E5, E6, and E7 proteins, which are early expressed during the viral infection. L1 and L2 proteins are expressed later during the productive infection and constitute the viral capsid [24]. HPV replication depends upon the host cell differentiation program [25]. *E1* and *E2* are expressed at the beginning of the infection, bind to the LCR and participate in viral replication and transcription. E4 is expressed mainly during later stages of infection, contributes to the amplification of the viral genome and disrupts the cellular keratin network, facilitating virus release [26]. E5 inhibits maturation of endocytic vesicles from early to late endosomes [27]. E6 and E7 are the main viral oncoproteins, that as multifunctional proteins, regulate the cell cycle through their interaction with a set of cellular proteins, being within the most relevant the tumor suppressor proteins p53 and pRB, respectively [28,29]. E6 and E7 oncoproteins target numerous cellular proteins and thus promote cell cycle progression, apoptosis inhibition, DNA damage and evasion of the host immune response [30].

For example, the PDZ-binding motif (PBM) of E6 oncoproteins interacts with PDZ domain-containing proteins, such as disk large homolog 1 (Dlg1), Scribble, and membrane-associated guanylate kinase (MAGI)-1, -2, -3, which, in many cases, lead to their proteasome-mediated degradation [30]. High-risk E6 oncoproteins interact with at least 14 identified cellular PDZ-containing proteins, whose degradation gives rise to changes in cell morphology, reorganization of microfilament network and loss of tight junctions, associated with the immortalization of keratinocytes [30,31].

E7 is the main transforming protein of HR-HPV due to its interactions with at least 20 protein partners including pRb, c-Jun, and multiple epigenetic elements such as histone deacetylase 1 (HDAC1), DNA methyltransferase (DNMT1), Enhancer of zeste homolog 2 (EZH2), and CBP/p300 [32]. Dimerization of E7 with CBP/p300 provokes the acetylation of pRb, causing cell cycle disruption [33].

Moreover, it has been demonstrated that E6 and E7 oncoproteins from HPV16 impair the function of diverse enzymes involved in the homologous-recombination pathway reducing its ability to complete the double-strand break (DSB) repair, which contributes to genomic instability [34].

4.2. HPV Replicative Cycle

The HPV replicative cycle begins in the basal layer of the squamous epithelia, where viral particles arrive through micro wounds and there they complete their life cycle, taking advantage of the differentiation program of the host epithelial cells [25]. After the viral entry, the HPV genome is replicated as an episome. At the initial infection in undifferentiated basal cells, low levels of the E1, E2, E6, and E7 proteins are expressed, delaying normal keratinocyte differentiation. E1 is the only viral protein with enzymatic activity: it is a helicase/ATPase which is recruited by E2 at the origin of viral replication, generating approximately 50 to 100 copies of viral episomes per cell [35]. These numbers increase to thousands of DNA viral copies throughout the differentiation of the epithelium [36]. From the middle to the upper differentiating epithelial layers, E6 and E7 are expressed in high amounts [37] and the viral life cycle is completed by the expression of L1 and L2 proteins in the uppermost layer of the epithelium. Finally, the viral genome is encapsidated and the mature virions are released [35]. A complete viral life cycle does not represent a risk for cancer development. Moreover, near 90% of HPV infections, even with the high-risk types, are eliminated within two years [38]. A persistent infection with HR-HPV is the main determinant for cervical cancer development; where the accumulation of damaged DNA, probably due to the interactions of E7 and E6 with their cellular targets, leads to genomic instability and, in most cases, to the integration of the viral genome into the host genome [39]. Viral genome integration precedes the overregulation of *E6* and *E7* oncogenes, since in most of the cases *E2* gene is disrupted, which causes the overexpression of *E6* and *E7* oncogenes [40]. It has been reported that the ratio of HPV integrated genome copies in relation to DNA episomal particles increases as the cervical lesion progresses to cancer [41]. Nevertheless, there are other possible mechanisms of E6/E7 deregulation [40]. For example, in some cases, HPV DNA copies are found integrated in tandem and, even when the *E2* gene is not broken, the E2 binding sites in the LCR are found methylated, which affects the binding of E2, avoiding its repressive effect [42]. Tandem integration of the HPV genome can also cause the formation of super-enhancer like elements which strongly activate *E6/E7* expression [43].

5. HPV Oncoproteins and Cellular Metabolism

The continuous expression of E6 and E7 are required to maintain the malignant phenotype [44–46]. Accumulating evidence suggests that HPV is involved in the reprogrammed metabolism of cervical cancer cells. Diverse interactions of E6 and E7 with cellular partners have been demonstrated, and many of those affect cell biological functions, leading to transformation. E6 interacts with the E6 associated protein (E6AP), an E3 ubiquitin ligase, targeting cellular proteins to be ubiquitinated for proteasomal degradation [47]. In this way HR-HPV E6 induces p53 degradation by forming a complex with E6AP [48] and, thus, p53 dependent apoptosis is blocked [49]. It has been reported that p53 regulates genes involved in metabolic processes, such as glycolysis and OXPHOS pathways, which are inhibited and activated by p53, respectively [50]. For instance, p53 induces the expression of the TP53-induced glycolysis and apoptosis regulator (TIGAR), a transcriptional factor that inhibits glycolysis. p53 also upregulates the cytochrome c oxidase assembly protein (SCO2), necessary for the assembly of the complex IV of the electron transport chain [51]. Therefore, HR-HPV E6 modulates metabolic pathways due, in part, to its interaction with p53 (Figure 3). E6 HPV16 interacts with c-Myc which is a transcription factor that promotes the expression of genes involved in the control of glycolysis, such as enolase A, hexokinase II, lactate dehydrogenase A, phosphofructokinase and glucose transporter [52,53]. Moreover, E6 also activates the mammalian target of the rapamycin complex 1 (mTORC1) pathway mediating the increase in protein kinase B (PKB, also known as Akt) activity through the phosphoinositide-dependent kinase-1 (PDK1) and mTORC2 pathways. mTORC1 signaling cascade serves as a metabolic sensor that responds to nutrient and growth factor availability and leads to accumulation of hypoxia-inducible factor 1 (HIF1), a transcription factor involved in the cellular adaptation to hypoxia and other hypoxia response elements regulated proteins, such as the glucose transporter 1 (GLUT1) [54,55]. Therefore, HPV16 E6 can also regulate glycolysis by its association with c-Myc and PI3K/AKT pathways (Figure 3).

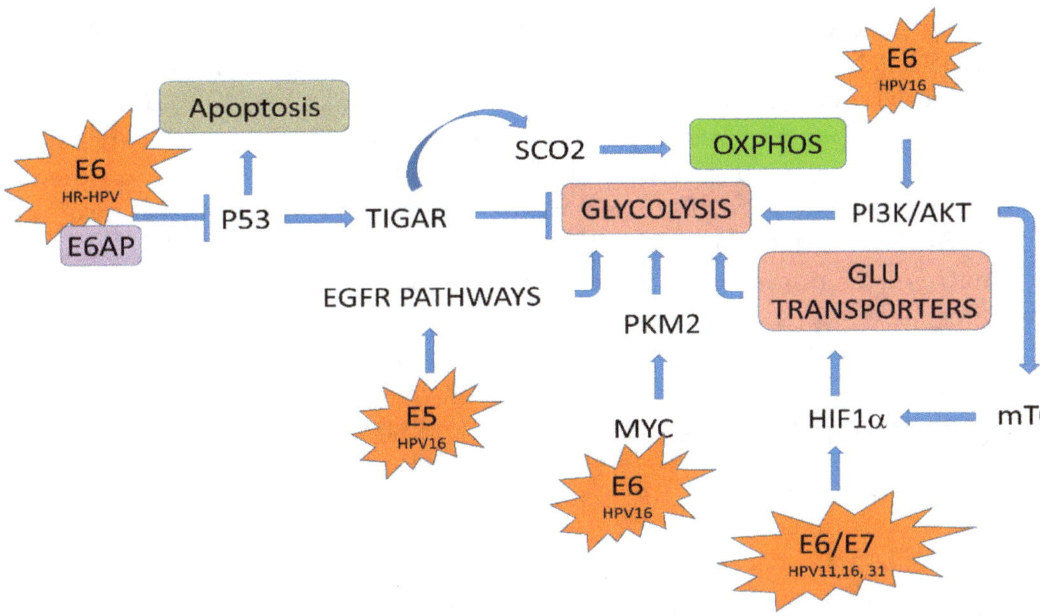

Figure 3. Modulation of the Warburg effect by HPV oncoproteins. E6, E7, and E5 viral proteins activate the glycolytic pathway (arrows). E6 with E6 associated protein (E6AP) inhibits repressors of the glycolytic pathway such as p53, which activates the TP53 induced glycolysis and apoptosis regulator (TIGAR), a glycolytic repressor protein (truncated arrow). TIGAR activates cytochrome c oxidase assembly protein (SCO2), activator of oxidative phosphorylation (OXPHOS). Hence, the absence of p53 promotes glycolysis. E5 activates epidermal grow factor receptor (EGFR) pathways; E6 activates Pyruvate kinase isoform 2 (PKM2), PI3K/AKT and rapamycin complex 1 (mTOR); E6 and E7 activate hypoxia-inducible factor (HIF1. Glu: glucose). The HPV type is indicated.

The E7 oncoprotein can also promote the glycolytic pathway by modulating HIF1. Rodolico et al. (2011) found a strong association between the expression of HPV16 E7 and HIF1 in oral squamous cell carcinoma in patients infected with HPV16 [56]. HIF1 has two subunits of which HIF1α is regulated by oxygen concentrations and dimerizes with HIF1β to form the transcription factor HIF1. HIF1 expression also enhances the transcription of genes involved in glucose metabolism [57]. The activation of HIF1α contributes to the Warburg effect through the upregulation of glycolysis and downregulation of oxidative phosphorylation [58] (Figure 3). On the other hand, the E5 oncoprotein from HPV16 could indirectly regulate the Warburg effect through the axis of the epidermal growth factor and its receptor (EGF–EGFR). E5 stimulates EGFR signaling pathways, promoting a prolonged activation of the extracellular signal-regulated kinases-1,2 (ERK1/2) and AKT in response to EGF [59]. It has been shown that EGFR pathway promotes an enhancement of the glycolytic metabolic program in oral cancer cells [60].

5.1. HPV E6/E7 Oncoproteins and Glucose Transporters

Most of the nutrients which are necessary for cells to synthesize macromolecules and carbon sources are hydrophilic and do not diffuse freely through the cell double layer membrane, requiring help for their entry into the cell. For instance, glucose enters the cell through two families of glucose transporters, the GLUT transporters (GLUTs) and SGLT Na+/glucose cotransporters (SGLT1 and SGLT2 in epithelial cells) [61–63].

GLUTs are membrane proteins that transport monosaccharides, polyols, and other small carbon compounds [64]. GLUT transporter family internalizes glucose by a mechanism of facilitated diffusion. In cancer cells, there is an upregulation of specific glucose transporters in the plasma membrane. Under hypoxia or nutrient deprivation, tumor cells overexpress at least one of the GLUT transporters, predominantly GLUT1. A high expression of this glucose transporter has been described in many types of cancer including cervical cancer [65,66]. It has been reported that GLUT1 is a transcriptional target of HIF1, which activates a panel of target genes for example GLUT-1, GLUT-3, carbonic anhydrase IX (CAIX), and vascular endothelial growth factor (VEGF), promoting the transcription of glycolytic enzymes [58,66,67]. Cancer cells have a high demand of nutrients due to their high rate of proliferation, so the entry of glucose by facilitated diffusion through GLUT transporters could be insufficient, being required other mechanisms of glucose uptake may be required. For example, SGLT Na+/glucose cotransporters internalize the glucose in the cell through a secondary active transport coupled to Na+ entry, mediated by the Na+/K+ ATPase. The energy produced by the consumption of glucose is higher than that required by the Na+/K+ ATPase [68]. SGLT1 which is present in HPV-positive cervical carcinoma cells helps cells to accumulate glucose even when there is a low concentration of extracellular glucose [63]. In addition, in several tumors EGFR is highly expressed and this receptor is physically associated with SGLT1, inducing its stabilization and promoting glucose entry into cancer cells [64–66,69].

The expression of GLUT1 is under the control of multiple transcriptional factors, one of which is p53 that directly represses the transcription of GLUT1 and GLUT4 and indirectly GLUT3, limiting glucose uptake in cancer cells, which slows down their growth. Therefore, the degradation of p53 mediated by HPV E6 oncoprotein could lead to overexpression of GLUT1 and to elevation of glucose uptake in cervical cancer cells [70].

It has been reported that HPV16 E6 contributes to the Warburg effect preventing the interaction and degradation of HIF1 by Von Hippel-Lindau (VHL) tumor suppressor [71]. Moreover, it has been proposed that HPV E6 and E7 increase GLUT1 expression through the upregulation of HIF1α in lung cancer [72]. Interestingly, in Xenopus oocytes expressing SGLT1 it was found that the co-expression of HPV18 E6 significantly increased the SGLT1 protein abundance in the cell membrane [68].

5.2. HPV in the Glycolytic Pathway

Glycolysis also named as the Embden-Meyerhof pathway, is the first step of the glucose metabolism. It covers ten reactions that occur in the cytosol, where two molecules of pyruvate are obtained from the metabolism of one molecule of glucose [4] (Figure 4).

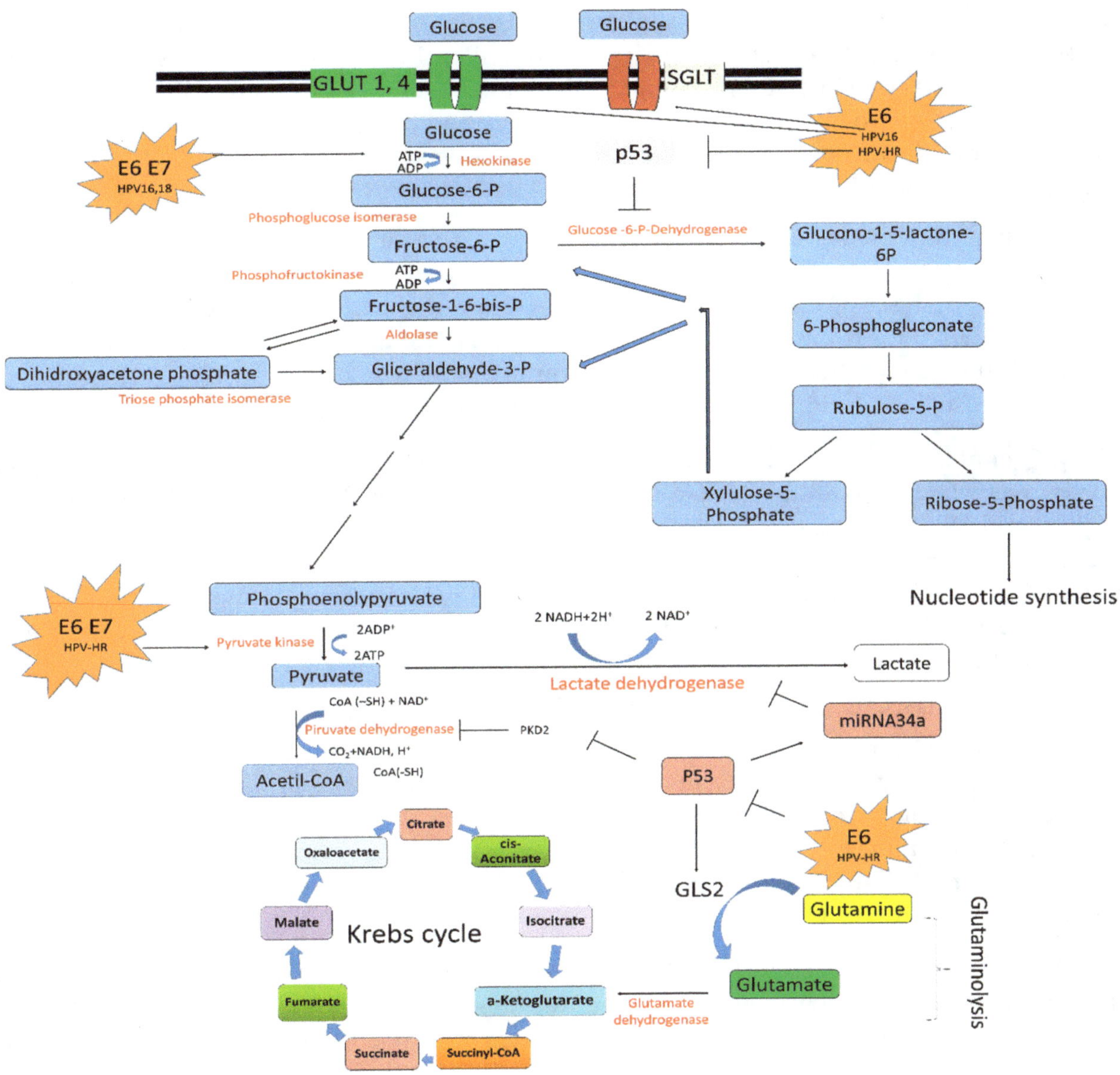

Figure 4. The modulation of glucose metabolism by HPV oncoproteins favor glycolysis instead of OXPHOS. HR-HPV E6 oncoproteins promote glucose uptake through the overexpression of glucose transporters GLUT and SGLT. Both E6 and E7 oncoproteins induce the expression of hexokinase, enzyme associated to the first step of glucose metabolism. Moreover, E6 can promote nucleotide synthesis through the degradation of p53, which is an antagonist of glucose 6-P-dehidrogenase, a key enzyme in the pentose phosphate pathway (PPP). The glycolytic pathway is carried out in two phases. First, in the preparatory or glucose activation phase, a six-carbon glucose molecule breaks down into two molecules with three carbons each: one glyceraldehyde-3-phosphate and the other is dihydroxyacetone phosphate, which is transformed into glyceraldehyde-3-phosphate. In these reactions, two molecules of ATP are consumed. Second, the energy extraction phase, involves the conversion of the two glyceraldehyde-3-phosphate molecules into two pyruvate molecules by pyruvate kinase, which is activated by E6 and E7 oncoproteins. This process results in the production of four ATP molecules through substrate-level phosphorylation. Pyruvate is metabolized to acetyl-CoA by pyruvate dehydrogenase, which, in turn, is negatively regulated by pyruvate dehydrogenase kinase 2 (PDK2). PDK2 is activated in the absence of p53, a process induced by E6, avoiding acetyl-CoA production. Moreover, p53 degradation avoids glutaminase 2 (GLS2) activation and glutaminolysis, which in turn decreases α-ketoglutarate levels. Thus, E6 does not permit an optimal function of the Krebs cycle, inducing lactate production by lactate dehydrogenase. E6 prevents the expression of miRNA34a and lactate dehydrogenase deactivation. HPV type is indicated. T arrows indicate inhibition. Black line arrows and thick blue arrows indicate the direction of the reaction.

Three key enzymes participate in glycolysis mediating irreversible reactions: hexokinase (HK), phosphofructokinase-1 (PFK-1) and pyruvate kinase (PK). In aerobic conditions the pyruvate is oxidized in the Krebs cycle into acetyl-coA and then into CO_2 and H_2O. Meanwhile, under anaerobic conditions pyruvate can follow the fermentation pathway and produce lactate [4,73]. Finally, the total energy in form of ATP produced in aerobic conditions consists of 36 ATP molecules, in contrast to anaerobic conditions which accumulates only two ATP molecules (Figure 4).

The enzyme hexokinase 2 (HK2) is an important regulatory protein of the "Warburg effect". In most normal tissues, it remains at very low levels but it is found frequently elevated in cancer [74,75]. The increased glycolysis in cancer cells promoted by the rise of HK2, provides energy and also precursors important for tumor growth [6]. In addition, HK2 supplies intermediaries for the Krebs cycle through the use of glutamine derived carbon in anaplerosis pathway [76]. When the cell is deficient in glucose, HK2 facilities autophagy by maintaining cell energy in homeostasis [77].

Zeng Q et al., in 2017 [78] reported that in mouse embryo fibroblasts, HK2 and consequently glycolysis, were increased in a c-Myc-dependent manner when HPV16 E6/E7 were ectopically overexpressed [79]. This suggests that E6/E7 oncoproteins directly activate the expression of HK2 and reprogram the HPV transformed cells to glycolysis as a new strategy to maintain the transformed phenotype [78,79]. Furthermore, it was demonstrated that E6/E7 silencing in the HeLa cell line caused a strong decrease of HK2 at mRNA and protein levels, mediated in part by the decrease in c-Myc expression and the increase in miR-143-3p [80].

Other evidence related to HPV in reprogrammed metabolism of cervical cancer cells is the interaction with pyruvate kinase (PK) protein (Figure 4). The PK protein has two isoforms, an embryological isoform (M2) with elevated enzymatic activity, and an adult isoform (M1) [81]. In NIH 3T3 cells, HPV-16 E7 oncoprotein directly binds to M2-PK inducing its dimerization, which promotes nucleic acid synthesis and cell proliferation [82]. In addition, malignant transformation of cells by HR-HPV E6 and E7 oncogenes induced a PK isoform switch from M1 to M2. This switch causes a shift from normal cellular metabolism to elevated glycolysis, beneficial for tumor growth [83].

The lactate dehydrogenase A (LDHA) enzyme is responsible for the processing of pyruvate to lactate in the absence or in low levels of oxygen. Recent studies revealed that microRNAs (miRNA) are involved in regulation of the Warburg effect by targeting specific enzymes [84,85]. miR-34a is a tumor suppressor upregulated by p53 [86]. Zhang et al. (2016) showed that miR-34a inhibited the lactate production by directly targeting LDHA. In cervical cancer, reduced levels of miR-34a has been reported as a result of p53 degradation by E6 [87]; therefore, metabolic reprogramming mediated by the absence of miR-34a could play an important role in cervical carcinogenesis (Figure 4).

5.3. HPV in the Krebs Cycle

p53 induces the expression of proteins, such as the cytochrome c oxidase 2 (SCO2) [51] (Figure 3), the apoptosis-inducing factor (AIF) [88] and ferredoxin reductase (FDXR) [89], which are associated with the maintenance of the mitochondrial integrity and oxidative phosphorylation. This transcription factor also induces OXPHOS promoting the inhibition of pyruvate dehydrogenase kinase 2 (PDK2), which is an enzyme that deactivates the pyruvate dehydrogenase complex (PDC) [90] (Figure 4). Thus, PDK2 deactivation induces the production of acetyl-CoA by PDC, an essential molecule in KC. Furthermore, p53 has a very important role in the glutamine metabolism pathway, which is an alternative via that feeds KC, since p53 induces the expression of glutaminase 2 (GLS2), an enzyme that converts glutamine to glutamate, which, in turn, is changed into α-ketoglutarate supplying the KC [91] (Figure 4). Thus, p53 has a crucial role in rising the KC and OXPHOS. However, in the presence of HR-HPV E6, p53 is degraded via proteasome and the conditions for the Warburg effect are then favorable (Figure 4).

5.4. HPV E2 Protein and the Oxidative Phosphorylation System

The OXPHOS system occurs in the mitochondrial inner membrane and consists of five enzyme complexes. The first four complexes: Complex I or NADH dehydrogenase, Complex II, or succinate dehydrogenase, Complex III, or ubiquinol-cytochrome *c* oxidoreductase, and Complex IV, or cytochrome, along with two electron carriers: ubiquinol and cytochrome C, form the electron transport chain which generates a proton gradient used by Complex V, or ATP synthase, to generate the majority of cellular ATP [92] (Figure 5).

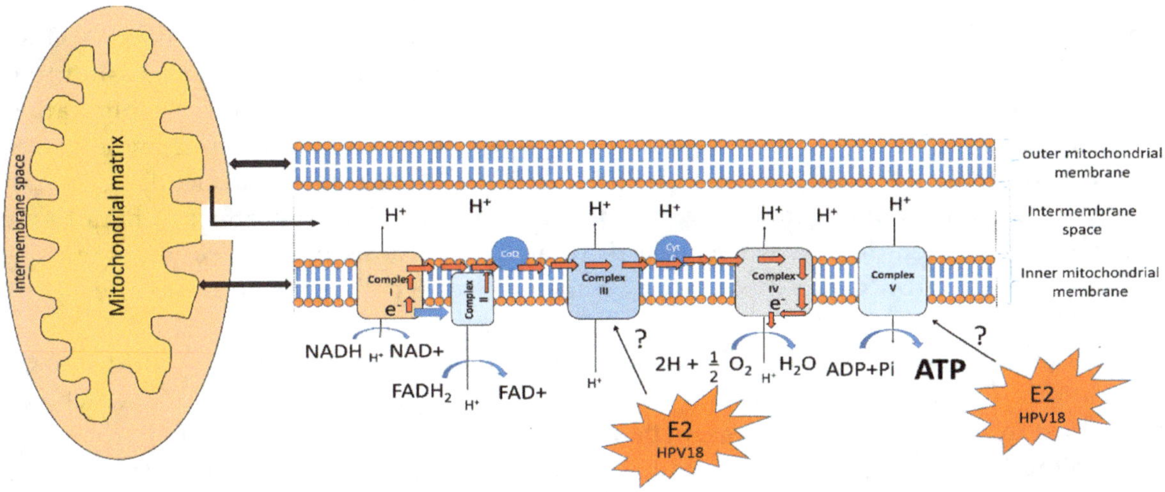

Figure 5. Modulation of the oxidative phosphorylation system by HPV18 E2 protein. Electrons are transferred from NADH or $FADH_2$ to O_2 using a series of electron carriers: Complex I, II, III, IV, and V favor the translocation of protons and generate an electrochemical gradient of protons in the intermembrane space. Red arrows show the electron flux. Electrons (e^-); protons (H^+); oxygen (O_2); water (H_2O); coenzyme Q (CoQ), and cytochrome C (Cyt C). Since E2 modifies the mitochondrial cristae morphology, releasing ROS, it could possibly modulate Complex III, which is a mediator of mitochondrial ROS production; and it could also modulate Complex V, which is a regulator of the mitochondrial cristae structure. The question mark (?) indicates the possible effect of E2 on the complexes III and V.

In aerobic conditions, 36 ATP molecules are produced from the complete oxidation of glucose into CO_2 and H_2O. It has been shown that mitochondrial respiration is also a major source of reactive oxygen species (ROS) that damage the cell and are markers of OXPHOS dysfunction [93] (Figure 5).

The E2 proteins from HR-HPV are negative regulators of E6 and E7 oncoproteins. E2 proteins actively shuttle their location between the cytoplasm and nucleus. In the cytoplasm, E2 can promote apoptosis [94] and, in the nucleus, it induces chromosomal instability and DNA breaks during mitosis which possibly facilitates the integration of the HPV genome into the cellular genome [95]. Lai et al., 2013 determined that E2 can also be found in the mitochondrial membrane where it modifies the cristae morphology increasing mitochondrial release of ROS and, therefore, modifying cellular respiration, which correlates with HIF1 stabilization and increased glycolysis [93]. As proteins from Complex III are central mediators of mitochondrial ROS production [96] and ATP synthase is a regulator of the cristae structure [97], E2 could modulate the mitochondrial structure and ROS release through the interaction with those proteins (Figure 5). During cellular homeostasis ROS are used as second messengers that influence cell proliferation and differentiation; however, their increase due to mitochondria alteration, or to processes such as the Warburg effect, could produce oxidative stress (OS) and, consequently, damage to DNA, lipids, or proteins. Cells have a battery of antioxidants, such as glutathione (GSH), superoxide dismutase (SOD) 1, 2 and 3, catalase (CAT), peroxidases (Prxs), and thioredoxins (Trxs), among others, which respond quickly and efficiently against OS [98]. It has been shown that the co-expression of E2 and E1 from HPV18 decreases GSH levels and SOD1 and 2 activity inducing an increase in ROS levels and DNA damage [99].

6. The Warburg Effect in the HPV Replicative Cycle

It has been proposed that the Warburg effect represents a metabolic strategy that allows cancer cells to optimally meet energy demands posed by stochastic or fluctuating tumor environments [100], so that oxidative phosphorylation and aerobic glycolysis work in a complementary way to satisfy ATP demands as required. Even though OXPHOS yields a maximal number of ATP molecules, its production is too slow for satisfying peaks of fluctuating ATP demands; in contrast, although aerobic glycolysis is less efficient, it can produce ATP with a very fast flux, more than 100-fold faster than oxidative phosphorylation [101]. Therefore, when tumor cells require fast accessibility to ATP, in order to temporarily face increased short-term energy demands, as in the case of proliferation [102], migration [103], or invasion [104], the energy production is optimized by glycolysis, which, in general, can represent a normal physiological function [105], which is increased in cancer cells.

In such a way, HPV-related cancer would benefit from the Warburg effect to maintain cancer hallmarks and it has been demonstrated that different interactions of HPV proteins with cellular proteins promote the Warburg effect. Now, the question arises whether the Warburg effect could also favor the HPV replicative cycle during the infectious process.

In HPV infected cells, a peak of ATP could be demanded for the synthesis of viral DNA during HPV replication. This process needs high amounts of energy in the form of ATP, which could be rapidly obtained from increased glycolysis, to be used by the E1-helicase/ATPase viral protein for DNA uncoiling and for the stimulation of DNA synthesis through the recruitment of cellular replicative DNA polymerases, which is dependent on the ability of the E1 helicase to hydrolyze ATP [106]. Replication of HPV also requires high amounts of nucleotides, which could be obtained from the processing of a glycolytic intermediary, the glucose-6-phospate (G6P), which is a substrate of the glucose-6-phosphate dehydrogenase (G6PD) in the pentose phosphate pathway (PPP) [13]. G6PD is a rate-limiting enzyme for PPP, which produce ribose and nicotinamide adenine dinucleotide phosphate (NADPH) (Figure 3). Ribose is an essential molecule in the synthesis of nucleotides, while NADPH is a cellular antioxidant and coenzyme of multiple cellular reactions. In such a way, HPV genome replication could be favored by the Warburg effect, which would provide the urgent needs of ATP and nucleotides, required in different steps of the viral life cycle.

7. Metabolism as a Therapeutic Target in HPV-Related Cancers

Oncologic patients undergo tumoral chemo- and radioresistance, due in part to the hypoxic tumor microenvironment. It has been shown that aerobic glycolysis favors radioresistance [107,108]. Liu and collaborators observed that HK2 enzyme is increased in HPV-infected cervical cancer cells, compared to those negative to HPV, and that the inhibition of HK2 activity in HPV-positive cells makes these cells radiosentive [79]. Hypoxia status has been related to radioresistance in tumors and cell lines. It is known that the human epidermal growth factor receptor 2 (HER-2), a receptor tyrosine kinase (RTK), is over-expressed in HPV-related cancers; although the role of this protein remains unclear for prognosis and therapy in cervical cancer. Clinical data have shown tumor resistance to treatment to receptor tyrosine kinase inhibitors (RTK inhibitors), related to HIF1α overexpression. It has been observed that when blocking metabolic mechanisms, including hypoxia and glycolysis as adjuvant treatments, tumoral cells become sensitive to conventional drugs and radiotherapy schemes, which represents a potential strategy for cancer therapy. There is some evidence that shows that cytotoxic activity can be enhanced when combining HER inhibitors (RTKs inhibitors) with inhibitors of glucose metabolism, in relation to the use of each drug alone [109,110]. Using SiHa-derived tumors as a model for HER2 and HPV positive cervical cancer, Martinho et al. (2017) observed a poor response to HER inhibitors, associated with an increase in lactate production [111]. HER inhibitors promote a glycolytic metabolism that is related with poor response in HPV-positive cancer cells, a phenomenon that was not observed in HPV-negative tumors. Interestingly, the combination of lepatinib, receptor tyrosine kinase (RTK) inhibitor, and 2-deoxy-D-glucose (2-DG), a non-metabolizable glucose analog, decreased lactate and HIF1α expression, resulting in a decrease in cell viability, tumor growth inhibition, as well as in a decrease in angiogenesis, compared to the cytotoxic effects of both drugs alone. These results suggest the use of glycolysis inhibitors as adjuvant therapy directed to achieve an increase in the cytotoxicity of conventional chemotherapy in cancer cells [111].

Since HPV could contribute to radio- and chemoresistance by the deregulation of key metabolic pathways, it is necessary to develop new strategies, including metabolic molecular targets, to enhance the cytotoxic effects of standard chemo- or radiotherapies, which will contribute to improve clinical responses in HPV-related cancers.

8. Conclusions

The Warburg effect is a metabolic change observed in practically all types of cancer. In HPV-related cancers, the E6 and E7 viral oncoproteins are responsible for maintaining the malignant phenotype and promote a metabolic switch, together with other viral proteins, such as E5 and E2, even though E6 and E7 regulate two critical enzymes of the glycolytic pathway, hexokinase and pyruvate kinase, as well as other proteins involved in lactate production, such as lactate dehydrogenase. Furthermore, the E2 viral protein also stimulates the Warburg effect, decreasing the respiratory chain and oxidative phosphorylation. The Warburg effect provides a rapid production of ATP, helping to satisfy the high-energy demands of HPV-related cancer cells during cell proliferation and, probably, Warburg also contributes to HPV productive infection during the replication of the viral genome. The identification of the effects of individual HPV proteins on elements that regulate the metabolic switch, strongly suggests that HPV regulates Warburg; nevertheless, how these individual effects of HPV proteins work together on metabolic cellular pathways is still rather uncertain. Nowadays, several studies have identified new cellular interacting partners of HPV proteins [112–114]. The detailed analysis of some of those interactions which could possibly be involved in the regulation of the Warburg effect, as well as the study of their functional implications, will enrich the knowledge that may contribute to improve the design of new therapeutic strategies for HPV-related cancers.

Author Contributions: Conceptualization: I.M.-R. and M.L.; investigation: I.M.-R., A.C.-G. (Adela Carrillo-García), A.C.-G. (Alfredo Cruz-Gregorio), A.C.-P., E.O.-S., and M.L.; writing—review and editing: I.M.-R., A.C.-G. (Adela Carrillo-García), A.C.-P., E.O.-S., A.C.-G. (Alfredo Cruz-Gregorio), and M.L.; visualization: I.M.-R. and M.L.; and supervision: I.M.-R. and M.L.

Acknowledgments: Imelda Martínez-Ramírez is a student from the Programa de Doctorado en Ciencias Bioquímicas, Universidad Nacional Autónoma de México (UNAM); and received a scholarship from CONACyT, México (45562).

Abbreviations

HPV	Human papillomavirus
KC	Krebs cycle
OXPHOS	Oxidative phosphorylation
TIGAR	TP53 induced glycolysis and apoptosis regulator
HIF1	Hypoxia-inducible factor 1
HK2	Hexokinase 2
PK	Pyruvate kinase
PFK-1	Phosphofructokinase-1
PBM	PDZ-binding motif
HDAC1	Histone deacetylase 1
DBMT1	DNA methyltransferase
EZH2	Enhancer of zeste homolog 2

References

1. Hanahan, D.; Weinberg, R.A. The hallmarks of cancer. *Cell* **2000**, *100*, 57–70. [CrossRef]
2. Hanahan, D.; Weinberg, R.A. Hallmarks of cancer: The next generation. *Cell* **2011**, *144*, 646–674. [CrossRef] [PubMed]
3. DeBerardinis, R.J.; Lum, J.J.; Hatzivassiliou, G.; Thompson, C.B. The biology of cancer: Metabolic reprogramming fuels cell growth and proliferation. *Cell Metab.* **2008**, *7*, 11–20. [CrossRef] [PubMed]
4. Berg, J.M.; Jeremy, M.; Tymoczko, J.L.; Stryer, L.; Stryer, L. *Biochemistry*; W.H. Freeman: New York, NY, USA, 2002; ISBN 0716730510.
5. Semenza, G.L.; Artemov, D.; Bedi, A.; Bhujwalla, Z.; Chiles, K.; Feldser, D.; Laughner, E.; Ravi, R.; Simons, J.; Taghavi, P.; et al. "The metabolism of tumours": 70 Years later. *Novartis Found. Symp.* **2001**, *240*, 251–260. [PubMed]
6. Liberti, M.V.; Locasale, J.W. The Warburg effect: How does it benefit cancer cells? *Trends Biochem. Sci.* **2016**, *41*, 211–218. [CrossRef] [PubMed]
7. de Martel, C.; Ferlay, J.; Franceschi, S.; Vignat, J.; Bray, F.; Forman, D.; Plummer, M. Global burden of cancers attributable to infections in 2008: A review and synthetic analysis. *Lancet Oncol.* **2012**, *13*, 607–615. [CrossRef]
8. Yoshida, M.; Miyoshi, I.; Hinuma, Y. Isolation and characterization of retrovirus from cell lines of human adult T-cell leukemia and its implication in the disease. *Proc. Natl. Acad. Sci. USA* **1982**, *79*, 2031–2035. [CrossRef] [PubMed]
9. Mesri, E.A.; Feitelson, M.A.; Munger, K. Human viral oncogenesis: A cancer hallmarks analysis. *Cell Host Microbe* **2014**, *15*, 266–282. [CrossRef] [PubMed]
10. Ferlay, J.; Soerjomataram, I.; Dikshit, R.; Eser, S.; Mathers, C.; Rebelo, M.; Parkin, D.M.; Forman, D.; Bray, F. Cancer incidence and mortality worldwide: Sources, methods and major patterns in GLOBOCAN 2012. *Int. J. Cancer* **2015**, *136*, E359–E386. [CrossRef] [PubMed]
11. Yabaluri, N.; Bashyam, M.D. Hormonal regulation of gluconeogenic gene transcription in the liver. *J. Biosci.* **2010**, *35*, 473–484. [CrossRef] [PubMed]
12. Ward, P.S.; Thompson, C.B. Metabolic reprogramming: A cancer hallmark even warburg did not anticipate. *Cancer Cell* **2012**, *21*, 297–308. [CrossRef] [PubMed]
13. Vander Heiden, M.G.; Cantley, L.C.; Thompson, C.B. Understanding the Warburg effect: The metabolic requirements of cell proliferation. *Science* **2009**, *324*, 1029–1033. [CrossRef] [PubMed]
14. Schurr, A.; West, C.A.; Rigor, B.M. Lactate-supported synaptic function in the rat hippocampal slice preparation. *Science* **1988**, *240*, 1326–1328. [CrossRef] [PubMed]
15. Koppenol, W.H.; Bounds, P.L.; Dang, C.V. Otto Warburg's contributions to current concepts of cancer metabolism. *Nat. Rev. Cancer* **2011**, *11*, 325–337. [CrossRef] [PubMed]
16. Vander Heiden, M.G.; Lunt, S.Y.; Dayton, T.L.; Fiske, B.P.; Israelsen, W.J.; Mattaini, K.R.; Vokes, N.I.; Stephanopoulos, G.; Cantley, L.C.; Metallo, C.M.; et al. Metabolic pathway alterations that support cell proliferation. *Cold Spring Harb. Symp. Quant. Biol.* **2011**, *76*, 325–334. [CrossRef] [PubMed]
17. Metallo, C.M.; Gameiro, P.A.; Bell, E.L.; Mattaini, K.R.; Yang, J.; Hiller, K.; Jewell, C.M.; Johnson, Z.R.; Irvine, D.J.; Guarente, L.; et al. Reductive glutamine metabolism by IDH1 mediates lipogenesis under hypoxia. *Nature* **2011**, *481*, 380–384. [CrossRef] [PubMed]
18. Bodily, J.; Laimins, L.A. Persistence of human papillomavirus infection: Keys to malignant progression. *Trends Microbiol.* **2011**, *19*, 33–39. [CrossRef] [PubMed]
19. de Villiers, E.-M. Cross-roads in the classification of papillomaviruses. *Virology* **2013**, *445*, 2–10. [CrossRef] [PubMed]
20. Muñoz, N.; Bosch, F.X.; de Sanjosé, S.; Herrero, R.; Castellsagué, X.; Shah, K.V.; Snijders, P.J.F.; Meijer, C.J.L.M. International Agency for Research on Cancer Multicenter Cervical Cancer Study Group Epidemiologic classification of human papillomavirus types associated with cervical cancer. *N. Engl. J. Med.* **2003**, *348*, 518–527. [CrossRef] [PubMed]
21. Clifford, G.M.; Rana, R.K.; Franceschi, S.; Smith, J.S.; Gough, G.; Pimenta, J.M. Human papillomavirus genotype distribution in low-grade cervical lesions: Comparison by geographic region and with cervical cancer. *Cancer Epidemiol. Biomark. Prev.* **2005**, *14*, 1157–1164. [CrossRef] [PubMed]
22. Tuna, M.; Amos, C.I. Next generation sequencing and its applications in HPV-associated cancers. *Oncotarget* **2017**, *8*, 8877–8889. [CrossRef] [PubMed]

23. Ang, K.K.; Harris, J.; Wheeler, R.; Weber, R.; Rosenthal, D.I.; Nguyen-Tân, P.F.; Westra, W.H.; Chung, C.H.; Jordan, R.C.; Lu, C.; et al. Human papillomavirus and survival of patients with oropharyngeal cancer. *N. Engl. J. Med.* **2010**, *363*, 24–35. [CrossRef] [PubMed]
24. Egawa, N.; Egawa, K.; Griffin, H.; Doorbar, J. Human papillomaviruses; epithelial tropisms, and the development of neoplasia. *Viruses* **2015**, *7*, 3863–3890. [CrossRef] [PubMed]
25. Doorbar, J.; Quint, W.; Banks, L.; Bravo, I.G.; Stoler, M.; Broker, T.R.; Stanley, M.A. The biology and life-cycle of human papillomaviruses. *Vaccine* **2012**, *30* (Suppl. 5), F55–F70. [CrossRef] [PubMed]
26. Doorbar, J. The E4 protein; structure, function and patterns of expression. *Virology* **2013**, *445*, 80–98. [CrossRef] [PubMed]
27. Thomsen, P.; van Deurs, B.; Norrild, B.; Kayser, L. The HPV16 E5 oncogene inhibits endocytic trafficking. *Oncogene* **2000**, *19*, 6023–6032. [CrossRef] [PubMed]
28. Vande Pol, S.B.; Klingelhutz, A.J. Papillomavirus E6 oncoproteins. *Virology* **2013**, *445*, 115–137. [CrossRef] [PubMed]
29. Roman, A.; Munger, K. The papillomavirus E7 proteins. *Virology* **2013**, *445*, 138–168. [CrossRef] [PubMed]
30. Yeo-Teh, N.S.L.; Ito, Y.; Jha, S. High-risk human papillomaviral oncogenes E6 and E7 target key cellular pathways to achieve oncogenesis. *Int. J. Mol. Sci.* **2018**, *19*, 1706. [CrossRef] [PubMed]
31. Massimi, P.; Shai, A.; Lambert, P.; Banks, L. HPV E6 degradation of p53 and PDZ containing substrates in an E6AP null background. *Oncogene* **2008**, *27*, 1800–1804. [CrossRef] [PubMed]
32. Balsitis, S.; Dick, F.; Lee, D.; Farrell, L.; Hyde, R.K.; Griep, A.E.; Dyson, N.; Lambert, P.F. Examination of the pRb-dependent and pRb-independent functions of E7 in vivo. *J. Virol.* **2005**, *79*, 11392–11402. [CrossRef] [PubMed]
33. Jansma, A.L.; Martinez-Yamout, M.A.; Liao, R.; Sun, P.; Dyson, H.J.; Wright, P.E. The high-risk HPV16 E7 oncoprotein mediates interaction between the transcriptional coactivator CBP and the retinoblastoma protein pRb. *J. Mol. Biol.* **2014**, *426*, 4030–4048. [CrossRef] [PubMed]
34. Wallace, N.A.; Khanal, S.; Robinson, K.L.; Wendel, S.O.; Messer, J.J.; Galloway, D.A. High-risk alphapapillomavirus oncogenes impair the homologous recombination pathway. *J. Virol.* **2017**, *91*, e01084-17. [CrossRef] [PubMed]
35. Moody, C. Mechanisms by which HPV Induces a replication competent environment in differentiating keratinocytes. *Viruses* **2017**, *9*, 261. [CrossRef] [PubMed]
36. Pinidis, P.; Tsikouras, P.; Iatrakis, G.; Zervoudis, S.; Koukouli, Z.; Bothou, A.; Galazios, G.; Vladareanu, S. Human papilloma virus' life cycle and carcinogenesis. *Maedica (Buchar)* **2016**, *11*, 48–54. [PubMed]
37. Coupe, V.M.; González-Barreiro, L.; Gutiérrez-Berzal, J.; Melián-Bóveda, A.L.; López-Rodríguez, O.; Alba-Domínguez, J.; Alba-Losada, J. Transcriptional analysis of human papillomavirus type 16 in histological sections of cervical dysplasia by in situ hybridisation. *J. Clin. Pathol.* **2012**, *65*, 164–170. [CrossRef] [PubMed]
38. Rodríguez, A.C.; Schiffman, M.; Herrero, R.; Wacholder, S.; Hildesheim, A.; Castle, P.E.; Solomon, D.; Burk, R. Proyecto Epidemiológico Guanacaste Group Rapid clearance of human papillomavirus and implications for clinical focus on persistent infections. *J. Natl. Cancer Inst.* **2008**, *100*, 513–517. [CrossRef] [PubMed]
39. Vinokurova, S.; Wentzensen, N.; Kraus, I.; Klaes, R.; Driesch, C.; Melsheimer, P.; Kisseljov, F.; Dürst, M.; Schneider, A.; von Knebel Doeberitz, M. Type-dependent integration frequency of human papillomavirus genomes in cervical lesions. *Cancer Res.* **2008**, *68*, 307–313. [CrossRef] [PubMed]
40. McBride, A.A.; Warburton, A. The role of integration in oncogenic progression of HPV-associated cancers. *PLoS Pathog.* **2017**, *13*, e1006211. [CrossRef] [PubMed]
41. Shukla, S.; Mahata, S.; Shishodia, G.; Pande, S.; Verma, G.; Hedau, S.; Bhambhani, S.; Kumari, A.; Batra, S.; Basir, S.F.; et al. Physical state & copy number of high risk human papillomavirus type 16 DNA in progression of cervical cancer. *Indian J. Med. Res.* **2014**, *139*, 531–543. [PubMed]
42. Chaiwongkot, A.; Vinokurova, S.; Pientong, C.; Ekalaksananan, T.; Kongyingyoes, B.; Kleebkaow, P.; Chumworathayi, B.; Patarapadungkit, N.; Reuschenbach, M.; von Knebel Doeberitz, M. Differential methylation of E2 binding sites in episomal and integrated HPV 16 genomes in preinvasive and invasive cervical lesions. *Int. J. Cancer* **2013**, *132*, 2087–2094. [CrossRef] [PubMed]
43. Dooley, K.E.; Warburton, A.; McBride, A.A. Tandemly integrated HPV16 can form a Brd4-dependent super-enhancer-like element that drives transcription of viral oncogenes. *mBio* **2016**, *7*, e01446-16. [CrossRef] [PubMed]

44. Androphy, E.J.; Hubbert, N.L.; Schiller, J.T.; Lowy, D.R. Identification of the HPV-16 E6 protein from transformed mouse cells and human cervical carcinoma cell lines. *EMBO J.* **1987**, *6*, 989–992. [PubMed]
45. Banks, L.; Spence, P.; Androphy, E.; Hubbert, N.; Matlashewski, G.; Murray, A.; Crawford, L. Identification of human papillomavirus type 18 E6 polypeptide in cells derived from human cervical carcinomas. *J. Gen. Virol.* **1987**, *68 Pt 5*, 1351–1359. [CrossRef] [PubMed]
46. Goodwin, E.C.; DiMaio, D. Repression of human papillomavirus oncogenes in HeLa cervical carcinoma cells causes the orderly reactivation of dormant tumor suppressor pathways. *Proc. Natl. Acad. Sci. USA* **2000**, *97*, 12513–12518. [CrossRef] [PubMed]
47. Scheffner, M.; Huibregtse, J.M.; Vierstra, R.D.; Howley, P.M. The HPV-16 E6 and E6-AP complex functions as a ubiquitin-protein ligase in the ubiquitination of p53. *Cell* **1993**, *75*, 495–505. [CrossRef]
48. Martinez-Zapien, D.; Ruiz, F.X.; Poirson, J.; Mitschler, A.; Ramirez-Ramos, J.; Forster, A.; Cousido-Siah, A.; Masson, M.; Vande Pol, S.; Podjarny, A.; et al. Structure of the E6/E6AP/p53 complex required for HPV-mediated degradation of p53. *Nature* **2016**, *529*, 541–545. [CrossRef] [PubMed]
49. Vogt, M.; Butz, K.; Dymalla, S.; Semzow, J.; Hoppe-Seyler, F. Inhibition of Bax activity is crucial for the antiapoptotic function of the human papillomavirus E6 oncoprotein. *Oncogene* **2006**, *25*, 4009–4015. [CrossRef] [PubMed]
50. Itahana, Y.; Itahana, K. Emerging roles of p53 family members in glucose metabolism. *Int. J. Mol. Sci.* **2018**, *19*, 776. [CrossRef] [PubMed]
51. Matoba, S.; Kang, J.-G.; Patino, W.D.; Wragg, A.; Boehm, M.; Gavrilova, O.; Hurley, P.J.; Bunz, F.; Hwang, P.M. p53 regulates mitochondrial respiration. *Science* **2006**, *312*, 1650–1653. [CrossRef] [PubMed]
52. Veldman, T.; Liu, X.; Yuan, H.; Schlegel, R. Human papillomavirus E6 and Myc proteins associate in vivo and bind to and cooperatively activate the telomerase reverse transcriptase promoter. *Proc. Natl. Acad. Sci. USA* **2003**, *100*, 8211–8216. [CrossRef] [PubMed]
53. Dang, C.V.; O'Donnell, K.A.; Zeller, K.I.; Nguyen, T.; Osthus, R.C.; Li, F. The c-Myc target gene network. *Semin. Cancer Biol.* **2006**, *16*, 253–264. [CrossRef] [PubMed]
54. Spangle, J.M.; Münger, K. The human papillomavirus type 16 E6 oncoprotein activates mTORC1 signaling and increases protein synthesis. *J. Virol.* **2010**, *84*, 9398–9407. [CrossRef] [PubMed]
55. Luo, W.; Semenza, G.L. Pyruvate kinase M2 regulates glucose metabolism by functioning as a coactivator for hypoxia-inducible factor 1 in cancer cells. *Oncotarget* **2011**, *2*, 551–556. [CrossRef] [PubMed]
56. Rodolico, V.; Arancio, W.; Amato, M.C.; Aragona, F.; Cappello, F.; Di Fede, O.; Pannone, G.; Campisi, G. Hypoxia inducible factor-1 α expression is increased in infected positive HPV16 DNA oral squamous cell carcinoma and positively associated with HPV16 E7 oncoprotein. *Infect. Agent. Cancer* **2011**, *6*, 18. [CrossRef] [PubMed]
57. Masoud, G.N.; Li, W. HIF-1α pathway: Role, regulation and intervention for cancer therapy. *Acta Pharm. Sin. B* **2015**, *5*, 378–389. [CrossRef] [PubMed]
58. Kappler, M.; Taubert, H.; Schubert, J.; Vordermark, D.; Eckert, A.W. The real face of HIF1α in the tumor process. *Cell Cycle* **2012**, *11*, 3932–3936. [CrossRef] [PubMed]
59. Crusius, K.; Auvinen, E.; Alonso, A. Enhancement of EGF- and PMA-mediated MAP kinase activation in cells expressing the human papillomavirus type 16 E5 protein. *Oncogene* **1997**, *15*, 1437–1444. [CrossRef] [PubMed]
60. Xu, Q.; Zhang, Q.; Ishida, Y.; Hajjar, S.; Tang, X.; Shi, H.; Dang, C.V.; Le, A.D. EGF induces epithelial-mesenchymal transition and cancer stem-like cell properties in human oral cancer cells via promoting Warburg effect. *Oncotarget* **2017**, *8*, 9557–9571. [CrossRef] [PubMed]
61. Nualart, F.; Los Angeles García, M.; Medina, R.A.; Owen, G.I. Glucose transporters in sex steroid hormone related cancer. *Curr. Vasc. Pharmacol.* **2009**, *7*, 534–548. [CrossRef] [PubMed]
62. Thorens, B.; Mueckler, M. Glucose transporters in the 21st Century. *Am. J. Physiol. Endocrinol. Metab.* **2010**, *298*, E141–E145. [CrossRef] [PubMed]
63. Wright, E.M.; Turk, E. The sodium/glucose cotransport family SLC5. *Pflugers Arch.* **2004**, *447*, 510–518. [CrossRef] [PubMed]
64. Ganapathy, V.; Thangaraju, M.; Prasad, P.D. Nutrient transporters in cancer: Relevance to Warburg hypothesis and beyond. *Pharmacol. Ther.* **2009**, *121*, 29–40. [CrossRef] [PubMed]
65. Medina, R.A.; Owen, G.I. Glucose transporters: Expression, regulation and cancer. *Biol. Res.* **2002**, *35*, 9–26. [CrossRef] [PubMed]

66. Gonzalez-Menendez, P.; Hevia, D.; Mayo, J.C.; Sainz, R.M. The dark side of glucose transporters in prostate cancer: Are they a new feature to characterize carcinomas? *Int. J. Cancer* **2018**, *142*, 2414–2424. [CrossRef] [PubMed]
67. Semenza, G.L. Targeting HIF-1 for cancer therapy. *Nat. Rev. Cancer* **2003**, *3*, 721–732. [CrossRef] [PubMed]
68. Leiprecht, N.; Munoz, C.; Alesutan, I.; Siraskar, G.; Sopjani, M.; Föller, M.; Stubenrauch, F.; Iftner, T.; Lang, F. Regulation of Na^+-coupled glucose carrier SGLT1 by human papillomavirus 18 E6 protein. *Biochem. Biophys. Res. Commun.* **2011**, *404*, 695–700. [CrossRef] [PubMed]
69. Weihua, Z.; Tsan, R.; Huang, W.-C.; Wu, Q.; Chiu, C.-H.; Fidler, I.J.; Hung, M.-C. Survival of cancer cells is maintained by EGFR independent of its kinase activity. *Cancer Cell* **2008**, *13*, 385–393. [CrossRef] [PubMed]
70. Schwartzenberg-Bar-Yoseph, F.; Armoni, M.; Karnieli, E. The tumor suppressor p53 down-regulates glucose transporters GLUT1 and GLUT4 gene expression. *Cancer Res.* **2004**, *64*, 2627–2633. [CrossRef] [PubMed]
71. Guo, Y.; Meng, X.; Ma, J.; Zheng, Y.; Wang, Q.; Wang, Y.; Shang, H. Human papillomavirus 16 E6 contributes HIF-1α induced Warburg effect by attenuating the VHL-HIF-1α interaction. *Int. J. Mol. Sci.* **2014**, *15*, 7974–7986. [CrossRef] [PubMed]
72. Fan, R.; Hou, W.-J.; Zhao, Y.-J.; Liu, S.-L.; Qiu, X.-S.; Wang, E.-H.; Wu, G.-P. Overexpression of HPV16 E6/E7 mediated HIF-1α upregulation of GLUT1 expression in lung cancer cells. *Tumour Biol.* **2016**, *37*, 4655–4663. [CrossRef] [PubMed]
73. Nolop, K.B.; Rhodes, C.G.; Brudin, L.H.; Beaney, R.P.; Krausz, T.; Jones, T.; Hughes, J.M. Glucose utilization in vivo by human pulmonary neoplasms. *Cancer* **1987**, *60*, 2682–2689. [CrossRef]
74. Wilson, J.E. Isozymes of mammalian hexokinase: Structure, subcellular localization and metabolic function. *J. Exp. Biol.* **2003**, *206*, 2049–2057. [CrossRef] [PubMed]
75. Mathupala, S.P.; Ko, Y.H.; Pedersen, P.L. Hexokinase II: Cancer's double-edged sword acting as both facilitator and gatekeeper of malignancy when bound to mitochondria. *Oncogene* **2006**, *25*, 4777–4786. [CrossRef] [PubMed]
76. Patra, K.C.; Wang, Q.; Bhaskar, P.T.; Miller, L.; Wang, Z.; Wheaton, W.; Chandel, N.; Laakso, M.; Muller, W.J.; Allen, E.L.; et al. Hexokinase 2 is required for tumor initiation and maintenance and its systemic deletion is therapeutic in mouse models of cancer. *Cancer Cell* **2013**, *24*, 213–228. [CrossRef] [PubMed]
77. Roberts, D.J.; Tan-Sah, V.P.; Ding, E.Y.; Smith, J.M.; Miyamoto, S. Hexokinase-II positively regulates glucose starvation-induced autophagy through TORC1 inhibition. *Mol. Cell* **2014**, *53*, 521–533. [CrossRef] [PubMed]
78. Zeng, Q.; Chen, J.; Li, Y.; Werle, K.D.; Zhao, R.-X.; Quan, C.-S.; Wang, Y.-S.; Zhai, Y.-X.; Wang, J.-W.; Youssef, M.; et al. LKB1 inhibits HPV-associated cancer progression by targeting cellular metabolism. *Oncogene* **2017**, *36*, 1245–1255. [CrossRef] [PubMed]
79. Liu, Y.; Murray-Stewart, T.; Casero, R.A.; Kagiampakis, I.; Jin, L.; Zhang, J.; Wang, H.; Che, Q.; Tong, H.; Ke, J.; et al. Targeting hexokinase 2 inhibition promotes radiosensitization in HPV16 E7-induced cervical cancer and suppresses tumor growth. *Int. J. Oncol.* **2017**, *50*, 2011–2023. [CrossRef] [PubMed]
80. Hoppe-Seyler, K.; Honegger, A.; Bossler, F.; Sponagel, J.; Bulkescher, J.; Lohrey, C.; Hoppe-Seyler, F. Viral E6/E7 oncogene and cellular hexokinase 2 expression in HPV-positive cancer cell lines. *Oncotarget* **2017**, *8*, 106342–106351. [CrossRef] [PubMed]
81. Mazurek, S. Pyruvate kinase type M2: A key regulator of the metabolic budget system in tumor cells. *Int. J. Biochem. Cell Biol.* **2011**, *43*, 969–980. [CrossRef] [PubMed]
82. Mazurek, S.; Zwerschke, W.; Jansen-Dürr, P.; Eigenbrodt, E. Effects of the human papilloma virus HPV-16 E7 oncoprotein on glycolysis and glutaminolysis: Role of pyruvate kinase type M2 and the glycolytic-enzyme complex. *Biochem. J.* **2001**, *356*, 247–256. [CrossRef] [PubMed]
83. Christofk, H.R.; Vander Heiden, M.G.; Harris, M.H.; Ramanathan, A.; Gerszten, R.E.; Wei, R.; Fleming, M.D.; Schreiber, S.L.; Cantley, L.C. The M2 splice isoform of pyruvate kinase is important for cancer metabolism and tumour growth. *Nature* **2008**, *452*, 230–233. [CrossRef] [PubMed]
84. Singh, P.K.; Mehla, K.; Hollingsworth, M.A.; Johnson, K.R. Regulation of aerobic glycolysis by microRNAs in cancer. *Mol. Cell. Pharmacol.* **2011**, *3*, 125–134. [PubMed]
85. Tomasetti, M.; Santarelli, L.; Neuzil, J.; Dong, L. MicroRNA regulation of cancer metabolism: Role in tumour suppression. *Mitochondrion* **2014**, *19 Pt A*, 29–38. [CrossRef] [PubMed]
86. Slabáková, E.; Culig, Z.; Remšík, J.; Souček, K. Alternative mechanisms of miR-34a regulation in cancer. *Cell Death Dis.* **2017**, *8*, e3100. [CrossRef] [PubMed]

87. Zhang, R.; Su, J.; Xue, S.-L.; Yang, H.; Ju, L.-L.; Ji, Y.; Wu, K.-H.; Zhang, Y.-W.; Zhang, Y.-X.; Hu, J.-F.; et al. HPV E6/p53 mediated down-regulation of miR-34a inhibits Warburg effect through targeting LDHA in cervical cancer. *Am. J. Cancer Res.* **2016**, *6*, 312–320. [PubMed]
88. Stambolsky, P.; Weisz, L.; Shats, I.; Klein, Y.; Goldfinger, N.; Oren, M.; Rotter, V. Regulation of AIF expression by p53. *Cell Death Differ.* **2006**, *13*, 2140–2149. [CrossRef] [PubMed]
89. Liu, G.; Chen, X. The ferredoxin reductase gene is regulated by the p53 family and sensitizes cells to oxidative stress-induced apoptosis. *Oncogene* **2002**, *21*, 7195–7204. [CrossRef] [PubMed]
90. Contractor, T.; Harris, C.R. p53 Negatively regulates transcription of the pyruvate dehydrogenase kinase Pdk2. *Cancer Res.* **2012**, *72*, 560–567. [CrossRef] [PubMed]
91. Suzuki, S.; Tanaka, T.; Poyurovsky, M.V.; Nagano, H.; Mayama, T.; Ohkubo, S.; Lokshin, M.; Hosokawa, H.; Nakayama, T.; Suzuki, Y.; et al. Phosphate-activated glutaminase (GLS2), a p53-inducible regulator of glutamine metabolism and reactive oxygen species. *Proc. Natl. Acad. Sci. USA* **2010**, *107*, 7461–7466. [CrossRef] [PubMed]
92. McDonald, A.E.; Pichaud, N.; Darveau, C.-A. "Alternative" fuels contributing to mitochondrial electron transport: Importance of non-classical pathways in the diversity of animal metabolism. *Comp. Biochem. Physiol. B Biochem. Mol. Biol.* **2017**. [CrossRef] [PubMed]
93. Lai, D.; Tan, C.L.; Gunaratne, J.; Quek, L.S.; Nei, W.; Thierry, F.; Bellanger, S. Localization of HPV-18 E2 at mitochondrial membranes induces ROS release and modulates host cell metabolism. *PLoS ONE* **2013**, *8*, e75625. [CrossRef] [PubMed]
94. Blachon, S.; Bellanger, S.; Demeret, C.; Thierry, F. Nucleo-cytoplasmic shuttling of high risk human Papillomavirus E2 proteins induces apoptosis. *J. Biol. Chem.* **2005**, *280*, 36088–36098. [CrossRef] [PubMed]
95. Bellanger, S.; Tan, C.L.; Nei, W.; He, P.P.; Thierry, F. The human papillomavirus type 18 E2 protein is a cell cycle-dependent target of the SCFSkp2 ubiquitin ligase. *J. Virol.* **2010**, *84*, 437–444. [CrossRef] [PubMed]
96. Nohl, H.; Gille, L.; Staniek, K. Intracellular generation of reactive oxygen species by mitochondria. *Biochem. Pharmacol.* **2005**, *69*, 719–723. [CrossRef] [PubMed]
97. Paumard, P.; Vaillier, J.; Coulary, B.; Schaeffer, J.; Soubannier, V.; Mueller, D.M.; Brèthes, D.; di Rago, J.-P.; Velours, J. The ATP synthase is involved in generating mitochondrial cristae morphology. *EMBO J.* **2002**, *21*, 221–230. [CrossRef] [PubMed]
98. Valko, M.; Leibfritz, D.; Moncol, J.; Cronin, M.T.D.; Mazur, M.; Telser, J. Free radicals and antioxidants in normal physiological functions and human disease. *Int. J. Biochem. Cell Biol.* **2007**, *39*, 44–84. [CrossRef] [PubMed]
99. Cruz-Gregorio, A.; Manzo-Merino, J.; Gonzaléz-García, M.C.; Pedraza-Chaverri, J.; Medina-Campos, O.N.; Valverde, M.; Rojas, E.; Rodríguez-Sastre, M.A.; García-Cuellar, C.M.; Lizano, M. Human Papillomavirus Types 16 and 18 Early-expressed Proteins Differentially Modulate the Cellular Redox State and DNA Damage. *Int. J. Biol. Sci.* **2018**, *14*, 21–35. [CrossRef] [PubMed]
100. Epstein, T.; Gatenby, R.A.; Brown, J.S. The Warburg effect as an adaptation of cancer cells to rapid fluctuations in energy demand. *PLoS ONE* **2017**, *12*, e0185085. [CrossRef] [PubMed]
101. Aw, T.Y. Intracellular compartmentation of organelles and gradients of low molecular weight species. *Int. Rev. Cytol.* **2000**, *192*, 223–253. [PubMed]
102. Stewart, M.P.; Helenius, J.; Toyoda, Y.; Ramanathan, S.P.; Muller, D.J.; Hyman, A.A. Hydrostatic pressure and the actomyosin cortex drive mitotic cell rounding. *Nature* **2011**, *469*, 226–230. [CrossRef] [PubMed]
103. Schwab, A.; Fabian, A.; Hanley, P.J.; Stock, C. Role of ion channels and transporters in cell migration. *Physiol. Rev.* **2012**, *92*, 1865–1913. [CrossRef] [PubMed]
104. Karlsson, T.; Bolshakova, A.; Magalhães, M.A.O.; Loitto, V.M.; Magnusson, K.-E. Fluxes of water through aquaporin 9 weaken membrane-cytoskeleton anchorage and promote formation of membrane protrusions. *PLoS ONE* **2013**, *8*, e59901. [CrossRef] [PubMed]
105. Epstein, T.; Xu, L.; Gillies, R.J.; Gatenby, R.A. Separation of metabolic supply and demand: Aerobic glycolysis as a normal physiological response to fluctuating energetic demands in the membrane. *Cancer Metab.* **2014**, *2*, 7. [CrossRef] [PubMed]
106. Chojnacki, M.; Melendy, T. The human papillomavirus DNA helicase E1 binds, stimulates, and confers processivity to cellular DNA polymerase epsilon. *Nucleic Acids Res.* **2018**, *46*, 229–241. [CrossRef] [PubMed]
107. Sattler, U.G.A.; Mueller-Klieser, W. The anti-oxidant capacity of tumour glycolysis. *Int. J. Radiat. Biol.* **2009**, *85*, 963–971. [CrossRef] [PubMed]

108. Trachootham, D.; Alexandre, J.; Huang, P. Targeting cancer cells by ROS-mediated mechanisms: A radical therapeutic approach? *Nat. Rev. Drug Discov.* **2009**, *8*, 579–591. [CrossRef] [PubMed]

109. Kim, S.M.; Yun, M.R.; Hong, Y.K.; Solca, F.; Kim, J.-H.; Kim, H.-J.; Cho, B.C. Glycolysis inhibition sensitizes non-small cell lung cancer with T790M mutation to irreversible EGFR inhibitors via translational suppression of Mcl-1 by AMPK activation. *Mol. Cancer Ther.* **2013**, *12*, 2145–2156. [CrossRef] [PubMed]

110. Komurov, K.; Tseng, J.-T.; Muller, M.; Seviour, E.G.; Moss, T.J.; Yang, L.; Nagrath, D.; Ram, P.T. The glucose-deprivation network counteracts lapatinib-induced toxicity in resistant ErbB2-positive breast cancer cells. *Mol. Syst. Biol.* **2012**, *8*, 596. [CrossRef] [PubMed]

111. Martinho, O.; Silva-Oliveira, R.; Cury, F.P.; Barbosa, A.M.; Granja, S.; Evangelista, A.F.; Marques, F.; Miranda-Gonçalves, V.; Cardoso-Carneiro, D.; de Paula, F.E.; et al. HER family receptors are important theranostic biomarkers for cervical cancer: blocking glucose metabolism enhances the therapeutic effect of HER Inhibitors. *Theranostics* **2017**, *7*, 717–732. [CrossRef] [PubMed]

112. Rozenblatt-Rosen, O.; Deo, R.C.; Padi, M.; Adelmant, G.; Calderwood, M.A.; Rolland, T.; Grace, M.; Dricot, A.; Askenazi, M.; Tavares, M.; et al. Interpreting cancer genomes using systematic host network perturbations by tumour virus proteins. *Nature* **2012**, *487*, 491–495. [CrossRef] [PubMed]

113. White, E.A.; Sowa, M.E.; Tan, M.J.A.; Jeudy, S.; Hayes, S.D.; Santha, S.; Münger, K.; Harper, J.W.; Howley, P.M. Systematic identification of interactions between host cell proteins and E7 oncoproteins from diverse human papillomaviruses. *Proc. Natl. Acad. Sci. USA* **2012**, *109*, E260–E267. [CrossRef] [PubMed]

114. White, E.A.; Kramer, R.E.; Tan, M.J.A.; Hayes, S.D.; Harper, J.W.; Howley, P.M. Comprehensive analysis of host cellular interactions with human papillomavirus E6 proteins identifies new E6 binding partners and reflects viral diversity. *J. Virol.* **2012**, *86*, 13174–13186. [CrossRef] [PubMed]

2

Genetic Diversity of the Noncoding Control Region of the Novel Human Polyomaviruses

Ugo Moens [1,*,†], Carla Prezioso [2,3] and Valeria Pietropaolo [2,†]

1 Department of Medical Biology, Faculty of Health Sciences, University of Tromsø—The Arctic University of Norway, 9037 Tromsø, Norway
2 Department of Public Health and Infectious Diseases, "Sapienza" University of Rome, 00185 Rome, Italy; carla.prezioso@uniroma1.it (C.P.); valeria.pietropaolo@uniroma1.it (V.P.)
3 IRCSS San Raffaele Pisana, Microbiology of Chronic Neuro-degenerative Pathologies, 00160 Rome, Italy
* Correspondence: ugo.moens@uit.no
† Both authors have contributed equally.

Academic Editors: John M. Lehman and Thomas D. Friedrich

Abstract: The genomes of polyomaviruses are characterized by their tripartite organization with an early region, a late region and a noncoding control region (NCCR). The early region encodes proteins involved in replication and transcription of the viral genome, while expression of the late region generates the capsid proteins. Transcription regulatory sequences for expression of the early and late genes, as well as the origin of replication are encompassed in the NCCR. Cell tropism of polyomaviruses not only depends on the appropriate receptors on the host cell, but cell-specific expression of the viral genes is also governed by the NCCR. Thus far, 15 polyomaviruses have been isolated from humans, though it remains to be established whether all of them are genuine human polyomaviruses (HPyVs). The sequences of the NCCR of these HPyVs show high genetic variability and have been best studied in the human polyomaviruses BK and JC. Rearranged NCCRs in BKPyV and JCPyV, the first HPyVs to be discovered approximately 30 years ago, have been associated with the pathogenic properties of these viruses in nephropathy and progressive multifocal leukoencephalopathy, respectively. Since 2007, thirteen novel PyVs have been isolated from humans: KIPyV, WUPyV, MCPyV, HPyV6, HPyV7, TSPyV, HPyV9, HPyV10, STLPyV, HPyV12, NJPyV, LIPyV and QPyV. This review describes all NCCR variants of the new HPyVs that have been reported in the literature and discusses the possible consequences of NCCR diversity in terms of promoter strength, putative transcription factor binding sites and possible association with diseases.

Keywords: disease; Merkel cell carcinoma; mutation; NCCR; novel human polyomaviruses; transcription factor binding sites

---

1. Introduction: Human Polyomaviruses

Polyomaviruses (PyVs) are non-enveloped viruses that are typically 40–45 nm in diameter, and that possess a double-stranded circular genome of around 5000 base-pairs. Birds and mammals, including humans, are natural hosts for PyVs [1,2]. Recently, PyVs have also been isolated from fish [3,4]. So far, 15 different polyomaviruses have been isolated from human samples. The first human polyomaviruses, BKPyV and JCPyV, were identified in 1971 [5,6]. In 2007, two new human polyomaviruses (Karolinska Institute PyV (KIPyV) [7] and Washington University PyV (WUPyV) [8] were detected, and in the following years, Merkel cell PyV (MCPyV) [9], HPyV6 [10], HPyV7 [10], Trichodisplasia spinulosa PyV (TSPyV) [11], HPyV9 [12], HPyV10 [13], Saint Louis PyV (STLPyV) [14], HPyV12 [15], New Jersey PyV (NJPyV) [16], Lyon IARC PyV (LIPyV) [17], and Quebec PyV [18] have been described. Their original source of isolation and association with human diseases is summarized in Table 1.

Table 1. The novel human polyomaviruses, their original source of isolation and their association with human diseases.

Virus	Original Source	Associated Disease	Reference
KIPyV	Nasopharyngeal aspirate	None	[7]
WUPyV	Bronchoavelar lavage	None	[8]
MCPyV	Merkel cell carcinoma	None	[9]
HPyV6	Healthy skin	Pruritic skin eruption in immunocompromised patients	[10]
HPyV7	Healthy skin	Pruritic skin eruption in immunocompromised patients	[10]
TSPyV	Trichodysplasia spinulosa spicules	Trichodysplasia spinulosa	[11]
HPyV9	Serum from renal transplant recipient	None	[12]
HPyV10	Condyloma specimens from a patient with WHIM * syndrome	None	[13]
STLPyV	Stool sample from a healthy 15-month-old child	None	[14]
HPyV12	Liver sample from patient with malignant disease	None	[15]
NJPyV	Muscle biopsy from a pancreatic transplant patient	None	[16]
LIPyV	Skin swab	None	[17]
QPyV	Stool sample from 85-year old hospital patient	None	[18]

* warts, hypogammaglobulinemia, infections, and myelokathexis.

Whether all of these are genuine human polyomaviruses (HPyVs) remains to be determined. BKPyV, JCPyV, KIPyV, WUPyV, MCPyV, HPyV6, HPyV, TSPyV, HPyV9, HPyV10, STLPyV, HPyV12, and NJPyV are classified as human polyomaviruses by the International Committee of Taxonomy of Viruses [19,20], LIPyV has only been very recently described, while LIPyV DNA was originally detected in human skin [17]. LIPyV seroreactivity in healthy individuals is ~5% in healthy individuals and much lower than the seroprevalence of the other HPyVs, which is between 50–100% [21]. Accordingly, LIPyV DNA was not detected or present in <2% of examined skin, eyebrow hair, gargle samples and tonsil samples [17,22,23]. Moreover, LIPyV DNA was frequently detected in the feces of cats [24], suggesting that it may be a feline PyV rather than a HPyV. QPyV DNA was detected in the feces of one patient [18], and the seroprevalence of this PyV has not been examined. Despite its original identification in human liver, gastro-intestinal tract and colon tissue and a VP1 seropositivity (respectively LT seropositivity) between ~20–90% (respectively 30–40%) in healthy adults or malignant and non-malignant gastro-intestinal tract patients [15,25], HPyV12 DNA could not be detected in numerous human samples from different sources [22,26–30]. Moreover, the group of Feltkamp reported that HPyV12 seroprevalence is only around 5% [21,31]. A nearly identical HPyV12 variant was isolated from shrew, suggesting that HPyV12 may be transmitted from shrew to humans, or that human HPyV12 positive samples were contaminated [32].

2. The Polyomavirus Genome: The Noncoding Control Region

Functionally, the PyV genome is tripartite consisting of the early region, the late region, and the noncoding control region (NCCR) (Figure 1A). The early region codes for regulatory proteins involved in replication and transcription of the viral genome. The major early proteins are large T-antigen (LT) and small t-antigen (sT). The late region codes for the structural proteins VP1, VP2 and VP3 that form the capsid. VP1 is the major capsid protein, while VP2 and VP3 are the minor capsid proteins [1,2]. However, not all PyVs express VP3 [33]. Interspersed between the early and late region are sequences that do not code for viral proteins, and is referred to as the NCCR.

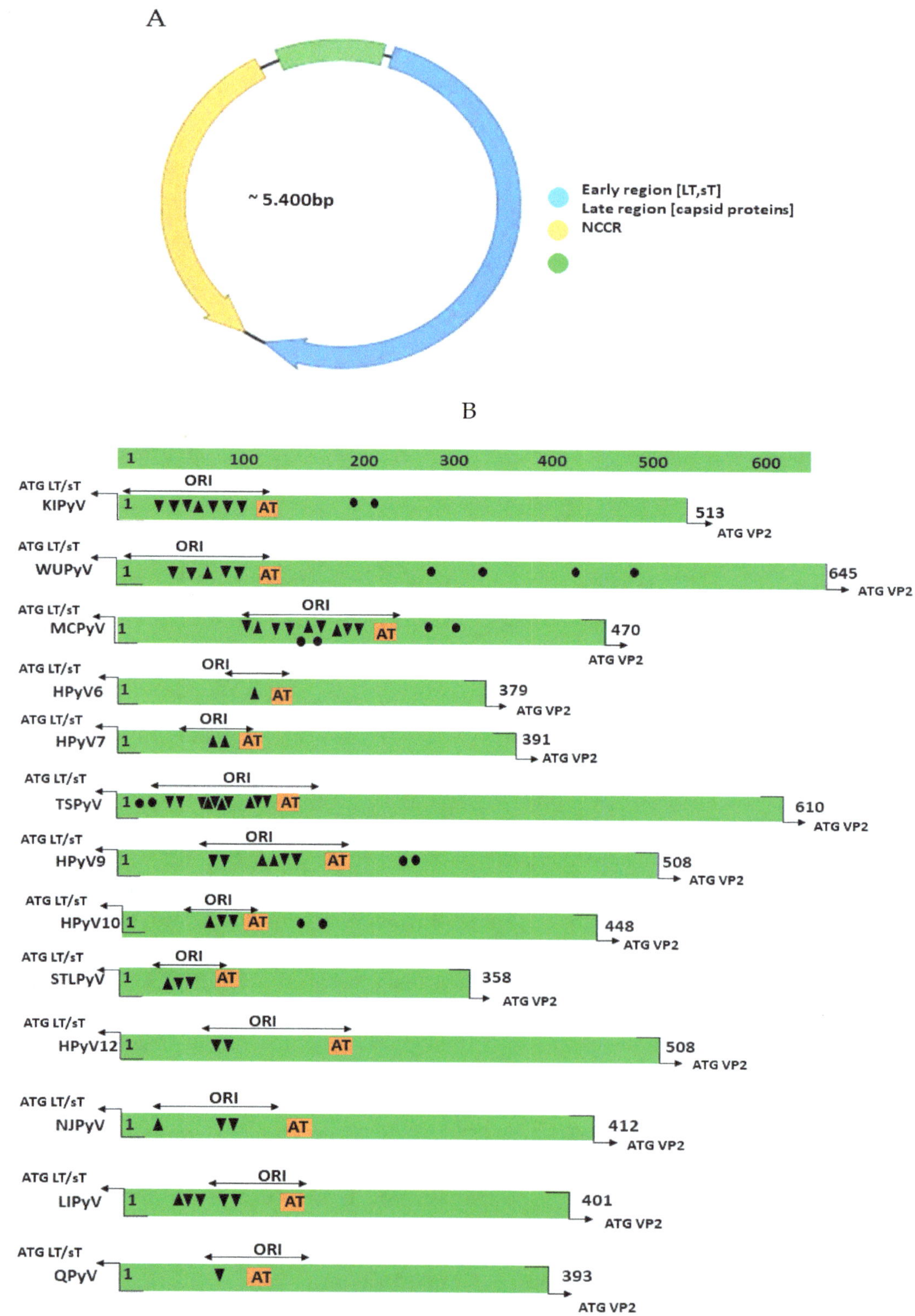

Figure 1. *Cont.*

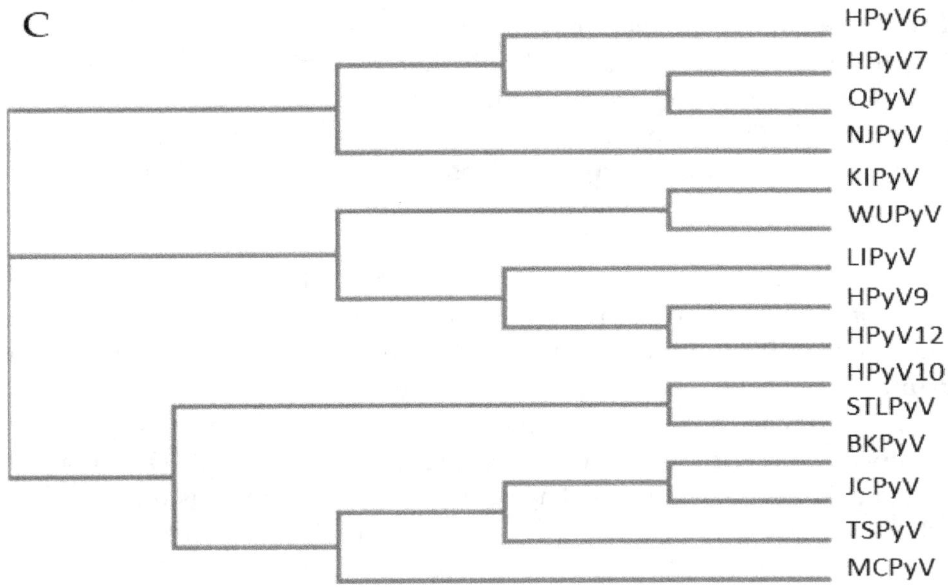

Figure 1. Genomic organization of the human polyomaviruses (HPyVs) genome and the structure of the noncoding control region (NCCR). (**A**) The circular dsDNA genome consists of the early and late regions that encode regulatory and structural proteins, respectively. Interspersed is the NCCR. (**B**) Schematic presentation of the NCCR of the novel HPyVs. The NCCR is the region between the start codon of Large T antigen (LT) and Small T antigen (sT) and the start codon of VP2. The AT-rich region (AT), repeated sequences (black dots), and LT binding motifs (upward pointing triangle = 5′-GRGGC-3′; downward pointing triangle = 5′-GCCYC-3′) are shown. (**C**) Phylogenetic tree bases on NCCR sequences of the different HPyVs. This is a neighbor-joining tree without distance corrections using Clustal Omega multiple sequence alignment [34].

Studies with simian virus 40 (SV40 or Macaca mulatta polyomavirus 1) and murine polyomaviruses have been pivotal in unveiling the functions of this region. The SV40 NCCR contains the origin of replication (ori), which consists of GRGGC motifs to which LT binds and is flanked by an AT-rich sequence and an easily denaturated imperfect palindrome [35,36]. Binding of LT to these motifs is also involved in regulation of viral transcription [37,38]. The NCCR also contains promoter and enhancer elements that control early and late transcription [39,40]. SV40 directly isolated from its natural host, rhesus monkey, has a NCCR that consists of an AT-rich region, triple GC-rich 21 base-pairs (bp) repeats, and a single 72 bp element. The 21 bp repeats contain the LT binding motif (GRGGC; [41,42]). This NCCR organization is known as the archetype. SV40 adapted to grow in cell culture has a duplication of this 72 bp element, with this type of NCCR referred to as prototype [43,44]. SV40 isolated from human tumors usually contain a single 72 bp repeat [43]. Rearrangements in the SV40 NCCR affect viral transcription and replication, as well as oncogenic properties of the virus [45,46]. The Mouse polyomavirus (Mus musculus polyomavirus 1; MPyV) NCCR encompasses the ori consisting of an AT-tract and a GC-rich (LT binding motifs) inverted repeat, and the transcription regulatory domains A (or α) and B (or β), C and D [47–50]. Alterations in the MPyV NCCR have an effect on viral replication in cell culture and in the host, the host range, and in vitro transformation [51–55].

The NCCR of the HPyVs varies between 267 bp (JCPyV CY-strain; accession number AB038249) to 645 bp (WUPyV prototype; accession number NC_009539) (see Figure S1 for the NCCR sequences of the novel HPyV), and similar to the NCCR of SV40 and MPyV, the NCCR of HPyVs also contain the origin of replication, LT binding motifs, and AT-rich region (Figure 1B). This region of the genome displays little or no sequence identity between the different HPyV species (Figure S2). A neighbor-joining tree without distance corrections shows which NCCRs are most closely related (Figure 1C).

The diversification based on the presence of a certain NCCR rearranged structure contributed to determining HPyVs strains as "archetype" or prototype". The importance of the NCCR rearrangements during HPyVs infection became obvious when different strains of JCPyV were examined. The archetype JCPyV NCCR strain (CY) is divided into six boxes named A (36 bp), B (23 bp), C (55 bp), D (66 bp), E (18 bp), and F (69 bp) and contains the origin of replication (ORI), the promoter and the enhancer elements [56]. The NCCR harbored transcription factor binding sites such as the nuclear transcription factor-1 (NF1), a JCPyV cell-specific regulator of promoter and enhancer activity [57,58], the activating protein 1 (AP1), involved in JCPyV early gene expression [57,59], and the specificity protein-1 (SP1) able to regulate JCPyV transcription [57,60]. The archetype NCCR is considered the transmissible form of the virus among the population, and could be released into the urine of healthy individuals due to periodic and subclinical reactivation in the kidney [61,62]. In contrast, in the context of immunosuppression or during immunomodulatory therapy or in AIDS patients, JCPyV can reactivate from latency to cause a fatal pathology of the central nervous system (CNS), known as progressive multifocal leukoencephalopathy (PML) [61]. JCPyV variants carrying rearranged NCCR were usually isolated from PML patients. The prototype Mad-1 strain is the most studied variant of JCPyV and is characterized by 98-bp tandem repeats in the NCCR late proximal region (arranged as ORI-A-C-E-A-C-E-F), and is able to increase viral gene expression in human glial cells, thereby indicating that it is involved in controlling cell gene expression [63–65]. The enhancer repeats found in the Mad-1 strain are lacking in the archetype JCPyV strains isolated from the urine of healthy individuals [64]. Additional NCCR rearrangements are implicated in the development of the JCPyV pathogenic strains. In fact, in a significant proportion of JCPyV archetype isolates, short deletions or duplications were observed, corroborating that this region is highly unstable [66,67]. Therefore, it is possible to assume that subsequent archetypal NCRR rearrangements could determine the onset of PML strains, such as Mad-1 [68].

Based on the occurrence in the NCRR of transcriptional enhancer repeat elements, BKPyV isolates can also be identified as archetype and prototype strains. The archetype BKPyV WW strain, characterized by five blocks named O (35 bp), which includes the origin of replication and a TATA-box, P (68 bp), Q (39 bp), R (63 bp), and S (63 bp), containing TATA-like elements and the regulatory regions for early and late genes expression, is considered the infectious strain, shed in the urine of immunocompetent individuals [69–71]. Approximately 30 transcription factor binding sites are in silico predicted: SP1 has been the most extensively studied [72–74], although the additional role played by other transcription factors such as NF1, ETS1, NFκB, the glucocorticoid and progesterone receptors, and CREB were evidenced in several studies [73,75–77].

Similarly to JCPyV, the plausible instability of the archetype BKPyV NCRR could contribute to the development of the prototype strains, which is able to cause polyomavirus-associated nephropathy in kidney transplant recipients and hemorrhagic cystitis in hematopoietic stem cell transplant recipients [78–81]. The Dunlop strain, the most salient prototype strain, was isolated from a kidney transplant recipient with ureteral stenosis [82]. This strain displays three 68-bp tandem repeat within the NCRR (O-P-P-P-S arrangement) with respect to the archetype strain, carrying a single 68-bp motif. This strain showed less enhancer activity than the prototype strain, thus confirming the significance of the triplicated motifs on transcriptional regulation, and on viral infectious activity [83]. In fact, BKPyV strains isolated from kidney transplant recipients with rearranged NCRR showed higher viral gene expression and viral loads with more extensive pathogenicity [84].

Additional NCCR structures have been described for both viruses [85–88]. In particular, the presence of a common pattern of JCPyV NCCR rearrangement, such as the D-box deletion, can be considered a hallmark needed for the initial NCCR rearrangements critical co-factor for the development of PML in immunosuppressed individuals [88,89]. Besides the triplication of the P region, rearrangements of BKPyV NCCR involve the adjacent O and Q blocks. Differently, the S block is always retained, hence highlighting the importance of these nucleotide sequences [70]. NCCR mutations were also observed during in vitro JCPyV and BKPyV cultivation, confirming that NCRR variants could arise after prolonged propagation of the viruses in cells [71,85,90,91]. The mechanisms by which both viruses determine relevant human diseases are not established, but it is accepted that the regulation of gene expression in HPyVs plays a role in determining the viral tropism, and in the promotion of pathogenesis progression [92].

Little is known about the genetic diversity of the NCCRs from the novel HPyVs and the biological relevance in terms of viral transcription, replication, and possible pathogenic properties. In this review, we provide an overview of the mutations in the NCCR, which is defined as the sequence between the start codon of the *LT/sT* gene and the start codon of the *VP2* gene, of the novel HPyVs and their known effect on promoter activity. We discuss how NCCR rearrangements may affect the binding of putative transcription factors, and whether specific NCCR configurations are associated with disease.

3. KI and WU NCCR Variants

KIPyV has been mostly isolated from oral and respiratory specimens from (pediatric) patients with respiratory diseases that suffer from other viral and bacterial infections (reviewed in [93]). Whether KIPyV is a genuine respiratory pathogen or an opportunistic co-infector has not been established [93,94]. Seventy-two full-length NCCR sequences have been deposited in GenBank so far (Table S1). They contain the LT binding motifs, an AT-rich stretch and repeated sequences (Figure 1B and Table 2).

Table 2. Repeat sequences in the noncoding regions of the novel human polyomaviruses.

HPyV	Sequence	Position	Length	Remarks
KIPyV	CGTGAAGCCAACTTCCTG-GGC CGTG-AGCCAGCTTCCTGCGGC	251–271; 272–292	21	Imperfect DR *
WUPyV	GCCCTTTGTA ATGTTGTGACATCTCC	319–328; 389–398 479–494, 548–563	10 16	-
MCPyV	CAGAGGCCTC AACTTTTTTTC	147–156; 170–179 328–338; 370–380	10 11	Palindrome
HPyV6	No repeats	-	-	
HPyV7	No repeats	-	-	-
TSPyV	GAAATGAA	34–41; 42–49	7	DR
HPyV9	CTGTGGTAT	275–283; 284–292	9	DR
HPyV10	GCTATTGTTGGCAA	168–181;182–195	14	DR
STLPyV	No repeats	-	-	-
HPyV12	GTTCC CAGGCAGACGGCCAAGTTCC	203–207; 229–233 208–227; 228–248	5 20	GTTCC is part of larger repeat DR separated by 1 nucleotide
NJPyV	No repeats	-	-	-
LIPyV	No repeats	-	-	-
QPyV	No repeats	-	-	-

* DR = direct repeat. Repeat sequences were identified using the programs Tandem Repeat finder program [95] and the repeat sequence finding tool from Novoprolabs (novoprolabs.com/tools/repeat-sequence-finder).

Most KIPyV NCCR sequences are obtained from nasopharyngeal swabs or aspirates, but also from blood form healthy blood donors [96] and from feces from a child with acute gastroenteritis [97]. The NCCR sequence of the Stockholm 60 isolate (Genbank accession number NC_009238; [7]; Figure S1) may represent the archetype NCCR because it is the most common sequence reported, and has been isolated from different biological samples in different parts of the world. Stockholm 60 KIPyV was originally isolated from respiratory tract specimens from a child. We found that 21 out of 48 isolates from nasopharyngeal aspirates of patients with respiratory symptoms or infections and 23 out of 38 isolates from healthy blood donors have the Stockholm 60 NCCR [96]. As described by us and others, NCCRs of other isolates, contain only minor point mutations scattered throughout the entire NCCR (Figure 2 and Table 3). Exceptions are the isolates Brisbane 001, Brisbane 005 and CU-255, whose NCCRs have the 10 bp AGGCGCTGCG insertion, and are clinical isolates obtained from respiratory tract (Table S1).

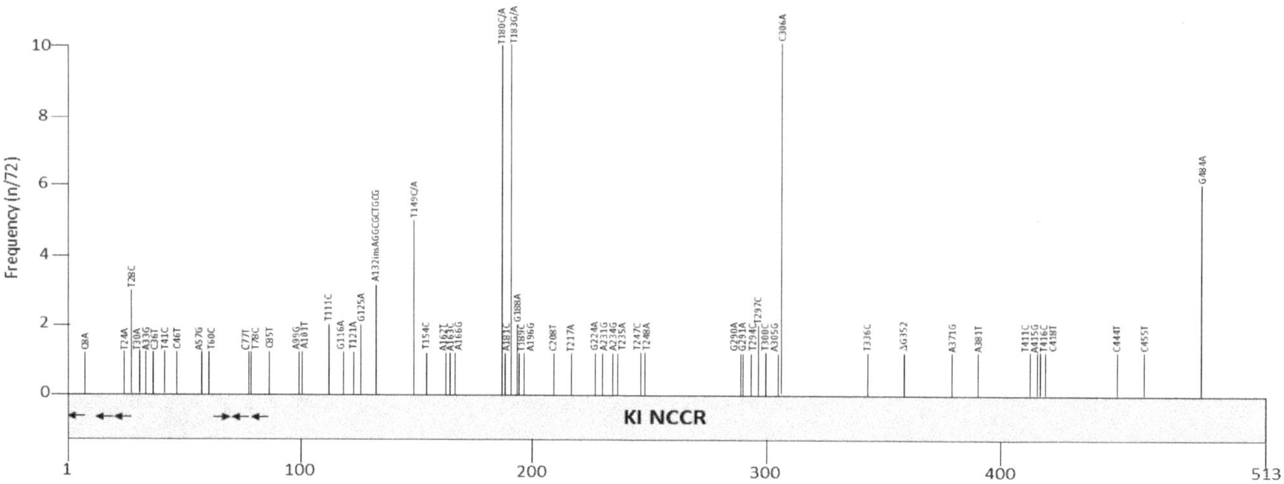

Figure 2. Mutations and their prevalence in variants of Karolinska Institute polyomavirus (KIPyV) noncoding control region (NCCR). The numbering of the NCCR is from early to late, with nucleotide 1 being the most proximal to the ATG start codon of the early genes and the most distal nucleotide, just upstream of the start codon of the *VP2* gene. The number of times a peculiar mutation is found in the different variants is given as frequency (with n the number of times the mutation was described/total NCCR variant sequences available). Putative Large T antigen (LT) binding motifs 5′-GRGGC-3′ (→) or 5′-GCCYC-3′ (←) are shown. The table summarizes the mutations, their location in the NCCR and their frequency. For details, see Table S1. Putative transcription factor binding sites are shown in Supplementary Table S2.

The KIPyV NCCR contains putative binding sites for several transcription factors (Table 4 and Table S2). The effect of this 10 bp AGGCGCTGCG insertion on KIPyV promoter activity or replication is not known, but the sequence contains a putative binding site for transcription factor AP4 [98]. AP4 is ubiquitously expressed, and can both activate and repress transcription [99,100]. Its effect on KIPyV NCCR has not been investigated. The point mutations remove or create putative binding sites for several transcription factors, including nuclear receptors, STAT proteins, HOXD, and POU the general transcription factors TBP and TFIID (see Table 3 in [101] for a detailed overview). We examined the effect of NCCR polymorphism in isolates from blood and nasopharyngeal samples on early and late promoter activity in HEK 293 cells [96]. These cells had previously been shown to give highest promoter activity of 10 different cell lines tested [102]. Eighteen isolates with a single nucleotide substitution were tested and revealed significant differences in early and late promoter activities for some of the isolates. One variant (NPA7d) had a mutation that destroyed a putative c-Myb binding motif compared to Stockholm 60 NCCR. Ectopic expression of c-Myb stimulated the early and late promoter activities of both Stockholm 60 and NPA7d, but there was no significant difference in c-Myb induced activation of the promoters [96]. Some of the mutations are located in putative LT binding sites and may therefore have an effect on promoter activity or/and viral DNA replication. It remains to be determined whether the NCCR may have an effect on the pathogenic properties of KIPyV because Stockholm 60 and Stockholm 60-like NCCRs have also been isolated from blood and respiratory specimens from healthy individuals, with no direct association between KIPyV and diseases having been established. Larger KIPyV NCCR rearrangements as seen for BKPyV and JCPyV NCCRs seem to be rare.

Table 3. Frequency of mutations in the noncoding control region of Karolinska Institute polyomavirus.

Mutation	Frequency *	Mutation	Frequency	Mutation	Frequency
C8A	1	A132insAGGCGCTGCG	3	T248A	1
T24A	1	T149C/A	4	G290A	1
T28C	3	T154C	1	G291A	1
T30A	1	A162T	1	T294C	1
A33G	1	A163C	1	T297C	2
C36T	1	A166G	1	T300C	1
T41C	1	T180C/A	10	A305G	1
C46T	1	A181C	1	C306A	10
A57G	1	T183G/A	10	T336C	1
T60C	1	G188A	2	ΔG352	1
C77T	1	T189C	1	A371G	1
T78C	1	A196G	1	A381T	1
C85T	1	C208T	1	T411C	1
A99G	1	T217A	1	A415G	1
A101T	1	G224A	1	T416C	1
T111C	2	A231G	1	C418T	1
G116A	1	A234G	1	C444T	1
T121A	1	T235A	1	C455T	1
G125A	2	T247C	1	G484A	6

* *n*/72.

A total of 185 partial or complete WUPyV NCCR sequences are available in the GenBank (Table S1). All strains have a NCCR of 645 bp, except variant J1, which has an insertion of one A at position 277 (Table S1), and contains an AT-rich stretch, GRGGC pentamers and repeated sequences of, respectively, 10 and 16 bp (Figure 1B and Table 2). Polymorphisms are predominantly in the NCCR part proximal to the early region (Figure 3 and Table 5). The most common point mutations are G54A and T59G. Both mutations are often present simultaneously. The substitution C52G is also common, but is always found in combination with the G54A mutation. The variants GD-WU709 and WU/Wuerzburg01/07 have C52T rather than C52G, whereas 12 variants have the triple mutation C52G/G54A/T59G. These three nucleotides are part of a sequence that is flanked, respectively, by 4 and 5 T residues; the triple substitution removes the putative binding site for transcription factor c-MYB, and creates motifs for TATA/TBP and retinoic acid receptor-related orphan receptor α [98]. The mutations A94G and C105G are also always simultaneously present, with the double mutations generating a remote sequence similarity with the binding motif of transcription factor AP1, though this does not seem to affect the binding of other putative factors (Table S2; [98]). Other common mutations include A284C and C285A, which are also found together except for the WU/Wuerzberg03/07 variant, which lacks the A284C substitution. A284/C285 are part of a putative site for RUNX1 (AML1; [98]), a transcription factor involved in hematopoiesis [103]. While G295A is found in 9 NCCR sequences, one strain (CQ6029/China_CQ/2014) had a G295C replacement. The CU_CHONBURI3 isolate from a patient with respiratory diseases had several unique point mutations. Overall, no typical mutations in specific specimens were detected, nor was an apparent correlation with a genotype and geographic regions. To the best of our knowledge, the effect of mutations on the WUPyV promoter activity has not been studied, nor have the consequences of viral replication been addressed. Whether mutations had an effect on putative transcription factor binding sites is also unknown, but because most mutations are single or few point mutations, they may not destroy or create novel binding sites.

Table 4. Number of putative transcription factor binding sites in the noncoding control region of the novel human polyomaviruses. Prediction based on TFBIND tool [98].

Transcription Factor	KIPyV	WUPyV	MCPyV	HPyV6	HPyV7	TSPyV	HPyV9	HPyV10	STLPyV	HPyV12	NJPyV	LIPyV	QPyV
AML1	6	3	3	6	4	6	6	3	4	2	7	4	3
ARNT	2	0	0	3	2	4	2	2	2	5	4	2	2
AP1	8	9	10	6	8	14	8	7	10	4	5	5	5
AP2	12	13	6	7	6	6	5	6	5	5	7	3	0
AP4	8	12	5	0	6	6	5	5	9	4	9	3	3
ATF/CREB	7	11	6	4	8	14	5	6	10	2	5	3	8
CAAT	7	7	3	5	5	7	8	9	4	3	6	4	5
C/EBP	7	9	8	9	9	12	13	9	4	7	7	11	0
CDP	2	2	0	1	1	1	2	4	3	3	2	1	1
E2F	9	10	4	4	5	14	9	6	8	6	7	5	2
E47	7	8	1	2	5	3	5	4	4	2	6	2	2
ELK1	10	8	9	5	4	12	8	8	3	6	8	6	5
EVI1	3	6	4	3	4	9	7	3	4	4	5	5	5
GATA-1	7	12	9	8	6	16	15	9	6	6	7	10	8
GATA-2	6	8	5	3	5	13	8	6	4	8	3	5	7
GATA-3	5	0	2	1	5	8	6	1	0	0	1	4	3
GR	2	0	0	1	2	3	0	2	1	3	0	1	3
HNF3	3	4	2	2	1	4	1	3	4	1	2	0	2
IRF	5	3	6	4	4	2	3	4	0	2	4	3	3
MYB	6	5	6	7	0	4	6	6	3	7	2	7	1
MYC/MAX	5	7	1	2	2	6	3	2	2	3	4	1	1

Table 4. *Cont.*

Transcription Factor	KIPyV	WUPyV	MCPyV	HPyV6	HPyV7	TSPyV	HPyV9	HPyV10	STLPyV	HPyV12	NJPyV	LIPyV	QPyV
MYOD	9	9	3	3	5	7	8	6	8	8	5	6	2
NF1	5	5	3	1	1	4	1	6	4	3	4	2	1
NFκB	2	3	5	2	1	5	7	5	0	1	1	1	1
NHLH1	2	2	2	1	1	1	0	0	1	0	1	1	0
OCT1	9	17	11	8	7	16	16	6	8	8	13	9	9
p53	7	8	5	4	3	8	9	5	1	11	10	5	3
PAX2	3	1	0	3	1	4	2	1	2	5	5	1	4
PAX5	7	4	5	2	3	3	8	1	5	2	3	1	3
PAX6	1	2	0	1	0	0	2	0	1	1	2	0	2
SOX5	3	7	4	2	4	8	2	5	7	2	5	4	4
RORα	7	4	3	4	6	5	4	1	2	6	2	0	3
SP1	12	12	14	6	8	13	10	9	7	11	8	11	11
SREBP	3	1	1	1	2	1	1	3	5	0	2	1	3
SRF	3	5	1	4	2	7	5	3	0	3	3	1	5
STAT	3	6	2	2	3	7	2	3	3	2	5	1	0
T3R	6	6	1	3	1	1	3	2	3	2	3	0	4
TBP	5	10	5	5	5	5	5	4	4	6	2	3	5
TST-1	3	5	4	2	0	4	2	2	1	0	1	2	1
USF	12	12	3	6	7	13	7	8	5	4	8	2	7
YY1	3	3	4	3	6	6	5	1	1	3	2	2	3

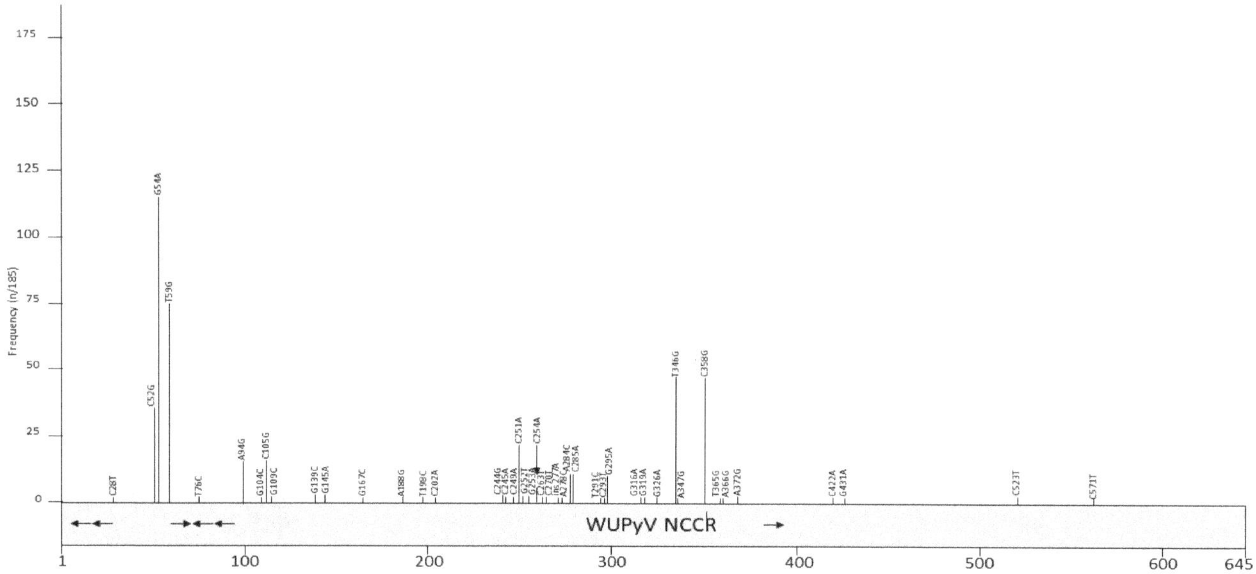

Figure 3. Mutations and their prevalence in variants of Washington University polyomavirus (WUPyV) noncoding control region (NCCR). The numbering of the NCCR is from early to late, with nucleotide being 1 the most proximal to the ATG start codon of the early genes and the most distal nucleotide, just upstream of the start codon of the *VP2* gene. The number of times a peculiar mutation is found in the different variants is given as frequency (with *n* the number of times the mutation was described/total NCCR variant sequences available). Putative Large T antigen (LT) binding motifs 5′-GRGGC-3′ (→) or 5′-GCCYC-3′ (←) are shown. The table summarizes the mutations, their location in the NCCR, and their frequency. For details see Table S1. Putative transcription factor binding sites are shown in Table S2.

Table 5. Frequency of mutations in the noncoding control region of Washington University polyomavirus.

Mutation	Frequency *	Mutation	Frequency	Mutation	Frequency
C28T	1	C245A	1	G316A	1
C52G	33	C249A	1	G319A	1
G54A	114	C251A	18	G326A	2
T76C	1	G252T	1	T346G	17
A94G	15	G253A	1	A347G	1
G104C	1	C254A	5	C358G	20
C105G	15	C263T	1	T365G	1
G109C	1	C270T	1	A366G	1
G139C	2	ins277A	1	A372G	5
G145A	3	A278C	1	C422A	1
G167C	1	A284C	10	G431A	1
A188G	1	C285A	10	C523T	1
T198C	1	T291C	1	C571T	1
C202A	1	C293T	1		
C244G	2	G295A	8		

* *n*/185.

4. MCPyV NCCR Variants

NCCR rearrangements are described as a pivotal event in the onset of HPyVs-related pathology, as demonstrated for JCPyV and BKPyV, in which NCCRs not only control gene expression, but also serve as the main determinants in viral replication, containing the origin of DNA replication and transcription factor binding sites [104,105]. MCPyV is a major causative agent of the skin cancer Merkel cell carcinoma [9], but whether the NCCR can influence the outcome of the infection remains elusive. More than 100 partial or complete NCCR sequences are available from MCC and non-MCC tissue (Table S1). Mutations have been described throughout the entire NCCR, but especially in the late promoter part region (Figure 4 and Table 6). Nucleotides 360–425 of the MCPyV NCCR contain putative binding sites for transcription factors AP1, AP2, C/EBPα and β, EVI1, NFκB, c-Myb, p53, SOX5, TST-1, and SP1 (Table 4; [98]), although their binding has not been proven so far. Some of the mutations affect putative LT binding motifs, and may therefore interfere with transcription and replication of the viral DNA. Indeed, studies by the group of Chang and Moore showed that mutations in nucleotides G143, C145A, A173 and C176 abolished the replication of MCC isolates MCV339 and MCV350 in the presence of full-length LT [106,107]. The NCCR from MCC isolates FraMerk22 and FraMerk24 both contain the mutations G143T and C176T, whereas MCC isolate MKT-23 has the mutations G143A, C145G, and A173A, with MKT-32 carrying the transversion C146G. Since all these isolates are derived from MCC, they are replication deficient due to the expression of truncated LT and integration. None of the mutations identified by the work of the Chang-Moore group that abrogate MCPyV replication have been reported in non-MCC PyV isolates (see Table S1).

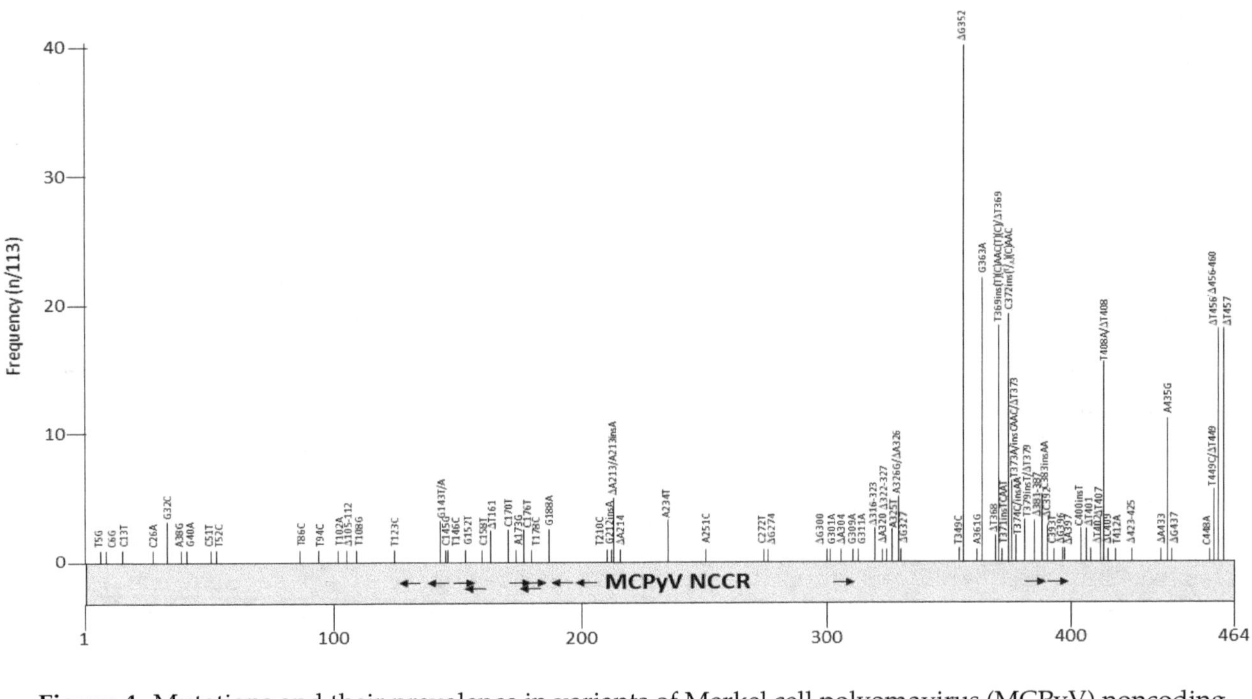

Figure 4. Mutations and their prevalence in variants of Merkel cell polyomavirus (MCPyV) noncoding control region (NCCR). The numbering of the NCCR is from early to late with nucleotide 1 being the most proximal to the ATG start codon of the early genes and the most distal nucleotide, just upstream of the start codon of the *VP2* gene. The number of times a peculiar mutation is found in the different variants is given as frequency (with *n* the number of times the mutation was described/total NCCR variant sequences available). Putative Large T antigen (LT) binding motifs 5′-GRGGC-3′ (→) or 5′-GCCYC-3′ (←) are shown. The table summarizes the mutations, their location in the NCCR, and their frequency. For details see Table S1. Putative transcription factor binding sites are shown in Supplementary Table S2.

Table 6. Frequency of mutations in the noncoding control region of Merkel cell polyomavirus.

Mutation	Frequency *	Mutation	Frequency	Mutation	Frequency
T5G	1	G188A	2	T371insCAAT	1
C6G	1	T210C	1	C372ins($^{T}/_{A}$)(C)AAC	19
C13T	1	G212insA	2	T373A/insCAAC/ΔT373	6
C26A	1	ΔA213/A213insA	5	T374C/insAA	2
G32C	3	ΔA214	1	T379insT/ΔT379	3
A38G	1	A234T	3	Δ381–387	3
G40A	1	A251C	1	C383insAA	5
C51T	1	C272T	1	ΔC392	3
T52C	1	ΔG274	1	C393T	1
T86C	1	ΔG300	1	ΔG396	1
T94C	1	G301A	1	ΔA397	1
T102A	1	ΔA304	1	C400insT	2
Δ105–112	2	G309A	1	ΔT401	2
T108G	1	G311A	1	ΔT402	1
T123C	1	Δ316–323	2	ΔT407	3
G143T/A	3	ΔA320	1	T408A/ΔT408	14
C145G	1	Δ322–327	1	ΔC409	1
T146C	1	A325T	2	T412A	1
G152T	2	A326G/ ΔA326	3	ΔT423–425	1
C158T	1	T349C	1	ΔA433	1
ΔT161	2	ΔG352	40	A435G	17
C170T	2	A361G	1	C448A	1
A173G	1	G363A	22	T449C/ΔT449	7
C176T	2	ΔT368	3	ΔT456/Δ456–460	18
T178C	1	T369ins(T)(C)AAC(T)(C)/ ΔT369	14	ΔT457	17

* *n*/113.

Whether a MCPyV variant with a particular NCCR architecture is associated with specific patient groups is not known. MCPyV with different NCCRs have been characterized in Merkel cell carcinoma samples (see Table S1), while for other diseases only a single sample of a particular individual was examined. Few studies have examined a larger cohort and/or different clinical samples from the same patient. Hashida, et. al. evaluated the genetic variability of MCPyV NCCR in skin swab specimens of healthy individuals with distinct ethnicities and geographic origins, identifying two major subtypes of MCPyV NCCR, subtypes I and II, with the presence or absence of a 25 bp tandem repeat (TGTCCTCCTCCCTTTGTAAGAGAAA) in the late promoter region. Based on the occurrences of two deletions (T368 and T369), and the 5 bp TCAAC insertion (compared to the consensus strain R17b), MCPyV strains were further assigned to five genotypes [108]. Delbue et.al. performed MCPyV NCCR molecular characterization on cerebrospinal fluid samples collected from patients affected by neurological disorders. The results obtained showed the presence of the MCPyV NCCR IIc strain, according to Hashida's NCCR classification [109]. Prezioso, et al., studying the MCPyV NCCR from urine, plasma and rectal swabs recovered from immunosuppressed population, observed, in plasma and rectal swabs, as well as the occurrence of the MCPyV NCCR IIa-2 strain, which contains the 5 bp insertion and represents the predominant strain among white persons of European descent [110]. The deletion of nucleotide G352 is unique for the MCPyV isolates in plasma, urine and rectal swab specimens from HIV-1 patients, and has not been described in MCPyV isolates from other patient groups. In addition to the NCCR genotypes circulating within a HIV-1-positive population in the same study, Prezioso et al. evaluated the MCPyV NCCR alterations focusing on putative binding sites of cellular transcription factors, in order to verify whether mutations and/or rearrangements could fall in some binding sites [110]. The analysis of distal NCCR sequences (nucleotides 302–464) and the analysis of the relative putative binding site, revealed a high degree of homology with R17b strain in urine samples, whereas transitions, transversions, and single or double deletions were observed in plasma and rectal swabs (Table S1). Differently from JCPyV and BKPyV, in which the early proximal side of NCCR is highly conserved and the late proximal side undergoes rearrangements [111], insertions and deletions occurred in both the early and late proximal side of the MCPyV NCCR. More specifically, representative TCAAT and AAC insertions (nucleotide positions 5210–5211) were observed in both plasma and rectal swabs. Analysis of the putative binding site showed that the MCC350 NCCR sequence contains putative NF1, NFκB, TST-1, OCT1, AP-1, and TATA sites, already described within the NCCRs of other HPyVs [98,112]. In several strains obtained from MCPyV-positive plasma and rectal swabs samples, deletions, insertions, or single base substitutions fell within these putative binding sites, thus making predictable that some of these changes would not allow the identification of putative binding motifs, such as SP1 and/or p53, already described in the NCCR of other HPyVs [112]. Further studies are warranted in order to define the importance of these NCCR binding sites and to understand how their changes (mutations, insertions, or deletions) may influence in vivo MCPyV pathogenicity. In contrast to NCCR analysis conducted on rectal swabs from an HIV-1-positive population, which were characterized by the onset of transitions, transversions, and single or double deletions [110], MCPyV NCCR in stool samples from patients with hematological disorders exhibited a high degree of sequence stability, thereby suggesting that sequence rearrangements occurred rarely in the gastrointestinal anatomical site [113]. To date, although it is well documented that MCPyV DNA has been detected in the upper and lower respiratory tract specimens of children and adults and in immunocompetent and immunocompromised patients [114–118] and that the detection of MCPyV DNA was also observed in cystic fibrosis patient respiratory secretions [119,120], the respiratory NCCR structure organization has not yet been investigated.

The relative early and late promoter strength of seven MCPyV NCCR variants was compared in human dermal fibroblasts, and in the non-classical MCPyV-positive MCC cell line MCC13 [121]. All variants that had mutations compared to the consensus strain R17b (GenBank accession number HM011556) had a 10–50% lower basal early and late activity in both cell lines. However, the I strain described by Hashida et al. ([108]) had an approximately 30% higher early and late promoter activity and the early promoter of isolate MKL1, a MCC isolate [122], was approximately 40% stronger in the fibroblasts. The promoter activity of other variants has not been compared, nor has the effect of mutations on the viral life cycle and transforming potential of this oncovirus been examined.

5. HPyV 6 and HPyV7 NCCR Variants

Although HPyV6 and HPyV7 DNA is commonly present in the normal skin of healthy persons [10,123], HPyV6 and HPyV7 are associated with rash and pruritic skin eruption [124–128], HPyV7 DNA was found in 19/35 cholangiocarcinomas [129], while HPyV6 DNA has been detected in a few cases of keratoacanthomas, basal cell carcinomas, squamous cell carcinomas and trichoblastomas [130,131]. HPyV6 DNA was detected in 1/234 cerebrospinal fluid samples and 1/1016 serum samples of healthy blood donors [109,132]. HPyV6 DNA prevalence was much higher in tonsil brushing samples from immunocompetent children and adults than HPyV7 DNA (113/689 versus 6/689). HPyV6 and HPyV7 DNA prevalence and copy number were significantly higher in skin swabs collected from lesional and non-lesional skins of 86 Japanese patients with inflammatory skin diseases and mycosis fungoides compared with specimens from 149 healthy control individuals [133].

HPyV6 and HPyV7 were detected in 1/55 skin specimens from cutaneous T-cell lymphoma patients [29]. Despite the presence of HPyV6 and HPyV7 DNA in samples of various disorders, it remains to be established whether these viruses play a direct role in causing such skin conditions.

Seventeen HPyV6 NCCR sequences are deposited in GenBank. Four of them are sequences obtained from HPyV6 DNA amplified in sewage (H6-cg-A2.f, B159.4, U43.1 and U43.3), six are from healthy skin (606b, 607a, 607b, 609a, 614a, and 627a), two are from bile samples (Bile-72 and Bile-81), and two are combined nose and throat samples from kidney transplant patients (QLD-49Br and QLD-61Br). One sample was obtained from pruritic skin lesion (UTSW6.1), one from a lymph node from a patient with an angiolymphoid hyperplasia with Kimura disease (LN1), and one from a nasopharyngeal aspirate of a child with respiratory tract infections (BJ376) (see Table S1 for details and references). Identical HPyV6 NCCRs were found in healthy skin, along with bile from patients with malignant biliary obstruction, combined nose and throat specimens from kidney transplant patients and a nasopharyngeal sample of a child with respiratory infection (Table S1). Two clinical samples (UTSW6.1 from pruritic skin and LN1 from the lymph node of a patient with Kimura disease) and the DNA amplified from sewage water had mutations compared to the reference strain. The mutation spectrum is shown in Figure 5 and Table 7. The UTSW6 isolate had a deletion of nucleotides 183–193 (CAAAGGTCAAA), a mutation of nucleotides 223–229 (except 228), and insertions of GGC and of TGGGCAGGGCCATTT distal of these point mutations. The 11 bp deletion removes binding motifs for AP1 and CREB, while the 15 bp insertion adds a putative SP1 and p53 binding site. Other putative binding sites are shown in Table 4 and Table S2. The point mutations affect an AT-rich region but no putative binding motifs are predicted in this sequence [98], which may affect viral replication, as this region is part of the predicted ori [134]. Based on the limited available HPyV6 NCCR sequences, no specific HPyV6 NCCR is associated with disease. The effect of mutations in the NCCR on the promoter activity and viral life cycle has not been tested.

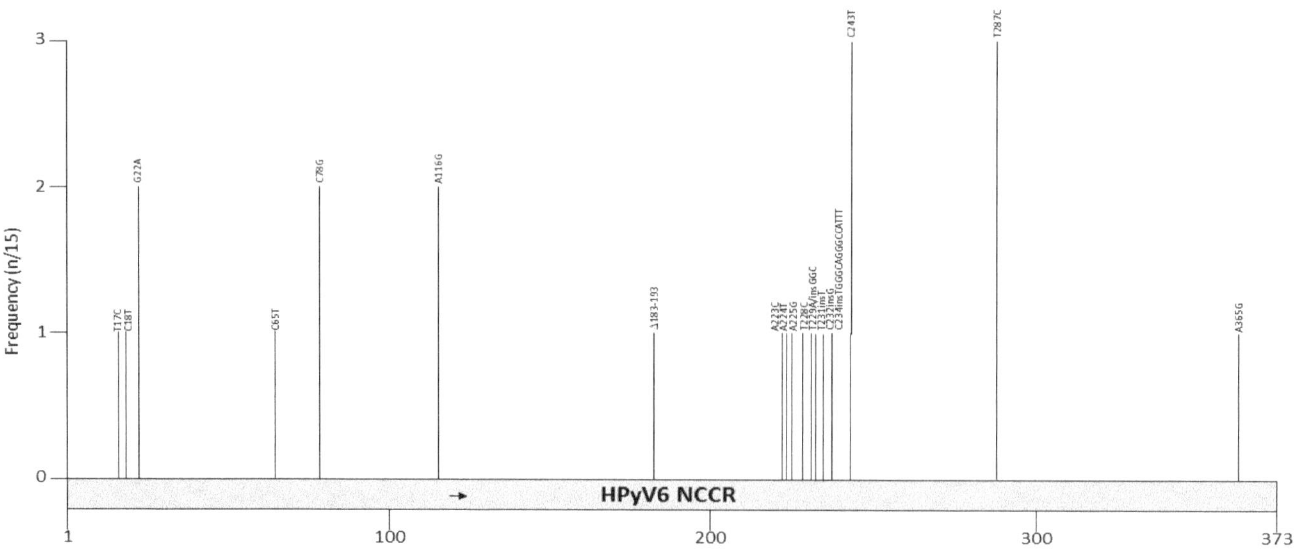

Figure 5. Mutations and their prevalence in variants of HPyV6 noncoding control region (NCCR). The numbering of the NCCR is from early to late with nucleotide 1 being the most proximal to the ATG start codon of the early genes and the most distal nucleotide, just upstream of the start codon of the *VP2* gene. The number of times a peculiar mutation is found in the different variants is given as frequency (with *n* the number of times the mutation was described/total NCCR variant sequences available). Putative Large T antigen (LT) binding motifs 5′-GRGGC-3′ (→) or 5′-GCCYC-3′ (←) are shown. The table summarizes the mutations, their location in the NCCR, and their frequency. For details see Table S1. Putative transcription factor binding sites are shown in Supplementary Table S2.

Table 7. Frequency of mutations in the noncoding control region of HPyV6.

Mutation	Frequency *	Mutation	Frequency	Mutation	Frequency
T17C	1	Δ183–193	1	T231insG	1
C18T	1	A223C	1	C232insG	1
G22A	2	A224T	1	C234insTGGGCAGGGCATTT	1
C65T	1	A225G	1	C243T	3
C78G	2	T228C	1	T287C	3
A116G	2	T229A/insGGC	1	A356G	1

* *n*/17.

Sequences of 15 different HPyV7 NCCR isolates are available in GenBank (variants 707a, 707b, 713a, 713b, 715b, 727a, CRC01, PITT1, PITT2, UTSW7.1, PLA1, PLA2, MUQ, URI, and BIO). Six are from healthy skin specimens, while nine are clinical samples from patients, including skin lesions from lung or renal transplant patients with rash and pruritus, and skin lesions from patients with pruritic and dyskeratotic dermatoses (Table S1 and references therein). The length of these NCCRs varies from 371 bp (PITT2 isolate) to 399 bp (PITT1 isolate). DNA of these two variants was isolated from the skin of lung transplant patients with a rash [124]. Five isolates from the same patient had an NCCR of 381 bp (BIO, MUQ, PLA1, PLA2, URI), five had a 383 bp NCCR (707a, 707b, 715b, 727a, UTSW7.1), two had a 385 bp NCCR (713a and 713b), and one had a NCCR of 387 bp (CRC01). No repeated sequences are present (Table 2). The mutations in the different HPyV7 variants are concentrated in the central part of the NCCR (Figure 6 and Table 8). The consensus is the nucleotide sequence that was present in the majority of the 15 available sequences, with the nucleotide numbering based on the HPyV7 reference strain R713a (GenBank accession number NC_014407=713). Most mutations are point mutations, whereas PITT1 also contains the insertion ACAGGATATGAT and PITT2 has a deletion removing nucleotides 150–161 (CTGGGTTACTGG). The insertion contains putative binding sites for the transcription factors ETS1, GATA1/2/3 and EVI1, whereas the deletion removes possible GATA2 and CDP binding motifs [98]. EVI1, CDP and GATA3 are expressed in the skin, while ETS1, GATA1, and GATA2 are not or weakly expressed in skin [100]. Putative binding sites for transcription factors in the HPyV7 NCCR are summarized in Table 4 and Table S2.

Table 8. Frequency of mutations in the noncoding control region of HPyV7.

Mutation	Frequency *	Mutation	Frequency	Mutation	Frequency
G99A/C	10	G161insTCA	1	ΔT205	10
A100T	1	G162C/insACAGGTATGAT	2	G228T	10
ΔT149	1	C164T	5	G230C/R	7
ΔC150/Δ150–161	2	ΔA180	5	C279G/insTT	6
C158G	5	ΔG181	5	G315T	11
T159(C)(A)/insA	7	G203A	10		
G160A	9	C204G/ΔC204	11		

* *n*/18.

The early promoter activity of the PITT1 and PITT2 variants was significantly higher than the activity of the reference strain in the colon adenocarcinoma cell line SW480, whereas a tendency to lower activity in human embryonal kidney HEK293 cells was observed [112]. The promoter activity was not examined in skin cells, although these variant were originally isolated from the skin [124]. Colon and kidney cells may not be authentic host cells because no HPyV7 LT expression was detected in 10 normal and 94 malignant colon samples, and in 10 normal and 65 renal cancers [135] and so far there are no reports of HPyV7 DNA in these organs. A transversion of A to T in the putative 5′-GAGGC-3′ LT motif was reported (Figure 6), although the effect on viral replication has not been exploited.

Interestingly, the NCCR of the recently isolated QPyV DNA shows >80% identity with the HPyV7 NCCR (Figure S3), while the complete genome is 81% identical with HPyV7 [18].

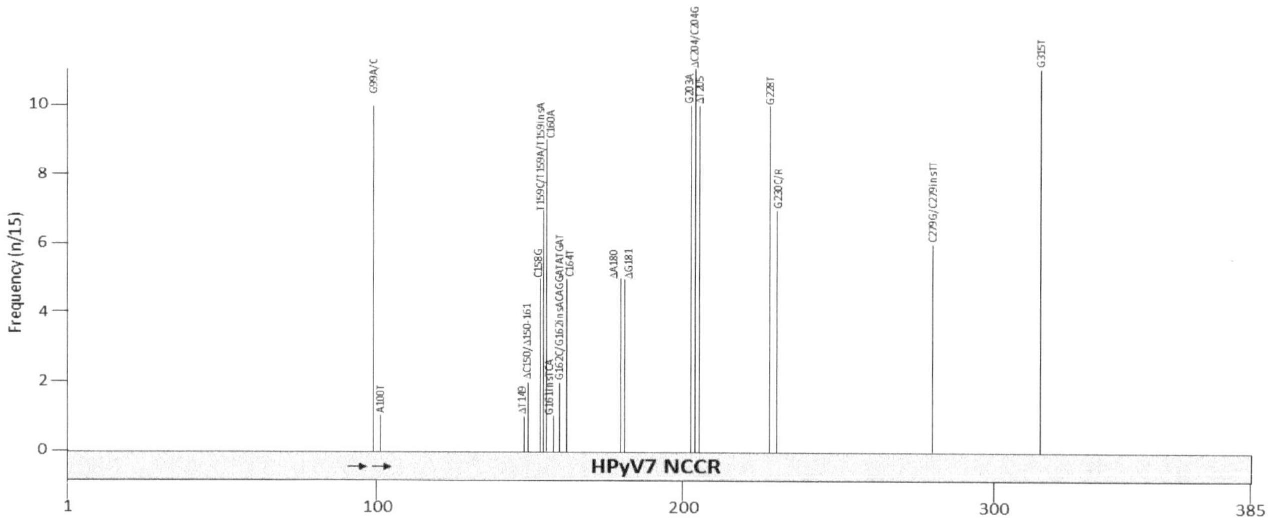

Figure 6. Mutations and their prevalence in variants of HPyV7 noncoding control region (NCCR). The numbering of the NCCR is from early to late with nucleotide 1 being the most proximal to the ATG start codon of the early genes and the most distal nucleotide, just upstream of the start codon of the *VP2* gene. The number of times a peculiar mutation is found in the different variants is given as frequency (with *n* the number of times the mutation was described/total NCCR variant sequences available). Putative Large T antigen (LT) binding motifs 5′-GRGGC-3′ (→) or 5′-GCCYC-3′ (←) are shown. The table summarizes the mutations, their location in the NCCR, and their frequency. For details see Table S1. Putative transcription factor binding sites are shown in Supplementary Table S2.

6. TSPyV NCCR Variants

Twenty-four TSPyV NCCR sequences are deposited in the GenBank (Table S1). Most samples are derived from skin spicules, but also a nasopharyngeal aspirate from a heart transplant patient, a heart from a myocarditis patient, and the CSF and serum of immunosuppressed patients contained TSPyV DNA. The NCCRs of the non-spicule isolates were identical or quasi identical with isolates from skin spicules. Most mutations are point mutations (Figure 7 and Table 9), but two skin spicule isolates (0602 and 1312) had deletions of 54 and 38 bp, respectively [136]. The relative promoter activity of these NCCR variants has not been examined, nor has the effect of mutations on the viral life cycle been investigated. The 39 bp deletion removes putative binding sites for AP1, SOX5, HNF3, OCT1, TATA/TBP, STAT, glucocorticoid receptor, retinoic acid receptor-related orphan receptor α, and CREB, whereas 54 bp deletion possesses possible binding sites for ARNT, AP1, AP2, AP4, CREB/ATF, CAAT, E2F, ELK, EVI1, GATA1/2/3, NHLH1, MYB, MYC, MYOD, NFκB, OCT1, PAX5, TST1, and USF [98]. While most of these factors are expressed in the skin, MYOD, PAX5, GATA1, and GATA2 seem to be absent in the skin [100]. However, the binding of these transcription factors and their possible role in regulating TSPyV transcription remain to be proven. The TSPyV NCCR contains several putative LT binding motifs, and mutations in some of them have been reported (Figure 7). Whether they have an effect on viral replication has not been tested.

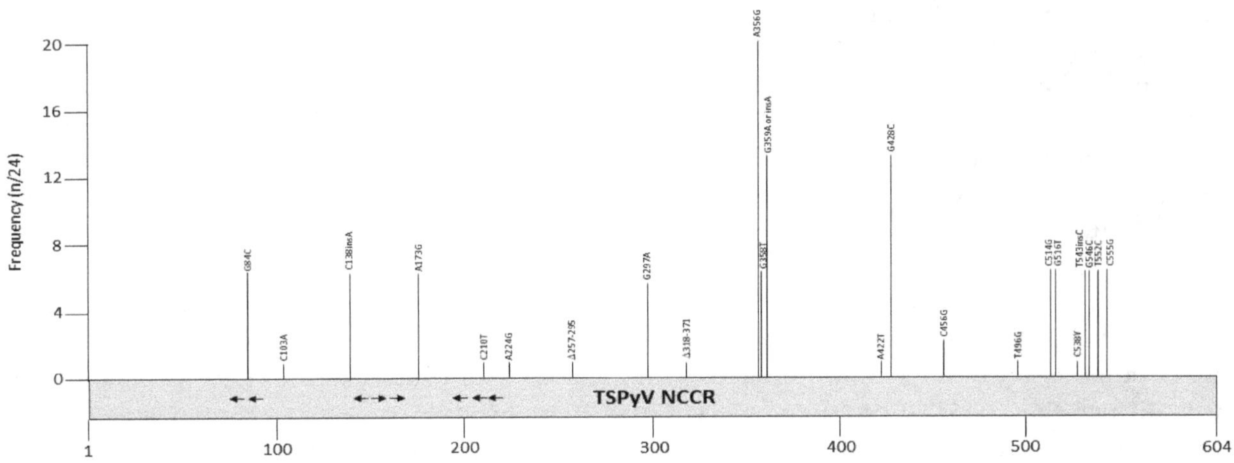

Figure 7. Mutations and their prevalence in variants of Trichodisplasia spinulosa polyomavirus (TSPyV) noncoding control region (NCCR). The numbering of the NCCR is from early to late with nucleotide 1 being the most proximal to the ATG start codon of the early genes and the most distal nucleotide, just upstream of the start codon of the *VP2* gene. The number of times a peculiar mutation is found in the different variants is given as frequency (with *n* the number of times the mutation was described/total NCCR variant sequences available). Putative Large T antigen (LT) binding motifs 5′-GRGGC-3′ (→) or 5′-GCCYC-3′ (←) are shown. The table summarizes the mutations, their location in the NCCR, and their frequency. For details see Table S1. Putative transcription factor binding sites are shown in Supplementary Table S2.

Table 9. Frequency of mutations in the noncoding control region of Trichodisplasia spinulosa polyomavirus.

Mutation	Frequency *	Mutation	Frequency *	Mutation	Frequency *
G84C	6	Δ318–371	1	C514G	6
C103A	1	A356G	20	G516T	6
C138insA	6	G358T	6	C538Y	1
A173G	6	G359A/G359insA	13	T543insC	6
C210T	1	A422T	1	G546C	6
A224G	1	G428C	13	T552C	6
Δ257–295	1	C456G	2	C555G	6
G297A	5	T496G			

* *n*/24.

7. HPyV 9 NCCR Variants

HPyV9 was originally detected in the serum and urine from a renal transplant patient under immunosuppressive treatment [12]. Shortly after, HPyV9 DNA was isolated from the facial surface of a Merkel cell carcinoma patient and tentatively named Institute Pasteur polyomavirus (IPPyV) [137]. The genome of IPPyV only differs by two nucleotides from HPyV9, hence IPPyV is a variant of HPyV9. Yet, none of these mutations are within the NCCR (Table S1; [137]). The HPyV9 isolate M149 from tonsils has an identical NCCR sequence as the original HPyV9 isolate (GenBank accession MH844627). An HPyV9 isolate (UF-1 isolate) from the blood of an AIDS patients displays an eight base-pair deletion, a 13 base-pair insertion and 24 point mutations in its NCCR [138]. These NCCR rearrangements created putative SP1 binding sites in the late promoter. We compared the basal early and late promoter activity of the original HPyV9 strain and the UF-1 clinical isolate in the human cell lines BEL7402, C33A, HEK293, HeLa, SK-N-BE, SW480, and U2OS. We found that the UF-1 early promoter was stronger in all cell lines except in U2OS, and the UF-1 late promoter was stronger in all cell lines except in C33A and HeLa cells [139]. The effect of LT on early and late promoter activity was monitored in BEL7402, HEK293 and HeLa cells. Whereas the UF-1 late promoter activity was more potently stimulated than the HPyV9 late promoter by LT in all three cell lines tested, a stronger LT-induced activation of the UF-1 early promoter compared to the HPyV9 early promoter was only observed in HEK293 cells. The mutations in the UF-1 NCCR generate two putative SP1 binding sites in the distal part of the late promoter. Mutating these two SP1 sites did not have an effect on the basal early promoter activity, but increased basal late promoter 2-fold. Disruption of these SP1 sites had also no effect on LT-induced early promoter activity, but reduced late promoter activity 7-fold compared to non-mutated late UF-1 promoter activity. Our results showed that the promoter activity of the clinical isolate UF-1 is stronger and more potently induced by LT compared with the promoter of the original HPyV9 isolate. A later study confirmed that the UF-1 promoter was stronger than the promoter of the original isolated HPyV9 in HEK293 and the lung carcinoma A549 cells [112]. Whether the rearrangements in the UF-1 NCCR may affect the life cycle and possible pathogenic properties of the virus remains to be determined. Additional putative transcription factor binding sites are summarized in Table 4 and Table S2.

8. HPyV10 NCCR Variants

Twenty NCCR sequences are available in GenBank (Table S1), with the length ranging from to 430 to 442 bp. The original isolates MWPyV (NC_018102) and MA095 (JQ898291), both from feces [140], are identical, but contain an 11 bp deletion compared to the other variants (Table S1, Figure 8 and Table 10). The NCCRs of isolates ww10, TEDDY-01, QLDMW04 and QLDMW10 are identical, although they were derived from different specimens from different patients. The ww10 isolate was detected in a condyloma specimen from a patient with warts, hypogammaglobulinemia, infections, and myelokathexis (WHIM) syndrome [13], QLDMW04 and 010 are from respiratory samples [141], and TEDDY_01 is from feces (direct submission to GenBank; accession number KC549591). Point mutations are dispersed throughout the NCCR for the other isolates. The 11bp deletion (ATTGTTGGCAA) contains possible binding sites for CDP and SOX5 [98]. CDP is ubiquitously expressed, but SOX5 is enriched in testis [100]. Other possible transcription factors that may bind the HPyV10 NCCR are given in Table 4 and Table S2. It is not known whether HPyV10 is associated with a disease, and the biological consequence of NCCR mutations remains elusive.

Table 10. Frequency of mutations in the noncoding control region of HPyV10.

Mutation	Frequency *	Mutation	Frequency	Mutation	Frequency
G21C	1	ΔA180	1	G263A	5
T22C	1	ΔG192	1	A287C	1
A24C	1	ΔA194	1	G311A	1
G28C/ΔG28	3	A195G/ΔA195	4	G338A	1
G40A	1	G196C	4	G345A	1
C53G	5	A198G/ΔA198	6	C354A	2
G57C	1	G201A/ΔG201	5	C356G	1
ΔC65	1	ΔT203	1	C363G	1
C131G	1	ΔA204	1	G372A	1
T132insG	1	A208G	1	C389G	1
A145G	1	A215T	6	A395C	1
C146T	1	T222C	1	G396T	3
G160A	7	G225A/C	2	A422	1
C166T	5	T256C/ΔT256	8	C433A	3
Δ171–181	1	A259A	1		

* *n*/20.

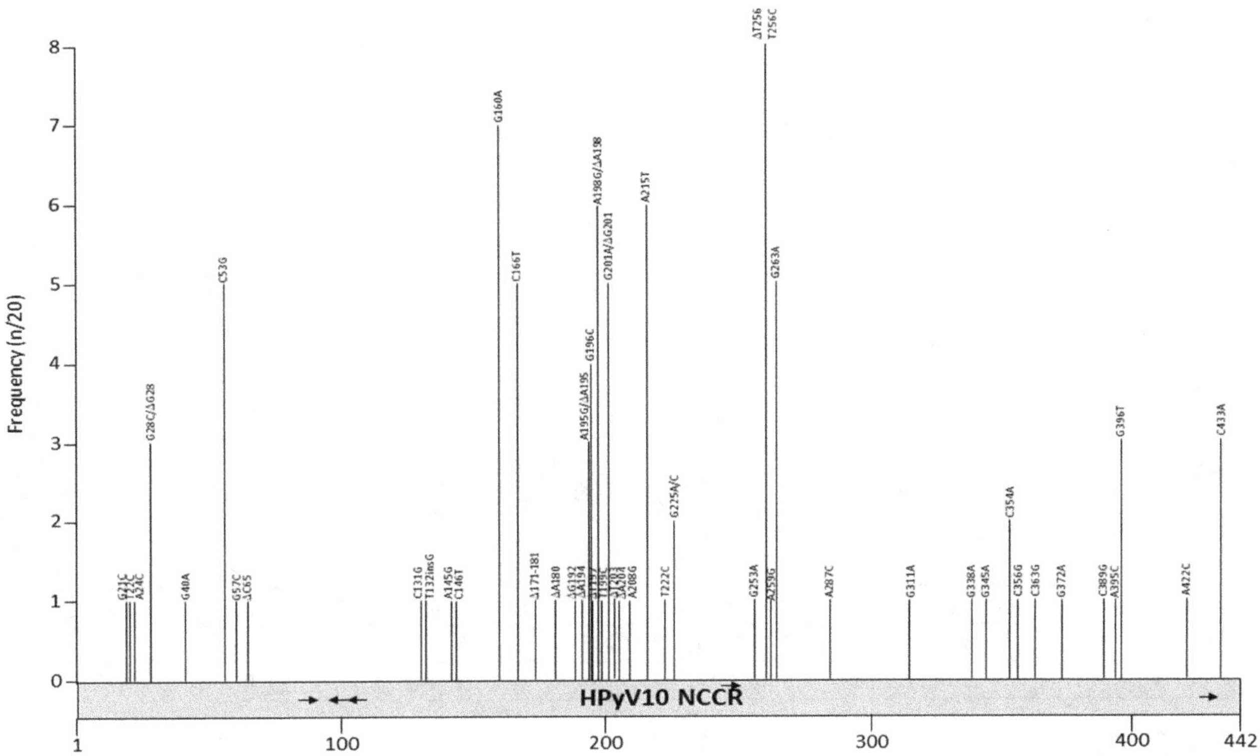

Figure 8. Mutations and their prevalence in variants of HPyV10 noncoding control region (NCCR). The numbering of the NCCR is from early to late with nucleotide 1 being the most proximal to the ATG start codon of the early genes and the most distal nucleotide, just upstream of the start codon of the *VP2* gene. The number of times a peculiar mutation is found in the different variants is given as frequency (with *n* the number of times the mutation was described/total NCCR variant sequences available). Putative Large T antigen (LT) binding motifs 5′-GRGGC-3′ (→) or 5′-GCCYC-3′ (←) are shown. The table summarizes the mutations, their location in the NCCR, and their frequency. For details see Table S1. Putative transcription factor binding sites are shown in Supplementary Table S2.

9. STLPyV NCCR Variants

Sequences of 7 STLPyV NCCRs are available in GenBank (Table S1). These variants were discovered in feces, respiratory swab or peri-anal warts [14,26,27,142,143]. Point mutations and one bp deletions have been observed (Figure 9 and Table S1), although the biological implications on promoter activity and viral replication have not been examined, nor has their possible role in disease been defined. So far, no variants with mutations in the possible LT binding motifs have been isolated. Putative binding sites for transcription factors in the STLPyV, NCCR are shown in Table 4 and Table S2.

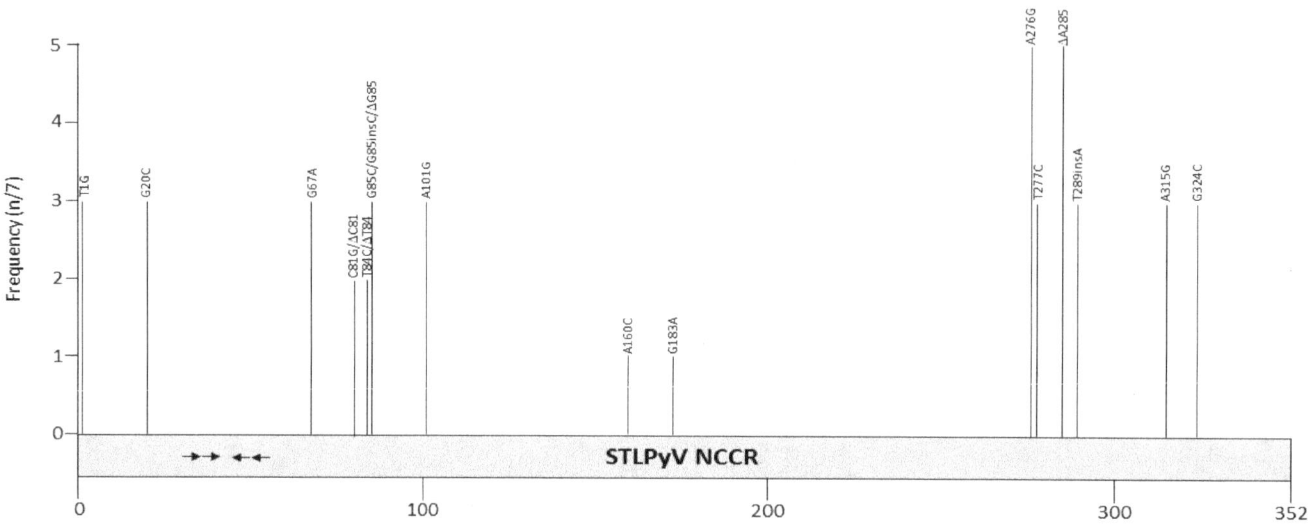

Mutation	Frequency *	Mutation	Frequency *	Mutation	Frequency *
T1G	3	G85C/G85insC/ΔG85	3	T277C	3
G20C	3	A101G	3	ΔA285	5
G67A	3	A160C	1	T289insA	3
C81G/ΔC81	2	G183A	1	A315G	3
T84C/ΔT84	3	A276G	5	G324C	3

* *n*/7.

Figure 9. Mutations and their prevalence in variants of Saint Louis polyomavirus (STLPyV) noncoding control region (NCCR). The numbering of the NCCR is from early to late with nucleotide 1 being the most proximal to the ATG start codon of the early genes and the most distal nucleotide, just upstream of the start codon of the *VP2* gene. The number of times a peculiar mutation is found in the different variants is given as frequency (with *n* the number of times the mutation was described/total NCCR variant sequences available). Putative Large T antigen (LT) binding motifs 5′-GRGGC-3′ (→) or 5′-GCCYC-3′ (←) are shown. The table summarizes the mutations, their location in the NCCR, and their frequency. For details see Table S1. Putative transcription factor binding sites are shown in Supplementary Table S2.

10. HPyV12, NJPyV, LIPyV and QPyV Variants

The NCCR sequence of two HPyV12 isolates is known. One of them carries a 26 bp deletion in the distal part of the late promoter (nucleotides 297–322) (Table S1; [102]). This deletion reduces the early and late promoter activity in 10 different human cell lines tested, except for the early promoter activity in BEL7402 and HEK293 cells, in which a significantly higher activity was measured compared with the early promoter activity of the original HPyV12 isolate [102]. The deletion eliminates putative c-MYB, CREB and AP4 binding sites, but it is not known whether these transcription factors actually bind the HPyV12 promoter.

Only one human and one feline LIPyV isolate have been reported, and they differ by four-point mutations (Table S1). No NJPyV and QPyV NCCR variants have been described thus far. Putative binding sites for transcription factors in the HPyV12, NJPyV and LIPyV NCCR are summarized in Table 4 and Table S2.

11. Conclusions

Similar to BKPyV and JCPyV, novel HPyV isolates with mutations in their NCCR are commonly detected in human samples. However, for the most recently described PyVs isolated from human specimens, none or very few isolates have been reported, and large deletions and/or duplication are lacking. Our knowledge of the effect of mutations in the NCCR on viral promoter activity and viral replication is incomplete because only a few studies have addressed the effect of NCCR mutations on the promoter activity, while the impact on viral replication has not been examined. Replication studies have been hampered by the lack of permissive cell systems for all novel HPyVs, except dermal fibroblasts which support productive MCPyV infection [144]. The HPyV NCCRs contain a plethora of binding motifs for host cell proteins, but their binding to the NCCR has not been confirmed. Chromatin immunoprecipitation assays at the early and late stages of infection may allow for the identification of transcription factors involved in early and late expression. Another unsolved question for most of the novel HPyVs is whether they are associated with specific diseases. Apart from MCPyV and its etiologic role in MCC, TSPyV as the causative agent of trichodysplasia spinulosa [9,145–147], and HPyV6 and HPyV7 with pruritic skin eruption in immunocompromised patients [124–128], firm evidence for pathogenic properties of the other novel HPyVs is lacking. So, far no specific MCPyV, HPyV6, HPyV7, and TSPyV NCCR variants seem to be associated with disease because virus variants with (quasi) identical NCCRs were also detected in samples from healthy individuals. Studies on different patient groups are required to unveil possible novel HPyV-associated diseases, as more NCCR sequences from larger and different patient cohorts are required to establish a possible connection between the genetic diversity of the NCCR and disease. The biological consequences of NCCR mutations for the viral life cycle warrants further investigation.

Author Contributions: Conceptualization, V.P., C.P. and U.M.; writing—original draft preparation, V.P., C.P. and U.M.; writing—review and editing, V.P., C.P. and U.M. All authors have read and agreed to the published version of the manuscript.

Abbreviations

bp	Base-pairs
HPyV	Human polyomavirus
IPPyV	Institute Pasteur polyomavirus
KIPy	Karolinska Institute polyomavirus
LIPyV	Lyon IARC polyomavirus
LT	Large T antigen
MCC	Merkel cell carcinoma
MCPyV	Merkel cell polyomavirus
MPyV	Mouse polyomavirus
NCCR	Non-coding control region
NJPyV	New Jersey polyomavirus

PML	Progressive multifocal leukoencephalopathy
QPyV	Quebec polyomavirus
sT	Small T antigen
STLPyV	Saint Louis polyomavirus
TSPyV	Trichodisplasia spinulosa polyomavirus
WUPyV	Washington University polyomavirus

References

1. DeCaprio, J.A.; Garcea, R.L. A cornucopia of human polyomaviruses. *Nat. Rev. Microbiol.* **2013**, *11*, 264–276. [CrossRef] [PubMed]
2. Moens, U.; Krumbholz, A.; Ehlers, B.; Zell, R.; Johne, R.; Calvignac-Spencer, S.; Lauber, C. Biology, evolution, and medical importance of polyomaviruses: An update. *Infect. Genet. Evol.* **2017**, *54*, 18–38. [CrossRef] [PubMed]
3. Peretti, A.; FitzGerald, P.C.; Bliskovsky, V.; Pastrana, D.V.; Buck, C.B. Genome Sequence of a Fish-Associated Polyomavirus, Black Sea Bass (Centropristis striata) Polyomavirus 1. *Genome Announc.* **2015**, *3*, e01476-14. [CrossRef] [PubMed]
4. Buck, C.B.; Van Doorslaer, K.; Peretti, A.; Geoghegan, E.M.; Tisza, M.J.; An, P.; Katz, J.P.; Pipas, J.M.; McBride, A.A.; Camus, A.C.; et al. The Ancient Evolutionary History of Polyomaviruses. *PLoS Pathog.* **2016**, *12*, e1005574. [CrossRef] [PubMed]
5. Gardner, S.D.; Field, A.M.; Coleman, D.V.; Hulme, B. New human papovavirus (B.K.) isolated from urine after renal transplantation. *Lancet* **1971**, *1*, 1253–1257. [CrossRef]
6. Padgett, B.L.; Walker, D.L.; ZuRhein, G.M.; Eckroade, R.J.; Dessel, B.H. Cultivation of papova-like virus from human brain with progressive multifocal leucoencephalopathy. *Lancet* **1971**, *1*, 1257–1260. [CrossRef]
7. Allander, T.; Andreasson, K.; Gupta, S.; Bjerkner, A.; Bogdanovic, G.; Persson, M.A.; Dalianis, T.; Ramqvist, T.; Andersson, B. Identification of a third human polyomavirus. *J. Virol.* **2007**, *81*, 4130–4136. [CrossRef]
8. Gaynor, A.M.; Nissen, M.D.; Whiley, D.M.; Mackay, I.M.; Lambert, S.B.; Wu, G.; Brennan, D.C.; Storch, G.A.; Sloots, T.P.; Wang, D. Identification of a novel polyomavirus from patients with acute respiratory tract infections. *PLoS Pathog.* **2007**, *3*, e64. [CrossRef]
9. Feng, H.; Shuda, M.; Chang, Y.; Moore, P.S. Clonal integration of a polyomavirus in human Merkel cell carcinoma. *Science* **2008**, *319*, 1096–1100. [CrossRef]
10. Schowalter, R.M.; Pastrana, D.V.; Pumphrey, K.A.; Moyer, A.L.; Buck, C.B. Merkel cell polyomavirus and two previously unknown polyomaviruses are chronically shed from human skin. *Cell Host Microbe* **2010**, *7*, 509–515. [CrossRef]
11. van der Meijden, E.; Janssens, R.W.; Lauber, C.; Bouwes Bavinck, J.N.; Gorbalenya, A.E.; Feltkamp, M.C. Discovery of a new human polyomavirus associated with trichodysplasia spinulosa in an immunocompromized patient. *PLoS Pathog.* **2010**, *6*, e1001024. [CrossRef] [PubMed]
12. Scuda, N.; Hofmann, J.; Calvignac-Spencer, S.; Ruprecht, K.; Liman, P.; Kühn, J.; Hengel, H.; Ehlers, B. A novel human polyomavirus closely related to the african green monkey-derived lymphotropic polyomavirus. *J. Virol.* **2011**, *85*, 4586–4590. [CrossRef] [PubMed]
13. Buck, C.B.; Phan, G.Q.; Raiji, M.T.; Murphy, P.M.; McDermott, D.H.; McBride, A.A. Complete genome sequence of a tenth human polyomavirus. *J. Virol.* **2012**, *86*, 10887. [CrossRef] [PubMed]
14. Lim, E.S.; Reyes, A.; Antonio, M.; Saha, D.; Ikumapayi, U.N.; Adeyemi, M.; Stine, O.C.; Skelton, R.; Brennan, D.C.; Mkakosya, R.S.; et al. Discovery of STL polyomavirus, a polyomavirus of ancestral recombinant origin that encodes a unique T antigen by alternative splicing. *Virology* **2013**, *436*, 295–303. [CrossRef]
15. Korup, S.; Rietscher, J.; Calvignac-Spencer, S.; Trusch, F.; Hofmann, J.; Moens, U.; Sauer, I.; Voigt, S.; Schmuck, R.; Ehlers, B. Identification of a novel human polyomavirus in organs of the gastrointestinal tract. *PLoS ONE* **2013**, *8*, e58021. [CrossRef]
16. Mishra, N.; Pereira, M.; Rhodes, R.H.; An, P.; Pipas, J.M.; Jain, K.; Kapoor, A.; Briese, T.; Faust, P.L.; Lipkin, W.I. Identification of a novel polyomavirus in a pancreatic transplant recipient with retinal blindness and vasculitic myopathy. *J. Infect. Dis.* **2014**, *210*, 1595–1599. [CrossRef]

17. Gheit, T.; Dutta, S.; Oliver, J.; Robitaille, A.; Hampras, S.; Combes, J.D.; McKay-Chopin, S.; Le Calvez-Kelm, F.; Fenske, N.; Cherpelis, B.; et al. Isolation and characterization of a novel putative human polyomavirus. *Virology* **2017**, *506*, 45–54. [CrossRef]
18. Ondov, B.D.; Starrett, G.J.; Sappington, A.; Kostic, A.; Koren, S.; Buck, C.B.; Phillippy, A.M. Mash Screen: High-throughput sequence containment estimation for genome discovery. *Genome Biol.* **2019**, *20*, 232. [CrossRef]
19. Calvignac-Spencer, S.; Feltkamp, M.C.; Daugherty, M.D.; Moens, U.; Ramqvist, T.; Johne, R.; Ehlers, B. A taxonomy update for the family Polyomaviridae. *Arch. Virol.* **2016**, *161*, 1739–1750. [CrossRef]
20. Moens, U.; Calvignac-Spencer, S.; Lauber, C.; Ramqvist, T.; Feltkamp, M.C.W.; Daugherty, M.D.; Verschoor, E.J.; Ehlers, B. ICTV Report Consortium. ICTV Virus Taxonomy Profile: Polyomaviridae. *J. Gen. Virol.* **2017**, *98*, 1159–1160. [CrossRef]
21. Kamminga, S.; van der Meijden, E.; Feltkamp, M.C.W.; Zaaijer, H.L. Seroprevalence of fourteen human polyomaviruses determined in blood donors. *PLoS ONE* **2018**, *13*, e0206273. [CrossRef] [PubMed]
22. Kourieh, A.; Combes, J.D.; Tommasino, M.; Dalstein, V.; Clifford, G.M.; Lacau St Guily, J.; Clavel, C.; Franceschi, S.; Gheit, T. For The Split Study, G. Prevalence and risk factors of human polyomavirus infections in non-malignant tonsils and gargles: The SPLIT study. *J. Gen. Virol.* **2018**, *99*, 1686–1698. [CrossRef] [PubMed]
23. Wang, Y.; Keinonen, A.; Koskenmies, S.; Pitkänen, S.; Fyhrquist, N.; Sadeghi, M.; Mäkisalo, H.; Söderlund-Venermo, M.; Hedman, K. Occurrence of newly discovered human polyomaviruses in skin of liver transplant recipients and their relation with squamous cell carcinoma in situ and actinic keratosis—a single-center cohort study. *Transpl. Int.* **2019**, *32*, 516–522. [CrossRef] [PubMed]
24. Fahsbender, E.; Altan, E.; Estrada, M.; Seguin, M.A.; Young, P.; Leutenegger, C.M.; Delwart, E. Lyon-IARC Polyomavirus DNA in Feces of Diarrheic Cats. *Microbiol. Resour. Announc.* **2019**, *8*. [CrossRef]
25. Gaboriaud, P.; Ferté, M.; Arnold, F.; Leblond, V.; Nicol, J.; Debare, H.; Le Meur, M.; Martini, F.; Tognon, M.; Touzé, A. Age-specific seroprevalence of human polyomavirus 12 and Saint Louis and New Jersey polyomaviruses. *Emerg. Microbes Infect.* **2018**, *7*, 22. [CrossRef] [PubMed]
26. Bialasiewicz, S.; Rockett, R.J.; Barraclough, K.A.; Leary, D.; Dudley, K.J.; Isbel, N.M.; Sloots, T.P. Detection of Recently Discovered Human Polyomaviruses in a Longitudinal Kidney Transplant Cohort. *Am. J. Transpl.* **2016**, *16*, 2734–2740. [CrossRef]
27. Li, K.; Zhang, C.; Zhao, R.; Xue, Y.; Yang, J.; Peng, J.; Jin, Q. The prevalence of STL polyomavirus in stool samples from Chinese children. *J. Clin.* **2015**, *66*, 19–23. [CrossRef]
28. Herberhold, S.; Hellmich, M.; Panning, M.; Bartok, E.; Silling, S.; Akgül, B.; Wieland, U. Human polyomavirus and human papillomavirus prevalence and viral load in non-malignant tonsillar tissue and tonsillar carcinoma. *Med. Microbiol. Immunol.* **2017**, *206*, 93–103. [CrossRef]
29. Bergallo, M.; Daprà, V.; Fava, P.; Ponti, R.; Calvi, C.; Montanari, P.; Novelli, M.; Quaglino, P.; Galliano, I.; Fierro, M.T. DNA from Human Polyomaviruses, MWPyV, HPyV6, HPyV7, HPyV9 and HPyV12 in Cutaneous T-cell Lymphomas. *Anticancer Res.* **2018**, *38*, 4111–4114. [CrossRef]
30. Daprà, V.; Galliano, I.; Rassu, M.; Calvi, C.; Montanari, P.; Merlino, C.; Bergallo, M. Lack of detection of HPyV12 DNA using real-time PCR in Italian infants with diarrhea. *Minerva Pediatr.* **2020**. [CrossRef]
31. Kamminga, S.; van der Meijden, E.; Wunderink, H.F.; Touzé, A.; Zaaijer, H.L.; Feltkamp, M.C.W. Development and Evaluation of a Broad Bead-Based Multiplex Immunoassay To Measure IgG Seroreactivity against Human Polyomaviruses. *J. Clin. Microbiol.* **2018**, *56*, e01566-17. [CrossRef] [PubMed]
32. Gedvilaite, A.; Tryland, M.; Ulrich, R.G.; Schneider, J.; Kurmauskaite, V.; Moens, U.; Preugschas, H.; Calvignac-Spencer, S.; Ehlers, B. Novel polyomaviruses in shrews (Soricidae) with close similarity to human polyomavirus 12. *J. Gen. Virol.* **2017**, *98*, 3060–3067. [CrossRef] [PubMed]
33. Schowalter, R.M.; Buck, C.B. The Merkel cell polyomavirus minor capsid protein. *PLoS Pathog.* **2013**, *9*, e1003558. [CrossRef] [PubMed]
34. Madeira, F.; Park, Y.M.; Lee, J.; Buso, N.; Gur, T.; Madhusoodanan, N.; Basutkar, P.; Tivey, A.R.N.; Potter, S.C.; Finn, R.D.; et al. The EMBL-EBI search and sequence analysis tools APIs in 2019. *Nucleic Acids Res.* **2019**, *47*, W636–W641. [CrossRef] [PubMed]
35. Fanning, E.; Zhao, K. SV40 DNA replication: From the A gene to a nanomachine. *Virology* **2009**, *384*, 352–359. [CrossRef]
36. Kelly, T.J. SV40 DNA replication. *J. Biol. Chem.* **1988**, *263*, 17889–17892. [PubMed]

37. Farmerie, W.G.; Folk, W.R. Regulation of polyomavirus transcription by large tumor antigen. *Proc. Natl. Acad. Sci. USA* **1984**, *81*, 6919–6923. [CrossRef]
38. Tjian, R. Regulation of viral transcription and DNA replication by the SV40 large T antigen. *Curr. Top Microbiol. Immunol.* **1981**, *93*, 5–24.
39. Zenke, M.; Grundström, T.; Matthes, H.; Wintzerith, M.; Schatz, C.; Wildeman, A.; Chambon, P. Multiple sequence motifs are involved in SV40 enhancer function. *EMBO J.* **1986**, *5*, 387–397. [CrossRef]
40. Rio, D.C.; Tjian, R. Multiple control elements involved in the initiation of SV40 late transcription. *J. Mol. Appl. Gen.* **1984**, *2*, 423–435.
41. Cowie, A.; Kamen, R. Multiple binding sites for polyomavirus large T antigen within regulatory sequences of polyomavirus DNA. *J. Virol.* **1984**, *52*, 750–760. [CrossRef] [PubMed]
42. Jones, K.A.; Tjian, R. Essential contact residues within SV40 large T antigen binding sites I and II identified by alkylation-interference. *Cell* **1984**, *36*, 155–162. [CrossRef]
43. Lednicky, J.A.; Butel, J.S. Simian virus 40 regulatory region structural diversity and the association of viral archetypal regulatory regions with human brain tumors. *Semin. Cancer Biol.* **2001**, *11*, 39–47. [CrossRef] [PubMed]
44. O'Neill, F.J.; Greenlee, J.E.; Carney, H. The archetype enhancer of simian virus 40 DNA is duplicated during virus growth in human cells and rhesus monkey kidney cells but not in green monkey kidney cells. *Virology* **2003**, *310*, 173–182. [CrossRef]
45. Lednicky, J.A.; Wong, C.; Butel, J.S. Artificial modification of the viral regulatory region improves tissue culture growth of SV40 strain 776. *Virus Res.* **1995**, *35*, 143–153. [CrossRef]
46. Sroller, V.; Vilchez, R.A.; Stewart, A.R.; Wong, C.; Butel, J.S. Influence of the viral regulatory region on tumor induction by simian virus 40 in hamsters. *J. Virol.* **2008**, *82*, 871–879. [CrossRef] [PubMed]
47. Muller, W.J.; Mueller, C.R.; Mes, A.M.; Hassell, J.A. Polyomavirus origin for DNA replication comprises multiple genetic elements. *J. Virol.* **1983**, *47*, 586–599. [CrossRef] [PubMed]
48. Pomerantz, B.J.; Hassell, J.A. Polyomavirus and simian virus 40 large T antigens bind to common DNA sequences. *J. Virol.* **1984**, *49*, 925–937. [CrossRef]
49. Herbomel, P.; Bourachot, B.; Yaniv, M. Two distinct enhancers with different cell specificities coexist in the regulatory region of polyoma. *Cell* **1984**, *39*, 653–662. [CrossRef]
50. Iacoangeli, A.; Melucci-Vigo, G.; Risuleo, G.; Santi, E. Role of mouse polyomavirus late region in the control of viral DNA replication: A review. *Biochimie* **1995**, *77*, 780–786. [CrossRef]
51. Jat, P.; Novak, U.; Cowie, A.; Tyndall, C.; Kamen, R. DNA sequences required for specific and efficient initiation of transcription at the polyoma virus early promoter. *Mol. Cell Biol.* **1982**, *2*, 737–751. [CrossRef] [PubMed]
52. Mueller, C.R.; Mes-Masson, A.M.; Bouvier, M.; Hassell, J.A. Location of sequences in polyomavirus DNA that are required for early gene expression in vivo and in vitro. *Mol. Cell Biol.* **1984**, *4*, 2594–2609. [CrossRef]
53. Rochford, R.; Campbell, B.A.; Villarreal, L.P. Genetic analysis of the enhancer requirements for polyomavirus DNA replication in mice. *J. Virol.* **1990**, *64*, 476–485. [CrossRef] [PubMed]
54. Sekikawa, K.; Levine, A.J. Isolation and characterization of polyoma host range mutants that replicate in nullipotential embryonal carcinoma cells. *Proc. Natl. Acad. Sci. USA* **1981**, *78*, 1100–1104. [CrossRef] [PubMed]
55. Veldman, G.M.; Lupton, S.; Kamen, R. Polyomavirus enhancer contains multiple redundant sequence elements that activate both DNA replication and gene expression. *Mol. Cell Biol.* **1985**, *5*, 649–658. [CrossRef] [PubMed]
56. White, M.K.; Khalili, K. Pathogenesis of progressive multifocal leukoencephalopathy—revisited. *J. Infect. Dis.* **2011**, *203*, 578–586. [CrossRef]
57. Assetta, B.; Atwood, W.J. The biology of JC polyomavirus. *Biol. Chem.* **2017**, *398*, 839–855. [CrossRef]
58. Chen, N.N.; Khalili, K. Transcriptional regulation of human JC polyomavirus promoters by cellular proteins YB-1 and Pur alpha in glial cells. *J. Virol.* **1995**, *69*, 5843–5848. [CrossRef]
59. Sadowska, B.; Barrucco, R.; Khalili, K.; Safak, M. Regulation of human polyomavirus JC virus gene transcription by AP-1 in glial cells. *J. Virol.* **2003**, *77*, 665–672. [CrossRef]
60. Romagnoli, L.; Sariyer, I.K.; Tung, J.; Feliciano, M.; Sawaya, B.E.; Del Valle, L.; Ferrante, P.; Khalili, K.; Safak, M.; White, M.K. Early growth response-1 protein is induced by JC virus infection and binds and regulates the JC virus promoter. *Virology* **2008**, *375*, 331–341. [CrossRef]

61. Pietropaolo, V.; Prezioso, C.; Bagnato, F.; Antonelli, G. John Cunningham virus: An overview on biology and disease of the etiological agent of the progressive multifocal leukoencephalopathy. *New Microbiol.* **2018**, *41*, 179–186. [PubMed]
62. Van Loy, T.; Thys, K.; Tritsmans, L.; Stuyver, L.J. Quasispecies analysis of JC virus DNA present in urine of healthy subjects. *PLoS ONE* **2013**, *8*, e70950. [CrossRef] [PubMed]
63. Kenney, S.; Natarajan, V.; Strike, D.; Khoury, G.; Salzman, N.P. JC virus enhancer-promoter active in human brain cells. *Science* **1984**, *226*, 1337–1339. [CrossRef] [PubMed]
64. Yogo, Y.; Kitamura, T.; Sugimoto, C.; Ueki, T.; Aso, Y.; Hara, K.; Taguchi, F. Isolation of a possible archetypal JC virus DNA sequence from nonimmunocompromised individuals. *J. Virol.* **1990**, *64*, 3139–3143. [CrossRef] [PubMed]
65. L'Honneur, A.S.; Leh, H.; Laurent-Tchenio, F.; Hazan, U.; Rozenberg, F.; Bury-Moné, S. Exploring the role of NCCR variation on JC polyomavirus expression from dual reporter minicircles. *PLoS ONE* **2018**, *13*, e0199171. [CrossRef] [PubMed]
66. Agostini, H.T.; Ryschkewitsch, C.F.; Stoner, G.L. Rearrangements of archetypal regulatory regions in JC virus genomes from urine. *Res. Virol.* **1998**, *149*, 163–170. [CrossRef]
67. Bofill-Mas, S.; Clemente-Casares, P.; Major, E.O.; Curfman, B.; Girones, R. Analysis of the excreted JC virus strains and their potential oral transmission. *J. Neurovirol.* **2003**, *9*, 498–507. [CrossRef]
68. Frisque, R.J.; Bream, G.L.; Cannella, M.T. Human polyomavirus JC virus genome. *J. Virol.* **1984**, *51*, 458–469. [CrossRef]
69. Markowitz, R.B.; Dynan, W.S. Binding of cellular proteins to the regulatory region of BK virus DNA. *J. Virol.* **1988**, *62*, 3388–3398. [CrossRef]
70. Moens, U.; Van Ghelue, M. Polymorphism in the genome of non-passaged human polyomavirus BK: Implications for cell tropism and the pathological role of the virus. *Virology* **2005**, *331*, 209–231. [CrossRef] [PubMed]
71. Rubinstein, R.; Pare, N.; Harley, E.H. Structure and function of the transcriptional control region of nonpassaged BK virus. *J. Virol.* **1987**, *61*, 1747–1750. [CrossRef] [PubMed]
72. Bethge, T.; Ajuh, E.; Hirsch, H.H. Imperfect Symmetry of Sp1 and Core Promoter Sequences Regulates Early and Late Virus Gene Expression of the Bidirectional BK Polyomavirus Noncoding Control Region. *J. Virol.* **2016**, *90*, 10083–10101. [CrossRef] [PubMed]
73. Bethge, T.; Hachemi, H.A.; Manzetti, J.; Gosert, R.; Schaffner, W.; Hirsch, H.H. Sp1 sites in the noncoding control region of BK polyomavirus are key regulators of bidirectional viral early and late gene expression. *J. Virol.* **2015**, *89*, 3396–3411. [CrossRef] [PubMed]
74. Helle, F.; Brochot, E.; Handala, L.; Martin, E.; Castelain, S.; Francois, C.; Duverlie, G. Biology of the BKPyV: An Update. *Viruses* **2017**, *9*, 327. [CrossRef]
75. Moens, U.; Sundsfjord, A.; Flaegstad, T.; Traavik, T. BK virus early RNA transcripts in stably transformed cells: Enhanced levels induced by dibutyryl cyclic AMP, forskolin and 12-O-tetradecanoylphorbol-13-acetate treatment. *J. Gen. Virol.* **1990**, *71*, 1461–1471. [CrossRef]
76. Moens, U.; Subramaniam, N.; Johansen, B.; Johansen, T.; Traavik, T. A steroid hormone response unit in the late leader of the noncoding control region of the human polyomavirus BK confers enhanced host cell permissivity. *J. Virol.* **1994**, *68*, 2398–2408. [CrossRef]
77. Gorrill, T.S.; Khalili, K. Cooperative interaction of p65 and C/EBPbeta modulates transcription of BKV early promoter. *Virology* **2005**, *335*, 1–9. [CrossRef]
78. Anselmo, A.; Prezioso, C.; Saccà, F.A.; Di Lella, F.M.; Palmieri, G.; Tisone, G.; Pietropaolo, V.; Ciotti, M. Kidney graft failure induced by BKPyV replication despite a strong reduction of the immunosuppressive therapy. *J. Med. Virol.* **2019**, *91*, 1698–1701. [CrossRef]
79. Arthur, R.R.; Shah, K.V.; Baust, S.J.; Santos, G.W.; Saral, R. Association of BK viruria with hemorrhagic cystitis in recipients of bone marrow transplants. *N. Engl. J. Med.* **1986**, *315*, 230–234. [CrossRef]
80. Rosen, S.; Harmon, W.; Krensky, A.M.; Edelson, P.J.; Padgett, B.L.; Grinnell, B.W.; Rubino, M.J.; Walker, D.L. Tubulo-interstitial nephritis associated with polyomavirus (BK type) infection. *N. Engl. J. Med.* **1983**, *308*, 1192–1196. [CrossRef]
81. McIlroy, D.; Halary, F.; Bressollette-Bodin, C. Intra-patient viral evolution in polyomavirus-related diseases. *Philos. Trans. R. Soc. B Biol. Sci.* **2019**, *374*, 20180301. [CrossRef] [PubMed]
82. Seif, I.; Khoury, G.; Dhar, R. The genome of human papovavirus BKV. *Cell* **1979**, *18*, 963–977. [CrossRef]

83. Yang, J.F.; You, J. Regulation of Polyomavirus Transcription by Viral and Cellular Factors. *Viruses* **2020**, *12*, 1072. [CrossRef] [PubMed]
84. Olsen, G.H.; Hirsch, H.H.; Rinaldo, C.H. Functional analysis of polyomavirus BK non-coding control region quasispecies from kidney transplant recipients. *J. Med. Virol.* **2009**, *81*, 1959–1967. [CrossRef]
85. Degener, A.M.; Pietropaolo, V.; Di Taranto, C.; Jin, L.; Ameglio, F.; Cordiali-Fei, P.; Trento, E.; Sinibaldi, L.; Orsi, N. Identification of a new control region in the genome of the DDP strain of BK virus isolated from PBMC. *J. Med. Virol.* **1999**, *58*, 413–419. [CrossRef]
86. Pietropaolo, V.; Videtta, M.; Fioriti, D.; Mischitelli, M.; Arancio, A.; Orsi, N.; Degener, A.M. Rearrangement patterns of JC virus noncoding control region from different biological samples. *J. Neurovirol.* **2003**, *9*, 603–611. [CrossRef] [PubMed]
87. Mischitelli, M.; Fioriti, D.; Videtta, M.; Degener, A.M.; Antinori, A.; Cinque, P.; Giordano, A.; Pietropaolo, V. Investigation on the role of cell transcriptional factor Sp1 and HIV-1 TAT protein in PML onset or development. *J. Cell Physiol.* **2005**, *204*, 913–918. [CrossRef] [PubMed]
88. Ciardi, M.R.; Zingaropoli, M.A.; Iannetta, M.; Prezioso, C.; Perri, V.; Pasculli, P.; Lichtner, M.; D'Ettorre, G.; Altieri, M.; Conte, A.; et al. JCPyV NCCR analysis in PML patients with different risk factors: Exploring common rearrangements as essential changes for neuropathogenesis. *Virol. J.* **2020**, *17*, 23. [CrossRef] [PubMed]
89. Prezioso, C.; Zingaropoli, M.A.; Iannetta, M.; Rodio, D.M.; Altieri, M.; Conte, A.; Vullo, V.; Ciardi, M.R.; Palamara, A.T.; Pietropaolo, V. Which is the best PML risk stratification strategy in natalizumab-treated patients affected by multiple sclerosis? *Mult. Scler. Relat. Disord.* **2020**, *41*, 102008. [CrossRef] [PubMed]
90. Prezioso, C.; Scribano, D.; Bellizzi, A.; Anzivino, E.; Rodio, D.M.; Trancassini, M.; Palamara, A.T.; Pietropaolo, V. Efficient propagation of archetype JC polyomavirus in COS-7 cells: Evaluation of rearrangements within the NCCR structural organization after transfection. *Arch. Virol.* **2017**, *162*, 3745–3752. [CrossRef]
91. Prezioso, C.; Scribano, D.; Rodio, D.M.; Ambrosi, C.; Trancassini, M.; Palamara, A.T.; Pietropaolo, V. COS-7-based model: Methodological approach to study John Cunningham virus replication cycle. *Virol. J.* **2018**, *15*, 29. [CrossRef] [PubMed]
92. Jiang, M.; Abend, J.R.; Johnson, S.F.; Imperiale, M.J. The role of polyomaviruses in human disease. *Virology* **2009**, *384*, 266–273. [CrossRef] [PubMed]
93. Babakir-Mina, M.; Ciccozzi, M.; Perno, C.F.; Ciotti, M. The human polyomaviruses KI and WU: Virological background and clinical implications. *APMIS* **2013**, *121*, 746–754. [CrossRef] [PubMed]
94. Jartti, T.; Jartti, L.; Ruuskanen, O.; Söderlund-Venermo, M. New respiratory viral infections. *Curr. Opin. Pulm. Med.* **2012**, *18*, 271–278. [CrossRef]
95. Gelfand, Y.; Rodriguez, A.; Benson, G. TRDB—the Tandem Repeats Database. *Nucleic Acids Res.* **2007**, *35*, D80–D87. [CrossRef]
96. Song, X.; Van Ghelue, M.; Ludvigsen, M.; Nordbø, S.A.; Ehlers, B.; Moens, U. Characterization of the non-coding control region of polyomavirus KI isolated from nasopharyngeal samples from patients with respiratory symptoms or infection and from blood from healthy blood donors in Norway. *J. Gen. Virol.* **2016**, *97*, 1647–1657. [CrossRef]
97. Li, K.; Guo, J.; Zhao, R.; Xue, Y.; Chen, L.; Yang, J.; Peng, J.; Jin, Q. Prevalence of 10 human polyomaviruses in fecal samples from children with acute gastroenteritis: A case-control study. *J. Clin. Microbiol.* **2013**, *51*, 3107–3109. [CrossRef]
98. Tsunoda, T.; Takagi, T. Estimating transcription factor bindability on DNA. *Bioinformatics* **1999**, *15*, 622–630. [CrossRef]
99. Jackstadt, R.; Röh, S.; Neumann, J.; Jung, P.; Hoffmann, R.; Horst, D.; Berens, C.; Bornkamm, G.W.; Kirchner, T.; Menssen, A.; et al. AP4 is a mediator of epithelial-mesenchymal transition and metastasis in colorectal cancer. *J. Exp. Med.* **2013**, *210*, 1331–1350. [CrossRef]
100. Uhlén, M.; Fagerberg, L.; Hallstrom, B.M.; Lindskog, C.; Oksvold, P.; Mardinoglu, A.; Sivertsson, Å.; Kampf, C.; Sjöstedt, E.; Asplund, A.; et al. Proteomics. Tissue-based map of the human proteome. *Science* **2015**, *347*, 1260419. [CrossRef]
101. Csoma, E.; Lengyel, G.; Bányai, K.; Takács, P.; Ánosi, N.; Marton, S.; Mátyus, M.; Pászti, E.; Gergely, L.; Szűcs, A. Study of Karolinska Institutet and Washington University polyomaviruses in tonsil, adenoid, throat swab and middle ear fluid samples. *Future Microbiol.* **2018**, *13*, 1719–1730. [CrossRef] [PubMed]

102. Moens, U.; Van Ghelue, M.; Ludvigsen, M.; Korup-Schulz, S.; Ehlers, B. Early and late promoters of BKPyV, MCPyV, TSPyV, and HPyV12 are among the strongest of all known human polyomaviruses in 10 different cell lines. *J. Gen. Virol.* **2015**, *96*, 2293–2303. [CrossRef] [PubMed]
103. Sood, R.; Kamikubo, Y.; Liu, P. Role of RUNX1 in hematological malignancies. *Blood* **2017**, *129*, 2070–2082. [CrossRef] [PubMed]
104. Gosert, R.; Rinaldo, C.H.; Funk, G.A.; Egli, A.; Ramos, E.; Drachenberg, C.B.; Hirsch, H.H. Polyomavirus BK with rearranged noncoding control region emerge in vivo in renal transplant patients and increase viral replication and cytopathology. *J. Exp. Med.* **2008**, *205*, 841–852. [CrossRef] [PubMed]
105. Gosert, R.; Kardas, P.; Major, E.O.; Hirsch, H.H. Rearranged JC virus noncoding control regions found in progressive multifocal leukoencephalopathy patient samples increase virus early gene expression and replication rate. *J. Virol.* **2010**, *84*, 10448–10456. [CrossRef]
106. Feng, H.; Kwun, H.J.; Liu, X.; Gjoerup, O.; Stolz, D.B.; Chang, Y.; Moore, P.S. Cellular and viral factors regulating Merkel cell polyomavirus replication. *PLoS ONE* **2011**, *6*, e22468. [CrossRef] [PubMed]
107. Kwun, H.J.; Guastafierro, A.; Shuda, M.; Meinke, G.; Bohm, A.; Moore, P.S.; Chang, Y. The minimum replication origin of merkel cell polyomavirus has a unique large T-antigen loading architecture and requires small T-antigen expression for optimal replication. *J. Virol.* **2009**, *83*, 12118–12128. [CrossRef] [PubMed]
108. Hashida, Y.; Higuchi, T.; Matsui, K.; Shibata, Y.; Nakajima, K.; Sano, S.; Daibata, M. Genetic Variability of the Noncoding Control Region of Cutaneous Merkel Cell Polyomavirus: Identification of Geographically Related Genotypes. *J. Infect. Dis.* **2018**, *217*, 1601–1611. [CrossRef]
109. Delbue, S.; Franciotta, D.; Giannella, S.; Dolci, M.; Signorini, L.; Ticozzi, R.; D'Alessandro, S.; Campisciano, G.; Comar, M.; Ferrante, P.; et al. Human Polyomaviruses in the Cerebrospinal Fluid of Neurological Patients. *Microorganisms* **2019**, *8*, 16. [CrossRef]
110. Prezioso, C.; Obregon, F.; Ambroselli, D.; Petrolo, S.; Checconi, P.; Rodio, D.M.; Coppola, L.; Nardi, A.; Vito, C.; Sarmati, L.; et al. Merkel Cell Polyomavirus (MCPyV) in the Context of Immunosuppression: Genetic Analysis of Noncoding Control Region (NCCR) Variability among a HIV-1-Positive Population. *Viruses* **2020**, *12*, 507. [CrossRef]
111. White, M.K.; Safak, M.; Khalili, K. Regulation of gene expression in primate polyomaviruses. *J. Virol.* **2009**, *83*, 10846–10856. [CrossRef] [PubMed]
112. Ajuh, E.T.; Wu, Z.; Kraus, E.; Weissbach, F.H.; Bethge, T.; Gosert, R.; Fischer, N.; Hirsch, H.H. Novel Human Polyomavirus Noncoding Control Regions Differ in Bidirectional Gene Expression according to Host Cell, Large T-Antigen Expression, and Clinically Occurring Rearrangements. *J. Virol.* **2018**, *92*, e02231-17. [CrossRef] [PubMed]
113. Prezioso, C.; Ciotti, M.; Obregon, F.; Ambroselli, D.; Rodio, D.M.; Cudillo, L.; Gaziev, J.; Mele, A.; Nardi, A.; Favalli, C.; et al. Polyomaviruses shedding in stool of patients with hematological disorders: Detection analysis and study of the non-coding control region's genetic variability. *Med. Microbiol. Immunol.* **2019**, *208*, 845–854. [CrossRef] [PubMed]
114. Bialasiewicz, S.; Lambert, S.B.; Whiley, D.M.; Nissen, M.D.; Sloots, T.P. Merkel cell polyomavirus DNA in respiratory specimens from children and adults. *Emerg. Infect. Dis.* **2009**, *15*, 492–494. [CrossRef]
115. Kantola, K.; Sadeghi, M.; Lahtinen, A.; Koskenvuo, M.; Aaltonen, L.M.; Möttönen, M.; Rahiala, J.; Saarinen-Pihkala, U.; Riikonen, P.; Jartti, T.; et al. Merkel cell polyomavirus DNA in tumor-free tonsillar tissues and upper respiratory tract samples: Implications for respiratory transmission and latency. *J. Clin. Virol.* **2009**, *45*, 292–295. [CrossRef]
116. Babakir-Mina, M.; Ciccozzi, M.; Lo Presti, A.; Greco, F.; Perno, C.F.; Ciotti, M. Identification of Merkel cell polyomavirus in the lower respiratory tract of Italian patients. *J. Med. Virol.* **2010**, *82*, 505–509. [CrossRef]
117. Abedi Kiasari, B.; Vallely, P.J.; Klapper, P.E. Merkel cell polyomavirus DNA in immunocompetent and immunocompromised patients with respiratory disease. *J. Med. Virol.* **2011**, *83*, 2220–2224. [CrossRef]
118. Shikova, E.; Emin, D.; Alexandrova, D.; Shindov, M.; Kumanova, A.; Lekov, A.; Moens, U. Detection of Merkel Cell Polyomavirus in Respiratory Tract Specimens. *Intervirology* **2017**, *60*, 28–32. [CrossRef]
119. Iaria, M.; Caccuri, F.; Apostoli, P.; Giagulli, C.; Pelucchi, F.; Padoan, R.F.; Caruso, A.; Fiorentini, S. Detection of KI WU and Merkel cell polyomavirus in respiratory tract of cystic fibrosis patients. *Clin. Microbiol. Infect.* **2015**, *21*, e609–e615. [CrossRef]

120. Prezioso, C.; Di Lella, F.M.; Rodio, D.M.; Bitossi, C.; Trancassini, M.; Mele, A.; de Vito, C.; Antonelli, G.; Pietropaolo, V. Merkel Cell Polyomavirus DNA Detection in Respiratory Samples: Study of a Cohort of Patients Affected by Cystic Fibrosis. *Viruses* **2019**, *11*, 571. [CrossRef]
121. Abdulsalam, I.; Rasheed, K.; Sveinbjørnsson, B.; Ehlers, B.; Moens, U. Promoter activity of Merkel cell Polyomavirus variants in human dermal fibroblasts and a Merkel cell carcinoma cell line. *Virol. J.* **2020**, *17*, 54. [CrossRef]
122. Shuda, M.; Feng, H.; Kwun, H.J.; Rosen, S.T.; Gjoerup, O.; Moore, P.S.; Chang, Y. T antigen mutations are a human tumor-specific signature for Merkel cell polyomavirus. *Proc. Natl. Acad. Sci. USA* **2008**, *105*, 16272–16277. [CrossRef] [PubMed]
123. Hashida, Y.; Higuchi, T.; Matsuzaki, S.; Nakajima, K.; Sano, S.; Daibata, M. Prevalence and Genetic Variability of Human Polyomaviruses 6 and 7 in Healthy Skin Among Asymptomatic Individuals. *J. Infect. Dis.* **2018**, *217*, 483–493. [CrossRef] [PubMed]
124. Ho, J.; Jedrych, J.J.; Feng, H.; Natalie, A.A.; Grandinetti, L.; Mirvish, E.; Crespo, M.M.; Yadav, D.; Fasanella, K.E.; Proksell, S.; et al. Human polyomavirus 7-associated pruritic rash and viremia in transplant recipients. *J. Infect. Dis.* **2015**, *211*, 1560–1565. [CrossRef]
125. Nguyen, K.D.; Lee, E.E.; Yue, Y.; Stork, J.; Pock, L.; North, J.P.; Vandergriff, T.; Cockerell, C.; Hosler, G.A.; Pastrana, D.V.; et al. Human polyomavirus 6 and 7 are associated with pruritic and dyskeratotic dermatoses. *J. Am. Acad. Derm.* **2017**, *76*, 932–940.e933. [CrossRef] [PubMed]
126. Canavan, T.N.; Baddley, J.W.; Pavlidakey, P.; Tallaj, J.A.; Elewski, B.E. Human polyomavirus-7-associated eruption successfully treated with acitretin. *Am. J. Transpl.* **2018**, *18*, 1278–1284. [CrossRef] [PubMed]
127. Smith, S.D.B.; Erdag, G.; Cuda, J.D.; Rangwala, S.; Girardi, N.; Bibee, K.; Orens, J.B.; Prono, M.D.; Toptan, T.; Loss, M.J. Treatment of human polyomavirus-7-associated rash and pruritus with topical cidofovir in a lung transplant patient: Case report and literature review. *Transpl. Infect. Dis.* **2018**, *20*, e12793. [CrossRef]
128. Rosenstein, R.K.; Pastrana, D.V.; Starrett, G.J.; Sapio, M.R.; Hill, N.T.; Jo, J.H.; Lee, C.R.; Iadarola, M.J.; Buck, C.B.; Kong, H.H.; et al. Host-Pathogen Interactions in Human Polyomavirus 7 (HPyV7)-associated Pruritic Skin Eruption. *J. Investig. Derm.* **2020**. [CrossRef]
129. Klufah, F.; Mobaraki, G.; Chteinberg, E.; Alharbi, R.A.; Winnepenninckx, V.; Speel, E.J.M.; Rennspiess, D.; Olde Damink, S.W.; Neumann, U.P.; Kurz, A.K.; et al. High Prevalence of Human Polyomavirus 7 in Cholangiocarcinomas and Adjacent Peritumoral Hepatocytes: Preliminary Findings. *Microorganisms* **2020**, *8*, 1125. [CrossRef]
130. Schrama, D.; Groesser, L.; Ugurel, S.; Hafner, C.; Pastrana, D.V.; Buck, C.B.; Cerroni, L.; Theiler, A.; Becker, J.C. Presence of human polyomavirus 6 in mutation-specific BRAF inhibitor-induced epithelial proliferations. *JAMA Dermatol.* **2014**, *150*, 1180–1186. [CrossRef]
131. Hampras, S.S.; Locke, F.L.; Chavez, J.C.; Patel, N.S.; Giuliano, A.R.; Miller, K.; Gheit, T.; Tommasino, M.; Rollison, D.E. Prevalence of cutaneous viral infections in incident cutaneous squamous cell carcinoma detected among chronic lymphocytic leukemia and hematopoietic stem cell transplant patients. *Leuk. Lymphoma* **2018**, *59*, 911–917. [CrossRef] [PubMed]
132. Kamminga, S.; van der Meijden, E.; de Brouwer, C.; Feltkamp, M.; Zaaijer, H. Prevalence of DNA of fourteen human polyomaviruses determined in blood donors. *Transfusion* **2019**, *59*, 3689–3697. [CrossRef] [PubMed]
133. Hashida, Y.; Higuchi, T.; Tanaka, M.; Shibata, Y.; Nakajima, K.; Sano, S.; Daibata, M. Prevalence and Viral Loads of Cutaneous Human Polyomaviruses in the Skin of Patients With Chronic Inflammatory Skin Diseases. *J. Infect. Dis.* **2019**, *219*, 1564–1573. [CrossRef] [PubMed]
134. Harrison, C.; Jiang, T.; Banerjee, P.; Meinke, G.; D'Abramo, C.M.; Schaffhausen, B.; Bohm, A. Polyomavirus large T antigen binds symmetrical repeats at the viral origin in an asymmetrical manner. *J. Virol.* **2013**, *87*, 13751–13759. [CrossRef]
135. Toptan, T.; Yousem, S.A.; Ho, J.; Matsushima, Y.; Stabile, L.P.; Fernández-Figueras, M.T.; Bhargava, R.; Ryo, A.; Moore, P.S.; Chang, Y. Survey for human polyomaviruses in cancer. *JCI Insight* **2016**, *1*, e85562. [CrossRef]
136. Kazem, S.; Lauber, C.; van der Meijden, E.; Kooijman, S.; Kravchenko, A.A.; Feltkamp, M.C.; Gorbalenya, A.E. Limited variation during circulation of a polyomavirus in the human population involves the COCO-VA toggling site of Middle and Alternative T-antigen(s). *Virology* **2016**, *487*, 129–140. [CrossRef]
137. Sauvage, V.; Foulongne, V.; Cheval, J.; Ar Gouilh, M.; Pariente, K.; Dereure, O.; Manuguerra, J.C.; Richardson, J.; Lecuit, M.; Burguière, A.; et al. Human polyomavirus related to African green monkey lymphotropic polyomavirus. *Emerg. Infect. Dis.* **2011**, *17*, 1364–1370. [CrossRef]

138. Lednicky, J.A.; Butel, J.S.; Luetke, M.C.; Loeb, J.C. Complete genomic sequence of a new Human polyomavirus 9 strain with an altered noncoding control region. *Virus Genes.* **2014**, *49*, 490–492. [CrossRef]
139. Moens, U.; Song, X.; Van Ghelue, M.; Lednicky, J.A.; Ehlers, B. A Role of Sp1 Binding Motifs in Basal and Large T-Antigen-Induced Promoter Activities of Human Polyomavirus HPyV9 and Its Variant UF-1. *Int. J. Mol. Sci.* **2017**, *18*, 2414. [CrossRef]
140. Siebrasse, E.A.; Reyes, A.; Lim, E.S.; Zhao, G.; Mkakosya, R.S.; Manary, M.J.; Gordon, J.I.; Wang, D. Identification of MW polyomavirus, a novel polyomavirus in human stool. *J. Virol.* **2012**, *86*, 10321–10326. [CrossRef]
141. Rockett, R.J.; Sloots, T.P.; Bowes, S.; O'Neill, N.; Ye, S.; Robson, J.; Whiley, D.M.; Lambert, S.B.; Wang, D.; Nissen, M.D.; et al. Detection of novel polyomaviruses, TSPyV, HPyV6, HPyV7, HPyV9 and MWPyV in feces, urine, blood, respiratory swabs and cerebrospinal fluid. *PLoS ONE* **2013**, *8*, e62764. [CrossRef]
142. Pastrana, D.V.; Fitzgerald, P.C.; Phan, G.Q.; Raiji, M.T.; Murphy, P.M.; McDermott, D.H.; Velez, D.; Bliskovsky, V.; McBride, A.A.; Buck, C.B. A divergent variant of the eleventh human polyomavirus species, saint louis polyomavirus. *Genome Announc.* **2013**, *1*, e00812-13. [CrossRef] [PubMed]
143. Peng, J.; Li, K.; Zhang, C.; Jin, Q. MW polyomavirus and STL polyomavirus present in tonsillar tissues from children with chronic tonsillar disease. *Clin. Microbiol. Infect.* **2016**, *22*, e91–e97. [CrossRef]
144. Liu, W.; Yang, R.; Payne, A.S.; Schowalter, R.M.; Spurgeon, M.E.; Lambert, P.F.; Xu, X.; Buck, C.B.; You, J. Identifying the Target Cells and Mechanisms of Merkel Cell Polyomavirus Infection. *Cell Host Microbe* **2016**, *19*, 775–787. [CrossRef] [PubMed]
145. Pietropaolo, V.; Prezioso, C.; Moens, U. Merkel Cell Polyomavirus and Merkel Cell Carcinoma. *Cancers* **2020**, *12*, 1774. [CrossRef] [PubMed]
146. Kazem, S.; van der Meijden, E.; Feltkamp, M.C. The trichodysplasia spinulosa-associated polyomavirus: Virological background and clinical implications. *APMIS* **2013**, *121*, 770–782. [CrossRef] [PubMed]
147. Narayanan, D.; Rady, P.L.; Tyring, S.K. Recent developments in trichodysplasia spinulosa disease. *Transpl. Infect. Dis.* **2020**, e13434. [CrossRef] [PubMed]

3

Viral Genomic Characterization and Replication Pattern of Human Polyomaviruses in Kidney Transplant Recipients

Lucia Signorini [1,*], Maria Dolci [1], Evaldo Favi [2,3], Caterina Colico [3], Mariano Ferraresso [2,3], Rosalia Ticozzi [1], Giuseppe Basile [4], Pasquale Ferrante [1] and Serena Delbue [1]

1 Biomedical, Surgical and Dental Sciences, University of Milano, 20133 Milano, Italy; dolci@unimi.it (M.D.); rosalia.ticozzi@unimi.it (R.T.); pasquale.ferrante@unimi.it (P.F.); serena.delbue@unimi.it (S.D.)

2 Department of Clinical Sciences and Community Health, University of Milano, 20122 Milano, Italy; evaldofavi@gmail.com (E.F.); mariano.ferraresso@unimi.it (M.F.)

3 Kidney Transplantation, Fondazione IRCCS Ca' Granda, Ospedale Maggiore Policlinico, 20122 Milano, Italy; caterina.colico@policlinico.mi.it

4 Service of Legal Medicine, San Siro Clinical Institute, 20148 Milano, Italy; basiletraumaforense@gmail.com

* Correspondence: lucia.signorini@unimi.it

Abstract: Human Polyomavirus (HPyV) infections are common, ranging from 60% to 100%. In kidney transplant (KTx) recipients, HPyVs have been associated with allograft nephropathy, progressive multifocal leukoencephalopathy, and skin cancer. Whether such complications are caused by viral reactivation or primary infection transmitted by the donor remains debated. This study aimed to investigate the replication pattern and genomic characterization of BK Polyomavirus (BKPyV), JC Polyomavirus (JCPyV), and Merkel Cell Polyomavirus (MCPyV) infections in KTx. Urine samples from 57 KTx donor/recipient pairs were collected immediately before organ retrieval/transplant and periodically up to post-operative day 540. Specimens were tested for the presence of BKPyV, JCPyV, and MCPyV genome by virus-specific Real-Time PCR and molecularly characterized. HPyVs genome was detected in 49.1% of donors and 77.2% of recipients. Sequences analysis revealed the archetypal strain for JCPyV, TU and Dunlop strains for BKPyV, and IIa-2 strain for MCPyV. VP1 genotyping showed a high frequency for JCPyV genotype 1 and BKPyV genotype I. Our experience demonstrates that after KTx, HPyVs genome remains stable over time with no emergence of quasi-species. HPyVs strains isolated in donor/recipient pairs are mostly identical, suggesting that viruses detected in the recipient may be transmitted by the allograft.

Keywords: Human Polyomavirus JC (JCPyV); Human Polyomavirus BK (BKPyV), Merkel Cell Polyomavirus (MCPyV); kidney transplantation (KTx); urine; molecular characterization

1. Introduction

Human Polyomaviruses (HPyVs) are ubiquitous DNA viruses with a reported seroprevalence in the general population ranging from 60% to 100%, depending on the series [1,2]. After primary infection, usually contracted during childhood and asymptomatic, they often establish life-long latency in renal tubular epithelial cells and to a lesser degree in other cell types. In immunocompromised hosts, HPyVs can start uncontrolled replication and eventually lead to cell lysis, necrosis, and irreversible tissue damage [1].

Due to their chronic exposure to powerful immunosuppressive agents, kidney transplant (KTx) recipients have a greater risk of clinically significant HPyVs infections than healthy controls [3,4].

To date, the most relevant HPyVs detected in this particular group of patients are BK Polyomavirus (BKPyV), JC Polyomavirus (JCPyV), and Merkel Cell Polyomavirus (MCPyV) [2,4,5]. In particular, BKPyV has been linked to Polyomavirus-associated nephropathy (PVAN), JCPyV to progressive multifocal leukoencephalopathy (PML), and MCPyV to some forms of non-melanoma skin cancer [6]. Whether such complications are actually caused by reactivations of latent viruses of the recipient or rather by viruses transmitted by the donor through the allograft remains unclear [5].

HPyVs double-stranded DNA genome is circular and covalently closed. It is divided into an Early, a Late, and a Non-Coding Control Region (NCCR) containing the origin of DNA replication as well as the promoter/enhancer elements directing viral genes transcription [7]. The Early region, normally expressed prior to the initiation of viral DNA replication, encodes the Large and Small Tumour Antigens (LTAg and STAg) that play a key role in DNA replication, late viral genes transcription, and virions assembly [8]. The Late region encodes three viral capsid proteins (VP1, VP2, and VP3) required for capsid assembly and sustaining the uncoating process of the virus when it enters the host cells [9]. HPyV Late region could also contain open reading frames (*agnogene*), encoding for a 66 amino acid protein named agnoprotein—a regulatory cytoplasmic protein involved in viral release. The presence of the *agnogene* was demonstrated for JCPyV and BKPyV while there is no evidence that MCPyV expresses proteins analogous to agnoprotein [10–12]. Sequence variations analyses of the VP1 gene are routinely performed to discriminate between different viral genotypes [13]. Nevertheless, while BKPyV and JCPyV genotypes have been extensively characterized, MCPyV genotyping is still in progress [14–24]. The hypervariability of the NCCR, especially considering MCPyV, represents another topic of interest as differences in the sequences constituting this specific region are currently used to distinguish archetypal strains from rearranged ones and to obtain prognostic information regarding the course of the infection [13,25–28].

The present study aimed to evaluate origins, urinary replication patterns, and molecular characteristics of BKPyV, JCPyV, and MCPyV infections after KTx.

2. Materials and Methods

2.1. Study Design

In this single-center prospective observational study with an 18-month follow-up, we enrolled 57 consecutive KTx donor/recipient pairs. Both deceased and living donor transplants were considered. There were no exclusion criteria. Organ retrievals were carried out in hospitals under the area of influence of the Nord Italia Transplant (NITp) Organization whereas transplants were performed at the Fondazione IRCCS Ca' Granda Ospedale Maggiore Policlinico (Milan) between June 2016 and December 2019.

The main demographic and clinical characteristics of the participants are summarized in Table 1.

Table 1. Demographic and clinical characteristics of kidney transplant donors and recipients.

	Donors (n = 57)	**Recipients (n = 57)**
Variables	Mean ± SD or n (%)	
Male:female	26:24 *	32:25
Country of origin ≠ Italy	4/53 ## (7.5)	7/50 ### (14)
Age (years)	55.2 ± 13.2 **	44.3 ± 17.9 ***
Living:deceased	23:29 #	-
Basiliximab	-	31/52 $ (59.6)
Tacrolimus	-	51/52 $ (98.1)
Mycophenolate mofetil	-	50/52 $ (96.1)

* 7 (12.3%) of 57 missing cases; ** 8 (14%) of 57 missing cases; *** 1 (1.8%) of 57 missing cases; # 5 (8.8%) of 57 missing cases; ## Foreign countries: Albania, Sri Lanka, and Colombia; 1 (25%) of 4 missing case; ### Foreign countries: Kenya, Ethiopia, Tunisia, Austria, Lebanon; 2 (28.6%) of 7 missing cases; $ 5 (8.8%) of 57 missing cases.

Baseline urine samples were collected immediately before organ retrieval (donors) and transplants (recipients). Follow-up specimens (recipients) were obtained according to the time-schedule reported in Table 2.

Table 2. Urinary specimen's collection time points.

Days After Transplantation	0	1	15	30	60	90	180	270	360	540
Code	T0	T1	T2	T3	T4	T5	T6	T7	T8	T9

The subjects involved signed a specific informed consent. The study was approved by the Fondazione IRCCS Ca' Granda Ospedale Maggiore Policlinico Ethical Committee (protocol 2191) and conducted according to the World Health Organization Declaration of Helsinki and applicable regulatory requirements.

2.2. *DNA Isolation and Quantitative Real-Time PCR Assay*

Urine samples from donors and recipients were collected in sterile containers and immediately transported to the laboratory where they were spun at 1500 rpm for 30′ and the supernatants were stored at −20 °C until use. Viral DNA was isolated from 150 μL of supernatant using the commercial kit NucleoSpin RNA virus kit (Macherey-Nagel GmbH & Co. KG, Dueren, Germany), according to manufacturer's instructions. Quantitative Real-Time PCR (Q-RT-PCR) was performed using the Applied Biosystems 7500 Real-Time PCR System (Applied Biosystems, Foster City, CA, USA) with specific primers and TaqMan probe technology targeting the VP1 sequence of BKPyV and MCPyV or the LTAg sequence of JCPyV, adopting reaction conditions and thermal profiles previously described [29]. Strict precautionary measures were taken to avoid cross-contamination, including the use of separate rooms for nucleic acid extraction, preparation of amplification mixtures, execution of Q-RT-PCR, and nested PCR assays. Negative controls of DNA extraction and amplification were added to each run. Results were expressed as copies/mL of the urine sample.

2.3. *HPyVs Molecular Characterization*

HPyVs strains were molecularly characterized using different sets of specific nested PCR reactions. Amplification products were analyzed on 2% agarose gel in 0.5X TBE buffer. Positive PCR products were purified utilizing the QIAquick Gel Extraction Kit (QIAGEN, Hilden, Germany), following the manufacturer's instructions.

2.3.1. Molecular Characterization of JCPyV

JCPyV genomes were molecularly characterized using nested PCR reactions for the amplification of the VP1 region and NCCR. Specific primer sets and thermal profiles were utilized for each nested PCR reaction (see Tables S1–S3 for details) [14,30–36]. The amplifications were carried out in a total volume of 50 μL containing 2 U of Taq DNA Polymerase (Solis BioDyne OÜ, Tartu, Estonia) in the presence of 10 μL of 5X Reaction Buffer supplied by the manufacturer and 3–10 μL of DNA template. For JCPyV VP1 external/internal amplification, the PCR reaction was finished adding 3 μL of 10 mM dNTPs solution, 5 μL of 25 mM $MgCl_2$, and 3 μL of 10 μM primers. For JCPyV NCCR external/internal amplification, the PCR reaction was finished adding 2 μL of 10 mM dNTPs solution, 4 μL of 25 mM $MgCl_2$, and 2 μL of 10 μM primers.

2.3.2. Molecular Characterization of BKPyV

BKPyV genomes were molecularly characterized using nested PCR reactions for the amplification of the VP1 region and NCCR. Specific primer sets and thermal profiles were utilized for each nested PCR reaction (see Tables S1–S3 for details) [14,30–36]. The amplifications were carried out in a total volume of 50 μL containing 2 U of Taq DNA Polymerase (Solis BioDyne OÜ, Tartu, Estonia) in the presence of 10 μL of 5X Reaction Buffer supplied by the manufacturer and 3–10 μL of DNA template. For BKPyV VP1 external/internal amplification, the PCR reaction was finished adding 4 μL of 10 mM dNTPs solution, 4 μL of 25 mM $MgCl_2$, and 2 μL of 10 μM primers. For BKPyV NCCR external/internal amplification, the PCR reaction was finished adding 2 μL of 10 mM dNTPs solution, 4 μL of 25 mM $MgCl_2$, and 2,5/2 μL of 10 μM primers.

2.3.3. Molecular Characterization of MCPyV

MCPyV genomes were characterized utilizing nested PCR reactions for the LTAg region and NCCR. Specific primer sets and thermal profiles were used for each nested PCR reaction (see Tables S1–S3 for details) [14,30–36]. The amplifications were carried out in a total volume of 50 μL containing 2 U of Taq DNA Polymerase (Solis BioDyne OÜ, Tartu, Estonia) in the presence of 10 μL of 5X Reaction Buffer supplied by the manufacturer and 3–10 μL of DNA template. For MCPyV VP1 external/internal amplification, the PCR reaction was finished adding 2 μL of 10 mM dNTPs solution, 2 μL of 25 mM $MgCl_2$, and 2.5 μL of 10 μM primers. For MCPyV NCCR external/internal amplification, the PCR reaction was finished adding 2 μL of 10 mM dNTPs solution, 4/2 μL of 25 mM $MgCl_2$, and 2 μL of 10 μM primers. Each reaction was completed with nuclease-free water up to a total volume of 50 μL.

2.4. DNA Automatic Sequencing

DNA sequencing of the inner fragment of the nested PCR was carried out at an external facility (Eurofins Genomics GmbH, Ebersberg, Germany) using both forward and reverse primers. As previously described by Agostini et al. [37], for JCPyV genotyping we performed BLAST searches on the NCBI website (https://blast.ncbi.nlm.nih.gov) to assess sequences homology. As reported by Ault et al. [19] and Jensen et al. [38], BLAST searches were also performed to identify JCPyV NCCR rearrangements. BKPyV genotypes were determined following the classification method proposed by Jin et al. [39]. MCPyV sequences were aligned with those of the prototype strain MCC350 deposited in GenBank under the accession number EU375803 and analyzed according to the classification method proposed by Hashida et al. [35].

2.5. Immunosuppression and Concomitant Medications

As induction, low-immunological risk recipients (primary transplant, last panel-reactive antibody status < 50%, undetectable donor-specific antibody, and kidney retrieved from a living donor or a standard-criteria donation after brain death donor) were given intravenous basiliximab (Simulect®, Novartis, Basel, Switzerland) 20 mg on days 0 and day 4 whereas high-immunological risk patients received intravenous rATG (Thymoglobulin®, Genzyme, Cambridge, MA, USA) 5 mg/kg total-dose from day 0 to day 4. Pre- and post-transplant plasmapheresis along with a seven-day course of intravenous IgG up to 2 g/kg total-dose were selectively administered in case of donor-specific antibody > 2000 MFI at the time of transplant. Participants were also given intravenous methylprednisolone 500 mg on day 0, 250 mg on day 1, and 125 mg on day 2. As for maintenance, patients received oral tacrolimus (Adoport®, Novartis, Basel, Switzerland) 0.10 mg/kg/day and oral mycophenolate mofetil (Myfenax®, Teva, Petach Tikva, Israel) 2000 mg/day from day 0. Tacrolimus dose was adjusted to achieve a trough level of 8–12 ng/mL during the first post-transplant month and 6–8 ng/mL thereafter. From day 3, patients received oral prednisone 20 mg/day, progressively tapered to 5 mg/day after 1 month. For pneumocystis jirovecii and cytomegalovirus prophylaxis we used oral trimethoprim/sulfamethoxazole 80/400 mg/day three times per week for three months and oral valganciclovir (dose titrated according to renal function) for 3 months, respectively.

2.6. Statistical Analysis

Categorical and numerical variables were described using proportions, medians, and 1st–3rd interquartile ranges (IQR). Results were compared using Fisher's exact test, Chi-square test or

Mann–Whitney U test as appropriate. The significance of statistical tests was taken at a p-value < 0.05. Analyses were performed using SPSS (version 23.0; IBM Corp., Armonk, NY, USA).

3. Results

3.1. *HPyVs Viruria in the Study Cohort*

As summarized in Table 3, HPyVs genomes were detected in the urine of 28/57 (49.1%) donors and 44/57 (77.2%) recipients ($p = 0.0033$). Among the donors with viruria, 25/28 (89.3%) tested positive for JCPyV, 3/28 (10.7%) for BKPyV, and 2/28 (7.1%) for MCPyV ($p < 0.00001$ comparing JCPyV vs. BKPyV and JCPyV vs. MCPyV). The proportions of viruric recipients positive for JCPyV, BKPyV, and MCPyV were 25/44 (56.8%), 21/44 (47.7%), and 17/44 (38.6%), respectively (p = ns). Notably, the differences observed in virus-specific prevalence between donors and recipients were all statistically significant ($p = 0.0039$ for JCPyV; $p = 0.0017$ for BKPyV, and $p = 0.0029$ for MCPyV). Coinfections were only recorded in the recipients' group (19/44, 43.2% vs. 0/28, 0%; $p < 0.00001$). More in details, the following combinations were observed: JCPyV plus BKPyV (6/44, 13.6%), JCPyV plus MCPyV (8/44, 18.2%), BKPyV plus MCPyV (2/44, 4.5%), and JCPyV plus BKPyV plus MCPyV (3/44, 6.8%). Specific coinfection rates were not significantly different (p = ns).

Table 3. Human Polyomavirus (HPyV) prevalence in kidney transplant donors and recipients.

	JCPyV +/tot (%)	BKPyV +/tot (%)	MCPyV +/tot (%)	HPyV/tot (%)
Donors *	25/28 (89.3) $^{\#,\$,\&}$	3/28 (10.7) $^{\hat{},\$}$	2/28 (7.1) $^{\circ,\&}$	28/57 (49.1) $^{+}$
Recipients **	25/44 (56.8) $^{\#}$	21/44 (47.7) $^{\hat{}}$	17/44 (38.6) $^{\circ}$	44/57 (77.2) $^{+}$
Total	50 $^{\alpha}$/72 (69.4) $^{\S,£}$	24 $^{\beta}$/72 (33.3) §	19 $^{\pi}$/72 (26.4) $^{£}$	72/114 (63.1)

* Two coinfection, ** Nineteen coinfection, $^{\alpha}$ Nineteen donor/recipient pairs, $^{\beta}$ Two donor/recipient pair, $^{\pi}$ One donor/recipient pair, $^{\#,\hat{},\circ,+,\$,\&,\S,£}$ $p < 0.05$.

3.2. *HPyVs Viruria in Urinary Samples*

As summarized in Table 4, HPyVs genomes were detected in 272/551 (49.4%) urinary samples. More in details, 172/551 (31.2%) specimens tested positive for JCPyV, 62/551 (11.2%) for BKPyV, and 38/551 (6.9%) for MCPyV ($p < 0.00001$ comparing JCPyV vs. BKPyV and JCPyV vs. MCPyV; $p = 0.016$ comparing BKPyV vs. MCPyV). Coinfection was observed in 38/551 (6.9%) samples: JCPyV plus BKPyV in 10/551 (1.8%), JCPyV plus MCPyV in 24/551 (4.3%) and BKPyV plus MCPyV in 4/551 (0.7%).

Table 4. HPyVs prevalence in urinary samples collected from kidney transplant donors and recipients.

JCPyV +/tot (%)	BKPyV +/tot (%)	MCPyV +/tot (%)	HPyV/tot (%)
172/551 (31.2) $^{*,\circ,\#}$	62/551 (11.2) $^{*,\hat{},\#,\$}$	38/551 (6.9) $^{\circ,\hat{},\#,\$}$	272/551 (49.4)

* Ten JCPyV/BKPyV coinfections, ° Twenty-four JCPyV/MCPyV coinfections, ˆ Four BKPyV/MCPyV coinfections. $^{\#}$ $p < 0.0001$, $^{\$}$ $p < 0.05$.

3.3. *HPyVs Urinary Load*

Median virus-specific urinary loads in the study cohort are reported in Table 5 and depicted in Figure 1. JCPyV showed the highest urinary load in both donors' and recipients' groups: 7.11×10^5 copies/mL (IQR $9.51 \times 10^4 - 1.26 \times 10^6$ copies/mL) and 1.29×10^6 copies/mL (IQR $5.66 \times 10^4 - 3.17 \times 10^7$), respectively ($p = 0.0043$ comparing JCPyV vs. BKPyV among donors; $p = 0.0166$ and $p = 0.0158$ comparing JCPyV vs. BKPyV and JCPyV vs. MCPyV among recipients. We also observed that BKPyV urinary load was significantly higher in the recipients' group than donors' one (1.02×10^5 copies/mL, IQR $1.30 \times 10^4 - 3.69 \times 10^6$ vs. 1.80×10^3 copies/mL, IQR $1.18 \times 10^3 - 2.76 \times 10^3$; $p = 0.0107$).

Table 5. HPyVs urinary loads in kidney transplant donors and recipients.

	Median Viral Load Copies/mL (IQR *)			
	JCPyV	**BKPyV**	**MCPyV**	**HPyVs**
Donors	$7.11 \times 10^{5\,\circ}$ (1.17×10^{6})	$1.80 \times 10^{3\,\circ,\$}$ (1.58×10^{3})	1.18×10^{5} (1.10×10^{5})	3.64×10^{5} (1.15×10^{6})
Recipients	$1.29 \times 10^{6\,\#,\&}$ (3.16×10^{7})	$1.02 \times 10^{5\,\#,\$}$ (3.67×10^{6})	$3.76 \times 10^{5\,\&}$ (1.47×10^{6})	5.65×10^{5} (1.54×10^{7})
Total	1.01×10^{6} (2.57×10^{7})	6.39×10^{4} (2.60×10^{6})	3.67×10^{5} (1.34×10^{6})	5.42×10^{5} (1.16×10^{7})

* Interquartile range = 75° percentile—25° percentile; $^{\circ,\#,\&,\$}$ $p < 0.05$.

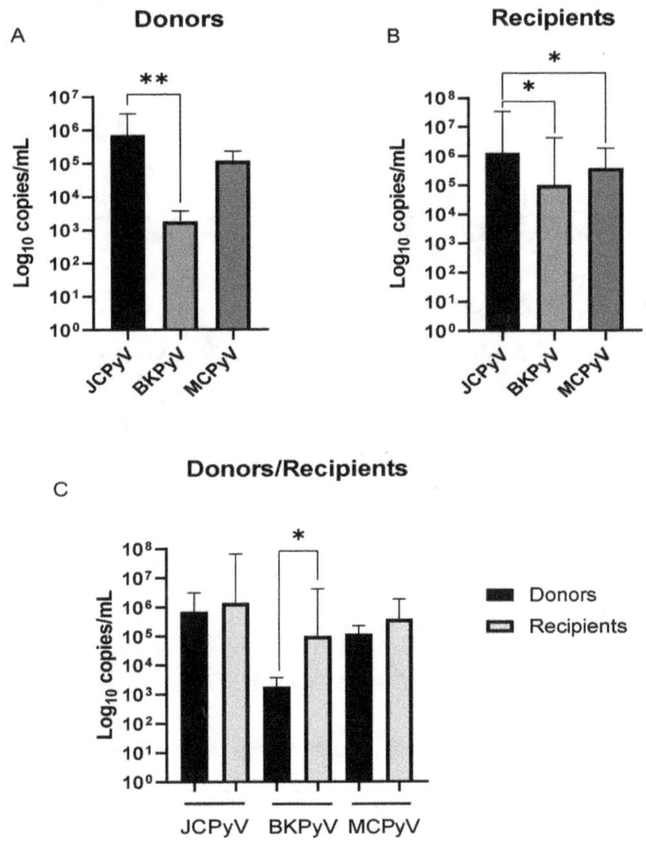

Figure 1. Median (IQR) virus-specific urinary loads in donors' (**A**) and recipients' (**B**) groups; direct comparison between donors' and recipients' median (IQR) virus-specific urinary loads (**C**). p values: ** $p = 0.0043$ (A), * $p = 0.0166$ (JCPyV vs. BKPyV and * $p = 0.0158$ JCPyV vs. MCPyV (B), * $p = 0.0107$ (C).

3.4. HPyVs Molecular Characterization

Donors' and recipients' urinary samples positive for HPyVs DNA as detected by Q-RT-PCR assay were amplified using HPyVs nested-PCR reactions and then sequenced for VP1 genes and NCCR to assess possible rearrangements. PCR reactions were performed on viral strains amplified in specimens collected from the donors at baseline and in specimens collected from the recipients on post-transplant day 1, 270, and 540.

3.4.1. JCPyV Molecular Characterization

JCPyV molecular characterization was carried out in 24 donors and 24 recipients. Sequence analysis of the NCCR demonstrated the presence of the archetypal strain, with several point mutations (data not shown), in all donors and recipients isolates, at every time points. As detailed in Table 6, VP1 gene sequence analysis in the donors' group allowed to identify genotype 1a (6/24, 25%), genotype 1b (7/24, 29.2%), genotype 2a (1/24, 4.2%), genotype 2c (3/24, 12.5%), genotype 4 (5/24, 20.8%), and genotype 6 (1/24, 4.2%). Strains and genotypes distributions in the recipients' group (Table 6) were concordant with those observed in the donor's group with the only exception of 2/24 (8.3%) recipients who experienced a shift from genotype 1 to genotype 4 at 270 and 540 days after KTx. In both cases, genotype 4 was transmitted by the donor (Table 6 and Figure S1).

Table 6. JC Polyomavirus (JCPyV) genotypes distribution in urine samples from kidney transplant donors and recipients.

JCPyV Genotype	Positive Donors (n = 24) (%)	Positive Recipients (n = 24) (%)
1a	6 (25.0%)	7* (25.0%)
1b	7 (29.2%)	7 (29.2%)
2a	1 (4.2%)	1 (4.2%)
2c	3 (12.5%)	2 (8.3%)
4	5 (20.8%)	6 * (25.0%)
6	1 (4.2%)	1 (4.2%)
Without identification	1 (4.2%)	-

* T1 genotype, switched to genotype 4 in the following time points in two recipients. *n* = number of donors/recipients.

3.4.2. BKPyV Molecular Characterization

BKPyV molecular characterization was performed in three donors and 13 recipients. Sequence analysis of the NCCR in the donors' group demonstrated the presence of the archetypal strain (1/3, 33.3%) and the TU strain (1/3, 33.3%); due to technical issues, results were only available for 2 out of 3 donors. In the recipients' group, we detected the WW strain (7/8, 87.5%) and the Dunlop strain (1/8; 12.5%). As described in Table 7 and depicted in Figure S2, VP1 gene sequence analysis among donors allowed to identify genotype Ia (1/3, 33.3%) and genotype Ib-1 (2/3, 66.6%). In the recipients' group, genotypes were as follow: genotype Ia (1/13, 7.7%), genotype Ib-1 (8/13, 61.6%), and genotype Ib-2 (4/13, 30.8%). Matching donors and recipients, we found that strains and genotypes were concordant with the only exception of a single donor/recipient pair showing the WW strain genotype Ib-1 in the donor and the Dunlop strain genotype Ia in the recipient (Table 7 and Figure S2).

Table 7. BK Polyomavirus (BKPyV) genotypes distribution in urine samples from kidney transplant donors and recipients.

BKPyV Genotype	Positive Donors (n = 3) (%)	Positive Recipients (n = 13) (%)
Ia	1 (33.3%)	1 (7.7%)
Ib-1	2 (66.7%)	8 (61.5%)
Ib-2	-	4 (30.8%)

3.4.3. MCPyV Molecular Characterization

MCPyV molecular characterization was carried out in two donors and 14 recipients. Sequence analysis of the NCCR region, performed in one donor and nine recipients, demonstrated the presence of the genotype IIa-2 in all subjects. VP1 gene sequence was analyzed in one donor and five recipients. With the exclusion of a few point mutations (data not shown), results demonstrated a complete alignment with the MCC350 strain.

3.5. HPyV-Associated Diseases

No episodes of PML or MCPyV-related skin cancer were recorded. One patient with BKPyV viruria (and BKPyV viremia) developed PVAN during the follow-up. This recipient successfully managed to reduce the net state of immunosuppression. Following treatment, allograft function remained consistently stable over time with no evidence of rejection.

4. Discussion

KTx is universally considered the golden standard treatment of end-stage renal disease [3]. Due to a widespread application of aggressive immunosuppressive protocols, we have witnessed a remarkable reduction of acute rejection-related allograft losses over the years [40,41]. The drawback of such a great achievement is that we are now facing a significant rise in post-transplant infectious diseases and drug-related adverse effects [42–44]. Among infectious complications, those caused by HPyVs represent a strenuous challenge for the transplant community. In fact, the ubiquitous distribution of these specific viruses in the general population, their ability to establish asymptomatic latency, the severity of the conditions caused under immunosuppression, and the lack of proper antiviral therapies make it extremely difficult to develop adequate prevention and treatment strategies [1,45,46]. In this context, a better understanding of HPyVs epidemiology, biology, and behavior in the immunocompromised host is paramount.

The main objectives of the present study were to evaluate the prevalence of JCPyV, BKPyV, and MCPyV in the actual KTx donors and recipients pool, to investigate whether episodes of viral replication during the various post-transplant phases are due to reactivations of previously acquired latent viruses or rather primary infections transmitted by the allograft, to assess possible interactions between different HPyVs, and to further characterize clinically relevant strains and genotypes.

Testing urinary samples collected from an Italian pool of KTx donor/recipient pairs, we found that the overall prevalence of HPyVs was about 50% among donors and 80% among recipients. Finding a higher percentage of positive recipients compared to donors might represent an ideal situation, since recipients would be more likely to develop a prompt immunological response against the HPyVs, in case of reactivation.

These results are in line with those from other reports demonstrating active replication of HPyVs in up to 60% of healthy individuals [2,47–52] and 75% of adult transplanted patients [53]. They also confirm the key role of immunosuppression in promoting HPyVs replication as the proportion of recipients with viruria was significantly higher than donors and progressively increased along with the duration of the exposure to immunosuppressants [2,47–52].

In this series, JCPyV was the most frequently detected HPyV in both donors' and recipients' groups. High JCPyV replication rates in immunocompetent subjects have been already demonstrated [2,47–52]. On the contrary, available data on KTx recipients remain conflicting [53,54]. Even though some authors have shown that JCPyV urinary prevalence can be as high as 75% and greater than BKPyV [53,54], the majority of reports suggest that BKPyV viruria is actually more frequent than JCPyV [55,56]. A possible reason behind this discrepancy is that most of the previously published studies evaluating BKPyV prevalence in urine samples focused on patients with documented viremia, symptomatic infection, or biopsy-proven PVAN and therefore more likely to have concomitant BKPyV viruria compared to our population of prospectively followed-up asymptomatic KTx recipients [55,56]. The fact that the cumulative incidence of clinically significant BKPyV infection observed during the present study was on the lower end of the values reported in the literature may also have contributed. Results obtained assessing MCPyV prevalence are more difficult to interpret as there is still a generalized lack of information regarding this specific virus in the KTx setting. Overall, they seem aligned with the few reports available showing relatively low percentages of viruric individuals among both donors and recipients [51,55,57].

In the present study, the relationship between immunosuppressive medications and HPyVs prevalence or replication pattern in urine was not investigated. In previous research, we observed that high-immunological risk characteristics such as Afro-Caribbean ethnicity, donor-recipient HLA mismatch > 4, and panel-reactive antibody test > 50% were predictors of BKPyV infection and viremia after KTx but no statistically significant associations between specific induction agents or maintenance drugs could be detected [58]. In-depth analysis of the effects of different anti-rejection treatments on HPyV infections susceptibility as well as targeted investigations aiming to determine a possible dose-effect or concentration-effect of tacrolimus and mycophenolate mofetil on HPyVs replication in blood and urine are certainly warranted.

Quantitative analysis of HPyVs viruria after transplantation demonstrated that JCPyV exhibited the highest viral load, followed by MCPyV and BKPyV. As previously reported by our group and others authors, these findings confirm high JCPyV urinary loads in patients receiving immunosuppressive therapy [51,59,60] and further support the hypothesis that, similarly to BKPyV and JCPyV, MCPyV may persist in renal tubular epithelial cells and eventually reactivate in case of transient alteration of the net state of immunosuppression [51,55].

JCPyV NCCR sequence analysis demonstrated that all tested subjects were infected by the archetypal strain. Data on JCPyV NCCR molecular characterization in KTx patients are extremely scarce. Therefore, direct comparison with other reports remains difficult if not impossible. In a similar study by Yin et al. [60], multiple JCPyV strains (namely, CY, TW1, and TW3) were detected with numerous episodes of simultaneous replication whereas strains with modified NCCR archetypal structure were identified by Seppälä and colleagues [59] performing a full-length JCPyV strain analysis on both plasma and urine samples.

Using JCPyV VP1 gene sequence analysis, we found that the most frequently represented genotypes were 1a and 1b, followed by genotypes 2, 4, and 6. As genotypes 1a and 1b represent the predominant genotypes in Europe [14,22], this result suggests that JCPyV genotypes distribution in KTx recipients may simply resemble JCPyV genotypes geographical distribution [53,54,59,61,62]. Importantly, a very strong concordance between donor and recipient genotypes was observed thus confirming the hypothesis that post-transplant JCPyV replication is mostly due to de novo acquired infections transmitted by the donor through the allograft [61,63].

BKPyV NCCR and VP1 sequences analysis showed that only half of our donor/recipient pairs with viruria actually shared the same viral strain and genotype. Considering the limited amount of BKPyV infections recorded in this series, no firm conclusions can be drawn. However, we can reasonably speculate that at least some episodes of BKPyV replication occurring during the post-transplant course may be due to latent viruses previously acquired by the recipient and reactivated under immunosuppression.

Interestingly, in one donor and one recipient, we detected rearranged forms of BKPyV NCCR already described in the literature and often linked to BKPyV-associated diseases [34,64]. Other studies assessing BKPyV NCCR genetic modifications in KTx recipients showed the presence of the archetypal and AS strains in renal allograft biopsies [55,56] and demonstrated BKPyV strains with few single-nucleotide substitutions within P, Q, R, and S blocks in blood and urine samples [34,65,66]. However, the clinical relevance of specific BKPyV NCCR rearrangements is still unclear and larger populations, as well as the longer follow-up, are needed to further address this topic.

In our experience, genotype 1b was the most commonly detected BKPyV genotype. This finding is overall in line with current evidence demonstrating the predominance of genotype 1 across the world. Nevertheless, considering the specific country of origin of the vast majority of the patients enrolled in the present study (Italy), it does not match the geographical distribution reported in the general population [52,62,67].

Finally, our sequences analysis of MCPyV amplicons showed the presence of the IIa-2 strain, already isolated from the skin of healthy individuals and patients with malignancy in several countries all around the world [35]. To date, there is only one report on MCPyV NCCR molecular characterization. According to it, there is a relationship between MCPyV genotypes and ethnicity [54].

5. Conclusions

This prospective observational study with an 18-month follow-up investigated the prevalence, viral load, and molecular characteristics of JCPyV, BKPyV, and MCPyV in urine samples collected from a cohort of KTx donor/recipient pairs. Our data show that: (a) HPyVs genome remains stable after transplantation with no de novo generation of quasispecies; (b) the vast majority of HPyVs strains isolated in donor/recipient pairs are identical thus suggesting that episodes of viral replication detected after transplantation may be more likely due to viruses transmitted by the donor through the allograft rather than reactivated viruses previously acquired by the recipient.

Author Contributions: Conceptualization, S.D., P.F., and M.F.; methodology, M.F., and S.D.; investigation, L.S., M.D., E.F., C.C., and R.T.; data curation, C.C., S.D., and L.S.; writing—original draft preparation, L.S. and S.D. writing—review and editing, E.F., S.D., G.B., and P.F.; supervision, C.C., M.F., and P.F.; funding acquisition, S.D. and P.F. All authors have read and agreed to the published version of the manuscript.

References

1. Feltkamp, M.C.W.; Kazem, S.; van der Meijden, E.; Lauber, C.; Gorbalenya, A.E. From Stockholm to Malawi: Recent developments in studying human polyomaviruses. *J. Gen. Virol.* **2013**, *94*, 482–496. [CrossRef]
2. Cook, L. Polyomaviruses. *Microbiol. Spectr.* **2016**, 4. [CrossRef]
3. Wolfe, R.A.; Ashby, V.B.; Milford, E.L.; Ojo, A.O.; Ettenger, R.E.; Agodoa, L.Y.; Held, P.J.; Port, F.K. Comparison of mortality in all patients on dialysis, patients on dialysis awaiting transplantation, and recipients of a first cadaveric transplant. *N. Engl. J. Med.* **1999**, *341*, 1725–1730. [CrossRef] [PubMed]
4. Hirsch, H.H.; Drachenberg, C.B.; Steiger, J.; Ramos, E. Polyomavirus-associated nephropathy in renal transplantation: Critical issues of screening and management. *Adv. Exp. Med. Biol.* **2006**, *577*, 160–173. [PubMed]
5. Cukuranovic, J.; Ugrenovic, S.; Jovanovic, I.; Visnjic, M.; Stefanovic, V. Viral infection in renal transplant recipients. *ScientificWorldJournal* **2012**, *2012*, 820621. [CrossRef] [PubMed]
6. Kartau, M.; Sipilä, J.O.; Auvinen, E.; Palomäki, M.; Verkkoniemi-Ahola, A. Progressive Multifocal Leukoencephalopathy: Current Insights. *Degener. Neurol. Neuromuscul. Dis.* **2019**, *9*, 109–121. [CrossRef] [PubMed]
7. DeCaprio, J.A.; Garcea, R.L. A cornucopia of human polyomaviruses. *Nat. Rev. Microbiol.* **2013**, *11*, 264–276. [CrossRef]
8. Moens, U.; Calvignac-Spencer, S.; Lauber, C.; Ramqvist, T.; Feltkamp, M.C.W.; Daugherty, M.D.; Verschoor, E.J.; Ehlers, B.; ICTV Report Consortium. ICTV Virus Taxonomy Profile: Polyomaviridae. *J. Gen. Virol.* **2017**, *98*, 1159–1160.
9. Schowalter, R.M.; Buck, C.B. The Merkel cell polyomavirus minor capsid protein. *PLoS Pathog.* **2013**, *9*, e1003558. [CrossRef]
10. Spurgeon, M.E.; Lambert, P.F. Merkel cell polyomavirus: A newly discovered human virus with oncogenic potential. *Virology* **2013**, *435*, 118–130. [CrossRef] [PubMed]
11. Gerits, N.; Moens, U. Agnoprotein of mammalian polyomaviruses. *Virology* **2012**, *432*, 316–326. [CrossRef]
12. Prado, J.C.M.; Monezi, T.A.; Amorim, A.T.; Lino, V.; Paladino, A.; Boccardo, E. Human polyomaviruses and cancer: An overview. *Clinics (Sao Paulo)* **2018**, *73* (Suppl. S1), e558s. [CrossRef]
13. Khalili, K.; Stoner, G.L. *Human Polyomaviruses: Molecular and Clinical Perspectives*; Wiley-Liss: Hoboken, NJ, USA, 2001; p. 688.
14. Agostini, H.T.; Deckhut, A.; Jobes, D.V.; Girones, R.; Schlunck, G.; Prost, M.G.; Frias, C.; Pérez-Trallero, E.; Ryschkewitsch, C.F.; Stoner, G.L. Genotypes of JC virus in East, Central and Southwest Europe. *J. Gen. Virol.* **2001**, *82*, 1221–1331. [CrossRef]
15. Sugimoto, C.; Kitamura, T.; Guo, J.; Al-Ahdal, M.N.; Shchelkunov, S.N.; Otova, B.; Ondrejka, P.; Chollet, J.Y.; El-Safi, S.; Ettayebi, M.; et al. Typing of urinary JC virus DNA offers a novel means of tracing human migrations. *Proc. Natl. Acad. Sci. USA* **1997**, *94*, 9191–9196. [CrossRef]
16. Agostini, H.T.; Yanagihara, R.; Davis, V.; Ryschkewitsch, C.F.; Stoner, G.L. Asian genotypes of JC virus in Native Americans and in a Pacific Island population: Markers of viral evolution and human migration. *Proc. Natl. Acad. Sci. USA* **1997**, *94*, 14542–14546. [CrossRef]
17. Jobe, D.V.; Friedlaender, J.S.; Mgone, C.S.; Agostini, H.T.; Koki, G.; Yanagihara, R.; Ng, T.C.N.; Chima, S.C.; Ryschkewitsch, C.F.; Stoner, G.L. New JC virus (JCV) genotypes from papua new guinea and micronesia (type 8 and type 2E) and evolutionary analysis of 32 complete JCV genomes. *Arch. Virol.* **2001**, *146*, 2097–2113. [CrossRef] [PubMed]

18. Ferenczy, M.W.; Marshall, L.J.; Nelson, C.D.; Atwood, W.J.; Nath, A.; Khalili, K.; Major, E.O. Molecular biology, epidemiology, and pathogenesis of progressive multifocal leukoencephalopathy, the JC virus-induced demyelinating disease of the human brain. *Clin. Microbiol. Rev.* **2012**, *25*, 471–506. [CrossRef]
19. Ault, G.S.; Stoner, G.L. Human polyomavirus JC promoter/enhancer rearrangement patterns from progressive multifocal leukoencephalopathy brain are unique derivatives of a single archetypal structure. *J. Gen. Virol.* **1993**, *74*, 1499–1507. [CrossRef]
20. Agostini, H.T.; Ryschkewitsch, C.F.; Singer, E.J.; Stoner, G.L. JC virus regulatory region rearrangements and genotypes in progressive multifocal leukoencephalopathy: Two independent aspects of virus variation. *J. Gen. Virol.* **1997**, *78*, 659–664. [CrossRef]
21. Jobes, D.V.; Chima, S.C.; Ryschkewitsch, C.F.; Stoner, G.L. Phylogenetic analysis of 22 complete genomes of the human polyomavirus JC virus. *J. Gen. Virol.* **1998**, *79*, 2491–2498. [CrossRef]
22. Pagani, E.; Delbue, S.; Mancuso, R.; Borghi, E.; Tarantini, L.; Ferrante, P. Molecular analysis of JC virus genotypes circulating among the Italian healthy population. *J. Neurovirol.* **2003**, *9*, 559–566. [CrossRef]
23. Zheng, H.Y.; Nishimoto, Y.; Chen, Q.; Hasegawa, M.; Zhong, S.; Ikegaya, H.; Ohno, N.; Sugimoto, C.; Takasaka, T.; Kitamura, T.; et al. Relationships between BK virus lineages and human populations. *Microbes Infect.* **2007**, *9*, 204–213. [CrossRef]
24. Martel-Jantin, C.; Pedergnana, V.; Nicol, J.T.; Leblond, V.; Trégouët, D.A.; Tortevoye, P.; Plancoulaine, S.; Coursaget, P.; Touzé, A.; Abel, L.; et al. Merkel cell polyomavirus infection occurs during early childhood and is transmitted between siblings. *J. Clin. Virol.* **2013**, *58*, 288–291. [CrossRef]
25. Gosert, R.; Rinaldo, C.H.; Funk, G.A.; Egli, A.; Ramos, E.; Drachenberg, C.B.; Hirsch, H.H. Polyomavirus BK with rearranged noncoding control region emerge in vivo in renal transplant patients and increase viral replication and cytopathology. *J. Exp. Med.* **2008**, *205*, 841–852. [CrossRef]
26. Olsen, G.H.; Hirsch, H.H.; Rinaldo, C.H. Functional analysis of polyomavirus BK non-coding control region quasispecies from kidney transplant recipients. *J. Med. Virol.* **2009**, *81*, 1959–1967. [CrossRef]
27. Gosert, R.; Kardas, P.; Major, E.O.; Hirsch, H.H. Rearranged JC virus noncoding control regions found in progressive multifocal leukoencephalopathy patient samples increase virus early gene expression and replication rate. *J. Virol.* **2010**, *84*, 10448–10456. [CrossRef]
28. Ajuh, E.T.; Wu, Z.; Kraus, E.; Weissbach, F.H.; Bethge, T.; Gosert, R.; Fischer, N.; Hirsch, H.H. Novel Human Polyomavirus Noncoding Control Regions Differ in Bidirectional Gene Expression according to Host Cell, Large T-Antigen Expression, and Clinically Occurring Rearrangements. *J. Virol.* **2018**, 92. [CrossRef] [PubMed]
29. Zanotta, N.; Delbue, S.; Signorini, L.; Villani, S.; D'Alessandro, S.; Campisciano, G.; Colli, C.; De Seta, F.; Ferrante, P.; Comar, M. Merkel Cell Polyomavirus Is Associated with Anal Infections in Men Who Have Sex with Men. *Microorganisms* **2019**, 7. [CrossRef]
30. Delbue, S.; Pagani, E.; Guerini, F.R.; Agliardi, C.; Mancuso, R.; Borghi, E.; Rossi, F.; Boldorini, R.; Veggiani, C.; Car, P.G.; et al. Distribution, characterization and significance of polyomavirus genomic sequences in tumors of the brain and its covering. *J. Med. Virol.* **2005**, *77*, 447–454. [CrossRef]
31. Gorelik, L.; Reid, C.; Testa, M.; Brickelmaier, M.; Bossolasco, S.; Pazzi, A.; Bestetti, A.; Carmillo, P.; Wilson, E.; McAuliffe, M.; et al. Progressive multifocal leukoencephalopathy (PML) development is associated with mutations in JC virus capsid protein VP1 that change its receptor specificity. *J. Infect. Dis.* **2011**, *204*, 103–114. [CrossRef]
32. Johnsen, J.I.; Seternes, O.M.; Johansen, T.; Moens, U.; Mäntyjärvi, R.; Traavik, T. Subpopulations of non-coding control region variants within a cell culture-passaged stock of BK virus: Sequence comparisons and biological characteristics. *J. Gen. Virol.* **1995**, *76*, 1571–1581. [CrossRef]
33. Flaegstad, T.; Sundsfjord, A.; Arthur, R.R.; Pedersen, M.; Traavik, T.; Subramani, S. Amplification and sequencing of the control regions of BK and JC virus from human urine by polymerase chain reaction. *Virology* **1991**, *180*, 553–560. [CrossRef]
34. Anzivino, E.; Bellizzi, A.; Mitterhofer, A.P.; Tinti, F.; Barile, M.; Colosimo, M.T.; Fioriti, D.; Mischitelli, M.; Chiarini, F.; Ferretti, G.; et al. Early monitoring of the human polyomavirus BK replication and sequencing analysis in a cohort of adult kidney transplant patients treated with basiliximab. *Virol. J.* **2011**, *8*, 407. [CrossRef] [PubMed]
35. Hashida, Y.; Higuchi, T.; Matsui, K.; Shibata, Y.; Nakajima, K.; Sano, S.; Daibata, M. Genetic Variability of the Noncoding Control Region of Cutaneous Merkel Cell Polyomavirus: Identification of Geographically Related Genotypes. *J. Infect. Dis.* **2018**, *217*, 1601–1611. [CrossRef]

36. Bofill-Mas, S.; Rodriguez-Manzano, J.; Calgua, B.; Carratala, A.; Girones, R. Newly described human polyomaviruses Merkel cell, KI and WU are present in urban sewage and may represent potential environmental contaminants. *Virol. J.* **2010**, *7*, 141. [CrossRef]
37. Agostini, H.T.; Ryschkewitsch, C.F.; Stoner, G.L. Genotype profile of human polyomavirus JC excreted in urine of immunocompetent individuals. *J. Clin. Microbiol.* **1996**, *34*, 159–164. [CrossRef]
38. Jensen, P.N.; Major, E.O. A classification scheme for human polyomavirus JCV variants based on the nucleotide sequence of the noncoding regulatory region. *J. Neurovirol.* **2001**, *7*, 280–287.
39. Jin, L.; Gibson, P.E.; Knowles, W.A.; Clewley, J.P. BK virus antigenic variants: Sequence analysis within the capsid VP1 epitope. *J. Med. Virol.* **1993**, *39*, 50–56. [CrossRef]
40. Van Loon, E.; Bernards, J.; Van Craenenbroeck, A.H.; Naesens, M. The Causes of Kidney Allograft Failure: More Than Alloimmunity. A Viewpoint Article. *Transplantation* **2020**, *104*, e46–e56. [CrossRef]
41. Favi, E.; James, A.; Puliatti, C.; Whatling, P.; Ferraresso, M.; Rui, C.; Cacciola, R. Utility and safety of early allograft biopsy in adult deceased donor kidney transplant recipients. *Clin. Exp. Nephrol.* **2020**, *24*, 356–368. [CrossRef] [PubMed]
42. Kotton, C.N.; Fishman, J.A. Viral infection in the renal transplant recipient. *J. Am. Soc. Nephrol.* **2005**, *16*, 1758–1774. [CrossRef] [PubMed]
43. Fishman, J.A. Infection in Organ Transplantation. *Am. J. Transplant.* **2017**, *17*, 856–879. [CrossRef]
44. Moroni, G.; Binda, V.; Quaglini, S.; Sacchi, L.; Raffiotta, F.; Cosa, F.; Montagnino, G.; Favi, E.; Messa, P.; Ponticelli, C. Causes of late transplant failure in cyclosporine-treated kidney allograft recipients. *Clin. Exp. Nephrol* **2019**, *23*, 1076–1086. [CrossRef]
45. Mishra, N.; Pereira, M.; Rhodes, R.H.; An, P.; Pipas, J.M.; Jain, K.; Kapoor, A.; Briese, T.; Faust, P.L.; Lipkin, W.I. Identification of a novel polyomavirus in a pancreatic transplant recipient with retinal blindness and vasculitic myopathy. *J. Infect. Dis.* **2014**, *210*, 1595–1599. [CrossRef]
46. Kwak, E.J.; Vilchez, R.A.; Randhawa, P.; Shapiro, R.; Butel, J.S.; Kusne, S. Pathogenesis and management of polyomavirus infection in transplant recipients. *Clin. Infect. Dis.* **2002**, *35*, 1081–1087. [CrossRef]
47. Knowles, W.A. Discovery and epidemiology of the human polyomaviruses BK virus (BKV) and JC virus (JCV). *Adv. Exp. Med. Biol.* **2006**, *577*, 19–45.
48. Jiang, M.; Abend, J.R.; Johnson, S.F.; Imperiale, M.J. The role of polyomaviruses in human disease. *Virology* **2009**, *384*, 266–273. [CrossRef]
49. Polo, C.; Pérez, J.L.; Mielnichuck, A.; Fedele, C.G.; Niubò, J.; Tenorio, A. Prevalence and patterns of polyomavirus urinary excretion in immunocompetent adults and children. *Clin. Microbiol. Infect.* **2004**, *10*, 640–644. [CrossRef]
50. Jeong, B.H.; Lee, K.H.; Choi, E.K.; Kim, K.; Kim, Y.S. Genotyping of the JC virus in urine samples of healthy Korean individuals. *J. Med. Virol.* **2004**, *72*, 281–289. [CrossRef]
51. Signorini, L.; Belingheri, M.; Ambrogi, F.; Pagani, E.; Binda, S.; Ticozzi, R.; Ferraresso, M.; Ghio, L.; Giacon, B.; Ferrante, P.; et al. High frequency of Merkel cell polyomavirus DNA in the urine of kidney transplant recipients and healthy controls. *J. Clin. Virol.* **2014**, *61*, 565–570. [CrossRef]
52. Chehadeh, W.; Kurien, S.S.; Nampoory, M.R. Molecular characterization of BK and JC viruses circulating among potential kidney donors in Kuwait. *Biomed. Res. Int.* **2013**, *2013*, 683464. [CrossRef] [PubMed]
53. Boukoum, H.; Nahdi, I.; Abid, N.; Foulongne, V.; Sahtout, W.; Zallema, D.; Skiri, H.; Aloui, S.; Achour, A.; Segondy, M.; et al. Distribution of JC polyomavirus genotypes in Tunisian renal transplant recipients between January 2008 and January 2011. *J. Med. Virol.* **2012**, *84*, 1818–1824. [CrossRef]
54. Comerlato, J.; Campos, F.S.; Oliveira, M.T.; Cibulski, S.P.; Corrêa, L.; Kulmann, M.I.; Arantes, T.S.; Hentges, L.P.; Spilki, F.R.; Roehe, P.M.; et al. Molecular detection and characterization of BK and JC polyomaviruses in urine samples of renal transplant patients in Southern Brazil. *J. Med. Virol.* **2015**, *87*, 522–528. [CrossRef] [PubMed]
55. Husseiny, M.I.; Anastasi, B.; Singer, J.; Lacey, S.F. A comparative study of Merkel cell, BK and JC polyomavirus infections in renal transplant recipients and healthy subjects. *J. Clin. Virol.* **2010**, *49*, 137–140. [CrossRef]
56. Drachenberg, C.B.; Hirsch, H.H.; Papadimitriou, J.C.; Gosert, R.; Wali, R.K.; Munivenkatappa, R.; Nogueira, J.; Cangro, C.B.; Haririan, A.; Mendley, S.; et al. Polyomavirus BK versus JC replication and nephropathy in renal transplant recipients: A prospective evaluation. *Transplantation* **2007**, *84*, 323–330. [CrossRef]

57. Bialasiewicz, S.; Rockett, R.J.; Barraclough, K.A.; Leary, D.; Dudley, K.J.; Isbel, N.M.; Sloots, T.P. Detection of Recently Discovered Human Polyomaviruses in a Longitudinal Kidney Transplant Cohort. *Am. J. Transplant.* **2016**, *16*, 2734–2740. [CrossRef]
58. Favi, E.; Puliatti, C.; Sivaprakasam, R.; Ferraresso, M.; Ambrogi, F.; Delbue, S.; Gervasi, F.; Salzillo, I.; Raison, N.; Cacciola, R. Incidence, risk factors, and outcome of BK polyomavirus infection after kidney transplantation. *World J. Clin. Cases* **2019**, *7*, 270–290. [CrossRef]
59. Seppälä, H.M.; Helanterä, I.T.; Laine, P.K.S.; Lautenschlager, I.T.; Paulín, L.G.; Jahnukainen, T.J.; Auvinen, P.O.V.; Auvinen, E. Archetype JC Polyomavirus (JCPyV) Prevails in a Rare Case of JCPyV Nephropathy and in Stable Renal Transplant Recipients With JCPyV Viruria. *J. Infect. Dis.* **2017**, *216*, 981–989. [CrossRef]
60. Yin, W.Y.; Lu, M.C.; Lee, M.C.; Liu, S.C.; Lin, T.Y.; Lai, N.S. A correlation between polyomavirus JC virus quantification and genotypes in renal transplantation. *Am. J. Surg.* **2010**, *200*, 53–58. [CrossRef]
61. Baksh, F.K.; Finkelstein, S.D.; Swalsky, P.A.; Stoner, G.L.; Ryschkewitsch, C.F.; Randhawa, P. Molecular genotyping of BK and JC viruses in human polyomavirus-associated interstitial nephritis after renal transplantation. *Am. J. Kidney Dis.* **2001**, *38*, 354–365. [CrossRef]
62. Chehadeh, W.; Nampoory, M.R. Genotypic diversity of polyomaviruses circulating among kidney transplant recipients in Kuwait. *J. Med. Virol.* **2013**, *85*, 1624–1631. [CrossRef]
63. Saundh, B.K.; Baker, R.; Harris, M.; Hale, A. A prospective study of renal transplant recipients reveals an absence of primary JC polyomavirus infections. *J. Clin. Virol.* **2016**, *77*, 101–105. [CrossRef]
64. Cubitt, C.L. Molecular genetics of the BK virus. *Adv. Exp. Med. Biol.* **2006**, *577*, 85–95.
65. Olsen, G.H.; Andresen, P.A.; Hilmarsen, H.T.; Bjørang, O.; Scott, H.; Midtvedt, K.; Rinaldo, C.H. Genetic variability in BK Virus regulatory regions in urine and kidney biopsies from renal-transplant patients. *J. Med. Virol.* **2006**, *78*, 384–393. [CrossRef] [PubMed]
66. Perets, T.T.; Silberstein, I.; Rubinov, J.; Sarid, R.; Mendelson, E.; Shulman, L.M. High frequency and diversity of rearrangements in polyomavirus bk noncoding regulatory regions cloned from urine and plasma of Israeli renal transplant patients and evidence for a new genetic subtype. *J. Clin. Microbiol.* **2009**, *47*, 1402–1411. [CrossRef] [PubMed]
67. Zalona, A.C.; Lopes, G.S.; Schrago, C.G.; Gonçalves, R.T.; Zalis, M.G.; Varella, R.B. Molecular characterization of BK polyomavirus subtypes in renal transplant recipients in Brazil. *J. Med. Virol.* **2011**, *83*, 1401–1405. [CrossRef] [PubMed]

4

Role of the DNA Damage Response in Human Papillomavirus RNA Splicing and Polyadenylation

Kersti Nilsson, Chengjun Wu and Stefan Schwartz *

Department of Laboratory Medicine, Lund University, 221 84 Lund, Sweden;
Kersti.Nilsson@med.lu.se (K.N.); troy_chengjun.wu@med.lu.se (C.W.)
* Correspondence: Stefan.Schwartz@med.lu.se

Abstract: Human papillomaviruses (HPVs) have evolved to use the DNA repair machinery to replicate its DNA genome in differentiated cells. HPV activates the DNA damage response (DDR) in infected cells. Cellular DDR factors are recruited to the HPV DNA genome and position the cellular DNA polymerase on the HPV DNA and progeny genomes are synthesized. Following HPV DNA replication, HPV late gene expression is activated. Recent research has shown that the DDR factors also interact with RNA binding proteins and affects RNA processing. DDR factors activated by DNA damage and that associate with HPV DNA can recruit splicing factors and RNA binding proteins to the HPV DNA and induce HPV late gene expression. This induction is the result of altered alternative polyadenylation and splicing of HPV messenger RNA (mRNA). HPV uses the DDR machinery to replicate its DNA genome and to activate HPV late gene expression at the level of RNA processing.

Keywords: papillomavirus; splicing; polyadenylation; SR proteins; hnRNP C; BRCA1; BCLAF1; TRAP150; DDR; U2AF65

---

1. Introduction

Human papillomaviruses (HPVs) are small DNA viruses that infect the keratinocytes of squamous and mucosal epithelia [1,2]. Thought to precede the amniotes (reptiles, birds and mammals), they are highly adapted to their host and most HPV infections are asymptomatic and resolve spontaneously. However, in rare cases, some HPV infections persist and cause disease such as warts and cancer. Approximately 50% of all virus-associated human cancers are caused by HPV [3]. This is largely attributed to a subset of sexually transmitted HPVs that cause anogenital and head and neck cancer. HPV16 is the most prevalent of the cancer-associated HPV types [4,5]. Knowledge of the HPV gene expression program is important to understand how HPV interacts with the infected cell in a manner that causes long-term persistence and cancer.

The Life Cycle of HPV

The HPV genome is about 8 kb in size and exists as an episome, a circular genome with independent replication [6,7]. The viral genome is associated with histones in a manner that is highly similar to human chromatin organization [8]. The HPV16 coding region contains at least six early (E) genes (Figure 1), which are expressed in the lower and mid layers of the infected epithelium. The HPV genome also encodes two late (L) genes, which encode the L1 and L2 structural proteins that are expressed only in terminally differentiated keratinocytes in the upper part of the epithelium (Figure 1) [9].

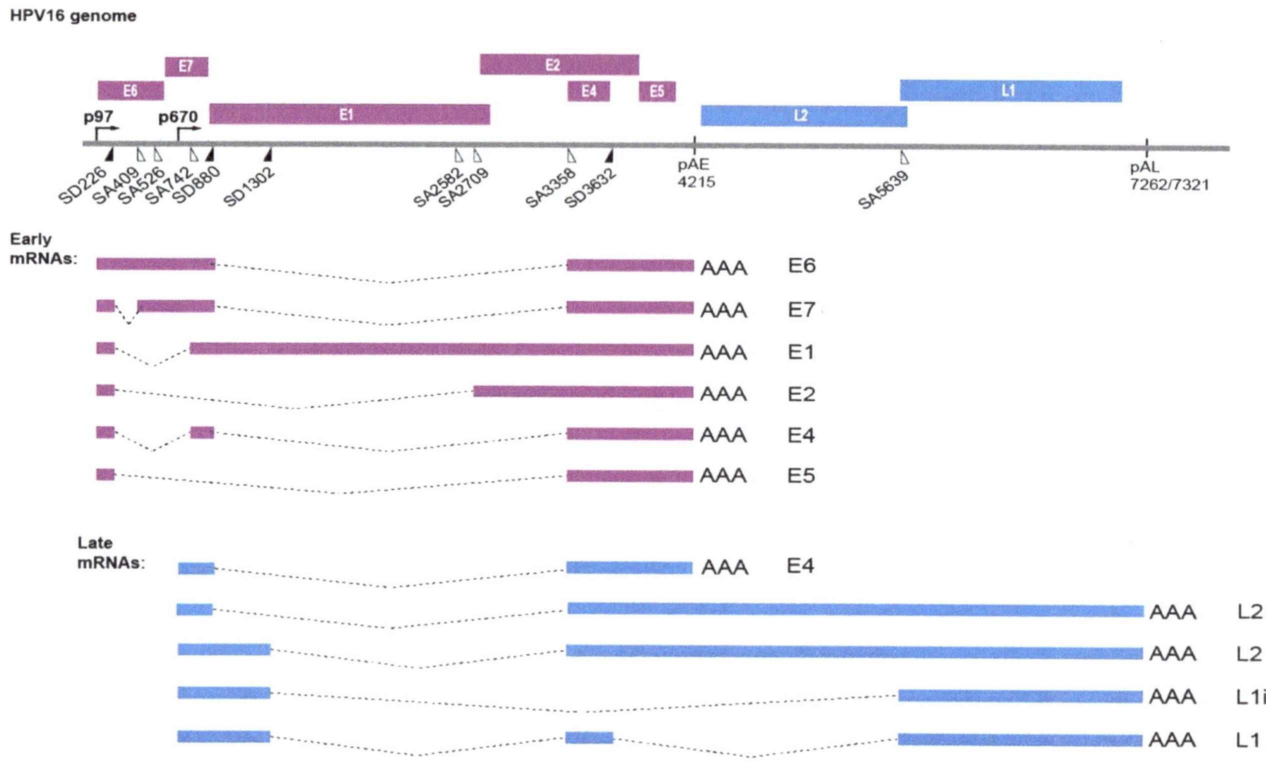

Figure 1. Schematic representation of the HPV16 genome. HPV16 early genes E1, E2, E4, E5, E6 and E7 and HPV16 late *L1* and *L2* genes are indicated. HPV16 early promoter (p97), late promoter (p670) and early (pAE) and late (pAL) polyadenylation signals are shown. Filled triangles represent 5′-splice sites and open triangles represent 3′-splice sites. Splice sites SD226, SA409, SA526 and SA742 are used exclusively by early mRNAs; SD3632 and SA5639 are used exclusively by late mRNAs; and splices sites SD880, SD1302, SA2582, SA2709 and SA3358 are used both by early and late mRNAs. A subset of HPV16 alternatively spliced early mRNAs and late mRNAs are shown.

The life cycle of HPV is coupled to the differentiation program of the keratinocyte, which results in an ordered expression of the viral genes [10]. HPV has no means of replicating its own DNA genome and is totally dependent on the DNA replication machinery of the host cell. Therefore, infection starts by HPV gaining access to the actively dividing cells in basal layer of the epithelium. Replication of the viral genome is divided into three phases; establishment-, maintenance- and productive-replication [7]. In the basal layer, the genome is amplified to a low copy number during establishment replication that is followed by maintenance amplification and HPV early gene expression. E6 and E7 promote cell cycle entry and prevent p53-mediated apoptosis to delay epithelial differentiation and maintain expression of cellular replication factors [11–13]. HPV E1 and E2 are directly involved in HPV genome amplification [14,15]. Downregulation of E6 and E7 expression eventually allows for terminal cell differentiation, expression of the HPV late genes L1 and L2 and production of progeny virus. The HPV gene expression program is dictated by the cellular differentiation program that controls HPV gene expression at the level of transcription [16,17] and at the level of RNA processing, including alternative splicing and polyadenylation [18–20]. HPVs produce a plethora of alternatively spliced and polyadenylated mRNAs that are controlled by cellular- [18–22] and viral factors (Figure 1) [18,23]. In this review, we discuss how DNA damage response (DDR) factors that are recruited to the HPV DNA to replicate the HPV genome can also be utilized to activate HPV late gene expression at the level of RNA splicing and polyadenylation. This review focus on the most common cancer-associated HPV types of the α-genus with emphasis on HPV type 16.

2. Human Papillomavirus (HPV) and the Cellular DNA Damage Response (DDR)

2.1. HPV Employs the Cellular DNA Damage Response for Genome Amplification

The integrity of the eukaryotic genome is maintained through a network collectively referred to as the DNA damage response (DDR) that senses and signals DNA damage arrests the cell cycle and activates repair mechanisms or eliminates the damaged cells through apoptosis (Figure 2). Different types of insult to the DNA are detected through unique sensors. DNA damage signals are then relayed to effector molecules in a manner similar to signal transduction pathways, including post-translational modifications such as phosphorylation [24]. The major upstream kinases in the signal transduction pathway that orchestrate the response to DNA damage are members of the phosphatidylinositol 3-kinase-related kinase (PIKKs) family and include Ataxia telangiectasia mutated kinase (ATM) and Ataxia telangiectasia and Rad3-related protein FRAP-related protein 1 (ATR) (Figure 2) [25]. ATM and ATR appear to regulate the broadest spectrum of downstream factors that contribute to the DDR (Figure 2) [26–28]. In addition, they induce further phosphorylation events through the activation of the Chk1 and Chk2 kinases (Figure 2) [29,30]. ATM is activated in response to double stranded breaks (DSBs) [31,32], whereas ATR is activated by the presence of single stranded DNA [25,33,34]. The downstream events in the DDR signal transduction chain include cell cycle check-points, apoptosis or DNA synthesis to restore the integrity of the DNA molecule. The latter feature of the DDR is exploited by some DNA viruses such as HPV that lacks a DNA polymerase and has evolved to employ the DDR for amplification of the viral genome.

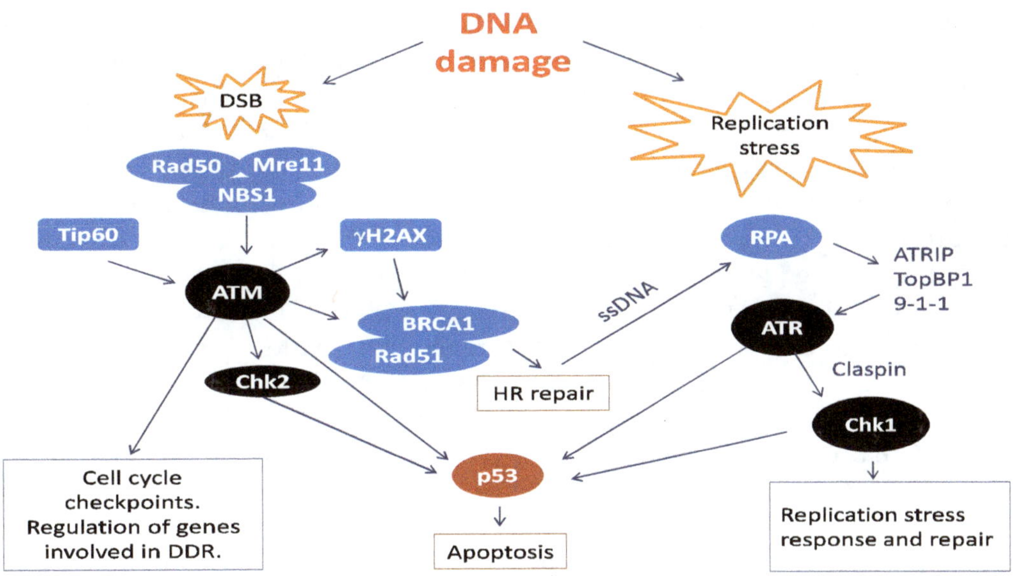

Figure 2. The Ataxia-Telangiectasia Mutated (ATM) and ATM and Rad3-related (ATR) signalling pathways in response to DNA damage. Double stranded breaks (DSBs) are detected by the sensory complex MRN (Mre11, Rad50, and Nbs1). The MRN complex and the acetyltransferase Tip60 activate ATM, which relays the damage signal to targets such as γH2AX, Chk2, p53, and Breast Cancer Susceptibility Gene 1 (BRCA1). γH2AX nucleates the site of damage, leading to the recruitment of several E3 Ubiquitin ligases that bring homologous repair factors (HR) such as BRCA1 and Rad51 to the site of damage. Downstream effects of the signal are cell cycle arrest, DNA repair, or apoptosis. ATR is activated in response to single stranded DNA (ssDNA) that arises when damaged DNA interfere with replication or transcription. ATR can also be activated in an ATM-dependent manner during repair of DSBs as intermediate structures during repair display ssDNA. The Replication Protein A (RPA) forms filaments on ssDNA and recruits ATRIP, the 9-1-1 complex (Rad9-Hus1-Rad1) and TopBP1 that all activate ATR. The damage signal is then passed on via Claspin and Chk1 and the DNA damage is repaired, if possible.

2.2. HPV Proteins Perturb Cell Differentiation to Allow for Replication of HPV DNA

Keratinocytes exit the cell cycle and differentiate as they leave the basal layer. To maintain an environment that supports viral replication, HPV E7 binds to the Rb family proteins to alleviate their suppression of the cellular transcription factor E2F [12]. The liberated E2F protein activates expression of cell cycle promoting proteins. Consequently, the HPV-infected cell enters a G2-like phase in which differentiation factors and replication factors required for productive viral replication can coexist [35,36]. Meanwhile, HPV E6 targets p53 for degradation to suppress p53-mediated apoptosis that would otherwise have been elicited by the unscheduled re-entry into the cell cycle [13]. The HPV proteins E1 and E2 support initial establishment and maintenance replication of the HPV genome. HPV E1 is a DNA helicase that separates the DNA strands at the HPV origin of replication, while E2 functions by positioning E1 and the cellular replication machinery onto the HPV DNA genome [14,15]. Efficient amplification of HPV genomes requires activation of the late, differentiation-dependent HPV promoter to provide high expression levels of the HPV E1, E2 and E4 proteins. Initially, the early promoter remains active upon differentiation that allows expression also of E6 and E7. However, the HPV early promoter is subsequently shut down by the accumulated levels of the E2 protein to allow for cell differentiation and differentiation-dependent expression of the HPV late *L1* and *L2* genes.

2.3. DDR Factors Contribute to HPV DNA Replication

In addition to HPV proteins, HPV genome amplification also requires cellular proteins of the ATM and ATR branches of the DDR [37–39]. ATR is active during all stages of the HPV life cycle [38,39], suggesting that this branch of the DDR is necessary for initial-, maintenance- and productive-replication [3,40,41]. Further, TopBP1 that acts upstream of ATR signalling is a required component of the viral replication loci [39]. The HPV E1 and E7 proteins can independently activate ATR and Chk1 [3,38,42]. Alternatively, this activation is a consequence of the replication stress that arises from replication of the HPV genome, the unspecific DNA helicase activity of E1, the aberrant cell cycle entry created by the viral proteins or the ssDNA generated during homologous recombination (HR)-mediated productive HPV replication [43]. However, different HPV types seem to have specific effects on the ATR signalling [44]. As the signalling from the ATM and ATR branch overlap, perhaps this reflects a variable ability of HPV proteins to interact with cellular components to elicit the DDR required for genome amplification [44]. The HPV infection activates the DDR with the purpose of exploiting the DDR DNA synthesis machinery for HPV genome replication (Figure 3). However, induction of the DDR is accompanied with a risk of inducing p53-mediated apoptosis. To prevent apoptosis, the HPV E6 protein binds and degrades cellular p53 (Figure 3). ATM is also active in HPV infected cells and contributes to the productive phase of HPV DNA replication [3,39]. As the levels of HPV E1 and E2 rise in the mid layers of the HPV-infected epithelium, E1 and E2 nucleate the viral origin of replication together with cellular HR factors Rad51, BRCA1 and the MRN (MRE11, Rad50and NBS1) complex (Figure 3). These factors are all required for productive HPV DNA replication. HR mediated repair creates a large area of ssDNA that invades a sister chromatid to use a homologues sequence as template for synthesis of new DNA. Thus, HPV may specifically activate ATM to recruit HR factors as they offer high fidelity replication in G2-arrested cells upon differentiation. Alternatively, ATM activation is a result of the rolling circle replication used for the productive amplification of the viral genome [45]. The modified histone γH2AX, a hallmark of DNA damage, is also found on HPV genomes at onset of productive replication [46]. It is aiding in the recruitment of DNA repair factors to the HPV genome. Additional proteins associated with the ATR branch of the DDR, such as CHK1 and TopBP1, are also found in the HPV replication foci [3,41,47]. HPV E7 appears to increase the abundance of these factors, partly through transcriptional activation by E2F [48], partly through protein stabilization [37,39]. Activation of the DDR by E7 is also mediated by interactions with signal transducer and trans activator 5 protein (STAT5) and the Tip60 acetyltransferase (Figure 3) [38,49,50]. In conclusion, several cellular DDR factors are required for replication of the HPV DNA genome.

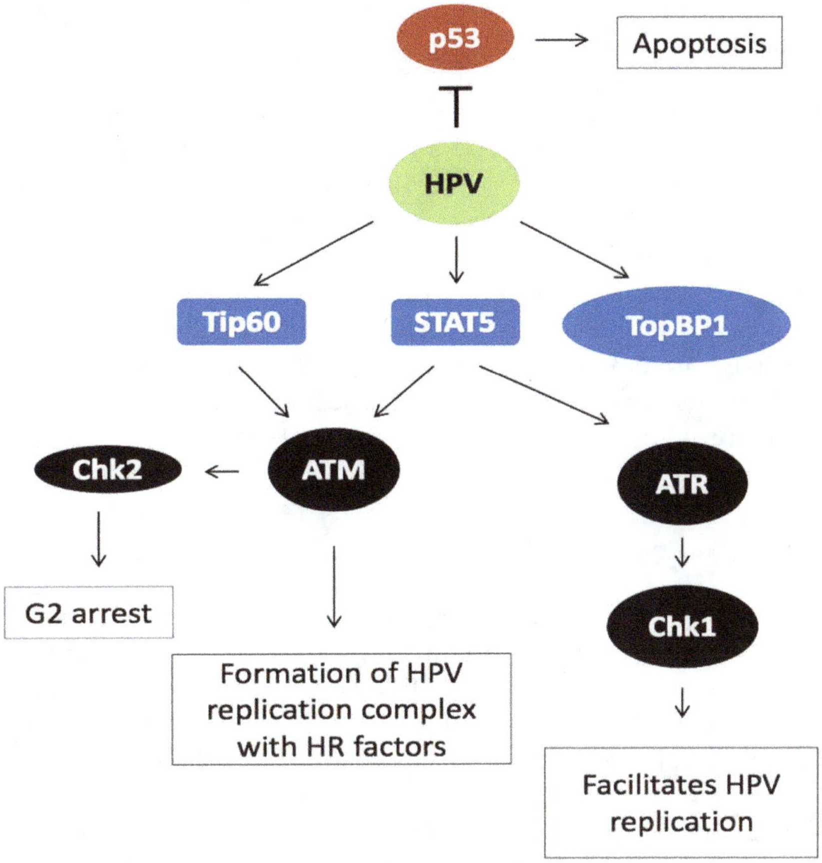

Figure 3. Interactions between HPV and the Ataxia-Telangiectasia Mutated (ATM) and ATM and Rad3-related (ATR) signalling during productive viral replication. HPV activates the ATM branch of the DDR to gain access to factors associated with homologous recombination. This activation occurs at least partially through the Tip60 acetyltransferase and signal transducer 5 (STAT5) that are both required for activation of ATM. In addition, the ATR branch of the DDR is activated by HPV. HPV replication foci contain TOPBP1, a protein necessary for ATR activation. The exact mechanism of ATR activation is unclear Downstream of ATM/ATR signal transduction are the kinases Chk1 and Chk2, both of which have been found in HPV replication foci and are known to be crucial for cell cycle arrest and regulation of genes needed for HPV genome amplification. To counteract the potential induction of apoptosis by the cellular DDR, HPV E6 targets p53 for degradation to inhibit apoptosis.

2.4. HPV Gene Regulation

The coding region of the HPV genome consists of at least two promoters, two polyadenylation signals and eight protein-coding genes (Figure 1). The early (E) genes are expressed from the early promoter and polyadenylated at the early polyadenylation signal (pAE) (Figure 1). However, early proteins E1, E2 and E4 can also be expressed from mRNAs initiated at the HPV late promoter but are polyadenylated at pAE (Figure 1). HPV late genes *L1* and *L2* are expressed from the late promoter and polyadenylated at the late polyadenylation signal (pAL) (Figure 1). To ensure efficient expression of each viral gene in a highly regulated fashion, HPV makes extensive use of alternative mRNA splicing and polyadenylation [18–22,51]. Although HPV uses the cellular splicing and polyadenylation machineries, the HPV genome differs from the cellular genome in that the vast majority of the HPV genome is protein coding (Figure 1). In addition, many of the HPV open reading frames (ORFs) overlap. The molecular anatomy of the HPV genome is therefore particularly challenging since RNA elements that control HPV splice sites and polyadenylation signals are likely to be situated in regions of the HPV genome that are constrained by a protein coding region, or even two overlapping protein coding regions (Figure 1) [18–20]. In addition, the 3′-untranslated regions of HPV encode RNA elements that control HPV mRNA stability and/or translation efficiency [52].

Expression of the HPV late *L1* and *L2* genes requires a switch to the differentiation-dependent late HPV promoter. The late promoter is located in the 5′-end of the genome, while the *L1* and *L2* genes are located in the 3′-end of the genome (Figure 1). Consequently, mRNA splicing and polyadenylation play major roles in the control of HPV late gene expression [18–20]. In addition to activation of the HPV late promoter, inhibition of the early polyadenylation signal pAE is required for production of pre-mRNAs encoding L1 and L2. Activation of the two suppressed, exclusively late splice sites SD3632 and SA5639 gives rise to the L1 mRNAs and is paramount for L1 and L2 expression [53,54]. High levels of the HPV16 E2 protein inhibit HPV16 early polyadenylation and E2 therefore contributes to activation of HPV16 late gene expression [55]. In addition to E2, recruitment of cellular splicing factors and RNA binding proteins is of vital importance for HPV late gene expression [22].

2.5. Induction of HPV Late Gene Expression by the DNA Damage Response

The HPV E2 protein binds to the HPV DNA genome and together with HPV E1 it is required for replication of the HPV genome [14,15]. As the E2 protein accumulates to high levels in the HPV infected cells, E2 binds to multiple sites in the HPV early promoter to shut it down [14], thereby inhibiting E6 and E7 expression and allowing the cell to resume differentiation. Cell differentiation activates the late, differentiation-dependent HPV promoter [16], thereby paving the way for late L1 and L2 expression. The HPV E2 protein also has an inhibitory effect on the HPV early polyadenylation signal, possibly through interactions with CPSF30, and can cause read-through into the HPV late region of the genome [55]. Thus, E2 has a dual role in the HPV life cycle: it functions in HPV DNA replication and in the regulation of HPV gene expression. Recruitment of E2 to the DNA genome is required for HPV DNA replication and HPV E2 contributes to induction of HPV late gene expression by inhibiting the HPV early polyadenylation signal pAE. Similar to HPV E2, DDR factors are recruited to the HPV DNA genome and they are required for replication of the HPV genome [56]. It has recently been shown that activation of the cellular DDR also involves recruitment of RNA processing factors [57–59]. Thus, it was reasonable to speculate that DDR factors already recruited to the HPV genome also contribute to induction of HPV late gene expression, especially since HPV late gene expression occurs immediately following HPV genome replication. Furthermore, it has been recently shown that the cellular DDR interacts with RNA processing factors [57–60] and that the cellular DDR affects alternative splicing of cellular mRNAs [61–64]. To test the idea that the DDR contributes to HPV late gene expression, we used reporter cell line C33A2 that is designed to study induction of HPV16 late gene expression to investigate if the DNA damage response could activate HPV16 late gene expression [53,65,66]. Addition of the DNA damaging agent melphalan to this reporter cell line efficiently induced the DNA damage response in the C33A2 cells, and efficiently activated the HPV16 late *L1* and *L2* gene expression [66]. We observed a several hundred-fold induction of HPV16 L1 and L2 mRNAs as a result of inhibition of HPV16 early polyadenylation and activation of HPV16 L1 mRNA splicing, while the effect at the level of transcription was relatively modest [66]. Figure 4 shows the striking shift from early polyA site usage in HPV16 to primarily late polyA signal usage in response to induction of the DDR (Figure 4). Thus, the DDR induced HPV16 late gene expression at the level of HPV16 RNA processing, primarily by altering HPV16 splicing and polyadenylation [66]. The DDR factors BRCA1, Chk1, Chk2 and ATM were phosphorylated in response to DNA damage, as expected. Inhibition of ATM- or Chk1/2-phosphorylation, but not ATR-phosphorylation, prevented induction of HPV16 late gene expression [66], demonstrating that activation of the DDR contributed to induction of HPV16 late gene expression at the level of RNA processing.

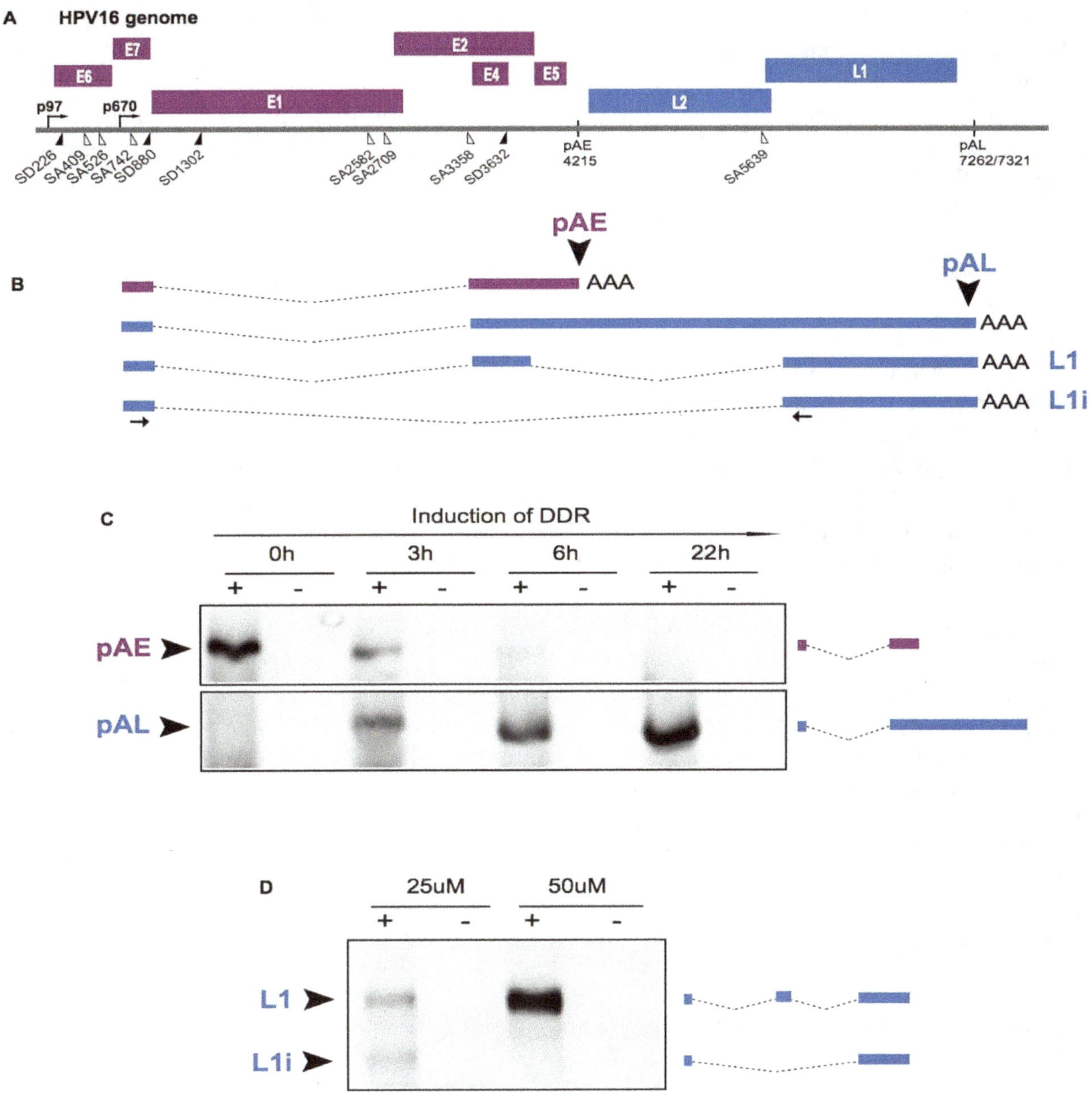

Figure 4. The DNA damage response alters HPV16 mRNA polyadenylation and splicing. (**A**) Schematic representation of the HPV16 genome. (**B**) Examples of alternatively polyadenylated and alternatively spliced HPV16 mRNAs. (**C**) 3′-RACE assay with primers specific for either the HPV16 early polyadenylation signal pAE, or HPV16 late polyadenylation signal pAL was performed on RNA extracted from HPV16 reporter cell line C33A2 treated with 100uM melphalan for the indicated time periods. Induction of the DNA damage response with melphalan in the HPV16 reporter cell line C33A2 inhibits HPV16 early polyadenylation and activates HPV16 late polyadenylation over time. (**D**) RT-PCR with primers that specifically detect the two alternatively spliced HPV16 L1 mRNAs named L1 and L1i. RT-PCR primers are indicated in (**B**). The DNA damage response induced with 50 uM melphalan alters HPV16 mRNA splicing and results in efficient inclusion of the central exon on the HPV16 L1 mRNA.

2.6. Cellular DNA Damage Response Factors Associate with HPV16 DNA and Recruit Cellular RNA Processing Factors

Inducing the DNA damage response in the C33A2 reporter cell line for HPV16 late gene expression resulted in recruitment of DDR factors BRCA1, in particular phosphorylated BRCA1, and BARD1 to the HPV16 DNA [66]. In addition, the more elusive BCLAF1 [67] and TRAP150 [68,69] proteins were also recruited to the HPV16 DNA [66]. Although BCLAF1 is bound to general splicing factor U2AF65 in both the absence and presence of DNA damage, it is associated with phosphorylated BRCA1 only in response to DNA damage [66]. These results suggested that interactions between BCLAF1 and phosphorylated BRCA1 occurred in response to DNA damage and resulted in recruitment of splicing factor U2AF65 to the HPV16 DNA. In addition, phosphorylated BRCA1 interacted with general splicing factor SF3b in the presence of DNA damage. The close relative of BCLAF1 named TRAP150 associated with HPV16 DNA and interacted with splicing factor U2AF65 only in the presence of DNA damage, suggesting that also TRAP150 recruits U2AF65 to HPV16 DNA. However, in contrast to BCLAF1, TRAP150 appeared to recruit U2AF65 independently of phosphorylated BRCA1. Indeed, the splicing factor U2AF65 was increasingly associated with HPV16 DNA in response to DNA damage [66]. Other studies indicate that TRAP150 binds U2AF65 directly [66]. We also observed an increased association between the HPV16 DNA of phosphorylated SR-proteins in response to DNA damage [66]. Serine and arginine-rich (SR) proteins are well known for their splicing regulatory functions [70,71] and several SR proteins have been shown to control HPV mRNA splicing [18–20,22,51]. The effect on HPV16 alternative splicing is best shown by the increased inclusion of the exon located between SA3358 and SD3632 in the L1 mRNAs (Figure 4). Taken together, DDR factors that are associated with HPV16 DNA may recruit splicing factors to the HPV16 DNA, thereby increasing the chances that they associate with de novo synthesized HPV16 mRNAs and affect HPV16 mRNA processing.

In addition to recruiting general splicing factors and SR proteins to the HPV16 DNA, the DDR factors also recruited other cellular RNA binding proteins, e.g., heterogenuos ribonuclearprotein C (hnRNP C) [66]. This protein has previously been shown to induce HPV16 late gene expression and affect L1 mRNA splicing in just the same way as induction of the DDR did [72]. Phosphorylated BRCA1 interacted with hnRNP C only in response to DNA damage and hnRNP C increasingly associated with HPV16 DNA in response to DNA damage [66]. hnRNP C has been shown to co-localize with sites of DNA damage as part of the BRCA1-, BRCA2- and PALB2-complex in response to DNA damage [57]. This hnRNP C-containing complex affected mRNA splicing. Combined, these results suggested that phosphorylated BRCA1 recruited hnRNP C to the HPV16 DNA and that this recruitment increased the chances that hnRNP C would bind newly synthesized HPV16 mRNAs and potentially alter HPV16 alternative splicing.

2.7. Increased Association between HPV16 mRNA-Binding Proteins and Cellular Polyadenylation Factors in Response to DNA Damage

The association between hnRNP C and polyadenylation factors CPSF30 and Fip1 increased in response to DNA damage, as did the binding of hnRNP C to the HPV16 early untranslated region [66]. Both hnRNP C and Fip1 binds to the U-rich region in the HPV16 early 3′-untranslated region (UTR) [72,73]. This suggested that hnRNP C contributed to inhibition of the HPV16 early polyadenylation signal pAE by binding to HPV16 mRNAs and negatively interfering with the polyadenylation factors Fip1 and CPSF30. Overexpression of hnRNP C with HPV16 subgenomic plasmids caused inhibition of the HPV16 early polyadenylation signal [66]. Knock-down or inhibition of CPSF30 inhibited the HPV16 early polyadenylation signal, but not the downstream late HPV16 polyadenylation signal [66]. In addition to hnRNP C, HuR binding to HPV16 early 3′-UTR increased in response to DNA damage [66]. HuR has been shown to inhibit HPV16 early polyadenylation and to contribute to export of HPV16 late mRNA from the nucleus [74]. Combined, these results support a model in which DDR factors assemble on HPV16 DNA and recruit RNA binding proteins including hnRNP C and HuR that bind to the HPV16 mRNAs. hnRNP C binds to polyadenylation factors CPSF30 and Fip1 to inhibit HPV16 early polyadenylation, thereby causing read-through into the late L1 and L2 coding region and activating HPV16 late gene expression (Figure 5).

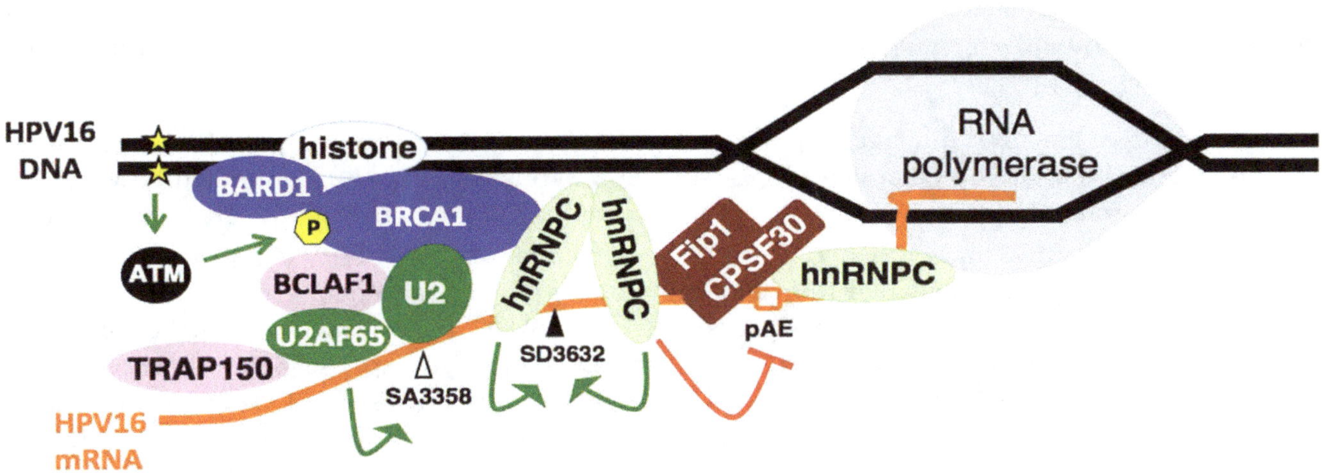

Figure 5. The DNA damage response activates HPV16 late gene expression by altering HPV16 mRNA splicing and polyadenylation. The DNA damage response induces ATM signalling. Activated ATM phosphorylates BRCA1, which leads to the formation of a pBRCA1-BCLAF1 complex that is recruited to the HPV DNA. The pBRCA1-BCLAF1 complex associates with splicing factors SF3b and U2AF65 and recruits them to the HPV16 DNA, thereby positioning the spliceosome in a strategic position for efficient detection of nascent HPV16 mRNA. hnRNP C is also recruited to both HPV16 DNA and mRNA through interactions with phosphorylated BRCA1. hnRNP C binds to the HPV16 early untranslated region and inhibits the HPV16 early polyadenylation signal pAE, possibly through interactions with Fip1 and CPSF30. This inhibition causes read-through at the early polyadenylation signal (pAE) and activates HPV16 late gene expression. hnRNP C also regulates HPV16 alternative splicing by activating late L1 splice site SD3632, contributing to the production of splices late L1 mRNAs. When the DNA damage response is activated, levels of the splicing regulatory protein TRAP150 increased in affected cells. The DNA damage response promotes the association of TRAP150 HPV16 DNA as well as with general splicing factor U2A65, thereby recruiting U2AF65 to HPV16 and contributing to the enhanced association of U2AF65 with HPV16 mRNAs. Taken together, the DNA damage response-induced associations of DNA damage response factors with RNA processing factors and with the HPV16 DNA and mRNAs alters HPV16 splicing and polyadenylation to induce HPV16 late gene expression.

2.8. DNA Damage Response Factors Recruit Splicing Factors to HPV16 DNA That Alter Splicing of HPV16 mRNAs

In addition to its role in HPV16 early polyadenylation, hnRNP C has also been shown to activate the suppressed HPV16 late splice site SD3632 to produce L1 mRNAs over the alternatively spliced L1i mRNA (see Figure 1 for structures of the HPV16 L1 and L1i mRNAs) [72]. This effect of hnRNP C is dependent on the HPV16 early UTR to which hnRNP C binds. Activation of HPV16 SD3632 results in L1 mRNAs in which the central exon between SA3358 and SD3632 is included on the mRNA as opposed to L1i mRNAs on which this exon is excluded. This effect of hnRNP C on HPV16 L1 mRNAs reproduced the effect of the DDR on alternative splicing of HPV16 L1 mRNAs (Figure 4) [66,72]. Thus, the hnRNP C proteins that were recruited to HPV16 DNA and to HPV16 mRNAs interacted with the HPV16 early UTR and inhibited HPV16 early polyadenylation and activated HPV16 L1 mRNA-specific late splice site SD3632. hnRNP C also suppresses polyadenylation of cellular mRNAs [75]. It is also of interest to note that hnRNP G, which is an RNA binding protein that plays an active role in the DDR [58], also controls HPV16 L1 mRNA splicing [76]. In conclusion, DDR factors recruit hnRNP C to the HPV16 DNA, thereby promoting association of hnRNP C with de novo synthesized HPV16 mRNAs. Consequently, splicing and polyadenylation of the HPV16 mRNAs are altered to favour HPV16 late gene expression.

Induction of the DNA damage response also resulted in enhanced splicing to HPV16 E2 splice site SA2709 and the HPV16 E4 splice site SA3358 [66]. While it is currently unknown how the E2 splice site is regulated, splice site SA3358 is controlled by splicing factors from the SR protein family including SRSF1, SRSF3 and SRSF9 [22,77–81]. The area at and around HPV16 splice sites SA3358 and SD3632 are hot-spots for cellular RNA binding proteins [82]. Enhanced splicing to SA3358 would explain the increase in the HPV16 E4 mRNAs spliced from SD880 to SA3358 as well as the enhanced production of the L2 mRNAs following activation of the DNA damage response [66]. It is reasonable to speculate that increased splicing to SA3358 is mediated by the enhanced association of phosphorylated SR proteins with the HPV16 mRNAs in response to DNA damage and/or the enhanced association of HPV16 mRNAs with general splicing factor U2AF65 [66]. In conclusion, activation of the DNA damage response results in the association of DDR factors with HPV16 DNA. These factors recruit various RNA binding proteins and RNA processing factor that alter HPV16 mRNA splicing and polyadenylation in a manner that favours HPV16 late gene expression. Thus, DNA damage response factors control HPV gene expression at the level of RNA processing in addition to their role in HPV DNA replication. Combined, the results suggest a model for activation of HPV16 late gene expression with the aid of the DDR that is presented in Figure 5.

2.9. The DNA Damage Response Affects Alternative Splicing of Cellular mRNAs

Given the ability of RNA binding proteins to interact with both chromatin and nascent mRNA, they could contribute to the response to DNA insult and to maintenance of the DDR signal. It has been shown that the BRCA1-BCLAF1 complex may position the spliceosome on genes for proper processing of transcripts in response to ATM/ATR signalling [83]. Indeed, we found that pBRCA1 and BCLAF1 were recruited to HPV16 chromatin and that they loaded splicing factors and RNA binding proteins onto HPV16 mRNAs [83]. Further, apart from activating HPV16 late gene expression at the level of RNA processing, induction of the DDR with melphalan also affected expression of many cellular genes as determined by an array analysis. Transcriptional changes in cellular genes not only affected DDR-genes, but also included genes coding for proteins involved in mRNA processing, RNA catabolic processes and RNA localization (Figure 6), suggesting that the DDR affected alternative splicing also of cellular mRNAs. As can be seen in Figure 7, up to 30% of the mRNAs in some gene groups showed changes in their alternative splicing in response to DDR activation (Figure 7).

Some of the mRNAs that were alternatively spliced in DDR-activated cells encoded DDR factors [66]. The HPV-infection alters the levels of many RNA binding proteins and splicing factors [84,85] and HPV16 E2 appears to indirectly affect splicing [23,86,87] as well as polyadenylation [55]. HPV infections may alter alternative splicing of cellular mRNAs through activation of the DDR. It has recently been shown that E6 and E7 increase transcription of HR-genes [88]. This effect could be due to the E6 and E7 effect on cellular transcription factors such as p53. p53 is one of the most well known examples of a mediator between the DDR response and RNA metabolism, effecting transcription and RNA turnover of many genes involved in the response to DNA damage. In addition, both E6 and E7 seem to induce cellular DSBs, independent of viral replication [88,89]. The amount of DSBs in viral and cellular DNA were the same until differentiation when active recruitment of HR-proteins to HPV DNA seemed to drive DSB repair, on the expense of cellular DSB repair [89]. As HPV has evolved to employ the DDR for genome amplification, it is possible that HPV gene expression has evolved in parallel to respond to the RNA processing factors brought to HPV DNA in complex with DDR factors.

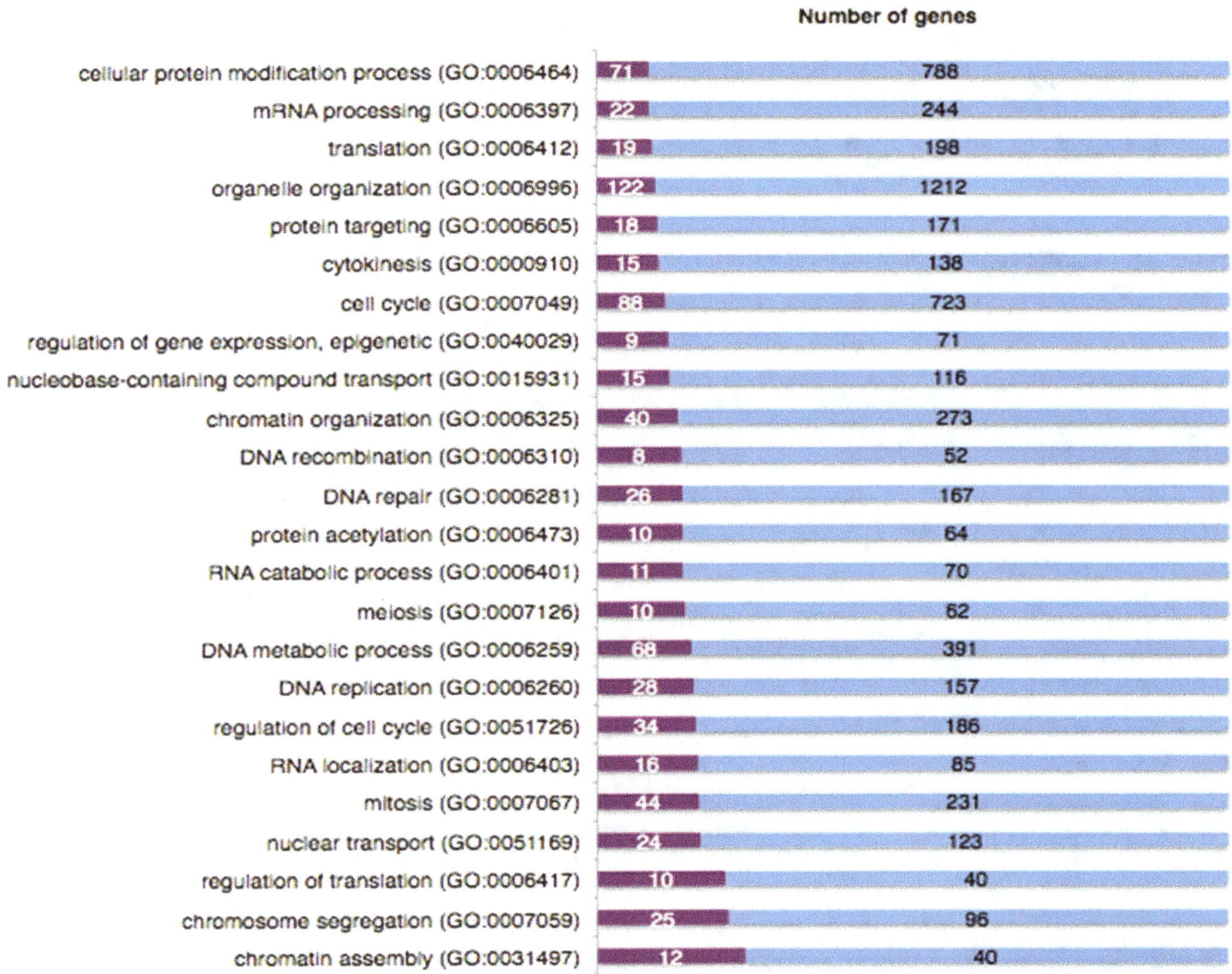

Figure 6. The DNA damage response affects mRNA levels of cellular genes with various biological functions, including genes encoding mRNA processing factors. Total RNA was harvested from HPV16 reporter cell line C33A2 after induction of the DNA damage response with 100 μM melphalan for 22 h. The RNA samples obtained from DMSO- or melphalan-treated C33A2 cells were subjected to microarray analysis to detect changes in mRNA levels throughout the genome. Total RNA was prepared using Qiagen RNeasy Mini Kit (Qiagen, Hilden, Germany) according to the manufacturer's protocol. The RNA quality was determined using a Bioanalyzer (Agilent, Santa Clara, CA, USA). In total, five RNA samples each from melphalan and DMSO treated cells were analysed on Affymetrix GeneChip Human Transcriptome array 2.0 at SCIBLU Genomics (Lund University, Lund, Sweden). Protein coding genes that displayed at least a 2-fold change in mRNA levels between melphalan and DMSO treated cells, were sorted in the Transcriptome Analysis Console (TAC) from Thermo Fisher, Waltham, MA, USA). Following sorting, these genes were exported to PANTHER version 13.1, Gene List Analysis tool (Available online: http://pantherdb.org) for an overrepresentation test based on their biological function. Results of the RNA array analysis of RNA from DMSO or melphalan treated C33A2 cells are displayed as percentage of genes in each category that were either up- or down-regulated more than two-fold. The blue area shows the total number of genes in each biological-function category, and the purple area the number of genes that displayed a higher than two-fold change in mRNA levels between DMSO and melphalan treated cells.

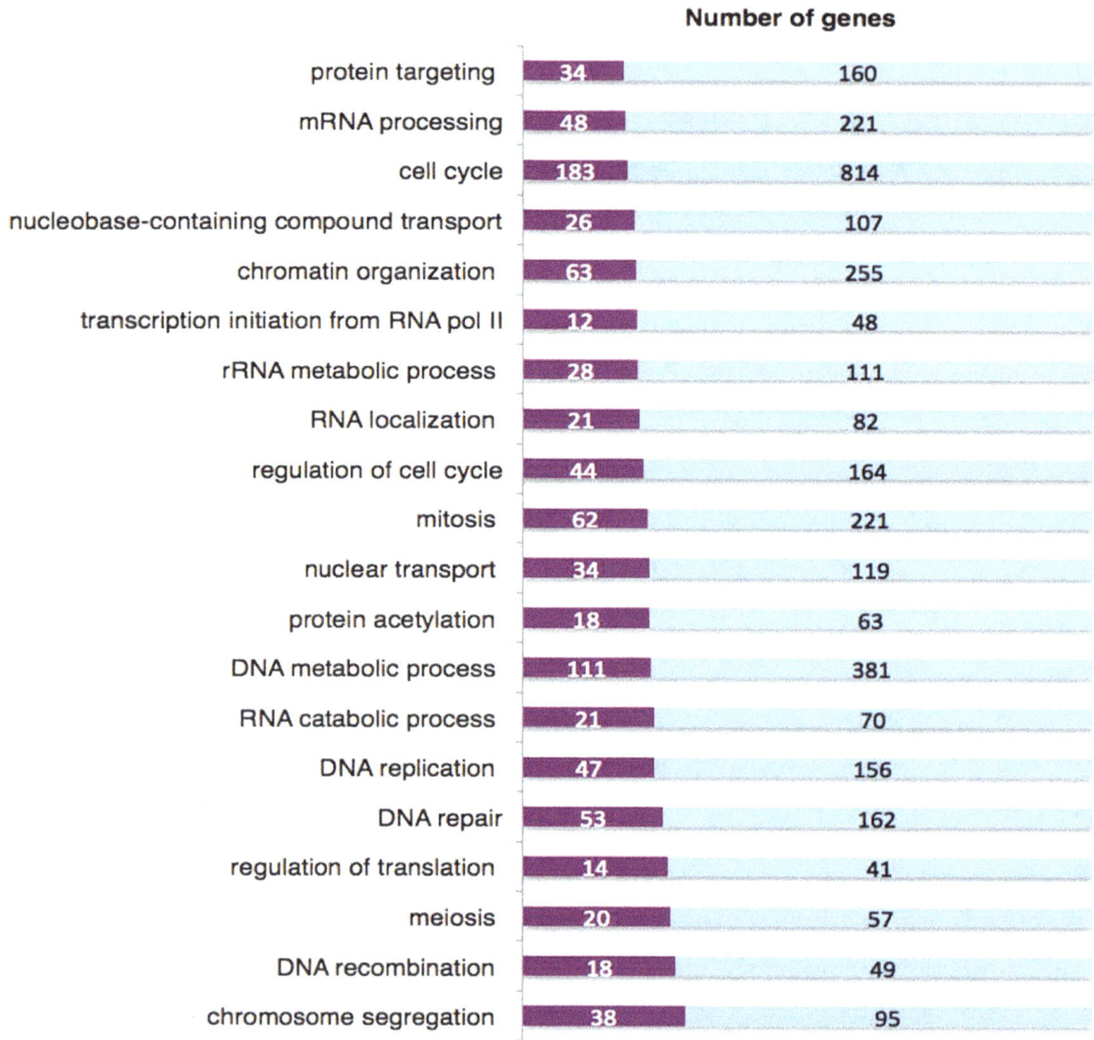

Figure 7. The DNA damage response induced by Melphalan affects splicing of cellular genes with various biological functions, including genes encoding mRNA processing factors. The data set obtained with the Affymetrix GeneChip Human Transcriptome array 2.0 and described in the legend of Figure 6 was analysed with the Transcriptome Analysis Console software (TAC) from Thermo Fisher. Protein coding genes with at least one two-fold change in the use of a splice junction or exon inclusion were exported into PANTHER, version 13.1, Gene List Analysis tool (Available online: http://pantherdb.org) for an overrepresentation test based on their biological function. Results of the RNA array analysis of DMSO or melphalan treated C33A2 cells are displayed as percentage of genes in each category that displayed altered splicing upon melphalan treatment. The blue area represents total number of genes in each biological category, and the purple area represents the number of genes producing mRNAs with altered alternative splicing in response to melphalan.

3. Future Perspective

It is intriguing that DNA damage response factors can recruit RNA processing factors to the HPV16 DNA and that these RNA processing factors alter HPV16 mRNA splicing and polyadenylation in such a way that HPV16 late gene expression is activated [66]. These results warrant investigations of the connection between the DNA damage response and RNA processing in experimental systems that better illustrate the cell-differentiation-dependent HPV life cycle [90,91]. HPV16 gene expression is complex and involves regulation at the levels of transcription, splicing and polyadenylation [18]. Given that there are at least 10 different splice sites and two different polyadenylation sites that all compete with each other and are regulated by several different cellular RNA processing factors, it is conceivable that there are additional connections between the DDR and HPV mRNA processing.

Author Contributions: K.N., C.W. and S.S. wrote the manuscript. K.N. designed figures, performed experiments and analysed the array data.

References

1. Zur Hausen, H. Papillomaviruses and cancer: From basic studies to clinical application. *Nat. Rev. Cancer* **2002**, *2*, 342–350. [CrossRef] [PubMed]
2. Howley, P.M.; Lowy, D.R. Papillomaviridae. In *Virology*, 5th ed.; Knipe, D.M., Howley, P.M., Eds.; Lippincott/The Williams & Wilkins Co.: Philadelphia, PA, USA, 2006; Volume 2, pp. 2299–2354.
3. Sakakibara, N.; Mitra, R.; McBride, A.A. The papillomavirus E1 helicase activates a cellular DNA damage response in viral replication foci. *J. Virol.* **2011**, *85*, 8981–8995. [CrossRef] [PubMed]
4. Walboomers, J.M.; Jacobs, M.V.; Manos, M.M.; Bosch, F.X.; Kummer, J.A.; Shah, K.V.; Snijders, P.J.; Peto, J.; Meijer, C.J.; Munoz, N. Human papillomavirus is a necessary cause of invasive cervical cancer worldwide. *J. Pathol.* **1999**, *189*, 12–19. [CrossRef]
5. Bouvard, V.; Baan, R.; Straif, K.; Grosse, Y.; Secretan, B.; El Ghissassi, F.; Benbrahim-Tallaa, L.; Guha, N.; Freeman, C.; Galichet, L.; et al. A review of human carcinogens—Part B: Biological agents. *Lancet Oncol.* **2009**, *10*, 321–322. [CrossRef]
6. Kadaja, M.; Silla, T.; Ustav, E.; Ustav, M. Papillomavirus DNA replication—From initiation to genomic instability. *Virology* **2009**, *384*, 360–368. [CrossRef] [PubMed]
7. McBride, A.A. Mechanisms and strategies of papillomavirus replication. *Biol. Chem.* **2017**, *398*, 919–927. [CrossRef] [PubMed]
8. Favre, M.; Breitburd, F.; Croissant, O.; Orth, G. Chromatin-like structures obtained after alkaline disruption of bovine and human papillomaviruses. *J. Virol.* **1977**, *21*, 1205–1209. [PubMed]
9. Chow, L.T.; Broker, T.R.; Steinberg, B.M. The natural history of human papillomavirus infections of the mucosal epithelia. *APMIS* **2010**, *118*, 422–449. [CrossRef] [PubMed]
10. Doorbar, J.; Quint, W.; Banks, L.; Bravo, I.G.; Stoler, M.; Broker, T.R.; Stanley, M.A. The biology and life-cycle of human papillomaviruses. *Vaccine* **2012**, *30* (Suppl. 5), F55–F70. [CrossRef] [PubMed]
11. Moody, C.A.; Laimins, L.A. Human papillomavirus oncoproteins: Pathways to transformation. *Nat. Rev. Cancer* **2010**, *10*, 550–560. [CrossRef] [PubMed]
12. Roman, A.; Munger, K. The papillomavirus E7 proteins. *Virology* **2013**, *445*, 138–168. [CrossRef] [PubMed]
13. Vande Pol, S.B.; Klingelhutz, A.J. Papillomavirus E6 oncoproteins. *Virology* **2013**, *445*, 115–137. [CrossRef] [PubMed]
14. McBride, A.A. The papillomavirus E2 proteins. *Virology* **2013**, *445*, 57–79. [CrossRef] [PubMed]
15. Bergvall, M.; Melendy, T.; Archambault, J. The E1 proteins. *Virology* **2013**, *445*, 35–56. [CrossRef] [PubMed]
16. Bernard, H.U. Regulatory elements in the viral genome. *Virology* **2013**, *445*, 197–204. [CrossRef] [PubMed]
17. Thierry, F. Transcriptional regulation of the papillomavirus oncogenes by cellular and viral transcription factors in cervical carcinoma. *Virology* **2009**, *384*, 375–379. [CrossRef] [PubMed]
18. Johansson, C.; Schwartz, S. Regulation of human papillomavirus gene expression by splicing and polyadenylation. *Nat. Rev. Microbiol.* **2013**, *11*, 239–251. [CrossRef] [PubMed]
19. Jia, R.; Zheng, Z.M. Regulation of bovine papillomavirus type 1 gene expression by RNA processing. *Front. Biosci.* **2009**, *14*, 1270–1282. [CrossRef]
20. Graham, S.V.; Faizo, A.A. Control of human papillomavirus gene expression by alternative splicing. *Virus Res.* **2017**, *231*, 83–95. [CrossRef] [PubMed]
21. Wu, C.; Kajitani, N.; Schwartz, S. Splicing and Polyadenylation of Human Papillomavirus Type 16 mRNAs. *Int. J. Mol. Sci.* **2017**, *18*, 366. [CrossRef] [PubMed]
22. Kajitani, N.; Schwartz, S. RNA Binding Proteins that Control Human Papillomavirus Gene Expression. *Biomolecules* **2015**, *5*, 758–774. [CrossRef] [PubMed]
23. Bodaghi, S.; Jia, R.; Zheng, Z.M. Human papillomavirus type 16 E2 and E6 are RNA-binding proteins and inhibit in vitro splicing of pre-mRNAs with suboptimal splice sites. *Virology* **2009**, *386*, 32–43. [CrossRef] [PubMed]

24. Marechal, A.; Zou, L. DNA damage sensing by the ATM and ATR kinases. *Cold Spring Harb. Perspect. Biol.* **2013**, *5*, 012716. [CrossRef] [PubMed]
25. Zhou, B.B.; Elledge, S.J. The DNA damage response: Putting checkpoints in perspective. *Nature* **2000**, *408*, 433–439. [CrossRef] [PubMed]
26. Matsuoka, S.; Ballif, B.A.; Smogorzewska, A.; McDonald, E.R., 3rd; Hurov, K.E.; Luo, J.; Bakalarski, C.E.; Zhao, Z.; Solimini, N.; Lerenthal, Y.; et al. ATM and ATR substrate analysis reveals extensive protein networks responsive to DNA damage. *Science* **2007**, *316*, 1160–1166. [CrossRef] [PubMed]
27. Smolka, M.B.; Albuquerque, C.P.; Chen, S.H.; Zhou, H. Proteome-wide identification of in vivo targets of DNA damage checkpoint kinases. *Proc. Natl. Acad. Sci. USA* **2007**, *104*, 10364–10369. [CrossRef] [PubMed]
28. Stokes, M.P.; Rush, J.; Macneill, J.; Ren, J.M.; Sprott, K.; Nardone, J.; Yang, V.; Beausoleil, S.A.; Gygi, S.P.; Livingstone, M.; et al. Profiling of UV-induced ATM/ATR signaling pathways. *Proc. Natl. Acad. Sci. USA* **2007**, *104*, 19855–19860. [CrossRef] [PubMed]
29. Matsuoka, S.; Huang, M.; Elledge, S.J. Linkage of ATM to cell cycle regulation by the Chk2 protein kinase. *Science* **1998**, *282*, 1893–1897. [CrossRef] [PubMed]
30. Reinhardt, H.C.; Aslanian, A.S.; Lees, J.A.; Yaffe, M.B. p53-deficient cells rely on ATM- and ATR-mediated checkpoint signaling through the p38MAPK/MK2 pathway for survival after DNA damage. *Cancer Cell* **2007**, *11*, 175–189. [CrossRef] [PubMed]
31. Lavin, M.F. Ataxia-telangiectasia: From a rare disorder to a paradigm for cell signalling and cancer. *Nat. Rev. Mol. Cell Biol.* **2008**, *9*, 759–769. [CrossRef] [PubMed]
32. Shiloh, Y. ATM and related protein kinases: Safeguarding genome integrity. *Nat. Rev. Cancer* **2003**, *3*, 155–168. [CrossRef] [PubMed]
33. Jazayeri, A.; Falck, J.; Lukas, C.; Bartek, J.; Smith, G.C.; Lukas, J.; Jackson, S.P. ATM- and cell cycle-dependent regulation of ATR in response to DNA double-strand breaks. *Nat. Cell Biol.* **2006**, *8*, 37–45. [CrossRef] [PubMed]
34. Branzei, D.; Foiani, M. Regulation of DNA repair throughout the cell cycle. *Nat. Rev. Mol. Cell Biol.* **2008**, *9*, 297–308. [CrossRef] [PubMed]
35. Nakahara, T.; Peh, W.L.; Doorbar, J.; Lee, D.; Lambert, P.F. Human papillomavirus type 16 E1E4 contributes to multiple facets of the papillomavirus life cycle. *J. Virol.* **2005**, *79*, 13150–13165. [CrossRef] [PubMed]
36. Stark, G.R.; Taylor, W.R. Control of the G2/M transition. *Mol. Biotechnol.* **2006**, *32*, 227–248. [CrossRef]
37. Anacker, D.C.; Gautam, D.; Gillespie, K.A.; Chappell, W.H.; Moody, C.A. Productive replication of human papillomavirus 31 requires DNA repair factor Nbs1. *J. Virol.* **2014**, *88*, 8528–8544. [CrossRef] [PubMed]
38. Hong, S.; Cheng, S.; Iovane, A.; Laimins, L.A. STAT-5 Regulates Transcription of the Topoisomerase IIbeta-Binding Protein 1 (TopBP1) Gene To Activate the ATR Pathway and Promote Human Papillomavirus Replication. *MBio* **2015**, *6*, e02006–e02015. [CrossRef] [PubMed]
39. Moody, C.A.; Laimins, L.A. Human papillomaviruses activate the ATM DNA damage pathway for viral genome amplification upon differentiation. *PLoS Pathog.* **2009**, *5*, e1000605. [CrossRef] [PubMed]
40. Anacker, D.C.; Aloor, H.L.; Shepard, C.N.; Lenzi, G.M.; Johnson, B.A.; Kim, B.; Moody, C.A. HPV31 utilizes the ATR-Chk1 pathway to maintain elevated RRM2 levels and a replication-competent environment in differentiating Keratinocytes. *Virology* **2016**, *499*, 383–396. [CrossRef] [PubMed]
41. Reinson, T.; Toots, M.; Kadaja, M.; Pipitch, R.; Allik, M.; Ustav, E.; Ustav, M. Engagement of the ATR-dependent DNA damage response at the human papillomavirus 18 replication centers during the initial amplification. *J. Virol.* **2013**, *87*, 951–964. [CrossRef] [PubMed]
42. Fradet-Turcotte, A.; Bergeron-Labrecque, F.; Moody, C.A.; Lehoux, M.; Laimins, L.A.; Archambault, J. Nuclear accumulation of the papillomavirus E1 helicase blocks S-phase progression and triggers an ATM-dependent DNA damage response. *J. Virol.* **2011**, *85*, 8996–9012. [CrossRef] [PubMed]
43. Moody, C. Mechanisms by which HPV Induces a Replication Competent Environment in Differentiating Keratinocytes. *Viruses* **2017**, *9*, 261. [CrossRef] [PubMed]
44. Hong, S.Y. DNA damage response is hijacked by human papillomaviruses to complete their life cycle. *J. Zhejiang Univ. Sci. B* **2017**, *18*, 215–232. [CrossRef] [PubMed]
45. Anacker, D.C.; Moody, C.A. Modulation of the DNA damage response during the life cycle of human papillomaviruses. *Virus Res.* **2017**, *231*, 41–49. [CrossRef] [PubMed]

46. Gillespie, K.A.; Mehta, K.P.; Laimins, L.A.; Moody, C.A. Human papillomaviruses recruit cellular DNA repair and homologous recombination factors to viral replication centers. *J. Virol.* **2012**, *86*, 9520–9526. [CrossRef] [PubMed]
47. Gauson, E.J.; Donaldson, M.M.; Dornan, E.S.; Wang, X.; Bristol, M.; Bodily, J.M.; Morgan, I.M. Evidence supporting a role for TopBP1 and Brd4 in the initiation but not continuation of human papillomavirus 16 E1/E2-mediated DNA replication. *J. Virol.* **2015**, *89*, 4980–4991. [CrossRef] [PubMed]
48. Johnson, B.A.; Aloor, H.L.; Moody, C.A. The Rb binding domain of HPV31 E7 is required to maintain high levels of DNA repair factors in infected cells. *Virology* **2017**, *500*, 22–34. [CrossRef] [PubMed]
49. Hong, S.; Laimins, L.A. The JAK-STAT transcriptional regulator, STAT-5, activates the ATM DNA damage pathway to induce HPV 31 genome amplification upon epithelial differentiation. *PLoS Pathog.* **2013**, *9*, e1003295. [CrossRef] [PubMed]
50. Hong, S.; Dutta, A.; Laimins, L.A. The acetyltransferase Tip60 is a critical regulator of the differentiation-dependent amplification of human papillomaviruses. *J. Virol.* **2015**, *89*, 4668–4675. [CrossRef] [PubMed]
51. Schwartz, S. Papillomavirus transcripts and posttranscriptional regulation. *Virology* **2013**, *445*, 187–196. [CrossRef] [PubMed]
52. Graham, S.V. Papillomavirus 3′UTR regulatory elements. *Front. Biosci.* **2008**, *13*, 5646–5663. [CrossRef] [PubMed]
53. Li, X.; Johansson, C.; Glahder, J.; Mossberg, A.K.; Schwartz, S. Suppression of HPV-16 late L1 5′-splice site SD3632 by binding of hnRNP D proteins and hnRNP A2/B1 to upstream AUAGUA RNA motifs. *Nucleic Acids Res.* **2013**, *22*, 10488–10508. [CrossRef] [PubMed]
54. Zhao, X.; Rush, M.; Schwartz, S. Identification of an hnRNP A1 dependent splicing silencer in the HPV-16 L1 coding region that prevents premature expression of the late L1 gene. *J. Virol.* **2004**, *78*, 10888–10905. [CrossRef] [PubMed]
55. Johansson, C.; Somberg, M.; Li, X.; Backström Winquist, E.; Fay, J.; Ryan, F.; Pim, D.; Banks, L.; Schwartz, S. HPV-16 E2 contributes to induction of HPV-16 late gene expression by inhibiting early polyadenylation. *EMBO J.* **2012**, *13*, 3212–3227. [CrossRef] [PubMed]
56. Hong, S.; Laimins, L.A. Regulation of the life cycle of HPVs by differentiation and the DNA damage response. *Future Microbiol.* **2013**, *8*, 1547–1557. [CrossRef] [PubMed]
57. Anantha, R.W.; Alcivar, A.L.; Ma, J.; Cai, H.; Simhadri, S.; Ule, J.; Konig, J.; Xia, B. Requirement of heterogeneous nuclear ribonucleoprotein C for BRCA gene expression and homologous recombination. *PLoS ONE* **2013**, *8*, e61368. [CrossRef] [PubMed]
58. Adamson, B.; Smogorzewska, A.; Sigoillot, F.D.; King, R.W.; Elledge, S.J. A genome-wide homologous recombination screen identifies the RNA-binding protein RBMX as a component of the DNA-damage response. *Nat. Cell Biol.* **2012**, *14*, 318–328. [CrossRef] [PubMed]
59. Marechal, A.; Li, J.M.; Ji, X.Y.; Wu, C.S.; Yazinski, S.A.; Nguyen, H.D.; Liu, S.; Jimenez, A.E.; Jin, J.; Zou, L. PRP19 transforms into a sensor of RPA-ssDNA after DNA damage and drives ATR activation via a ubiquitin-mediated circuitry. *Mol. Cell* **2014**, *53*, 235–246. [CrossRef] [PubMed]
60. Kai, M. Roles of RNA-Binding Proteins in DNA Damage Response. *Int. J. Mol. Sci.* **2016**, *17*, 310. [CrossRef] [PubMed]
61. Ciccia, A.; Elledge, S.J. The DNA damage response: Making it safe to play with knives. *Mol. Cell* **2010**, *40*, 179–204. [CrossRef] [PubMed]
62. Choi, H.H.; Choi, H.K.; Jung, S.Y.; Hyle, J.; Kim, B.J.; Yoon, K.; Cho, E.J.; Youn, H.D.; Lahti, J.M.; Qin, J.; et al. CHK2 kinase promotes pre-mRNA splicing via phosphorylating CDK11(p110). *Oncogene* **2014**, *33*, 108–115. [CrossRef] [PubMed]
63. Shkreta, L.; Chabot, B. The RNA Splicing Response to DNA Damage. *Biomolecules* **2015**, *5*, 2935–2977. [CrossRef] [PubMed]
64. Dutertre, M.; Vagner, S. DNA-Damage Response RNA-Binding Proteins (DDRBPs): Perspectives from a New Class of Proteins and Their RNA Targets. *J. Mol. Biol.* **2017**, *429*, 3139–3145. [CrossRef] [PubMed]
65. Johansson, C.; Jamal Fattah, T.; Yu, H.; Nygren, J.; Mossberg, A.K.; Schwartz, S. Acetylation of intragenic histones on HPV16 correlates with enhanced HPV16 gene expression. *Virology* **2015**, *482*, 244–259. [CrossRef] [PubMed]

66. Nilsson, K.; Wu, C.; Kajitani, N.; Yu, H.; Tsimtsirakis, E.; Gong, L.; Winquist, E.B.; Glahder, J.; Ekblad, L.; Wennerberg, J.; et al. The DNA damage response activates HPV16 late gene expression at the level of RNA processing. *Nucleic Acids Res.* **2018**, *46*, 5029–5049. [CrossRef] [PubMed]
67. Sarras, H.; Alizadeh Azami, S.; McPherson, J.P. In search of a function for BCLAF1. *Sci. World J.* **2010**, *10*, 1450–1461. [CrossRef] [PubMed]
68. Lee, K.M.; Hsu Ia, W.; Tarn, W.Y. TRAP150 activates pre-mRNA splicing and promotes nuclear mRNA degradation. *Nucleic Acids Res.* **2010**, *38*, 3340–3350. [CrossRef] [PubMed]
69. Yarosh, C.A.; Tapescu, I.; Thompson, M.G.; Qiu, J.; Mallory, M.J.; Fu, X.D.; Lynch, K.W. TRAP150 interacts with the RNA-binding domain of PSF and antagonizes splicing of numerous PSF-target genes in T cells. *Nucleic Acids Res.* **2015**, *43*, 9006–9016. [CrossRef] [PubMed]
70. Shepard, P.J.; Hertel, K.J. The SR protein family. *Genome Biol.* **2009**, *10*, 242. [CrossRef] [PubMed]
71. Long, J.C.; Caceres, J.F. The SR protein family of splicing factors: Master regulators of gene expression. *Biochem. J.* **2009**, *417*, 15–27. [CrossRef] [PubMed]
72. Dhanjal, S.; Kajitani, N.; Glahder, J.; Mossberg, A.K.; Johansson, C.; Schwartz, S. Heterogeneous Nuclear Ribonucleoprotein C Proteins Interact with the Human Papillomavirus Type 16 (HPV16) Early 3′-Untranslated Region and Alleviate Suppression of HPV16 Late L1 mRNA Splicing. *J. Biol. Chem.* **2015**, *290*, 13354–13371. [CrossRef] [PubMed]
73. Zhao, X.; Öberg, D.; Rush, M.; Fay, J.; Lambkin, H.; Schwartz, S. A 57 nucleotide upstream early polyadenylation element in human papillomavirus type 16 interacts with hFip1, CstF-64, hnRNP C1/C2 and PTB. *J. Virol.* **2005**, *79*, 4270–4288. [CrossRef] [PubMed]
74. Cumming, S.A.; Chuen-Im, T.; Zhang, J.; Graham, S.V. The RNA stability regulator HuR regulates L1 protein expression in vivo in differentiating cervical epithelial cells. *Virology* **2009**, *383*, 142–149. [CrossRef] [PubMed]
75. Gruber, A.J.; Schmidt, R.; Gruber, A.R.; Martin, G.; Ghosh, S.; Belmadani, M.; Keller, W.; Zavolan, M. A comprehensive analysis of 3′ end sequencing data sets reveals novel polyadenylation signals and the repressive role of heterogeneous ribonucleoprotein C on cleavage and polyadenylation. *Genome Res.* **2016**, *26*, 1145–1159. [CrossRef] [PubMed]
76. Yu, H.; Gong, L.; Wu, C.; Nilsson, K.; Li-Wang, X.; Schwartz, S. hnRNP G prevents inclusion on the HPV16 L1 mRNAs of the central exon between splice sites SA3358 and SD3632. *J. Gen. Virol.* **2018**. [CrossRef] [PubMed]
77. Jia, R.; Liu, X.; Tao, M.; Kruhlak, M.; Guo, M.; Meyers, C.; Baker, C.C.; Zheng, Z.M. Control of the papillomavirus early-to-late switch by differentially expressed SRp20. *J. Virol.* **2009**, *83*, 167–180. [CrossRef] [PubMed]
78. Somberg, M.; Li, X.; Johansson, C.; Orru, B.; Chang, R.; Rush, M.; Fay, J.; Ryan, F.; Schwartz, S. SRp30c activates human papillomavirus type 16 L1 mRNA expression via a bimodal mechanism. *J. Gen. Virol.* **2011**, *92*, 2411–2421. [CrossRef] [PubMed]
79. Somberg, M.; Schwartz, S. Multiple ASF/SF2 sites in the HPV-16 E4-coding region promote splicing to the most commonly used 3′-splice site on the HPV-16 genome. *J. Virol.* **2010**, *84*, 8219–8230. [CrossRef] [PubMed]
80. Rush, M.; Zhao, X.; Schwartz, S. A splicing enhancer in the E4 coding region of human papillomavirus type 16 is required for early mRNA splicing and polyadenylation as well as inhibition of premature late gene expression. *J. Virol.* **2005**, *79*, 12002–12015. [CrossRef] [PubMed]
81. Li, X.; Johansson, C.; Cardoso-Palacios, C.; Mossberg, A.; Dhanjal, S.; Bergvall, M.; Schwartz, S. Eight nucleotide substitutions inhibit splicing to HPV-16 3′-splice site SA3358 and reduce the efficiency by which HPV-16 increases the life span of primary human keratinocytes. *PLoS ONE* **2013**, *8*, e72776. [CrossRef] [PubMed]
82. Kajitani, N.; Glahder, J.; Wu, C.; Yu, H.; Nilsson, K.; Schwartz, S. hnRNP L controls HPV16 RNA polyadenylation and splicing in an Akt-kinase-dependent manner. *Nucleic Acids Res.* **2017**, *45*, 9654–9678. [CrossRef] [PubMed]
83. Savage, K.I.; Gorski, J.J.; Barros, E.M.; Irwin, G.W.; Manti, L.; Powell, A.J.; Pellagatti, A.; Lukashchuk, N.; McCance, D.J.; McCluggage, W.G.; et al. Identification of a BRCA1-mRNA splicing complex required for efficient DNA repair and maintenance of genomic stability. *Mol. Cell* **2014**, *54*, 445–459. [CrossRef] [PubMed]
84. Fay, J.; Kelehan, P.; Lambkin, H.; Schwartz, S. Increased expression of cellular RNA-binding proteins in HPV-induced neoplasia and cervical cancer. *J. Med. Virol.* **2009**, *81*, 897–907. [CrossRef] [PubMed]

85. Mole, S.; McFarlane, M.; Chuen-Im, T.; Milligan, S.G.; Millan, D.; Graham, S.V. RNA splicing factors regulated by HPV16 during cervical tumour progression. *J. Pathol.* **2009**, *219*, 383–391. [CrossRef] [PubMed]
86. Klymenko, T.; Hernandez-Lopez, H.; MacDonald, A.I.; Bodily, J.M.; Graham, S.V. Human Papillomavirus E2 Regulates SRSF3 (SRp20) to Promote Capsid Protein Expression in Infected Differentiated Keratinocytes. *J. Virol.* **2016**, *90*, 5047–5058. [CrossRef] [PubMed]
87. Gauson, E.J.; Windle, B.; Donaldson, M.M.; Caffarel, M.M.; Dornan, E.S.; Coleman, N.; Herzyk, P.; Henderson, S.C.; Wang, X.; Morgan, I.M. Regulation of human genome expression and RNA splicing by human papillomavirus 16 E2 protein. *Virology* **2014**, *468–470*, 10–18. [CrossRef] [PubMed]
88. Wallace, N.A.; Khanal, S.; Robinson, K.L.; Wendel, S.O.; Messer, J.J.; Galloway, D.A. High-Risk Alphapapillomavirus Oncogenes Impair the Homologous Recombination Pathway. *J. Virol.* **2017**, *91*, e01084-17. [CrossRef] [PubMed]
89. Mehta, K.; Laimins, L. Human Papillomaviruses Preferentially Recruit DNA Repair Factors to Viral Genomes for Rapid Repair and Amplification. *MBio* **2018**, *9*. [CrossRef] [PubMed]
90. Wilson, R.; Laimins, L.A. Differentiation of HPV-containing cells using organotypic "raft" culture or methylcellulose. *Methods Mol. Med.* **2005**, *119*, 157–169. [CrossRef] [PubMed]
91. Lambert, P.F.; Ozbun, M.A.; Collins, A.; Holmgren, S.; Lee, D.; Nakahra, T. Using an imortalised cell line to study the HPV life cycle in organotypic "raft" cultures. *Methods Mol. Med.* **2005**, *119*, 141–155. [PubMed]

5

Coinfection with Epstein–Barr Virus (EBV), Human Papilloma Virus (HPV) and Polyoma BK Virus (BKPyV) in Laryngeal, Oropharyngeal and Oral Cavity Cancer

Bartłomiej Drop [1], Małgorzata Strycharz-Dudziak [2,*], Ewa Kliszczewska [3] and Małgorzata Polz-Dacewicz [3]

1. Department of Information Technology and Medical Statistics, Medical University of Lublin, 20-059 Lublin, Poland; bartlomiej.drop@umlub.pl
2. Chair and Department of Conservative Dentistry with Endodontics, Medical University of Lublin, 20-059 Lublin, Poland
3. Department of Virology, Medical University of Lublin, 20-059 Lublin, Poland; ewakliszczewska@gmail.com (E.K.); m.polz@umlub.pl (M.P.-D.)

* Correspondence: malgorzata.strycharz-dudziak@umlub.pl

Abstract: Most research providing evidence for the role of oncogenic viruses in head and neck squamous cell carcinoma (SCC) development is focused on one type of virus without analyzing possible interactions between two or more types of viruses. The aim of this study was to analyse the prevalence of co-infection with human papillomavirus (HPV), Epstein–Barr virus (EBV) and polyoma BK virus (BKPyV) in oral, oropharyngeal and laryngeal squamous cell carcinomas in Polish patients. The correlations between viral infection, SCC, demographic parameters, evidence of metastases and grading were also investigated. Fresh-frozen tumour tissue samples were collected from 146 patients with laryngeal, oropharyngeal and oral cancer. After DNA extraction, the DNA of the studied viruses was detected using polymerase chain rection (PCR) assay. Males (87.7%) with a history of smoking (70.6%) and alcohol abuse (59.6%) prevailed in the studied group. Histological type G2 was recognized in 64.4% cases. The patients were most frequently diagnosed with T2 stage (36.3%) and with N1 stage (45.8%). Infection with at least two viruses was detected in 56.2% of patients. In this group, co-infection with HPV/EBV was identified in 34.1% of cases, EBV/BKV in 23.2%, HPV/BKV in 22.0%, and HPV/EBV/BKV in 20.7%. No difference of multiple infection in different locations of cancer was observed. The prevalence of poorly differentiated tumours (G3) was more frequent in co-infection with all three viruses than EBV or BKV alone. A significant correlation was observed between tumour dimensions (T) and lymph-node involvement (N) in co-infected patients compared to single infection. Further studies are necessary to clarify whether co-infection plays an important role in the initiation and/or progression of oncogenic transformation of oral, oropharyngeal and laryngeal epithelial cells.

Keywords: squamous cell carcinoma (SCC); laryngeal cancer; oropharyngeal cancer; oral cancer; Epstein–Barr virus (EBV); human papillomavirus (HPV); BK virus (BKV); co-infection

---

1. Introduction

Head and neck cancer accounts for more than 550,000 cases and 380,000 deaths annually worldwide [1]. In Europe, in 2012, there were approximately 250,000 cases (an estimated 4% of cancer incidence) and 63,500 deaths due to malignancy in this area [2]. According to the Polish

National Cancer Registry, oral and oropharyngeal cancer accounts for 3.8% cancers in men and 1.6% cancers in women, while laryngeal cancer constitutes 2.5% cancers in men and 0.4% in women [3].

It is estimated that approximately 90% of all head and neck cancers are squamous cell carcinoma (SCC). As many studies proved, the etiology of SCC is complex and involves many factors. Molecular and epidemiological research has provided evidence of the role of oncogenic viruses in SCC development [4–7]. Human papilloma virus (HPV) was established by the International Agency for Research on Cancer to be an important human carcinogen causing cancer in the head and neck area [8,9]. However, the first human virus with attributable oncogenic potential was Epstein–Barr Virus (EBV) [10]. The relationship between infection with EBV and the development of cancer in the head and neck region was reported by many researchers [11–14].

The human polyoma BK virus (BKPyV) belonging to the *Polyomaviridae* family is associated with human tumors and is classified in group 2B as possibly carcinogenic to humans [15–17]. It is estimated that 90% of the population may be infected with BK virus (BKV) during childhood. Moreover, BKV DNA has been found in many types of tumors, e.g., human brain tumors, in neuroblastoma, in urinary tract tumors, in uterine cervix vulva, lips and tongue carcinomas, as well as in Kaposi's sarcoma [18–21]. The correlation between BKV and prostate and bladder carcinoma and between BKV and metastatic bladder carcinoma among immunosuppressed transplant recipients has been described as a result of possible BKV latency in the kidneys [22]. Another potential location of the virus are salivary glands, as BKV DNA has been detected in saliva [23]. Initial viral exposure often leads to latent infections. Latent episomal BK virus can be reactivated and it can then cause productive viral infections [24]. The etiologic contribution of BK Polyoma Virus (BKPyV) is suggested to represent mechanistically a 'driver' role to a given cancer [25].

In our previous study, BKPyV DNA was detected in 18.5% of patients with oral squamous cell carcinoma but only in 3.3% of the controls [26]. The present research investigated the prevalence of co-infection of human papillomavirus, Epstein–Barr virus and polyoma BK virus in fresh-frozen samples from patients with laryngeal, oropharyngeal and oral cavity cancer and analysed the effects of these co-infections on clinico-pathological and epidemiological features.

2. Results

Males (87.7%) with smoking (70.6%) and alcohol abuse (59.6%) problems prevailed in the studied group. A moderately differentiated (G2) histological type was recognized in 64.4% of cases. The patients were diagnosed most frequently with T2 stage (36.3%), and with N1 stage (45.8%) conditions. No distant metastasis was observed (M0 in 100% patients). Characteristics of patients with oral, oropharyngeal and laryngeal cancer are shown in Table 1.

Single infection was detected in 43.8% of patients and multiple infection in 56.2% (Table 2). Among 146 infected patients, HPV/EBV co-infection was observed in 34.1% of cases, HPV/BKV in 22.0% of cases, and EBV/BKV in 23.2%. Co-infection with all three viruses was detected in 17 cases (20.7%).

The relative frequencies of single infection with HPV, EBV and BKV among oral cavity, laryngeal and oropharyngeal cancers were significantly different ($p = 0.0102$). Single infection with HPV was most frequent in oral cavity cancer (44.5%), while EBV infection was most frequent in oropharyngeal cancer (57.1%) (Table 3). No difference of multiple infection in different locations was observed.

The frequency of co-infection was not dependent on the sex and age of the patients (Table 4). The relative frequency of HPV and EBV co-infection in males and females did not differ significantly from the relative frequency of each sex and age group in the study population. Co-infection was detected mainly in patients living in urban areas. No difference in the frequency of co-infection with the examined viruses in particular locations was found. In cases of co-infection, the histological grade G3 predominated ($p < 10^{-4}$). Stages N1 and N2 were significantly more frequent in co-infection, while T1 and T2 were more frequent in co-infection with three types of viruses.

Logistic regression analysis showed a correlation between histological grade and T, N parameters and type of co-infection (Table 5). The prevalence of poorly differentiated tumours (G3) was more than four times more frequent in HPV/EBV co-infection (OR = 4.14; $p = 0.0160$), five times more frequent in EBV/BKV co-infection (OR = 5; $p = 0.0080$), and 10 times more frequent in patients co-infected with HPV/EBV/BKV (OR = 10.5; $p = 0.0010$) compared to infection with EBV alone.

Another significant relationship was found between histological grading and the type of virus while analyzing single BKV infection and co-infection with HPV/EBV/BKV. Poorly differentiated tumours (G3) were more than four times more common in patients co-infected with HPV/EBV/BKV (OR = 4.5; $p = 0.0427$) compared to infection with single BKV.

A significant correlation was observed between tumour dimensions (T) and lymph-node involvement (N) in co-infected patients compared with single infection. The T3–T4 stages were more frequent in co-infection of all three viruses than BKV alone (OR = 3.5; $p = 0.0457$). On the other hand, N3–N4 was detected more frequently in co-infection of all viruses than HPV alone (OR = 11.5; $p = 0.0067$), than EBV alone (OR = 8.8; $p = 0.0160$), and then BKV alone (OR = 14.4; $p = 0.0204$).

Table 1. Epidemiological and clinical characteristics of patients.

		Total *n* = 146	
		n	%
Sex	Female	18	12.3
	Male	128	87.7
Age	<50	27	18.5
	50–69	92	63.0
	≥70	27	18.5
Place of residence	Urban	95	65.1
	Rural	51	34.9
Smoking	Yes	103	70.6
	No	43	29.4
Alcohol abuse	Yes	87	59.6
	No	59	40.4
Histological grading	G1	28	19.2
	G2	94	64.4
	G3	24	16.4
T stage	T1	29	19.9
	T2	53	36.3
	T3	27	18.5
	T4	37	25.3
N stage	N1	67	45.8
	N2	33	22.6
	N3	23	15.8
	N4	23	15.8
M stage	M0	146	100.0
	M1	0	0
Location of cancer	Oropharynx	53	36.3
	Larynx	40	27.4
	Oral cavity	53	36.3

T—tumour, dimensions; N—lymph nodes involvement; M—distant metastasis.

Table 2. Prevalence of human papillomavirus (HPV), Epstein–Barr virus (EBV) and BK virus (BKV) in infected patients.

Single Infection				
HPV 18 (28.1%)	EBV 35 (54.7%)		BKV 11 (17.2)	64 (43.8%)
Multiple Infection				
HPV + EBV 28 (34.1%)	HPV + BKV 18 (22%)	EBV + BKV 19 (23.2%)	HPV + EBV + BKV 17 (20.7%)	82 (56.2%)

Table 3. Prevalence of single and multiple infection according to location of cancer.

	Single Infection					
Location of Cancer	**HPV**	**EBV**	**BKV**	**Total**		***p***
Oropharynx	4 (22.2%)	20 (57.1%)	4 (36.4%)	28	43.7%	
Larynx	6 (33.3%)	12 (34.3%)	4 (36.4%)	22	34.4%	0.0102
Oral cavity	8 (44.5%)	3 (8.6%)	3 (27.2%)	14	21.9%	*
Total	18	35	11	64	100%	

	Multiple Infection						
Location of Cancer	**HPV + EBV**	**HPV + BKV**	**EBV + BKV**	**HPV + EBV + BKV**	**Total**		***p***
Oropharynx	8 (28.6%)	8 (44.5%)	6 (31.6%)	3 (17.7%)	25	30.5%	
Larynx	6 (21.4%)	4 (22.2%)	4 (21.0%)	4 (23.5%)	18	21.9%	0.7123
Oral cavity	14 (50%)	6 (33.3%)	9 (47.4%)	10 (58.8%)	39	47.6%	
Total	28	18	19	17	82	100%	

* Statistically significant.

Table 4. Epidemiological and clinical characteristics of coinfected patients.

		HPV + EBV n **= 28**		**HPV + BKV** n **= 18**		**EBV + BKV** n **= 19**		**HPV + EBV + BKV** n **= 17**	
		n	%	n	%	n	%	n	%
Sex	female	2	7.1	2	11.1	2	10.5	1	5.9
	male	26	92.9	16	88.9	17	89.5	16	94.1
p		0.9030		0.9010		0.9402		0.6601	
Age	<50	4	14.3	2	11.1	3	15.8	1	5.9
	50–69	19	67.8	14	77.8	13	68.4	14	82.3
	≥70	5	17.9	2	11.1	3	15.8	2	11.8
p		0.9211		0.6831		0.9880		0.5910	
Place of residence	Urban	23	82.1	15	83.3	16	84.2	16	94.1
	Rural	5	17.9	3	16.7	3	15.8	1	5.9
p		0.0400 *		0.0640		0.0330 *		0.0440 *	
Smoking	Yes	22	78.6	13	72.2	15	78.9	13	76.5
	No	6	21.4	5	27.8	4	21.1	4	23.5
p		0.6010		0.7280		0.5691		0.7581	
Alcohol abuse	Yes	26	92.9	15	83.3	16	84.2	15	88.2
	No	2	7.1	3	16.7	3	15.8	2	11.8
p		0.0004 *		0.0890		0.0420 *		0.0800	
Histological grading	G1–G2	12	42.8	5	27.8	7	36.8	3	17.7
	G3	16	57.2	13	72.2	12	63.2	14	82.3
p		10^{-4} *		10^{-4} *		2×10^{-4} *		10^{-4} *	
T stage	T1–T2	20	71.4	14	77.8	13	68.4	15	88.2
	T3–T4	8	28.6	4	22.2	6	31.6	2	11.8
p		0.3310		0.4260		0.7180		0.0430 *	
N stage	N1–N2	22	78.6	16	88.9	16	84.2	17	100.0
	N3–N4	6	21.4	2	11.1	3	15.8	0	0
p		0.0349 *		0.0235 *		0.0274 *		0.0009 *	

* Statistically significant.

Table 5. Odds ratio of predictive variables.

	OR (95% CI)	**OR (95% CI)**	**OR (95% CI)**
Variable	**HPV/HPV + EBV**	**HPV/HPV + BKV**	**HPV/HPV + EBV + BKV**
G1–G2	0.8 (0.14–4.46)	0.61 (0.18–2.06)	0.4 (0.1–1.69)
G3	1.45 (0.54–3.87)	2.46 (0.73–8.25)	3.7 (0.88–15.27)
p	0.7580	0.2250	0.1440
T1–T2	0.91 (0.59–4.23)	1.7 (0.48–6.13)	2.46 (0.65–9.32)
T3–T4	1.57 (0.59–4.2)	0.6 (0.12–3.1)	0.44 (0.05–3.1)
p	0.7800	0.7950	0.4320

Table 5. *Cont.*

	OR (95% CI)	OR (95% CI)	OR (95% CI)
Variable	**HPV/HPV + EBV**	**HPV/HPV + BKV**	**HPV/HPV + EBV + BKV**
N1–N2	0.65 (0.21–2.12)	2.86 (0.81–10.24)	0.23 (0.03–1.93)
N3–N4	1.67 (0.03–2.25)	0.80 (0.08–7.79)	11.45 (1.37–95.6)
p	0.4330	0.4120	0.0067 *
Variable	**EBV/HPV + EBV**	**EBV/EBV + BKV**	**EBV/HPV + EBV + BKV**
G1–G2	0.54 (0.11–2.73)	0.40 (0.14–1.21)	0.19 (0.05–0.78)
G3	4.14 (1.52–11.23)	5.0 (1.61-15.54)	10.5 (2.49–44.11)
p	0.0160 *	0.0080 *	0.0010 *
T1–T2	1.12 (0.39–3.59)	1.78 (0.52–5.98)	3.04 (0.82–11.23)
T3–T4	0.61 (0.16–2.37)	0.91 (0.23–3.68)	0.31 (0.04–2.66)
p	0.8700	0.8280	0.3060
N1–N2	1.33 (0.52–3.42)	1.92 (0.61–6.09)	0.35 (0.04–2.96)
N3–N4	1.14 (0.27–4.84)	0.5 (0.06–4.36)	8.8 (1.07–72.34)
p	0.7060	0.6930	0.0160 *
Variable	**BKV/HPV + BKV**	**BKV/EBV + BKV**	**BKV/HPV + EBV + BKV**
G1–G2	0.38 (0.11–1.39)	0.53 (0.16–1.79)	0.25 (0.06–1.13)
G3	3 (0.82–10.9)	2.2 (0.64–7.19)	4.5 (1.01–20.1)
p	0.2160	0.3870	0.0427 *
T1–T2	2.5 (0.59–10.55)	2.1 (0.5–8.6)	1.43 (0.36–5.65)
T3–T4	0.36 (0.06–1.93)	0.8 (0.18–2.17)	3.5 (0.8–15.9)
p	0.4300	0.6780	0.0457 *
N1–N2	3.59 (0.9–13.9)	3.1 (0.88–11.61)	0.3 (0.03–2.74)
N3–N4	0.28 (0.03–2.62)	0.5 (0.09–2.99)	14.4 (1.6–126.1)
p	0.2920	0.8950	0.0204 *

* Statistically significant.

3. Discussion

Malignancies associated with infectious agents may result from prolonged latency as a consequence of chronic infections [27]. Pathogenic infections are necessary but not sufficient for cancer initiation or progression [8,10,17,25,27]. In patients infected with one virus, secondary co-infection with another virus may serve as an important co-factor that may cause initiation and/or progression of tumors.

A number of studies available in the literature concern the role of viruses in the development of head and neck squamous cell carcinoma (HNSCC). However, they are mainly focused on one type of virus, while only some recent reports analyze the possible correlation between the infection of two oncogenic viruses and carcinogenesis. Our research is the first original observation that implicates HPV, EBV, BKPyV co-infection in laryngeal, pharyngeal and oral cancer in the Polish population. The role of HPV virus, especially HPV16 in head and neck squamous cell carcinoma is well established [28–38].

In our study, a vast majority of patients co-infected with HPV/EBV smoked cigarettes. HPV/EBV coinfection was also detected statistically more often in patients who had problems with alcohol abuse. The results of many researchers suggest a possible synergy between tobacco components and viral oncogenes, especially HPV16 E6/E7 in transformation of oral epithelial cells [39–41].

The role of EBV in oral squamous cell carcinoma (OSCC) development was first observed by zur Hausen [42]. Other authors also emphasize the role of EBV in the development of OSCC [13,14,43,44]. Jaloluli et al. [13] detected the presence of EBV in 55% of samples from eight different countries. Primary infection with EBV mostly occurs at an early age. The virus, together with saliva, gets to the squamous epithelium and the lymphoid organs, primarily B-cells of the pharynx. There, the virus can survive in a latent form. EBV can reactivate periodically without symptoms and can be detected in the saliva of the infected patients [45].

A number of studies point to co-infection with HPV and EBV in oral squamous cell carcinoma [11,46,47]. Moreover, several researchers indicate that co-infection by multiple oncogenic viruses may be an important risk factor in the development of OSCC [11,42,47,48]. Co-infections occur much more frequently in the areas of high prevalence of infectious agents, especially in developing countries [27]. The presence of HPV/EBV co-infection in the presented study was found in 34.1% of patients. Deng et al. [49], in research carried out in Japan, revealed HPV/EBV co-infection in 1% of patients with head and neck cancer (HNC) and in 10% of patients with nasopharyngeal carcinoma (NPC).

Infection with a number of pathogens very often causes inflammation of tissues or organs, which can lead to the initiation of carcinogenesis. Al Moustafa et al. [46] proposes that high-risk HPV and EBV co-infections play an important role in initiating neoplastic transformation of human oral epithelial cells. Jiang et al. [47] hypothesized that oropharyngeal tumors might be associated with both HPV and EBV rather than HPV alone, and co-infected cells can have a higher tumorigenic potential than normal cells.

According to the research, it is not clear which virus, HPV or EBV, contributes to the first infection in co-infected patients [50]. However, the study performed by Makielski et al. [51] indicated that infection with HPV in the oral cavity may increase the capacity of epithelial cells to support the EBV life cycle, which could in turn increase EBV-mediated pathogenesis in the oral cavity. Guidry and Scott [52] suggested that HPV/EBV co-infection increases EBV persistence either through latency or enhanced viral replication and by extending HPV oncogene expression.

Apart from HPV and EBV, BKV DNA was also detected (20.7%) in the studied material. Recent data has suggested a correlation between BK virus and various types of human cancers [53].

The presence of BKV DNA was confirmed in high-grade squamous intraepithelial cervical lesions (precancerous lesions) [54]. Burger-Calderon et al. [55] suggest a connection between BKPyV and the oral cavity. Several studies have suggested BKPyV to be oral-tropic [23]. BKPyV binds to cellular receptors such as N-linked glycoprotein with a 2,3-linkedsialic acids and gangliosides GD1b and Gt1b, which is true for kidney (Vero) and oral (HSG) cells in vitro. It has not been determined whether the viruses undergo true latency, expressing only a subset of specific viral genes. It is unclear whether polyomavirus DNA is commonly integrated into the host genome or whether integration is a rare event specific to HPyV subtypes [17]. The non-coding control region (NCCR) is a hypervariable region and comparative studies suggested that it may regulate host cell tropism [55].

BKV DNA was detected in tonsilar biopsy specimens and nasopharyngeal aspirates. Moreover, replication of BKPyV laboratory strain in human submandibular and parotid salivary gland cell lines (HSG and HSY) was also demonstrated. Besides, Moens et al. [56] suggest that polyomaviruses, including those induced by other oncogenic viruses, may be a co-factor in the development of cancer. Some authors suggest that BK virus may be a potential co-factor for HPV in the development of cervical neoplasia [57], especially together with the HPV genotype 16 [54]. In our studies HPV/BKV co-infection was detected in 22.0% cases, while EBV/BKV in 23.2%. EBV/BKV was statistically more often detected among urban inhabitants. In light of our and other authors' research results, we cannot exclude the role of BKV in SCC, considering the fact that the genetic material of BKV was detected in saliva [23]. The 'hit and run' hypothesis is a mechanism deemed valid to justify a co-factorial role of BKPyV in cancer onset and progression in humans [17]. Tag gene expression, leading to inactivation of p53, without evidence of a productive infection (i.e., viral protein expression, genome replication, etc.), leads to host cell transformation [58].

In our study the OR for low-differentiated tumours (G3) was about four times greater in patients with HPV/EBV co-infection, five times greater in patients with EBV/BKV co-infection, and more than 10 times greater in patients with HPV/EBV/BKV co-infection than in patients with only EBV infection. Gonzales-Moles et al. [59] found a correlation between EBV and poor differentiation of cancerous lesion in OSCC. Some researchers have revealed experimentally that EBV infection may delay epithelial differentiation and enhance the invasiveness of epithelial cells expressing *HPV16 E6* and *E7* oncogenes. However, delayed differentiation and greater invasiveness were still present in epithelial cells after loss of EBV, which may suggest that EBV infection led to epigenetic reprogramming [60].

Double or mixed infection with other oncogenic viruses may induce transformation. A limitation of our study is, however, only the epidemiological character of the research carried out. As DNA was extracted from resected tumour tissues, the sample might contain DNA from normal tissues, including infiltrating lymphocytes, which might be infected with viruses. Performing immunohistochemical analysis could have provide more conclusive data. Thus, further studies are needed to clarify whether BKPyV plays a role in oral squamous cell carcinoma or is a co-factor for cancers induced by other oncoviruses. It is well-known that chronic infection affects the immunological response of the host. Primary infection with a non-oncogenic virus may promote superinfection with an oncogenic virus capable of tumor transformation. The oncogenic potential of HPV is related to the expression of *E6* and *E7*, whereas the oncogenic potential of EBV to the expression of LMP-1 (latent membrane protein 1) and LMP-2 and of the BKV-LTag (large tumour antigen). These oncoproteins can be in cooperation and they can lead to the transformation of the oral epithelium [17,25,27]. Toll-like receptors (TLRs) play a critical role in the early innate immune response to invading pathogens by sensing microorganisms and they are involved in sensing endogenous danger signals. The LTag of the virus BKPyV as well as the protein of LMP-1 of the virus EBV lowers the expression of TLR9 [61,62]. EBV latent membrane protein 1 is a negative regulator of TLR9. These observations may contribute to future studies. There are few studies examining the association between HPV, EBV and/or BKV in the progression of oral cancers. A detailed understanding of co-infection will enable the targeting of new methods for the early detection, prevention and treatment of viral-associated cancers.

4. Materials and Methods

4.1. Patients

The present study involved 146 patients with a diagnosed and histopathologically confirmed SCC of larynx, oropharynx and the oral cavity who were infected with at least one virus—HPV, EBV, or BKV. The patients were hospitalized at the Otolaryngology Division of the Masovian Specialist Hospital in Radom, Poland. The patients had not received radiotherapy or chemotherapy before. The samples were collected during surgery, but TNM was calculated during primary diagnosis. TNM classification was done according to the criteria of the Union Against Cancer (UICC) [63]. Histological grading was performed according to World Health Organization criteria, which divide tumors into three types: well differentiated (G1), moderately differentiated (G2), and poorly differentiated (G3) [64].

The research was approved by the Medical University of Lublin Ethics Committee and is in accordance with the GCP regulations (no. KE-0254/133/2013, 23 May 2013).

4.2. DNA Extraction from Fresh Frozen Tumour Tissue; Detection of EBV DNA

DNA extraction from fresh frozen tumour tissue, detection of EBV DNA, and amplification of the *EBNA-2* gene (the nested PCR) were performed as previously described [65].

4.3. HPV Detection and Genotyping

HPV genotyping was performed using the INNO-LiPA HPV Genotyping Extraassay (Innogenetics, Gent, Belgium). The kit is based on the amplification of a 65 bp fragment from the L1 region of the HPV genome with a SPF10 primer set. PCR products are subsequently typed with the reverse hybridization assay. This kit identifies 28 HPV genotypes: HPV 6, 11, 16, 18, 26, 31, 33, 35, 39, 40, 43, 44, 45, 52, 52, 53, 54, 56, 58, 59, 66, 68, 69, 70,71, 73, 74, and 82.

4.4. Detection of BKV

The polymerase chain reaction (PCR) method was used to detect BKPyV in the specimens. With the aim of detecting the genetic material of the BKPyV, the primers described for the first time by Arthur et al. [66], namely PEP-1 (5′-AGTCTTTAGGGTCTTCTACC-3′) and PEP-2 (5′-GGTGCCAACCTATGGAACAG-3′) were used. The oligonucleotides attach to a highly conservative region of early coding T-Ag. Primers amplify a 176-bp fragment of BKV genetic material. The final concentrations of the PCR reaction mixture were as follows: 2.0 mM $MgCl_2$, 200 µM dNTPs, 0.25 µM of each primer, 0.5 U Hot Start Taq DNA polymerase (Qiagen, Hilden, Germany). Amplification was performed under the following conditions: initial denaturation 94 °C 15 min, followed by 40 cycles: 94 °C 1 min, 55 °C 1 min, 72 °C 1 min; final extension: 72 °C 10 min. During each PCR run, the samples were tested, together with one negative and one positive control. DNA from the urine of a kidney transplant patient was used as a positive PCR control to assess the success of amplification (ATCC VR-837). PCR reagents without template DNA served as a negative control. The PCR products were analyzed using electrophoresis in 2% agarose gel.

4.5. Statistical Analysis

Statistical analysis was performed to investigate the relationship between the presence of HPV, EBV and BKV. The clinical and demographic characteristics of patients were determined by means of Pearson's chi-square test and with Fisher's exact test for small groups. Stepwise logistic regression was used to assess the effect of co-infection of HPV, EBV, BKV on the risk of the occurrence G and TN variables. The odds ratio with 95% confidence intervals was calculated. Statistical significance was defined as $p < 0.05$.

5. Conclusions

In Polish patients with oral, oropharyngeal and laryngeal cancer, co-infection with at least one virus was detected in 56.2% of cases. In this group, co-infection with HPV/EBV was identified in 34.1% of cases, EBV/BKV in 23.2%, HPV/BKV in 22.0%, and HPV/EBV/BKV in 20.7%. No difference of multiple infection in different locations of cancer was observed.

The prevalence of poorly differentiated tumours (G3) was more frequent in co-infection of all three viruses than EBV or BKV alone. T3–T4 and N3–N4 was more frequent in co-infection than in single viral infection.

Future epidemiological studies regarding the relationship between infection and gender, tobacco, alcohol and chronic inflammation in the development of oral cancer as well as studies on the mechanisms of co-infection and/or superinfection and their role in oral squamous cell carcinoma are necessary. Knowledge about the pathways of these viruses may provide targets for therapy and for devising diagnostic methods.

Acknowledgments: This study was supported by a Research Grant from the Medical University of Lublin, Lublin, Poland (DS 233). We are thankful to Sylwia Fołtyn for help in clinical material collection.

Author Contributions: Bartłomiej Drop: statistical and epidemiological analysis; Małgorzata Strycharz-Dudziak: data analysis, manuscript preparation; Ewa Kliszczewska: carried out serological and molecular identification, data analysis, manuscript preparation; Małgorzata Polz-Dacewicz: conceived the study, data analysis, coordination and help in drafting the manuscript; all authors read and approved the final manuscript.

References

1. Fitzmaurice, C.; Allen, C.; Barber, R.M.; Barregard, L.; Bhutta, Z.A.; Brenner, H.; Dicker, D.J.; Chimed-Orchir, O.; Dandona, R.; Dandona, L.; et al. Global, Regional, and National Cancer Incidence, Mortality, Years of Life Lost, Years Lived With Disability, and Disability-Adjusted Life-years for 32 Cancer Groups, 1990 to 2015: A Systematic Analysis for the Global Burden of Disease Study. *JAMA Oncol.* **2017**, *3*, 524–548. Available online: https://jamanetwork.com/journals/jamaoncology/fullarticle/2588797 (accessed on 7 November 2017). [CrossRef] [PubMed]
2. Gatta, G.; Botta, L.; Sánchez, M.J.; Anderson, L.A.; Pierannunzio, D.; Licitra, L.; EUROCARE Working Group. Prognoses and improvement for head and neck cancers diagnosed in Europe in early 2000s: The EUROCARE-5 population-based study. *Eur. J. Cancer* **2015**, *51*, 2130–2143. Available online: http://www.ejcancer.com/article/S0959-804900749-2/fulltext (accessed on 7 November 2017). [CrossRef] [PubMed]

3. Wojciechowska, U.; Olasem, P.; Czauderna, K.; Didkowska, J. *Cancer in Poland in 2014*; Ministerstwo Zdrowia: Warszawa, Poland, 2016; pp. 42–45.
4. Fakhry, C.; Westra, W.H.; Li, S.; Cmelak, A.; Ridge, J.A.; Pinto, H.; Forastiere, A.; Gillison, M.L. Improved survival of patients with human papillomavirus-positive head and neck squamous cell carcinoma in a prospective clinical trial. *J. Natl. Cancer Inst.* **2008**, *100*, 261–269. Available online: https://academic.oup.com/jnci/article-lookup/doi/10.1093/jnci/djn011 (accessed on 7 November 2017). [CrossRef] [PubMed]
5. Gillison, M.L.; Koch, W.M.; Capone, R.B.; Spafford, M.; Westra, W.H.; Wu, L.; Zahurak, M.L.; Daniel, R.W.; Viglione, M.; Symer, D.E.; et al. Evidence for a causal association between human papillomavirus and a subset of head and neck cancers. *J. Natl. Cancer Inst.* **2000**, *92*, 709–720. Available online: https://academic.oup.com/jnci/article/92/9/709/2906131/Evidence-for-a-Causal-Association-Between-Human (accessed on 7 November 2017). [CrossRef] [PubMed]
6. Hillbertz, N.S.; Hirsch, J.M.; Jalouli, J.; Jalouli, M.M.; Sand, L. Viral and molecular aspects of oral cancer. *Anticancer Res.* **2012**, *32*, 4201–4212. Available online: http://ar.iiarjournals.org/content/32/10/4201.long (accessed on 7 November 2017). [PubMed]
7. Scully, C. Oral cancer aetiopathogenesis; past, present and future aspects. *Medicina Oral Patologia Oral Y Cirugia Bucal* **2011**, *16*, e306–e311. Available online: https://www.researchgate.net/publication/50868315_Oral_cancer_aetiopathogenesis_past_present_and_future_aspects (accessed on 7 November 2017). [CrossRef]
8. *IARC Monographs on the Evaluation of Carcinogenic Risks to Humans*; International Agency for Research on Cancer: Lyon, France, 2007; pp. 222–230. ISBN 978-92-832-1290-4.
9. *IARC Monographs on the Evaluation of Carcinogenic Risks to Humans. A Review of Human Carcinogens. Biological Agents*; International Agency for Research on Cancer: Lyon, France, 2012; p. 255. ISBN 978-92-832-1319-2.
10. *IARC Monographs on the Evaluation of Carcinogenic Risks to Humans. Epstein-Barr virus and Kaposi' Sarcoma Herpesvirus/Human Herpesvirus 8*; International Agency for Research on Cancer: Lyon, France, 1997; pp. 47–262. ISBN 92832-12703.
11. Acharya, S.; Ekalaksananan, T.; Vatanasapt, P.; Loyha, K.; Phusingha, P.; Promthet, S.; Kongyingyoes, B.; Pientong, C. Association of Epstein-Barr virus infection with oral squamous cell carcinoma in a case-control study. *J. Oral Pathol. Med.* **2015**, *44*, 252–257. Available online: http://onlinelibrary.wiley.com/doi/10.1111/jop.12231/full (accessed on 7 November 2017). [CrossRef] [PubMed]
12. Jalouli, J.; Ibrahim, S.O.; Mehrotra, R.; Jalouli, M.M.; Sapkota, D.; Larson, P.A.; Hirsch, J.M. Prevalence of viral (HPV, EBV, HSV) infections in oral submucous fibrosis and oral cancer from India. *Acta Oto-Laryngol.* **2010**, *130*, 1306–1311. Available online: http://www.tandfonline.com/doi/full/10.3109/00016481003782041?needAccess=true (accessed on 7 November 2017). [CrossRef] [PubMed]
13. Jalouli, J.; Jalouli, M.M.; Sapkota, D.; Ibrahim, S.O.; Larson, P.A.; Sand, L. Human papilloma virus, herpes simplex and Epstein-Barr virus in oral squamous cell carcinoma from eight different countries. *Anticancer Res.* **2012**, *32*, 571–580. Available online: http://ar.iiarjournals.org/content/32/2/571.long (accessed on 7 November 2017). [PubMed]
14. Kis, A.; Feher, K.; Gall, T.; Tar, I.; Boda, R.; Toth, E.D.; Méhes, G.; Gergely, L.; Szarka, K. Epstein-Barr virus prevalence in oral squamous cell cancer and potentially malignant oral disorders in an eastern Hungarian population. *Eur. J. Oral Sci.* **2009**, *117*, 537–540. Available online: http://onlinelibrary.wiley.com/wol1/doi/10.1111/j.1600-0722.2009.00660.x/full (accessed on 7 November 2017). [CrossRef] [PubMed]
15. Bennett, S.; Broekema, N.; Imperiale, M. BK polyomavirus: Emerging pathogen. *Microbes Infect.* **2012**, *14*, 672–683. Available online: https://www.ncbi.nlm.nih.gov/pmc/articles/PMC3568954/pdf/nihms-360495.pdf (accessed on 7 November 2017). [CrossRef] [PubMed]
16. IARC Monographs on the Evaluation of Carcinogenic Risks to Humans. *Malaria and Some Polyomaviruses (SV40, BK, JC, and Merkel Cell Viruses)*; IARC Monographs: Lyon, France, 2014; Volume 104, pp. 215–251. ISBN 978-92-832-0142-7.
17. Burger-Calderon, R.; Webster-Cyriaque, J. Human BK Polyomavirus—The Potential for Head and Neck Malignancy and Disease. *Cancers* **2015**, *7*, 1244–1270. Available online: https://www.ncbi.nlm.nih.gov/pmc/articles/PMC4586768/pdf/cancers-07-00835.pdf (accessed on 7 November 2017). [CrossRef] [PubMed]
18. Neirynck, V.; Claes, K.; Naesens, M.; De Wever, L.; Pirenne, J.; Kuypers, D.; Vanrenterghem, Y.; van Poppel, H.; Kabanda, A.; Lerut, E. Renal cell carcinoma in the allograft: What is the role of Polyomavirus?

Case Rep. Nephrol. Urol. **2012**, *2*, 125–134. Available online: https://www.karger.com/Article/Pdf/341917 (accessed on 7 November 2017). [CrossRef] [PubMed]

19. Schowalter, R.; Reinhold, W.; Buck, C. Entry tropism of BK and Merkel cell polyomaviruses in cell culture. *PLoS ONE* **2012**, *7*, e42181. Available online: http://journals.plos.org/plosone/article?id=10.1371/journal.pone.0042181 (accessed on 7 November 2017). [CrossRef] [PubMed]
20. Raeesi, N.; Gheissari, A.; Akrami, M.; Moghim, S. Urinary BK virus excretion in children newly diagnosed with acute lymphoblastic leukemia. *Int. J. Prev. Med.* **2012**, *6*, 402–407. Available online: https://www.ncbi.nlm.nih.gov/pmc/articles/PMC3389437/ (accessed on 7 November 2017).
21. Konietzny, R.; Fischer, R.; Ternette, N.; Wright, C.; Turney, B.; Chakera, A. Detection of BK virus in urine from renal transplant subjects my mass spectrometry. *Clin. Proteomics* **2012**, *9*, 4–13. Available online: https://clinicalproteomicsjournal.biomedcentral.com/articles/10.1186/1559-0275-9-4 (accessed on 7 November 2017). [CrossRef] [PubMed]
22. Tognon, M.; Corallini, A.; Martini, F.; Negrini, M.; Barbanti-Brodano, G. Oncogenic transformation by BK virus and association with human tumors. *Oncogene* **2003**, *22*, 5192–5200. Available online: http://www.nature.com/onc/journal/v22/n33/full/1206550a.html (accessed on 7 November 2017). [CrossRef] [PubMed]
23. Jeffers, L.K.; Madden, V.; Webster-Cyriaque, J. BK virus has tropism for human salivary gland cells in vitro: Implications for transmission. *Virology* **2009**, *394*, 183–193. Available online: http://www.sciencedirect.com/science/article/pii/S0042682209004267?via%3Dihub (accessed on 7 November 2017). [CrossRef] [PubMed]
24. Kenan, D.J.; Mieczkowski, P.A.; Latulippe, E.; Côté, I.; Singh, H.K.; Nickeleit, V. BK Polyomavirus Genomic Integration and Large T Antigen Expression: Evolving Paradigms in Human Oncogenesis. *Am. J. Transplant.* **2017**, *17*, 1674–1680. Available online: http://onlinelibrary.wiley.com/doi/10.1002/path.4584/full (accessed on 7 November 2017). [CrossRef] [PubMed]
25. Dalianis, T.; Hirsch, H.H. Human polyomaviruses in disease and cancer. *Virology* **2013**, *437*, 63–72. Available online: http://www.sciencedirect.com/science/article/pii/S0042682213000044?via%3Dihub (accessed on 7 November 2017). [CrossRef] [PubMed]
26. Polz, D.; Morshed, K.; Jarzyński, A.; Polz-Dacewicz, M. Prevalence of Polyoma BKVirus (BKPyV), Epstein-Barr Virus (EBV) and Human Papilloma Virus (HPV) in Oropharyngeal Cancer. *Pol. J. Microbiol.* **2015**, *64*, 323–328. Available online: http://www.pjm.microbiology.pl/archive/vol6442015323.pdf (accessed on 7 November 2017). [CrossRef]
27. Vedham, V.; Divi, R.L.; Starks, V.L.; Verma, M. Multiple infections and cancer: Implications in epidemiology. *Technol. Cancer Res. Treat.* **2014**, *13*, 177–194. Available online: http://journals.sagepub.com/doi/abs/10.7785/tcrt.2012.500366 (accessed on 7 November 2017). [CrossRef] [PubMed]
28. Sand, L.; Wallström, M.; Hirsch, J.M. Smokeless tobacco, viruses and oral cancer. *Oral Health Dent. Manag.* **2014**, *13*, 372–378. Available online: https://www.omicsonline.org/open-access/smokeless-tobacco-viruses-and-oral-cancer-2247-2452.1000594.pdf (accessed on 7 November 2017). [PubMed]
29. Metgud, R.; Astekar, M.; Verma, M.; Sharma, A. Role of viruses in oral squamous cell carcinoma. *Oncol. Rev.* **2012**, *6*, 164–170. Available online: https://www.ncbi.nlm.nih.gov/pmc/articles/PMC4419625/ (accessed on 7 November 2017). [CrossRef] [PubMed]
30. Sathish, N.; Wang, X.; Yuan, Y. Human Papillomavirus (HPV)-associated oral cancers and treatment strategies. *J. Dent. Res.* **2014**, *93*, 29S–36S. Available online: https://www.ncbi.nlm.nih.gov/pmc/articles/PMC4107541/ (accessed on 7 November 2017). [CrossRef] [PubMed]
31. Gillison, M.L. Current topics in the epidemiology of oral cavity and oropharyngeal cancers. *Head Neck* **2007**, *29*, 779–792. Available online: http://onlinelibrary.wiley.com/wol1/doi/10.1002/hed.20573/full (accessed on 7 November 2017). [CrossRef] [PubMed]
32. Gillison, M.L.; Castellsagué, X.; Chaturvedi, A.; Goodman, M.T.; Snijders, P.; Tommasino, M.; Arbyn, M.; Franceschi, S. Eurogin Roadmap: Comparative epidemiology of HPV infection and associated cancers of the head and neck and cervix. *Int. J. Cancer* **2014**, *134*, 497–507. Available online: http://onlinelibrary.wiley.com/doi/10.1002/ijc.28201/full (accessed on 7 November 2017). [CrossRef] [PubMed]
33. Syrjänen, K.; Syrjänen, S. Detection of human papillomavirus in sinonasal carcinoma: Systematic review and meta-analysis. *Hum. Pathol.* **2013**, *44*, 983–991. Available online: http://www.humanpathol.com/article/S0046-817700320-6/fulltext (accessed on 7 November 2017). [CrossRef] [PubMed]
34. Syrjänen, K.J.; Syrjänen, S.M.; Lamberg, M.A.; Pyrhönen, S. Human papillomavirus (HPV) involvement in squamous cell lesions of the oral cavity. *Proc. Finn. Dent. Soc.* **1983**, *79*, 1–8. [PubMed]

35. Anantharaman, D.; Gheit, T.; Waterboer, T.; Abedi-Ardekani, B.; Carreira, C.; McKay-Chopin, S.; Gaborieau, V.; Marron, M.; Lagiou, P.; Ahrens, W.; et al. Human Papillomavirus infections and upper aero-digestive tract cancers. The ARCAGE study. *J. Natl. Cancer Int.* **2013**, *105*, 536–545. Available online: https://academic.oup.com/jnci/article-lookup/doi/10.1093/jnci/djt053 (accessed on 7 November 2017). [CrossRef] [PubMed]
36. Fakhry, C.; D'Souza, G. Discussing the diagnosis of HPV-OSCC: Common questions and answers. *Oral Oncol.* **2013**, *49*, 863–871. Available online: https://www.ncbi.nlm.nih.gov/pmc/articles/PMC4264664/pdf/nihms647259.pdf (accessed on 7 November 2017). [CrossRef] [PubMed]
37. Fakhry, C.; Psyrri, A.; Chaturvedi, A. HPV and head and neck cancers: State-of-the-science. *Oral Oncol.* **2014**, *50*, 353–355. Available online: http://www.sciencedirect.com/science/article/pii/S1368837514001006?via%3Dihub (accessed on 7 November 2017). [CrossRef] [PubMed]
38. Benson, E.; Li, R.; Eisele, D.; Fakhry, C. The clinical impast of HPV tumor status upon head and neck squamous cell carcinomas. *Oral Oncol.* **2014**, *50*, 565–574. Available online: http://www.sciencedirect.com/science/article/pii/S136883751300691X?via%3Dihub (accessed on 7 November 2017). [CrossRef] [PubMed]
39. Haukioja, A.; Asunta, M.; Söderling, E.; Syrjänen, S. Persistent oral human papillomavirus infection is associated with smoking and elevated salivary immunoglobulin G concentration. *J. Clin. Virol.* **2014**, *61*, 101–106. Available online: http://www.journalofclinicalvirology.com/article/S1386-653200243-1/fulltext (accessed on 7 November 2017). [CrossRef] [PubMed]
40. Rautava, J.; Syrjänen, S. Human papillomavirus infections in the oral mucosa. *J. Am. Dent. Assoc.* **2011**, *142*, 905–914. Available online: http://jada.ada.org/article/S0002-817762065-1/fulltext (accessed on 7 November 2017). [CrossRef] [PubMed]
41. Gillison, M.L.; Zhang, Q.; Jordan, R.; Xiao, W.; Westra, W.H.; Trotti, A.; Spencer, S.; Harris, J.; Chung, C.H.; Ang, K.K. Tobacco smoking and increased risk of death and progression for patients with p16 posotive and p16-negative oropharyngeal cancer. *J. Clin. Oncol.* **2012**, *30*, 2102–2111. Available online: https://www.ncbi.nlm.nih.gov/pmc/articles/PMC3397696/ and http://ascopubs.org/doi/full/10.1200/JCO.2011.38.4099 (accessed on 7 November 2017). [CrossRef] [PubMed]
42. Zur Hausen, H. Biochemical approaches to detection of Epstein-Barr virus in human tumors. *Cancer Res.* **1976**, *36*, 678–680. Available online: https://pdfs.semanticscholar.org/5f61/2ed773454019a9eb509e542e82f4f3a2a93e.pdf (accessed on 7 November 2017). [PubMed]
43. Jalouli, J.; Ibrahim, S.; Sapkota, D.; Jalouli, M.M.; Vasstrand, E.N.; Hirsch, J.M.; Larson, P.A. Presence of human papilloma virus, herpes simplex virus and Epstein-Barr virus DNA in oral biopsies from Sudanese patients with regard to toombak use. *J Oral Pathol. Med.* **2011**, *19*, 599–604. Available online: http://onlinelibrary.wiley.com/wol1/doi/10.1111/j.1600-0714.2010.00910.x/full (accessed on 7 November 2017). [CrossRef] [PubMed]
44. Senyuta, N.; Yakovleva, L.; Goncharova, E.; Scherback, L.; Diduk, S.; Smirova, K.; Maksimovich, D.; Gurtsevitch, V. Epstein-Barr virus latent membrane protein 1 polymorphism in nasopharyngeal carcinoma and other oral cavity tumors in Russia. *J. Med. Virol.* **2013**, *86*, 290–300. Available online: http://onlinelibrary.wiley.com/wol1/doi/10.1002/jmv.23729/full (accessed on 7 November 2017). [CrossRef] [PubMed]
45. Gulley, M.L.; Tang, W. Laboratory Assays for Epstein-Barr Virus-Related Disease. *J. Mol. Diagn.* **2008**, *10*, 279–292. Available online: https://www.ncbi.nlm.nih.gov/pmc/articles/PMC2438195/ (accessed on 7 November 2017). [CrossRef] [PubMed]
46. Al Moustafa, A.E.; Chen, D.; Ghabreau, L.; Akil, N. Association between human papillomavirus and Epstein-Barr virus infections in human oral carcinogenesis. *Med. Hypotheses* **2009**, *73*, 184–186. Available online: http://www.sciencedirect.com/science/article/pii/S0306987709001686?via%3Dihub (accessed on 7 November 2017). [CrossRef] [PubMed]
47. Jiang, R.; Ekshyyan, O.; Moore-Medlin, T.; Rong, X.; Nathan, S.; Gu, X.; Abreo, F.; Rosenthal, E.L.; Shi, M.; Guidry, J.T.; et al. Association between human papilloma virus/Epstein-Barr virus coinfection and oral carcinogenesis. *J. Oral Pathol. Med.* **2015**, *44*, 28–36. Available online: https://www.ncbi.nlm.nih.gov/pmc/articles/PMC4286485/ and http://onlinelibrary.wiley.com/wol1/doi/10.1111/jop.12221/full (accessed on 7 November 2017). [CrossRef] [PubMed]
48. Sand, L.; Jalouli, J. Viruses and oral cancer. Is there a link? *Microbes Infect.* **2014**, *16*, 371–378. Available online: http://www.sciencedirect.com/science/article/pii/S1286457914000239?via%3Dihub (accessed on 7 November 2017). [CrossRef] [PubMed]

49. Deng, Z.; Uehara, T.; Maeda, H.; Hasegawa, M.; Matayoshi, S.; Kiyuna, A.; Agena, S.; Pan, X.; Zhang, C.; Yamashita, Y.; et al. Epstein-Barr virus and human papillomavirus infections and genotype distribution in head and neck cancers. *PLoS ONE* **2014**, *9*, e113702. Available online: http://journals.plos.org/plosone/article?id=10.1371/journal.pone.0113702 (accessed on 7 November 2017). [CrossRef] [PubMed]
50. Ying, S.; Song-Ling, P.; Li-Fang, Y.; Xue, C.; Yong-Guang, T.; Ya, C. Co-infection of Epstein-Barr virus and human papillomavirus in human tumorigenesis. *Chin. J. Cancer* **2016**, *35*, 16. Available online: https://cjcjournal.biomedcentral.com/articles/10.1186/s40880-016-0079-1 (accessed on 7 November 2017). [CrossRef]
51. Makielski, K.R.; Lee, D.; Lorenz, L.D.; Nawandar, D.M.; Chiu, Y.F.; Kenney, S.C.; Lambert, P.F. Human papillomavirus promotes Epstein–Barr virus maintenance and lytic reactivation in immortalized oral keratinocytes. *Virology* **2016**, *495*, 52–62. Available online: http://www.sciencedirect.com/science/article/pii/S0042682216301052?via%3Dihub and https://www.ncbi.nlm.nih.gov/pmc/articles/PMC4912861/ (accessed on 7 November 2017). [CrossRef] [PubMed]
52. Guidry, J.T.; Scott, T.S. The interaction between human papillmavirus and other viruses. *Virus Res.* **2017**, *231*, 130–147. Available online: http://www.sciencedirect.com/science/article/pii/S0168170216306633?via%3Dihub (accessed on 7 November 2017). [CrossRef] [PubMed]
53. Hachana, M.; Amara, K.; Ziadi, S.; Gacem, R.B.; Korbi, S.; Trimeche, M. Investigation of human JC and BK polyomaviruses in breast carcinomas. *Breast Cancer Res. Treat.* **2012**, *133*, 969–977. Available online: https://link.springer.com/article/10.1007%2Fs10549-011-1876-5 (accessed on 7 November 2017). [CrossRef] [PubMed]
54. Comar, M.; Bonifacio, D.; Zanconati, F.; Di Napoli, M.; Isidoro, E.; Martini, F.; Torelli, L.; Tognon, M. High prevalence of BK poliyomavirus sequences in Human papillomavirus-16-positive precancerous cervical lesions. *J. Med. Virol.* **2011**, *83*, 1770–1776. Available online: http://onlinelibrary.wiley.com/doi/10.1002/jmv.22184/full (accessed on 7 November 2017). [CrossRef] [PubMed]
55. Burger-Calderon, R.; Madden, V.; Hallett, R.A.; Gingerich, A.D.; Nickeleit, V.; Webster-Cyriaque, J. Replication of oral BK virus in human salivary gland cells. *J. Virol.* **2014**, *88*, 559–573. Available online: http://jvi.asm.org/content/88/1/559.full (accessed on 7 November 2017). [CrossRef] [PubMed]
56. Moens, U.; van Ghule, M.; Ehlers, B. Are human polyomaviruses co-factors for cancers induced by other oncoviruses? *Rev. Med. Virol.* **2014**, *24*, 343–360. Available online: http://onlinelibrary.wiley.com/wol1/doi/10.1002/rmv.1798/full (accessed on 7 November 2017). [CrossRef] [PubMed]
57. Fraase, K.; Hart, J.; Wu, H.; Pang, X.; Ma, L.; Grant, F.; Li, A.; Lennon, A.; Hu, P.C.; Dong, J. BK virus as a potential co-factor for HPV in the development of cervical neoplasia. *Ann. Clin. Lab. Sci.* **2012**, *42*, 130–134. Available online: http://www.annclinlabsci.org/content/42/2/130.long (accessed on 7 November 2017). [PubMed]
58. Delbue, S.; Ferrante, P.; Provenzano, M. Polyomavirus BK and prostate cancer: An unworthy scientific effort? *Oncoscience* **2014**, *1*, 296–303. Available online: https://www.ncbi.nlm.nih.gov/pmc/articles/PMC4278296/ (accessed on 7 November 2017). [CrossRef] [PubMed]
59. Gonzalez-Moles, M.; Gutiérrez, J.; Ruiz, I.; Fernández, J.A.; Rodriguez, M.; Aneiros, J. Epstein-Barr virus and oral squamous cell carcinoma in patients without HIV infection: Viral detection by polymerase chain reaction. *Microbios* **1998**, *96*, 23–31. [PubMed]
60. Guidry, J.T.; Birdwell, C.E.; Scott, R.S. Epstein-Barr virus in the pathogenesis of oral cancers. *Oral Dis.* **2017**. Available online: http://onlinelibrary.wiley.com/doi/10.1111/odi.12656/full (accessed on 7 November 2017). [CrossRef] [PubMed]
61. Fathallah, I.; Parroche, P.; Gruffat, H.; Zannetti, C.; Johansson, H.; Yue, J.; Manet, E.; Tommasino, M.; Sylla, B.S.; Hasan, U.A. EBV latent membrane protein 1 is a negative regulator of TLR9. *J. Immunol.* **2010**, *185*, 6439–6447. Available online: http://www.jimmunol.org/content/185/11/6439.long (accessed on 7 November 2017). [CrossRef] [PubMed]
62. Shahzad, N.; Shuda, M.; Gheit, T.; Kwun, H.J.; Cornet, I.; Saidj, D.; Zannetti, C.; Hasan, U.; Chang, Y.; Moore, P.S.; et al. The T Antigen Locus of Merkel Cell Polyomavirus Downregulates Human Toll-Like Receptor 9 Expression. *J. Virol.* **2013**, *87*, 13009–13019. Available online: http://jvi.asm.org/content/87/23/13009 (accessed on 30 November 2017). [CrossRef] [PubMed]
63. Sobin, L.H.; Gospodarowicz, M.K.; Wittekind, C. *TNM Classification of Malignant Tumours*, 7th ed.; Wiley-Blackwell: Washington, DC, USA, 2009; pp. 22–45. ISBN 978-1-4443-3241-4.

64. Cardesa, A.; Gale, N.; Nadal, A.; Zidor, N. Squamous cell carcinoma. In *World Health Organization Classifiation of Tumours. Pathology and Genetics of Head and Neck Tumours*; Barnes, L., Eveson, J.W., Reichart, P., Sidransky, D., Eds.; IARC Press: Lyon, France, 2005; pp. 118–121. ISBN 92-832-2417-5.
65. Polz-Dacewicz, M.; Strycharz-Dudziak, M.; Dworzański, J.; Stec, A.; Kocot, J. Salivary and serum IL-10, TNF-α, TGF-β, VEGF levels in oropharyngeal squamous cell carcinoma and correlation with HPV and EBV infection. *Infect. Agent. Cancer* **2016**, *11*, 45–53. Available online: https://infectagentscancer.biomedcentral.com/articles/10.1186/s13027-017-0141-x (accessed on 7 November 2017). [CrossRef] [PubMed]
66. Arthur, R.R.; Dagostin, S.; Shah, K.V. Detection of BK virus and JC virus in urine and brain tissue by the polymerase chain reaction. *J. Clin. Microbiol.* **1989**, *27*, 1174–1179. Available online: https://www.ncbi.nlm.nih.gov/pmc/articles/PMC267522/pdf/jcm00066-0050.pdf (accessed on 7 November 2017). [PubMed]

6

Viral-Cellular DNA Junctions as Molecular Markers for Assessing Intra-Tumor Heterogeneity in Cervical Cancer and for the Detection of Circulating Tumor DNA

Katrin Carow [1,†], Mandy Gölitz [1,†], Maria Wolf [1], Norman Häfner [1], Lars Jansen [1], Heike Hoyer [2], Elisabeth Schwarz [3], Ingo B. Runnebaum [1] and Matthias Dürst [1,*]

1 Department of Gynecology, Jena University Hospital—Friedrich Schiller University Jena, 07747 Jena, Germany; katrin.carow@med.uni-jena.de (K.C.); mandy.goelitz@med.uni-jena.de (M.G.); mariawolf1990@gmx.de (M.W.); norman.haefner@med.uni-jena.de (N.H.); lars.jansen@med.uni-jena.de (L.J.); ingo.runnebaum@med.uni-jena.de (I.B.R.)
2 Institute of Medical Statistics, Information Sciences and Documentation, Jena University Hospital—Friedrich Schiller University Jena, 07743 Jena, Germany; heike.hoyer@med.uni-jena.de
3 Research Program Infection and Cancer, DKFZ, 69120 Heidelberg, Germany; e.schwarz@dkfz-heidelberg.de
* Correspondence: matthias.duerst@med.uni-jena.de
† These authors contributed equally to this work.

Abstract: The development of cervical cancer is frequently accompanied by the integration of human papillomaviruses (HPV) DNA into the host genome. Viral-cellular junction sequences, which arise in consequence, are highly tumor specific. By using these fragments as markers for tumor cell origin, we examined cervical cancer clonality in the context of intra-tumor heterogeneity. Moreover, we assessed the potential of these fragments as molecular tumor markers and analyzed their suitability for the detection of circulating tumor DNA in sera of cervical cancer patients. For intra-tumor heterogeneity analyses tumors of 8 patients with up to 5 integration sites per tumor were included. Tumor islands were micro-dissected from cryosections of several tissue blocks representing different regions of the tumor. Each micro-dissected tumor area served as template for a single junction-specific PCR. For the detection of circulating tumor-DNA (ctDNA) junction-specific PCR-assays were applied to sera of 21 patients. Samples were collected preoperatively and during the course of disease. In 7 of 8 tumors the integration site(s) were shown to be homogenously distributed throughout different tumor regions. Only one tumor displayed intra-tumor heterogeneity. In 5 of 21 analyzed preoperative serum samples we specifically detected junction fragments. Junction-based detection of ctDNA was significantly associated with reduced recurrence-free survival. Our study provides evidence that HPV-DNA integration is as an early step in cervical carcinogenesis. Clonality with respect to HPV integration opens new perspectives for the application of viral-cellular junction sites as molecular biomarkers in a clinical setting such as disease monitoring.

Keywords: HPV; viral-cellular junction; molecular marker; tumor heterogeneity; cell-free tumor DNA

---

1. Introduction

Genetic diversity is not only characteristic among tumors of the same entity but also within the tumor itself. The latter is often referred to as intra-tumor heterogeneity and reflects the co-existence of different tumor sub-clones. Indicators for intra-tumor heterogeneity are somatic copy number alterations [1], point mutations [2,3], differences in ploidy [4] or different transcript expression patterns [5]. The phenomenon needs to be considered when using individualized biomarkers in a clinical setting, for example during postoperative monitoring of cancer patients based on the detection of circulating tumor-derived DNA in serum or plasma. The potential value of this strategy has been demonstrated for a variety of cancer types [6–9], but so far, the impact of intra-tumor heterogeneity on biomarker performance has not been assessed systematically. Navin and colleagues report the existence of totally independent sub-clones in breast tumors [4]. Due to Darwinian selection, these sub-clones could grow into local relapse or distant metastases that differ in their genetic make-up from the primary tumor. In the absence of a common genetic aberration it would be necessary to perform multiple biopsies to identify a biomarker specific for each sub-clone in order to reliably detect persistent or recurrent disease based on liquid biopsies. This illustrates clearly the importance of a profound knowledge of the tumor's genomic landscape and mutations common to all cells for optimal biomarker design.

The clonal composition of cervical cancer has been addressed in several studies. A frequently applied technique for assessing clonal relation of cancer cells in women is the analysis of the X-chromosome inactivation pattern [10]. Using this strategy, Hu and colleagues have analyzed 24 preparations of the same cervical carcinoma and revealed its polyclonal origin [11]. Similar studies report polyclonality in 2 of 8 cases [12] as well as a monoclonal history of 6 examined adenosquamous cervical cancers [13]. However, since there is a 50% probability that two independent subpopulations carry the same inactivated X-chromosome, additional biomarkers are required to rule out heterogeneity. Another informative approach suitable for the investigation of intra-tumor heterogeneity is comparative genomic hybridization. Lyng and colleagues could demonstrate heterogeneity in chromosomal aberrations in 11 of 20 analyzed advanced cervical cancers [14]. However, a polyclonal history was ruled out due to the fact that the majority of the detected aberrations were present in all analyzed tumor regions.

Evidence is cumulating that human papillomaviruses (HPV) DNA integration may be particularly interesting in evaluating intra-tumor heterogeneity. A prerequisite for cervical cancer development is a persisting infection with an oncogenic type of HPV [15,16]. During carcinogenesis, HPV DNA is frequently integrated into the host genome. This is discussed as being an early event [17]. Integration is significantly associated with lesion progression [18,19], and evidently contributes to clonal expansion [17]. Integration frequencies in cervical carcinomas up to 81.7% have been reported [20]. Early studies used the physical state of the HPV genome in different tumor regions as a marker for clonality and provide evidence for monoclonal [13] as well as polyclonal lesions [21,22]. However, these studies were based on 2D-gel electrophoresis or HPV genome fragment ratios, which only provide indirect evidence for virus integration. In this context, the detection of HPV DNA integrates on sequence level is a far more specific and innovative approach: The uniqueness of each integration event renders the viral-cellular junctions as perfect molecular markers for tracing tumor cells. A first proof of feasibility was provided by Vinokurova and colleagues who used junction sequences as clonal markers to demonstrate a common history of vulvar and previous cervical lesions in 4 of 5 patients [23]. Should viral-cellular-junction sequences turnout to be representative of the whole tumor cell population, their detection may become a powerful tool in patient monitoring. In this context, Campitelli and colleagues provided a first prove of strategy value and demonstrated that viral-cellular junction sequences can be used to detect circulating tumor-derived DNA (ctDNA) in sera from 11 of 13 patients [24]. By analyzing sequential serum samples, they could correlate their findings with tumor recurrence.

However, thus far the distribution of viral-cellular junction sequences throughout the tumor mass has not been analyzed. Prior to a clinical implementation of HPV-integrates as biomarkers for disease monitoring it is essential to clarify the issue of intra-tumor heterogeneity. Our study addresses this topic systematically by analyzing DNA from micro-dissected tissue of tumors by amplification of the viral-cellular junction (vcj-PCR). Tumors with single and multiple integration sites were included. To assess the potential value of viral cellular junction sequences for disease monitoring, we performed vcj-PCRs with cell-free DNA (cfDNA) from sera of cervical cancer patients to identify circulating tumor DNA (ctDNA). The data were analyzed with respect to clinical parameters and patient outcome.

2. Results

To test our hypothesis that viral-cellular junction sequences are suited to serve as highly specific tumor markers for cervical carcinomas, we first addressed cervical cancer clonality by assessing the distribution of viral-cellular junction sequences in eight cervical cancers. We aimed to evaluate the influence of intra-tumor heterogeneity on biomarker performance. In the second part, we tested the suitability of viral-cellular junction sequences to serve as molecular markers for the detection of circulating tumor DNA by analyzing cell-free DNA from serum samples by vcj-PCR. Overall, we included tissue and serum samples of 21 patients with integrated HPV DNA and established (semi)-nested vcj-PCR-assays for 32 viral-cellular junctions (Table 1).

2.1. The Majority of Tumors Show Intra-Tumor Homogeneity with Respect to Junction Distribution

The specimens available for assessing intra-tumor heterogeneity comprised tumor tissue which was dissected into 3–11 (I-XI) individual blocks of up to 0.5 cm^3 in size. On basis of HE- and p16-staining of corresponding sections, tumor and stroma islands were identified and micro-dissected by laser capture technology for PCR analyses. For tumors with multiple junctions, areas in close proximity to each other or the corresponding areas of a subsequent section were selected for micro-dissection.

Tumors harboring single HPV integrates (n = 4) invariably tested positive for the respective junction sequence in all analyzed areas (Figure 1).

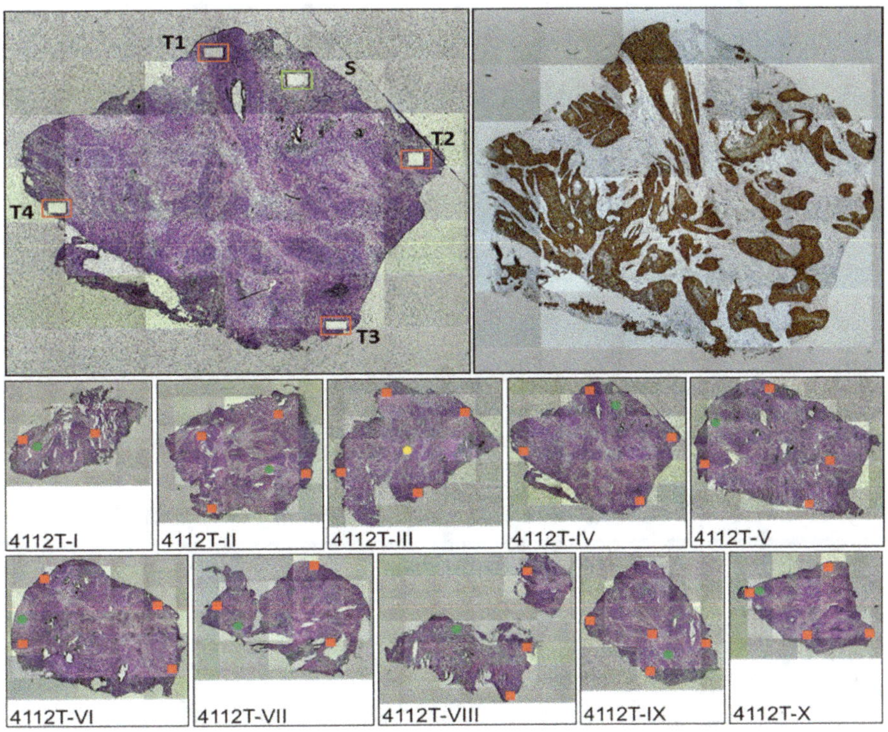

Figure 1. Intra-tumor heterogeneity analyses. Ten tissue blocks (I-X) of tumor 4112 were available. Left upper panel: Hematoxylin and eosin (HE)-stained section of block IV after micro-dissection; S: stroma, T1 to T4: tumor. (**Right upper**) panel: p16-staining of block IV to guide micro-dissection; (**Middle and lower**) panel: HE stained sections of all blocks. Squares and circles refer to micro-dissected tumor and stroma areas, respectively. Red, and green colors indicate the presence and absence of the viral junction, respectively. Yellow indicates a marginally positive result.

Table 1. Patient material.

Patient-ID	Age at Diagnosis	Histologic Subtype	Tumor Type	HPV Type	TNM-Classification			HPV-Integrates	3′ Integration Site	Tumor Blocks for ITH Analyses	Serum Samples for ctDNA Analyses
					T	N	M				
841	58	SCC	P	16	2a	0	1	1	2q33.3	6	3
								2	13q22.2		
								3	13q22.1		
								4	Xp22.11		
								5	Xp22.11		
1509	70	SCC	P	16	2b	1	0	1	9p13.3	6	1
								2	8q23.3		
1907	33	SCC	P	16	4			1	14q23.2	-	3
2349	58	SCC	P	16	1b1	0	0	1	Xp22.31	-	4
2555	48	SCC	P	16	2a	1	0	1	1p36.22	3	2
2723	52	SCC	P	16	2b	0	x	1	17q23.1	-	3
3817	42	SCC	P	16	1b1	1	0	1	16q23.1	-	1
3986/4112 *	47	SCC	P	18	3b	1		1	17p13.1	10	2
4154	46	SCC	P	18	1b	0	x	1	2q24.2	-	1
4338	60	SCC	P	16	2	0	x	1	9p24.1	11	1
4497	49	ADC	P	18	2b	0	x	1	3p24.2	-	2
4502	37	ADC	P	18	3b	1	x	1	7q34	-	1
4749	40	SCC	P	16	1b1	0	0	1	18p11.32	-	1
								2	18p11.32		
								3	18q12.2		
4977	67	SCC	P	16	2b	0	0	1	6q22.32	4	1
								2	7p15.1		
5234	33	SCC	P	16	1b2	0		1	3p21.31	5	2
5254	29	SCC	P	18	2b2	0		1	9p21.3	-	1
5613	37	ADC	P	18	1b1			1	8p12	-	1
3719	38	SCC	R	16	1b	1	0	1	2p22.3	-	1
								2	4q21.1		
4601	63	SCC	R	16	4	0	x	1	9p23	-	1
4995	52	SCC	R	18	2b	0	0	1	19p13.3	-	1
5189	48	SCC	R	16	1b2	0	0	1	8q24.21	3	1
								2	8q23.2		
								3	5p14.1		

HPV: human papillomaviruses; ITH: intra-tumor heterogeneity; ADC: adenocarcinoma; SCC: squamous cell carcinoma; P: primary tumor; R: relapse; TNM-Classification: Tumor, Node, Metastasis; * patient received two primary tumor surgeries yielding samples 3986 and 4112, sample 4112 served for heterogeneity analyses.

Likewise, 3 of 4 carcinomas harboring between 2 and 5 integrates showed a homogenous distribution of the viral integrates (Figure S1). Irregular junction distribution was observed in tumor 4977 only. We analyzed 4 blocks of this tumor and identified numerous regions in all blocks which displayed exclusively junction 2. Only 6 of 22 analyzed areas harbored junction 1, in four of them in coexistence with junction 2. Moreover, multiple areas were negative for both junction fragments (Figure 2 and Figure S2). vcj-PCRs performed with DNA from whole tissue sections of all blocks showed no evidence for presence of junction 1, except for block IV. For a subset of junction-negative tumor areas we additionally tested for the presence of HPV E6 DNA and obtained a positive result indicative for viral episomes or further unidentified integrates. All stroma areas ($n = 8$) were invariable negative for the junction sites. Thus, tumor 4977 clearly constitutes a case of intra-tumor heterogeneity with respect to junction distribution.

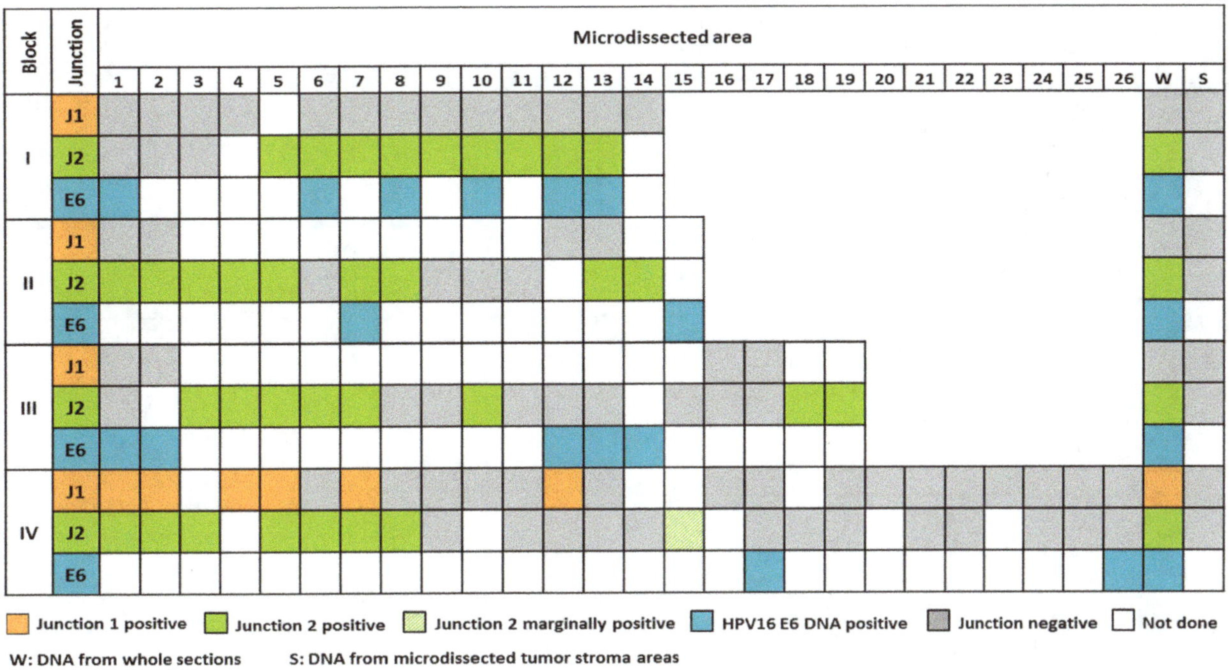

Figure 2. Varying numbers of areas were micro-dissected from four blocks of tumor 4977 and submitted to junction-specific PCR. Areas tested for the presence of both junctions received identical numbers. Successful detection is indicated by color. Both junctions are heterogeneously detected throughout the four blocks. Junction 1 is present in block IV only whereas junction 2 is detectable in all blocks, but not in all areas.

2.2. Detection of Viral-Cellular Junctions Fragments in Cell-Free Serum DNA

After demonstrating that viral-cellular junction sequences are in most cases representative for the respective tumor, we analyzed the suitability of the fragments to serve as marker for the detection of circulating tumor DNA. We tested pre-operative sera of all 8 patients included in the intra-tumor heterogeneity analyses as well as pre-operative sera of 13 additional cervical cancer patients (Table 2). Circulating cell-free DNA was isolated from 200 μL serum resulting in 50 μL of purified serum DNA.

Junction detection by (semi-) nested PCR was performed in triplicates with 4 μL template each and verified by Southern Blot hybridization using an internal oligonucleotide as probe. Cell-free serum-DNA of five of 21 patients was tested positive, providing clear evidence for the presence of tumor DNA (Table 2). Among cases displaying multiple integrates, junction detection was only successful in serum of patient 5189. Notably, all 3 junction fragments could be amplified. The serum corresponding to tumor 4977, which displayed intra-tumor heterogeneity, was negative for both known junctions.

Table 2. Junction detection in primary serum samples and follow-up of patients.

Tissue	Patient	Follow-Up (Days)	Status at End of Follow-Up	Junctions	Junction-Detection in Primary Serum	Junction-Detection in Sequential Sera
Primary Tumor	841	2549	deceased	J1	No	No
				J2	No	No
				J3	No	No
				J4	No	No
				J5	No	No
	1509	2663	deceased	J1	No	x
				J2	No	x
	1907	257	Deceased *	J1	No	No
	2349	2462	tumor-free	J1	No	No
	2555	4831	tumor-free	J1	No	No
	2723	3605	tumor-free	J1	No	No
	3817	643	Deceased *	J1	No	x
	3986	360	Deceased *	J1	Yes	Yes
	4154	4420	tumor-free	J1	No	x
	4338	1350	Deceased *	J1	Yes	Yes
	4497	4055	tumor-free	J1	No	No
	4502	1376	deceased	J1	No	x
	4749	0	tumor-free	J1	No	x
				J2	No	x
				J3	No	x
	4977	3980	tumor-free	J1	No	x
				J2	No	x
	5234	2681	tumor-free	J1	Yes	Yes
	5254	2923	tumor-free	J1	No	x
	5613	2740	tumor-free	J1	No	x
Relapse	3719	-	Deceased *	J1	No	x
				J2	No	x
	4601	-	Deceased *	J1	Yes	x
	4995	-	Deceased *	J1	No	x
	5189	-	Deceased *	J1	Yes	x
				J2	Yes	x
				J3	Yes	x

* tumor-related death; x: not done.

No correlation was found between the detection of viral-cellular junctions and TNM-stage (Fisher's exact test). However, Kaplan Meier analyses of the patients with primary tumors only revealed a significant association between the detection of junction fragments in pre-operative sera and a reduced recurrence free survival ($p = 0.03$) (Figure 3a). In the same cohort there was no correlation between reduced recurrence free survival and established risk factors like lymph node metastasis ($p = 0.43$), an advanced tumor stage ($p = 0.67$) or a combination thereof ($p = 0.52$) (Figure 3b–d).

Of 9 patients sequential serum samples were available which were collected during second surgery or at one of the follow up visits (Table S1). Samples of all 3 patients whose primary serum samples displayed junction fragments remained positive. Two of these patients had second surgery within 82 days. The third positive case was patient 4338 with one serum sample taken during surgery of the primary tumor and a second one taken 1.5 years later during relapse surgery. Both sera displayed junction presence. In follow up serum samples of the 6 initially negative patients no junctions could be detected despite one case of therapy failure (patient 1907).

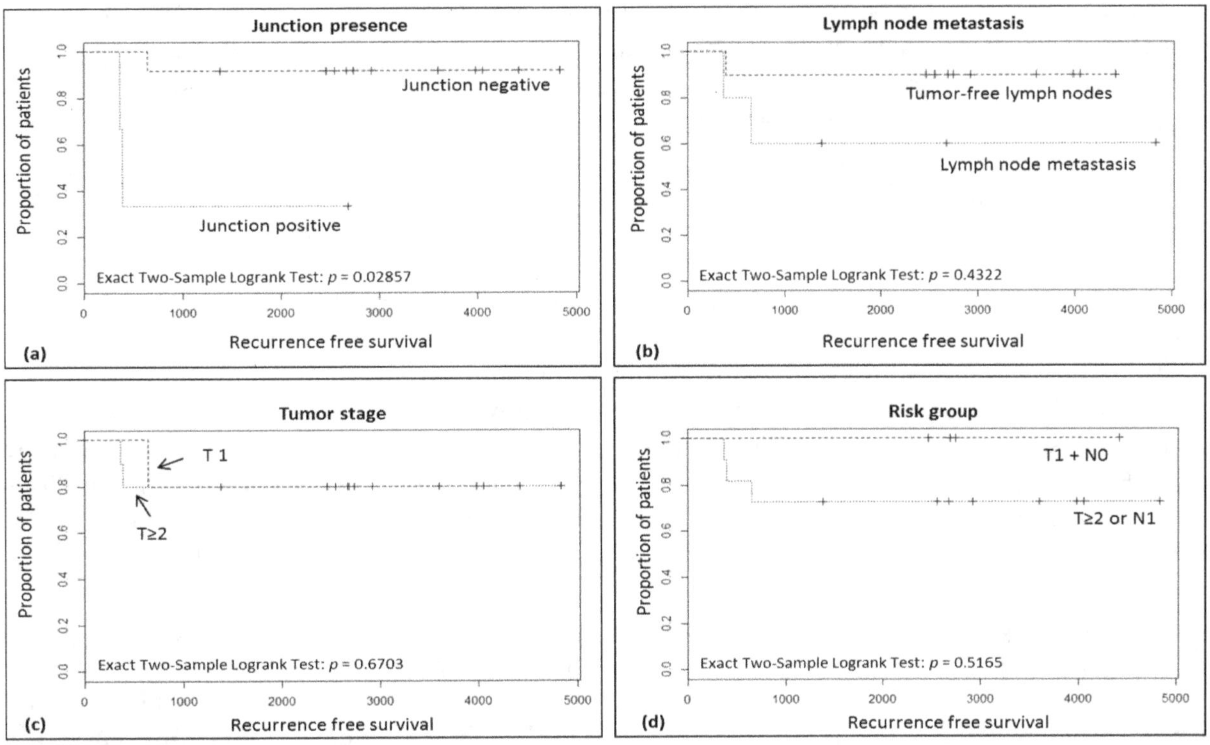

Figure 3. Kaplan Meier curves to assess the impact of (**a**) junction presence, (**b**) lymph node metastasis, (**c**) tumor stage and (**d**) the combined presence of risk factors on recurrence free survival. Censored subjects are indicated with a tick mark: (+).

3. Discussion

Viral-cellular junction sequences frequently arise during cervical carcinogenesis as a result of HPV DNA integration into the host genome. These integration sites are unique and highly specific for a patient's tumor and may thus serve as versatile tumor makers. Ideally, a tumor marker is representative for the entire tumor cell population. Intra-tumor heterogeneity typically observed in the course of clonal expansion complicates tumor biomarker research immensely. To determine the potential value of HPV DNA integration as a tumor biomarker, it is essential to assess whether HPV integrates are stable over time and space, i.e., whether they are representative for the tumor cell population. To address this, we analyzed 8 cervical cancers with regard to intra-tumor heterogeneity concerning the distribution of viral-cellular junction sequences. In this context, all 4 cervical carcinomas harboring single HPV integrates showed homogeneity. In case of tumors with multiple integrates, we found that 3 of 4 tumors harbored all tumor specific integrates in every micro-dissected tissue area in each tumor block, thereby implying monoclonal origin. Tumor 841 is of particular interest as it harbors 5 integrates on 3 different chromosomes (Table 1). On two chromosomes the integrates are in proximity of each other, possibly as a result of local rearrangements [25]. Since all 5 integrates were detected in all micro-dissected areas at least 3 independent integration events and two chromosomal rearrangements must have taken place in one common precursor cell. Both forms of integrates are also evident for tumor 5189 which harbors 3 integrates on 2 chromosomes. In contrast, tumor 1509 exclusively displays 2 integrates on 2 different chromosomes, which are likewise distributed homogenously. This extensive homogeneity in integrate distribution provides striking evidence for the importance of integration in an early phase of carcinogenesis and for its presumptive driving effect during clonal expansion. Moreover, it would appear that integrates that may have resulted from chromosomal rearrangements occurred at the same point in time. The only exception is tumor 4977. This tumor exhibits 2 integrates on different chromosomes and was the only case of irregular junction distribution. Junction 1 was highly underrepresented and detected in only 14% (6/43) of the tested areas. Junction 2 was present in 54% (34/63) of the tested areas. A co-detection of both junctions was evident in 4 tumor areas (Figures 2 and S2). Areas without detectable junctions, either harbor unidentified integrates, viral episomes only or both. The evident irregular integration pattern in this tumor and its relevance for carcinogenesis is open for all kinds of speculations including gain or loss of integrates at a later stage of tumor development.

Evidently, HPV DNA integration can be an early event in cervical carcinogenesis. In 7 of 8 examined tumors HPV integration has most likely driven clonal expansion. The homogeneous integrate distribution is also a strong indication that the continuous presence of the integrated viral DNA in the tumor is required for the maintenance of the carcinogenic phenotype. These observations, in particular the presence of specific integration site(s) representative for all tumor cells, highlight their usefulness as molecular markers for disease monitoring.

This extensive homogeneity of the driver event is highly remarkable in the light of the ambiguous situation of driver genetic event distribution in other tumor entities: human epidermal growth factor receptor 2 *HER2* gene amplification is an accepted driver event in breast cancer development. However, this aberration is not always detectable throughout the whole tumor tissue [26,27]. In contrast, Gerlinger and colleagues analyzed ten clear cell renal carcinomas and report that chromosome 3p loss and von Hippel–Lindau gene (*VHL*) aberrations were the only ubiquitous events, whereas 73–75% of identified driver aberrations were of subclonal origin [28]. In non-small cell lung cancer (NSCLC), de Briun et al. report that the majority of driver events seem to be clonal, although there are cases of heterogeneity [29]. On the other hand, a study by Zhang et al. describe heterogeneity in *TP53* mutations status in *TP53* driven NSCLC [30]. Compared to the complex genomic landscape of other tumor entities the unique virus induced etiology of cervical cancer allows a facilitated and precise analysis of clonal architecture. Our results clearly demonstrate a high selective pressure for integrate maintenance and the suitability of viral-cellular junction sequences for biomarker research.

Limitations of our study include the small number of patients and the use of micro-dissected tissue areas rather than single cells. However, clear draw-backs of single cell analyses are false negative results, especially if only one copy of the target molecule is present. The micro-dissected tissue areas used in our study represent 100 to 200 cells, which allow reliable analyses. In our study, intra-tumor homogeneity implies that the tumor-specific integrate(s) were detected in all micro-dissected areas taken within a tumor comprising up to several cm^2. To be scored positive the ΔC_t values had to correspond to a tumor cell fraction of $\geq$25% within a dilution series (Figure S3). Moreover, our approach does not allow drawing general conclusions about intra-tumor heterogeneity in cervical cancer as it focusses on viral integration sites only. Clearly, as in any other tumor entity, subclones evolve with distinct genetic alterations such as chromosomal aberration, DNA copy number changes and gene mutations.

To explore the use of integration sites as molecular markers for disease monitoring or prognosis we developed patient specific vcj-PCRs for the detection of circulating tumor DNA (ctDNA) in serum. We tested pre-operative sera of all 8 individuals included in above intra-tumor heterogeneity analyses as well as pre-operative sera of 13 additional cervical cancer patients (Tables 1 and 2). For 9 of 21 patients post-operative samples were also available (Table S1). Junction fragments were detectable in pre-operative 5 of five of 21 patients (23%). A similar study by Campitelli and colleagues demonstrated a detection rate of 68% [24]. Our low frequency of junction presence in sera may be explained by the fact that only 3 replicates corresponding to 48 μL serum were analyzed. In the former study 10 replicates corresponding to 160 μL serum were analyzed, with less than 50% of replicates being positive in the majority of cases. These stochastic effects seen in both studies clearly indicate that the analyses of some sera were done at the detection limit. Only 4 of the 7 sera of patients with proven intra-tumor homogeneity displayed junction fragments. Of note is that all 3 junctions characteristic for tumor 5189 could be detected in the corresponding serum. This co-detection of multiple viral-cellular junction sequences in serum has not been demonstrated before and implies that each junction is equally suited to serve as tumor marker. Not surprisingly, neither of the 2 junction fragments was detected in the serum of patient with tumor 4977 which showed intra-tumor heterogeneity. For more comprehensive results DNA from larger serum volumes need to be analyzed and quantified with respect to junction presence. This can be achieved by digital PCR [31].

Among the cases of relapse, we found a higher, albeit non-significant, detection rate of junction fragments in comparison to the sera of patients with primary tumors (60% vs. 21%). Also Campitelli and colleagues hypothesize that ctDNA is more easily released from a relapse than from a primary tumor. The detection of a relapse during monitoring would benefit from such a phenomenon.

Another novel finding of this study is that among the group of patients with primary tumors the detection rate of junction fragments in serum correlated significantly with a reduced recurrence-free survival ($p = 0.03$) (Figure 3a). A prognostic value of ctDNA presence before therapy has also been reported for other tumor entities such as colorectal [32,33], metastatic breast cancer [6] or metastatic uveal melanoma [34]. A further limitation of our study is the restricted number of patients, the different group size of junction positive and junction negative patients. As a consequence, the absence of a prognostic value of known established risk factors (Figure 3b–d) could be related to the limited number of subjects.

Finally, we evaluated the suitability of vcj-PCR based ctDNA detection to follow tumor dynamics. We analyzed 15 sequential serum samples of 9 patients. Three patients had junction fragments in pre-operative serum collected at primary surgery. Two of these patients required second surgery within 82 days at which a second serum sample was collected. Not surprisingly, the junction fragments were also detectable in the respective second serum sample. For the 6 patients without junction fragments in the sera taken at primary surgery, none of the follow up sera tested positive (Table S1). These patients were also free from tumor recurrence. Of particular interest is patient 4338 who relapsed close to two years after primary tumor resection. The junction fragment identified in the primary tumor was detectable in both sera of the patient illustrating the stability of the virus integration site over time.

In conclusion, we were able to use the unique viral etiology of cervical cancer to address tumor clonality in the context of intra-tumor heterogeneity and to open new perspectives for biomarker design.

4. Materials and Methods

4.1. Samples and Sample Processing

Tissue and serum samples of this retrospective study were obtained from a total of 21 cervical carcinoma patients undergoing treatment at Jena University Hospital between 1996 and 2006. Approval was given by the Ethics Committee of the Friedrich-Schiller University Jena (reference numbers 0175-02/00 and 2174-12/07, 20 September 2017). Written informed consent was obtained from all participants. All patients received an ID, which constitutes the number under which the first sample was archived. Further samples from the same patient were continuously numbered. Each tumor had been analyzed in previous studies for presence and site of HPV integrates [35,36] Tumors were HPV 16 ($n = 13$) or HPV 18 ($n = 8$) positive, of squamous cell ($n = 18$) or of glandular ($n = 3$) origin and ranged between FIGO (Fédération Internationale de Gynécologie et d'Obstétrique) stages IB and IVA (Table 1). Large tumors were cut into smaller blocks. For heterogeneity analyses, 3–11 blocks of each tumor representing different tumor areas were available. Frozen sections of these blocks were used for micro-dissection and DNA isolation.

Tumor cell content of sections was assessed after p16 and Ki67 immuno-staining. p16 overexpression and Ki67 are characteristic for HPV-transformed cells and cell proliferation, respectively. Cryosections of 7 μm thickness were stained on glass slides (Thermo Fisher Scientific, Waltham, MA, USA) with hematoxylin and eosin using standard protocols. For immunohistochemistry 7 μm cryosections (Thermo Fisher Scientific, Waltham, MA, USA) were fixed on Superfrost Plus slides with 4% paraformaldehyde for 10 min. Slides were washed with tris-buffered saline (50 mM Tris, 150 mM NaCl) and 0.1% Tween-20 (TBST) and incubated in 0.6% H_2O_2 for 7 min. After another washing step, blocking was performed with goat serum (1:5 dilution in TBST) for 20 min. Staining was performed over night with a Ki67 mouse monoclonal antibody clone mib-1 (DAKO, 80 mg/L), diluted at a ratio of 1:100 or with a p16 mouse monoclonal antibody clone E6H4 ready for use (CINtec Histology, Roche, Basel, Switzerland). For detection, the DAKO EnVision System was used according to the manufacturer's protocol. Finally, slides were counterstained with hematoxylin and covered with coverslips in gelatin.

Tumor sections for micro-dissection were stained with kresyl violet to enable tumor and stroma island identification while minimizing DNA damage. For this purpose, 15 μm-sections were placed on membrane slides (MembraneSlides NF, PALM Microlaser Technologies, Zeiss, Jena, Germany) and dried at room temperature. Staining was performed with 1% kresyl violet for 2 min. Afterwards, samples were washed twice with 70% ethanol and dried at room temperature.

DNA isolated from whole sections was used for evaluating assay sensitivity and as positive control. Ten consecutive 10 μm cryosections of each tissue specimens were digested with proteinase K overnight. DNA was isolated with QIAamp DNA Mini Kit (Qiagen, Hilden, Germany) following manufacturer's instructions.

Serum samples were used for ctDNA analyses. Blood was collected pre-operatively at the time point of surgical resection of primary tumor ($n = 17$) or of relapse ($n = 4$). From 9 patients, additional samples were available, taken pre-operatively at further surgeries or during follow up (Table 1 and Table S1). For coagulation the blood was stored for 30 min at room temperature. Afterwards it was centrifuged at 3000 rpm for 10 min and serum was stored at −30 °C. For isolation of cell-free DNA, 200 μL aliquots of sera were processed with the High Pure Viral Nucleic Acid Kit (Roche Applied Science, Mannheim, Germany) according to manufacturer's instructions. The 50 μL-eluat was purified by NaCl-EtOH-precipitation. The pellet was dissolved in 50 μL aqua dest. Quality and quantity of serum-DNA was controlled in β-Actin qPCR.

4.2. Intra-Tumor Heterogeneity Analyses: Assay Design and Optimization

For the detection of HPV-integrates by vcj-PCR, primers were designed to flank the junction of human and viral DNA. For heterogeneity analyses, a duplex nested PCR was performed (Table S2). In the first reaction the 3′junction and the housekeeping gene beta globin (HBB) were amplified in parallel in a standard Eppendorf PCR cycler. The first reaction step (50 °C, 2 min, for UDG-reaction) was followed by initial denaturation at 95 °C for 10 min. After 15 cycles (95 °C, 15 s denaturation; 20 s 58–63 °C annealing; 30 s, 72 °C elongation) and 2 min at 72 °C for, 1 μL of the reaction was transferred to two separate reactions, respectively, for amplification of the two templates with internal primer pairs. The second PCR was performed with 40 cycles using the same cycling conditions but in a real-time format (Rotor Gene Q 5plex HRM, Qiagen, Hilden, Germany). Finally, product identity was confirmed by melting curve analyses. The first duplex reaction mixture consisted of 10 ng template (sensitivity tests) or of one micro-dissected tissue area (heterogeneity analyses), 10 μL 2× Platinum® Quantitative PCR SuperMix-UDG (Thermo Fisher Scientific, Waltham, MA, USA), 0.25 μM external primers for vcj-PCR, 0.006–0.25 μM primers for HBB-amplification and was adjusted to 20 μL with aqua dest. Micro-dissected tissues were directly lysed in the entire PCR reaction mixture. The second simplex PCR consisted of 5 μL Fast Start Universal SYBRGreen-Master Mix (Roche Applied Science, Mannheim, Germany), 0.25 μM of each internal primer, 1 μL template and was adjusted to 10 μL with aqua dest.

Sensitivity and specificity of vcj-primers was tested in serial dilutions of DNA isolated from tumor sections in a background of DNA derived in equal parts from HeLa and Caski cells. The combination of both cell lines was chosen to illustrate the absence of any cross reactivity with unspecific HPV integrates. Each dilution step contained 5 ng/μL DNA with decreasing amounts of tumor-DNA (100%, 50%, 25%, 5%, 1%, 0.2%, 0.02%, 0.002% and 0%). The resulting C_t-values of the second PCR enabled a semi-quantitative analysis by calculating ΔC_t values between HBB- and vcj-PCR Ct. Based on the dilution series ΔC_t values were calculated for each assay to score the micro-dissected tissue as "clearly positive" (tumor cell fraction of $\geq$25%), "marginally positive" (tumor cell fraction of 0.02–5%) and "negative" for virus integrate presence (Figure S3).

4.3. *Intra-Tumor Heterogeneity Analyses: Laser Micro-Dissection*

Micro-dissection was performed with the help of the PALM laser capture microscope (Zeiss, Jena, Germany) and the PALM-ROBO-Software. For heterogeneity analyses tumor and stroma areas (20.000 μm^2) equivalent to 100–200 cells were dissected. Stroma served as control for specificity. In case of multiple integrates, areas in close proximity to each other or identical areas in consecutive sections were selected. Each dissected tissue area was transferred with a sterile needle to a reaction tube, which was directly supplemented with the complete PCR mix containing all necessary reagents. The PCR was performed as described above without further DNA-isolation. The results were evaluated according to ΔC_t-values.

4.4. *ctDNA Analyses: Detection of Viral-Cellular Junctions in Cell-Free Tumor DNA from Sera*

For ctDNA-detection, (semi-) nested-PCR primers were designed to flank all 33 junctions of 21 patients (Table S3). In contrast to DNA deriving from tissue, cell-free serum DNA is highly fragmented. Consequently ctDNA-pimers were designed to minimize product size. Performance of junction-specific PCR (vcj-PCR) assays was optimized in serial dilutions of tumor tissue-DNA in a background of HPV-negative genomic DNA from C33A cells. All reactions were performed in a 20 µL volume using the Fast Start Universal SYBR Green PCR Mastermix (Roche Applied Science, Mannheim, Germany) with 0.25 µM of forward and reverse primer, respectively. After initial denaturation (95 °C, 10 min) 15 amplification cycles (95 °C, 15 s; 55–66 °C, 40 s) were performed in a standard Eppendorf cycler. From this first reaction 2 µL were transferred to the second PCR comprising (semi-) nested primers. Amplification was done in 45 cycles under identical conditions as in the first PCR, but in a real-time format (Rotor Gene Q 5plex HRM, Qiagen, Hilden, Germany) and was followed by a dissociation stage for product identification.

Optimized vcj-primers were used for serum analyses with 4 µL serum-DNA as template. The (semi-) nested PCR approach was performed as described above. The results were verified by Southern Blot with an internal oligonucleotide as hybridization probe (Table S4). For this purpose, PCR products were separated by agarose gelelectrophoresis, denatured by treatment with blot buffer (0.6 N NaCl/0.4 N NaOH) for 3 $\times$ 20 min and transferred onto a nylon membrane. Membrane was washed with 2$\times$ SSC (sodium chloride sodium citrate for 2 $\times$ 10 min and crosslinked using a GS Gene Linker (Bio-Rad, Hercules, CA, USA) Prehybridization was done for two hours at 37 °C in 6$\times$ SSC, 5$\times$ Denhardt's, 0.02 M NaPP, 0.5% SDS and 100 µg/mL yeast total RNA. Hybridization was done in 6$\times$ SSC, 1$\times$ Denhardt's, 0.02 M NaPP, 100 µg/mL yeast total RNA and y^{32}-ATP labeled internal oligonucleotide. Labelling reaction was performed with Polynucleotide Kinase (Roche Applied Science, Mannheim, Germany) following manufacturer's instructions with 6 pmol of internal oligonucleotide. Using mini Quick Spin Columns (Roche Applied Science, Mannheim, Germany) product was purified following manufacturer's instructions. Hybridization was done overnight. Hybridization temperature was calculated from internal oligonucleotide melting temperature: T(Hybridisation) = T_M (Oligo) − 7 °C. Afterwards, the membrane was washed in 6$\times$ SSC and 0.02 M NaPP for 3 $\times$ 10 min at 37 °C and for 2 $\times$ 30 min at T_M (Oligo) − 5 °C. Exposition was done with Kodak X-OMAT radiographic films.

4.5. *Statistical Analyses*

The association between junction-detection and Tumor, Node, Metastasis-stage was analyzed by Fisher exact test. Survival analyses were performed with the statistical software R (package "coin"). Significance was evaluated by log-rank exact test with a two-sided significance level of 0.05.

Acknowledgments: This research work was financed by intramural funds (LoM) of Jena University Hospital. These funds also cover the costs to publish in open access.

Author Contributions: Katrin Carow, Mandy Gölitz, Elisabeth Schwarz, Ingo B. Runnebaum and Matthias Dürst conceived and designed the experiments; Katrin Carow, Mandy Gölitz, Maria Wolf and Lars Jansen performed the experiments; Katrin Carow, Mandy Gölitz, Norman Häfner, Heike Hoyer and Matthias Dürst analyzed the data; Matthias Dürst and Ingo B. Runnebaum contributed reagents/materials/analysis tools; Katrin Carow and Matthias Dürst wrote the paper.

Abbreviations

ADC	Adenocarcinoma
cfDNA	Circulating cell-free DNA
Ct	Cycle threshold
ctDNA	Circulating tumor DNA
FIGO	Fédération internationale de gynécologie et d'obstétrique
HE	Hematoxylin and eosin stain
HPV	Human papillomavirus
ITH	Intra-tumor heterogeneity
NSCLC	Non-small cell lung cancer
SCC	Squamous cell carcinoma
T_M	Melting temperature
vcj-PCR	Viral-cellular junction PCR

References

1. Kovac, M.; Navas, C.; Horswell, S.; Salm, M.; Bardella, C.; Rowan, A.; Stares, M.; Castro-Giner, F.; Fisher, R.; de Bruin, E.C.; et al. Recurrent chromosomal gains and heterogeneous driver mutations characterise papillary renal cancer evolution. *Nat. Commun.* **2015**, *6*, 6336. [CrossRef] [PubMed]
2. Cooper, C.S.; Eeles, R.; Wedge, D.C.; van Loo, P.; Gundem, G.; Alexandrov, L.B.; Kremeyer, B.; Butler, A.; Lynch, A.G.; Camacho, N.; et al. Analysis of the genetic phylogeny of multifocal prostate cancer identifies multiple independent clonal expansions in neoplastic and morphologically normal prostate tissue. *Nat. Genet.* **2015**, *47*, 367–372. [CrossRef] [PubMed]
3. Shah, S.P.; Roth, A.; Goya, R.; Oloumi, A.; Ha, G.; Zhao, Y.; Turashvili, G.; Ding, J.; Tse, K.; Haffari, G.; et al. The clonal and mutational evolution spectrum of primary triple-negative breast cancers. *Nature* **2012**, *486*, 395–399. [CrossRef] [PubMed]
4. Navin, N.; Krasnitz, A.; Rodgers, L.; Cook, K.; Meth, J.; Kendall, J.; Riggs, M.; Eberling, Y.; Troge, J.; Grubor, V.; et al. Inferring tumor progression from genomic heterogeneity. *Genome Res.* **2010**, *20*, 68–80. [CrossRef] [PubMed]
5. Bachtiary, B.; Boutros, P.C.; Pintilie, M.; Shi, W.; Bastianutto, C.; Li, J.H.; Schwock, J.; Zhang, W.; Penn, L.Z.; Jurisica, I.; et al. Gene expression profiling in cervical cancer: An exploration of intratumor heterogeneity. *Clin. Cancer Res.* **2006**, *12*, 5632–5640. [CrossRef] [PubMed]
6. Dawson, S.J.; Tsui, D.W.; Murtaza, M.; Biggs, H.; Rueda, O.M.; Chin, S.F.; Dunning, M.J.; Gale, D.; Forshew, T.; Mahler-Araujo, B.; et al. Analysis of circulating tumor DNA to monitor metastatic breast cancer. *N. Engl. J. Med.* **2013**, *368*, 1199–1209. [CrossRef] [PubMed]
7. Reinert, T.; Scholer, L.V.; Thomsen, R.; Tobiasen, H.; Vang, S.; Nordentoft, I.; Lamy, P.; Kannerup, A.S.; Mortensen, F.V.; Stribolt, K.; et al. Analysis of circulating tumour DNA to monitor disease burden following colorectal cancer surgery. *Gut* **2015**, *65*, 625–634. [CrossRef] [PubMed]
8. Wang, Z.; Chen, R.; Wang, S.; Zhong, J.; Wu, M.; Zhao, J.; Duan, J.; Zhuo, M.; An, T.; Wang, Y.; et al. Quantification and dynamic monitoring of EGFR T790M in plasma cell-free DNA by digital PCR for prognosis of EGFR-TKI treatment in advanced NSCLC. *PLoS ONE* **2014**, *9*, e110780. [CrossRef] [PubMed]

9. Wimberger, P.; Roth, C.; Pantel, K.; Kasimir-Bauer, S.; Kimmig, R.; Schwarzenbach, H. Impact of platinum-based chemotherapy on circulating nucleic acid levels, protease activities in blood and disseminated tumor cells in bone marrow of ovarian cancer patients. *Int. J. Cancer* **2011**, *128*, 2572–2580. [CrossRef] [PubMed]
10. Enomoto, T.; Fujita, M.; Inoue, M.; Tanizawa, O.; Nomura, T.; Shroyer, K.R. Analysis of clonality by amplification of short tandem repeats. Carcinomas of the female reproductive tract. *Diagn. Mol. Pathol.* **1994**, *3*, 292–297. [CrossRef] [PubMed]
11. Hu, X.; Pang, T.; Asplund, A.; Ponten, J.; Nister, M. Clonality analysis of synchronous lesions of cervical carcinoma based on X chromosome inactivation polymorphism, human papillomavirus type 16 genome mutations, and loss of heterozygosity. *J. Exp. Med.* **2002**, *195*, 845–854. [CrossRef] [PubMed]
12. Guo, Z.; Wu, F.; Asplund, A.; Hu, X.; Mazurenko, N.; Kisseljov, F.; Ponten, J.; Wilander, E. Analysis of intratumoral heterogeneity of chromosome 3p deletions and genetic evidence of polyclonal origin of cervical squamous carcinoma. *Mod. Pathol.* **2001**, *14*, 54–61. [CrossRef] [PubMed]
13. Ueda, Y.; Miyatake, T.; Okazawa, M.; Kimura, T.; Miyake, T.; Fujiwara, K.; Yoshino, K.; Nakashima, R.; Fujita, M.; Enomoto, T. Clonality and HPV infection analysis of concurrent glandular and squamous lesions and adenosquamous carcinomas of the uterine cervix. *Am. J. Clin. Pathol.* **2008**, *130*, 389–400. [CrossRef] [PubMed]
14. Lyng, H.; Beigi, M.; Svendsrud, D.H.; Brustugun, O.T.; Stokke, T.; Kristensen, G.B.; Sundfor, K.; Skjonsberg, A.; de Angelis, P.M. Intratumor chromosomal heterogeneity in advanced carcinomas of the uterine cervix. *Int. J. Cancer* **2004**, *111*, 358–366. [CrossRef] [PubMed]
15. Walboomers, J.M.; Jacobs, M.V.; Manos, M.M.; Bosch, F.X.; Kummer, J.A.; Shah, K.V.; Snijders, P.J.; Peto, J.; Meijer, C.J.; Munoz, N. Human papillomavirus is a necessary cause of invasive cervical cancer worldwide. *J. Pathol.* **1999**, *189*, 12–19. [CrossRef]
16. Zur Hausen, H. Papillomaviruses and cancer: From basic studies to clinical application. *Nat. Rev. Cancer* **2002**, *2*, 342–350. [CrossRef] [PubMed]
17. Jeon, S.; Allen-Hoffmann, B.L.; Lambert, P.F. Integration of human papillomavirus type 16 into the human genome correlates with a selective growth advantage of cells. *J. Virol.* **1995**, *69*, 2989–2997. [PubMed]
18. Hudelist, G.; Manavi, M.; Pischinger, K.I.; Watkins-Riedel, T.; Singer, C.F.; Kubista, E.; Czerwenka, K.F. Physical state and expression of HPV DNA in benign and dysplastic cervical tissue: Different levels of viral integration are correlated with lesion grade. *Gynecol. Oncol.* **2004**, *92*, 873–880. [CrossRef] [PubMed]
19. Badaracco, G.; Venuti, A.; Sedati, A.; Marcante, M.L. HPV16 and HPV18 in genital tumors: Significantly different levels of viral integration and correlation to tumor invasiveness. *J. Med. Virol.* **2002**, *67*, 574–582. [CrossRef] [PubMed]
20. Hu, Z.; Zhu, D.; Wang, W.; Li, W.; Jia, W.; Zeng, X.; Ding, W.; Yu, L.; Wang, X.; Wang, L.; et al. Genome-wide profiling of HPV integration in cervical cancer identifies clustered genomic hot spots and a potential microhomology-mediated integration mechanism. *Nat. Genet.* **2015**, *47*, 158–163. [CrossRef] [PubMed]
21. Badaracco, G.; Venuti, A. Physical status of HPV types 16 and 18 in topographically different areas of genital tumours and in paired tumour-free mucosa. *Int. J. Oncol.* **2005**, *27*, 161–167. [CrossRef] [PubMed]
22. Galehouse, D.; Jenison, E.; DeLucia, A. Differences in the integration pattern and episomal forms of human papillomavirus type 16 DNA found within an invasive cervical neoplasm and its metastasis. *Virology* **1992**, *186*, 339–341. [CrossRef]
23. Vinokurova, S.; Wentzensen, N.; Einenkel, J.; Klaes, R.; Ziegert, C.; Melsheimer, P.; Sartor, H.; Horn, L.C.; Hockel, M.; von Knebel Doeberitz, M. Clonal history of papillomavirus-induced dysplasia in the female lower genital tract. *J. Natl. Cancer Inst.* **2005**, *97*, 1816–1821. [CrossRef] [PubMed]
24. Campitelli, M.; Jeannot, E.; Peter, M.; Lappartient, E.; Saada, S.; de la Rochefordiere, A.; Fourchotte, V.; Alran, S.; Petrow, P.; Cottu, P.; et al. Human papillomavirus mutational insertion: Specific marker of circulating tumor DNA in cervical cancer patients. *PLoS ONE* **2012**, *7*, e43393. [CrossRef] [PubMed]
25. Dall, K.L.; Scarpini, C.G.; Roberts, I.; Winder, D.M.; Stanley, M.A.; Muralidhar, B.; Herdman, M.T.; Pett, M.R.; Coleman, N. Characterization of naturally occurring HPV16 integration sites isolated from cervical keratinocytes under noncompetitive conditions. *Cancer Res.* **2008**, *68*, 8249–8259. [CrossRef] [PubMed]
26. Hanna, W.; Nofech-Mozes, S.; Kahn, H.J. Intratumoral heterogeneity of HER2/neu in breast cancer—A rare event. *Breast J.* **2007**, *13*, 122–129. [CrossRef] [PubMed]

27. Ng, C.K.; Martelotto, L.G.; Gauthier, A.; Wen, H.C.; Piscuoglio, S.; Lim, R.S.; Cowell, C.F.; Wilkerson, P.M.; Wai, P.; Rodrigues, D.N.; et al. Intra-tumor genetic heterogeneity and alternative driver genetic alterations in breast cancers with heterogeneous HER2 gene amplification. *Genome Biol.* **2015**, *16*, 107. [CrossRef] [PubMed]
28. Gerlinger, M.; Horswell, S.; Larkin, J.; Rowan, A.J.; Salm, M.P.; Varela, I.; Fisher, R.; McGranahan, N.; Matthews, N.; Santos, C.R.; et al. Genomic architecture and evolution of clear cell renal cell carcinomas defined by multiregion sequencing. *Nat. Genet.* **2014**, *46*, 225–233. [CrossRef] [PubMed]
29. De Bruin, E.C.; McGranahan, N.; Mitter, R.; Salm, M.; Wedge, D.C.; Yates, L.; Jamal-Hanjani, M.; Shafi, S.; Murugaesu, N.; Rowan, A.J.; et al. Spatial and temporal diversity in genomic instability processes defines lung cancer evolution. *Science* **2014**, *346*, 251–256. [CrossRef] [PubMed]
30. Zhang, L.L.; Kan, M.; Zhang, M.M.; Yu, S.S.; Xie, H.J.; Gu, Z.H.; Wang, H.N.; Zhao, S.X.; Zhou, G.B.; Song, H.D.; et al. Multiregion sequencing reveals the intratumor heterogeneity of driver mutations in TP53-driven non-small cell lung cancer. *Int. J. Cancer* **2017**, *140*, 103–108. [CrossRef] [PubMed]
31. Vogelstein, B.; Kinzler, K.W. Digital PCR. *Proc. Natl. Acad. Sci. USA* **1999**, *96*, 9236–9241. [CrossRef] [PubMed]
32. Bettegowda, C.; Sausen, M.; Leary, R.J.; Kinde, I.; Wang, Y.; Agrawal, N.; Bartlett, B.R.; Wang, H.; Luber, B.; Alani, R.M.; et al. Detection of circulating tumor DNA in early- and late-stage human malignancies. *Sci. Transl. Med.* **2014**, *6*, 224ra24. [CrossRef] [PubMed]
33. Lecomte, T.; Berger, A.; Zinzindohoue, F.; Micard, S.; Landi, B.; Blons, H.; Beaune, P.; Cugnenc, P.H.; Laurent-Puig, P. Detection of free-circulating tumor-associated DNA in plasma of colorectal cancer patients and its association with prognosis. *Int. J. Cancer* **2002**, *100*, 542–548. [CrossRef] [PubMed]
34. Bidard, F.C.; Madic, J.; Mariani, P.; Piperno-Neumann, S.; Rampanou, A.; Servois, V.; Cassoux, N.; Desjardins, L.; Milder, M.; Vaucher, I.; et al. Detection rate and prognostic value of circulating tumor cells and circulating tumor DNA in metastatic uveal melanoma. *Int. J. Cancer* **2014**, *134*, 1207–1213. [CrossRef] [PubMed]
35. Xu, B.; Chotewutmontri, S.; Wolf, S.; Klos, U.; Schmitz, M.; Durst, M.; Schwarz, E. Multiplex Identification of Human Papillomavirus 16 DNA Integration Sites in Cervical Carcinomas. *PLoS ONE* **2013**, *8*, e66693. [CrossRef] [PubMed]
36. Schmitz, M.; Driesch, C.; Beer-Grondke, K.; Jansen, L.; Runnebaum, I.B.; Durst, M. Loss of gene function as a consequence of human papillomavirus DNA integration. *Int. J. Cancer* **2012**, *131*, E593–E602. [CrossRef] [PubMed]

7

Human Papilloma Virus and Autophagy

Domenico Mattoscio [1,2,*], Alessandro Medda [3] and Susanna Chiocca [3,*]

1 Department of Medical, Oral, and Biotechnology Science, University of Chieti-Pescara, 66100 Chieti, Italy
2 Center on Aging Science and Translational Medicine (CeSI-MeT), University of Chieti-Pescara, 66100 Chieti, Italy
3 Department of Experimental Oncology, European Institute of Oncology, 20139 Milan, Italy; alessandro.medda@ieo.it
* Correspondence: d.mattoscio@unich.it (D.M.); susanna.chiocca@ieo.it (S.C.)

Abstract: Human papilloma viruses (HPVs) are a group of double-stranded DNA viruses known to be the primary cause of cervical cancer. In addition, evidence has now established their role in non-melanoma skin cancers, head and neck cancer (HNC), and the development of other anogenital malignancies. The prevalence of HPV-related HNC, in particular oropharyngeal cancers, is rapidly increasing, foreseeing that HPV-positive oropharyngeal cancers will outnumber uterine cervical cancers in the next 15–20 years. Therefore, despite the successful advent of vaccines originally licensed for cervical cancer prevention, HPV burden is still very high, and a better understanding of HPV biology is urgently needed. Autophagy is the physiological cellular route that accounts for removal, degradation, and recycling of damaged organelles, proteins, and lipids in lysosomal vacuoles. In addition to this scavenger function, autophagy plays a fundamental role during viral infections and cancers and is, therefore, frequently exploited by viruses to their own benefit. Recently, a link between HPV and autophagy has clearly emerged, leading to the conceivable development of novel anti-viral strategies aimed at restraining HPV infectivity. Here, recent findings on how oncogenic HPV16 usurp autophagy are described, highlighting similarities and differences with mechanisms adopted by other oncoviruses.

Keywords: HPV; autophagy; cervical cancer; head and neck cancer; viral tumorigenesis; oncoviral proteins

1. Introduction

Oncovirus infection represents one of the leading causes of cancer worldwide, accounting for approximately 12% of total cancer burden [1]. Among them, human papilloma virus (HPV) infection accounts for approximately 2% and 7% of the total cancer burden in more developed and less developed countries, respectively [2]. In particular, high risk (HR) HPVs, such as HPV16, are the main cause of cervical cancer and are also involved in other anogenital tumors, in non-melanoma skin cancers and head and neck cancer (HNC) [3]. Even if vaccines against common tumorigenic HPVs have been established, the proportion of HPV-related malignancies, in particular HNC, is steadily increasing [4]. For these reasons, and for the absence of targeted therapies to treat HPV-tumors, a deep understanding of the molecular mechanisms underlying HPV tumorigenesis is fundamental.

Autophagy is a self-consumption mechanism occurring inside the cell mediating degradation of cargoes from different origins by lysosomal activity. In particular, autophagy is involved in degradation of long-lived proteins, but also of organelles like mitochondria (mitophagy), or endoplasmic reticulum (ER-phagy). Aggrephagy encompasses the degradation of aggregate-prone intracytoplasmic proteins, such as α-synuclein, mutant huntingtin, and protein tau. Xenophagy entails the degradation of pathogens, such as bacteria and viruses, that escape or cause ruptures in the vacuoles where they are confined (for recent reviews on these topics see [5–8]).

Depending on the delivery route carrying materials for digestion, autophagy can be distinguished in chaperone-mediated autophagy, microautophagy, and macroautophagy. In chaperone-mediated autophagy molecular chaperones translocate unfolded proteins containing a specific sequence in the cytosol into lysosomes by interaction with lysosome-associated membrane protein (LAMP) type 2A, which acts as a receptor (reviewed in [9]). Microautophagy is characterized by a direct invagination of lysosomal membrane with engulfment of cytoplasmic material (reviewed in [10]). In macroautophagy (hereafter simply referred as autophagy) a double-membraned organelle, the autophagosome, sequesters cytosolic materials, then fuses with a lysosome, resulting in an autolysosome, where cargoes are digested by lysosomal hydrolases (reviewed in [11]). Autophagy is a complex mechanism that can be schematically divided into phagophore initiation, elongation, and fusion with lysosomes. In mammals, during the initiation phase, the unc-51 like autophagy activating kinase 1 (ULK1) kinase complex, which consists of ULK1 (or ULK2), autophagy-related protein 13 (ATG13), FAK family kinase-interacting protein of 200 kDa (FIP200), and ATG101, is recruited either on cellular membranes at the ER-mitochondrial junction [12] or at the ER-Golgi intermediate compartment [13]. The activity of the complex is tightly regulated by nutrient conditions through the mammalian target of rapamycin (mTOR) [14]. In particular, during starvation, low energy levels are sensed by the AMP activated protein kinase (AMPK), which phosphorylates mTOR and inactivates the mTOR1 complex (mTORC1). Inactive mTORC1 dissociates from the ULK1 kinase complex, resulting in activation of ULK1 which phosphorylates mammalian ATG13 (mATG13) and FIP200 leading to phagophore initiation. The sequential recruitment of autophagy factors and membranes produces the phagophore, a curved, double-membrane sheet that detaches from the membrane it originates. Then, during elongation, the class III phosphatidylinositol 3 kinase (PI3K) complex formed by the vacuolar protein sorting 34 (VPS34) PI3K, its regulatory subunits ATG14L, VPS15, and Beclin 1, mediate phagophore membrane expansion. In particular, Beclin 1 dissociation from B-cell lymphoma/leukemia-2 (Bcl-2) activates the class III PI3K complex, leading to the recruitment of two ubiquitin-like conjugation systems that promote the conjugation of lipid molecules, namely phosphatidylethanolamine (PE), to the microtubule-associated protein 1 light chain 3 (LC3) (for a detailed review of autophagosome biogenesis see [15]). Lipidated LC3 (LC3-II), localized on both sides of the nascent autophagosome, is a specific marker of autophagosome formation widely used as an indicator of autophagic flux, and it regulates membrane elongation and autophagosome maturation [16]. Finally, the activity of the class III PI3K complex II, consisting of VPS34, VPS15, Beclin 1, and UV radiation resistance-associated gene protein (UVRAG), stimulates the Ras-associated protein-7 (Rab7), inducing its GTPase activity and leading to autophagosome fusion with late lysosomes [17]. This results in autolysosomes formation, with degradative properties given by lysozymes.

Although initially believed to have a role only in response to starvation and in degradation and recycling of macromolecules and organelles, it is now clear that autophagy impacts a wide range of cellular functions both in health and disease. For example, autophagy has important pathogenetic roles in muscular disorders, neurodegeneration, aging, and in inflammatory conditions (reviewed in [11]). Altered autophagy is also a key driver of malignant transformation. Indeed, autophagy could have a dual role in tumor development: tumor suppressor in early stages by restricting inflammation and intracellular reactive oxygen species (ROS) levels and supporting genomic stability and integrity, whilst tumor promoter in later stages of tumorigenesis through the increase in cancer cells' ability to live in nutrient starvation and hypoxic environment (recently reviewed in [18]). Therefore, in many cancer types, autophagy is inhibited in pre-cancerous cells while activated in growing tumors. In addition, given the pleiotropic role for autophagy in regulating a variety of cellular activities, it is not surprising that some pathogens evolved different mechanisms to usurp autophagic machinery to completely subvert the host milieu to their own advantage. Viruses, in particular, can evade autophagy in different ways, including prevention of autophagy induction, autophagosome maturation, pathogen recognition, and using autophagy components to increase their ability to survive or to replicate (reviewed in [19]).

Here, we focus on the role of autophagy during HPV infection and tumorigenesis and its potential function as a target for development of novel anti-viral and anti-neoplastic agents, also discussing similarities and differences on mechanisms adopted by other oncoviruses in order to modulate the host autophagic response.

2. HPV Manipulation of Host Autophagy

In recent years, a close interplay between autophagy and HPV has emerged. Like many other viruses, HPV also manipulates the autophagic machinery to promote its lifecycle inside host-infected cells. Additionally, oncogenic HPVs exploit cellular autophagy to support proliferation of infected epithelial cells, significantly contributing to cancer progression. In the following sections we will discuss how autophagy affects HPV pathogenesis and the strategy adopted by the virus to exploit this process in order to improve its persistence in the host.

2.1. Autophagy Inhibition Promotes HPV Infectivity

HPV is a double-stranded DNA virus with a circular genome that encodes early-, including *E1, E2, E4, E5, E6,* and *E7* essential for replication, transcription and transformation, and late-genes *L1* and *L2,* encoding for viral capsid proteins. The replication cycle of HPV is tightly linked to differentiation of the infected epithelium. Indeed, while HPV binding and infection arise exclusively in basal keratinocytes at microtrauma sites likely due to sexual intercourse, viral protein production and virus assembly occurs only in the upper differentiated layers of the epithelia. In wound basal layer, HPV particles initially interact with the basement membrane mostly through heparan sulfate proteoglycans (HSPGs)—capsid L1 contacts, and subsequently bind to HSPGs present on basal keratinocytes cell surface. This attachment triggers conformational changes in the L2 capsid protein, resulting in exposure of a consensus cleavage site in L2 N-terminus, whose proteolysis facilitates further interaction of viral capsid with secondary receptor(s) present on keratinocytes membrane. After such binding, HPVs are generally internalized by clathrin-dependent endocytosis, which is initially reliant on actin-rich cell protrusions, acting as the transport mechanism along the endocytic machinery (reviewed in [20]).

Both binding and internalization of HPV particles are processes intimately linked to manipulation of host autophagy. Interaction with HPSGs triggers rapid activation of several signaling pathways in host cells that benefit HPV infection and, among them, HPV entry is associated with autophagy suppression through activation of the mTOR pathway. In particular, upon HPSGs binding, HPV16 interacts with epidermal growth factor receptors (EGFRs) present on the plasma membrane of target cells [21], resulting in protein Kinase B (Akt) phosphorylation and inactivation of phosphatase and tensin homolog (PTEN), culminating with phosphorylation and activation of mTOR [22]. Once activated, mTOR promotes phosphorylation and activation of the mTOR complex 1 (mTORC1), substrates 4E-BP1 (elongation initiation factor (EIF)-4E binding protein 1) and S6K1 (ribosomal protein S6 kinase 1) [14], key components of the translation machinery, while it phosphorylates and de-activates ULK-1, the kinase localized on isolation membranes that is involved in autophagosome nucleation [23]. Thus, activation of the mTORC1 complex through manipulation of the PI3K/Akt/mTOR pathway, promoted by HPV binding, increases protein synthesis and inhibits cellular autophagy at very early stages of HPV16 cycle, namely before virus entry in target cells. Both of these functions are needed for successful infection of human keratinocytes [22]. Consistently, autophagy impairment obtained with the early inhibitor 3-methyladenine (3-MA) [24], or through genetic ablation of essential autophagic genes, greatly increases HPV16 infectivity in human keratinocytes [22,25,26], strongly indicating the importance of host autophagy in patrolling the early steps of HPV lifecycle. Mechanistically, rather than modulating virus attachment or internalization, autophagy inhibition protects HPV16 capsid degradation inside the autophagosome [26] through delaying the rate of L1 digestion [25].

Therefore, HPV16 diminishes the host-autophagic response, exploiting molecular events promoted by its binding and internalization as a defense mechanism to protect incoming virions from rapid degradation, thus extending HPV's lifespan inside infected cells (Figure 1).

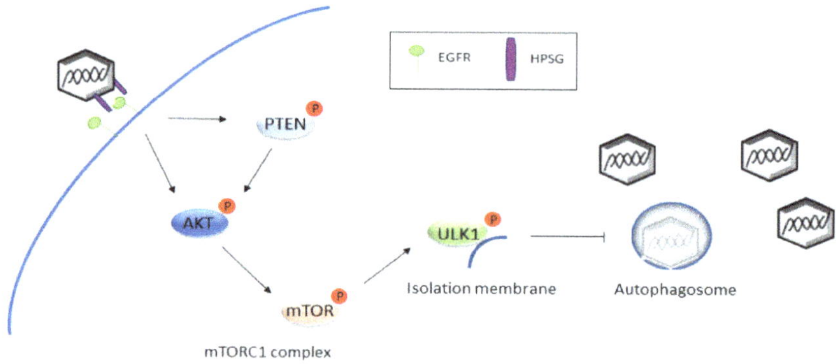

Figure 1. HPV16 binding and internalization inhibits autophagosome formation. HPV virions decorated with HPSGs interact with EGFRs present on the plasma membrane of target cells, resulting in Akt and PTEN phosphorylation, and in phosphorylation and activation of mTOR. Activated mTOR phosphorylates and inactivates ULK1, present on isolation membranes, inhibiting autophagosome nucleation and, therefore, delays L1 digestion and capsid degradation inside autophagosomes. Arrows indicate activating pathways; T-bar indicates inhibitory pathway; blue lines represent cellular and isolation membranes.

2.2. HPV16 Dampens Autophagy to Promote Cancer Progression

Once internalized in basal cells, HPV traffics via the endosomal compartment where capsid proteins are degraded inside acidified endosomes, and the viral genome enters the nucleus (reviewed in [27]). Here, HPV DNA is amplified and maintained as episomes in basal cells of the epithelium by E1- and E2-mediated mechanisms. Infection may latently persist or the virus may be triggered to proliferate, relying on keratinocyte replication proteins that mediate viral DNA synthesis. The infected basal cell initially enters into the suprabasal layer where viral episomal gene expression is maintained at low levels, and finally in the granular and cornified layers where the keratinocyte differentiates, exits the cell cycle, drives viral DNA replication, and promotes the expression of *E4*, *E5*, *E6*, *E7*, *L1*, and *L2* genes, leading to HPV virion formation and release (recently reviewed in [28]).

Although HR HPV infection is an important etiological factor for several cancer types, its presence is necessary, but not sufficient, for cancerogenesis, indicating that other events (such as genomic instabilities or failure of the immune system) are needed for cancer to occur. To this end, a fundamental role in cell transformation is mediated by E5, E6, and E7 oncoproteins, able to modify cell-cycle regulation, induce human telomerase reverse transcriptase (hTERT) expression with consequent telomere maintenance, and block apoptosis [29]. As a result, DNA damage and mutations accumulate, leading to the formation of a transformed cell, koilocyte [30], and then to the development of cancers. In particular, E5 is a multifunctional protein that contributes to cellular transformation mainly associating with and enhancing signaling network of growth factor receptors (reviewed in [31]). E6 promotes the proteasomal degradation of the tumor suppressor p53, through the ubiquitin ligase E6AP [32]. E7 interacts with the pocket proteins (retinoblastoma-RB, p107, and p130) family involved in cell cycle progression [33]. In addition, viral oncogenes can alter a large number of other cellular proteins (reviewed in [29]), modifying their normal function and driving transformation of epithelial basal and parabasal cells in koilocytes. However, before malignant transformation, the persistent activity of HPV oncoproteins can establish a number of precancerous lesions classified as low-grade squamous intraepithelial lesions (LSILs), and high-grade SILs (HSILs) during cervical cancer progression. LSILs are characterized by abnormalities in the lower third of epithelium, and lesions can spontaneously regress or evolve into HSILs. Transformed cells are present in HSILs throughout the epithelium and HPV more likely persists in the host also integrating its DNA in the cellular genome, contributing to cancer progression [34].

In addition to the well-characterized pathways that ultimately mediate HPV-promoted cellular transformation, a significant role for autophagy during the carcinogenic process is now emerging. Notably, the fact that E5, E6, and E7 oncoproteins evolved diverse mechanisms to affect the host autophagic pathway to induce cellular transformation underlines the importance of autophagy during each step of viral-mediated tumorigenesis (Figure 2). In particular, ectopic expression of HPV16 E5 in an HPV-negative keratinocytic cell line reduced levels of the autophagosome marker LC3-II, prevented the degradation of the autophagic susbstrate p62, and diminished the number of autophagosomes in the keratinocyte growth factor (KGF) and serum-starved triggered cells, indicating failure in autophagosome assembly. Consistently, depletion of E5 from HPV-positive cells abrogated these effects confirming that E5 expression drives autophagy inhibition acting at very early steps of the autophagic pathway. Mechanistically, HPV16 E5 interferes with the transcriptional activation of the autophagic machinery, down-regulating mRNA levels of key autophagic genes, such as *Beclin 1, ATG5, LC3, ULK1, ULK2, ATG4a,* and *ATG7* [35], suggesting inhibition of phagophore assembly. Notably, opposite to E5, which prevents autophagosome formation, HPV16 E6/E7 dampens autophagy, affecting one of the later steps (autophagosome-lysosome fusion), thus providing another mechanism against autophagy progression during HPV tumorigenesis. HPV16 E6/E7 overexpression in primary human keratinocytes increased the expression of both the lipidated LC3 and p62, indicating autophagosome accumulation (increase in LC3-II) in spite of decreased degradation capability (increased p62). Additionally, confocal and electron microscopic analyseis revealed reduced autolysosome formation in E6/E7 cells, pinpointing the fusion between autophagosomes and lysosomes as the defective stage affected by these viral oncoproteins. Impairment in later steps of autophagy was mainly due to an E6/p53-dependent mechanisms since HPV16 E6 expression alone increased both LC3-II and p62, and, compared to E6 wild type, E6 mutants defective in p53 degradation have a diminished ability to increase p62 expression in combination with E7 [36]. In addition, HPV16 E7 expression is also able to increase LC3-II levels in keratinocytes [37], suggestive of autophagic defects. Consistently with these results, depletion of the bicistronic HPV16 *E6/E7* mRNA using siRNA targeted against the E7 sequence induced significant enrichment of autophagic genes and phenotypic evidence of autophagy activation, such as the appearance of autophagosomes, punctate expression of LC3, conversion of LC3-I to LC3-II, and reduced levels of the autophagy substrate p62 [38].

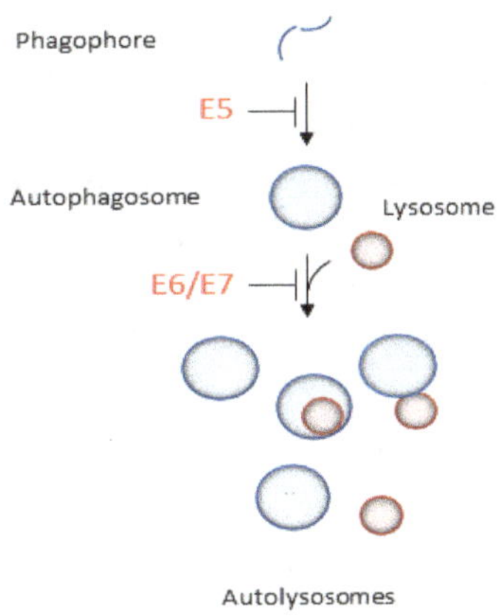

Figure 2. HPV16 oncoproteins dampen the host autophagic response acting at different levels of the autophagic pathway. E5 interferes with the transcriptional activation of the autophagic machinery down-regulating *Beclin 1, ATG5, LC3, ULK1, ULK2, ATG4a,* and *ATG7* mRNAs, thus suggesting inhibition of phagophore assembly, while E6 and E7 inhibit autophagosome/lysosome fusion may be due to depletion of autophagic genes. Arrows indicate activating pathways; T-bars indicate inhibitory pathways.

In support of the above described in vitro results collectively indicating that HPV16 oncoviral proteins impair autophagy through different mechanisms, a number of other observations obtained in vivo or ex vitro confirmed that autophagy manipulation is an important co-factor of several HPV-derived malignancies. These results also highlight several autophagic biomarkers that could be used for diagnosis and to monitor the evolution of diseases (Table 1). In fact, anuses of dysplastic

HPV16 transgenic mice (K14-E6/E7) are characterized by an amplified punctate expression of LC3 and p62, and an increase in autophagosomes without evidence of autolysosomes, as shown by transmission microscopy and immunofluorescence analysis. This peculiar pattern of autophagic markers expression, indicating autophagy inhibition at late stages, was also found in human samples, confirming that autophagic dysregulation is a key determinant of HPV-associated anal carcinogenesis [39]. In support of this hypothesis, treatment of both WT and K14-E6/E7 mice with the late autophagic inhibitor chloroquine significantly increases anal cancer development [39,40], while autophagy activation reduces tumor onset [40].

Table 1. Autophagy alterations during HPV-mediated tumor progression. Up and down arrows indicate increased and decreased expression of the reported proteins, respectively. ATAD3A: ATPase family AAA domain containing 3A.

Tumor	Autophagic Protein Altered	Putative Use as Biomarker	Refs.
Anal dysplasia	LC3↑, p62↑	Diagnosis, progression	[39]
Cervical dysplasia	p62↑	Diagnosis, progression	[36]
	Beclin 1↓	Diagnosis, progression, prognosis	[41,42]
	miR-224-3p↑, FIP200↓	Diagnosis, progression	[43]
Cervical cancer	LC3↓	HPV infection, diagnosis, prognosis	[44]
	ATAD3A↑	HPV infection, diagnosis, prognosis, therapeutic predictivity	[45]
	miR-224-3p↑, FIP200↓	Diagnosis	[43]
	miR-155-5p↓	Diagnosis	[46]

Similarly to anal cancer, autophagy impairment is also relevant for cervical cancer, the most common HPV-induced malignancy. Compared to adjacent normal tissues, p62 expression gradually increased through dysplastic transformation of cervical tissues from LSIL to HSIL, indicating that autophagy impairment during lesion progression could be clinically relevant to the HPV-mediated carcinogenic process [36]. Consistently, opposite to p62, Beclin 1 expression decreases during cancer progression since its levels were found to be significantly lower in dysplastic cervix and even less in cervical cancer specimens, still confirming a progressive dampening of the autophagic response during HPV-mediated cervical transformation. Notably, Beclin 1 levels, as determined by immunohistochemistry, negatively correlated with cervical cancer differentiation, lymph node metastasis, recurrence, death [41], and histological grade [42], indicating the fundamental role of autophagy in cancer development. In addition to Beclin 1, LC3 amount also negatively correlates with HR HPVs infection and higher clinical tumor node and lymph node metastasis [44], again indicative of a reduced autophagic flux in HPV-infected cervical tissues and worse clinicopathological features in affected patients. Moreover, the expression of the anti-autophagy protein ATAD3A in cervical cancer specimens is positively associated with persistent HPV infection, FIGO stage, lymph node involvement, proinflammatory and metastasis-related cytokines, and patient survival. In addition, ATAD3A silencing increases autophagosomes number and markedly decreases drug resistance in uterine cervical cancer cells, further underlying the importance of autophagy in HPV-mediated cancer development [45].

In addition, to hijack host genes and proteins for their own benefit, HR HPVs impact on micro RNAs (miRNAs) expression to dampen cellular autophagy and facilitate tumor progression. In fact, miRNAs profile of cervical tissues infected by HR HPVs showed increased expression of miR-224-3p, in dysplastic and cancer tissues. Gain and loss of function approaches in cervical cancer cell lines evidenced that miR-224-3p regulates autophagy through alteration of FIP200 levels, a protein involved in autophagosome formation [47]. Therefore, HPVs manipulate host autophagy by reducing the expression of FIP200 through the overexpressed miR-224-3p [43]. Similarly, miR-155-5p expression was recently found to be significantly down-regulated in cervical cancer tissues. Suppression of miR-155-5p levels in HPV positive cell lines inhibit LC3 while promotes p62 protein expression, indicative of autophagic defects, through a pyruvate dehydrogenase kinase 1 (PDK1)/mTOR-dependent mechanism [46].

Collectively, the above described findings highlighted that HR HPVs extensively impact on autophagy through many different mechanisms with the common endpoint of quenching cellular autophagic response. Therefore, manipulation of host autophagy is a crucial regulator of HPV-mediated malignant transformation and could be rationally targeted to develop innovative drugs, and exploited to generate useful biomarkers to monitor the clinical outcome of HPV-related diseases.

2.3. Tumor Viruses Usurp Host Autophagy during Their Infection Cycle

Similarly to HPV16, other oncoviruses affect the cellular autophagic response, suggesting the importance of the autophagic route in mediating oncovirus persistence and tumorigenesis. For example, the Epstein-Barr herpes virus (EBV), associated with several tumors, such as Burkitt's lymphoma, nasopharyngeal carcinoma, and Hodgkin's disease, decreases autophagy during its lytic replication cycle, and this activity promotes viral infectivity. Autophagy inhibition also increased EBV lytic gene expression, intracellular viral DNA and viral progeny yield [48]. Likewise to HPV16 E6/E7 [36], EBV promotes autophagosomes accumulation due to a reduction in autophagosome-lysosome fusion in a Rab7-dependent manner [49]. The block in the final degradative step allows the virus to prevent autophagy-mediated viral degradation, thus extending EBV infection and cancer progression. Moreover, the human T-cell leukemia virus type 1 (HTLV-1), a member of the deltaretrovirus family known to cause adult T-cell leukemia mainly due to the action of the oncoprotein Tax [50], also manipulates autophagy during its tumorigenic cycle to increase its persistence, promote proliferation, and inhibit apoptosis of infected T lymphocytes [51,52]. HTLV-1 infection induced autophagosome accumulation through the action of the viral protein Tax. Tax protein acts on autophagosome accumulation by blocking autophagosome-lysosome fusion [51], through the exploitation of peculiar cellular mechanisms involving recruitment of autophagic molecules to lipid rafts in a NF-κB inhibitor kinase (IKK) complex-dependent manner [52,53].

In addition, another oncogenic virus belonging to the herpes virus family, the Kaposi sarcoma herpesvirus (KSHV) associated with Kaposi sarcoma, a form of multicentric Castleman disease, and primary effusion lymphoma [54], also suppresses host autophagy through a mechanism involving the viral protein FADD-like interferon converting enzyme or caspase 8 (FLICE) inhibitory protein (vFLIP) -mediated dysregulation of the IKK/NF-κB (nuclear factor kappa-light-chain-enhancer of activated B cells) signaling axis [55,56]. This effect, occuring during virus latency, blocks senescence, and facilitates the proliferation of KSHV-infected cells [56]. During lytic reactivation of latent virus, the replication and transcription activator protein (RTA) potently increases autophagy of infected cells, and autophagy inhibition at this stage affects RTA-mediated lytic gene expression and viral DNA replication [57]. Therefore, the timely and coordinated usurpation of host cellular autophagy is fundamental to trigger completion of KSHV infection cycle.

Opposite to HPV, Hepatitis B virus (HBV), a hepatotropic virus that can cause hepatocellular carcinoma (HCC) [58], enhances the autophagic response of infected liver cells, transgenic mouse models, and tissue patients [59,60] to positively modulate HBV DNA replication. Mechanistically, the multifunctional Hepatitis B virus X protein (HBx) interacts with VPS34 leading to the activation of the PI3K complex III and promotion of autophagosome formation [59], and via Beclin-1 phosphorylation through a death-associated protein kinase (DAPK)-dependent manner [61]. However, despite autophagy induction, HBV infection does not alter the rate of autophagic degradation, indicating a later impairment of the lysosome-digestive properties. In fact, HBx could alter Rab7 expression and the V1D subunit of the proton-pumping V-type ATPase (V-ATPase) activity, resulting in a reduced maturation and function of lysosomes [62,63]. In addition to the induced production of HBV virions through autophagy activation, HBx also protects HBV-infected cells from apoptosis exploiting autophagy. To this end, HBx acts as an autophagic adaptor that facilitates the recruitment of dead receptor tumor necrosis factor receptor superfamily member 10B (TNFRSF10B) to the autophagic machinery, promoting its degradation and, thus, dampening the recognition of infected cells by immune cells. In addition, silencing of essential autophagic proteins, such as ATG5, ATG12, and ATG16L1, impairs HBV formation and release, altering the intracellular distribution of HBV. HBV core interaction with ATG12, a component of the autophagic elongation complex, drives HBV association with cellular membranes and allows the formation of mature viral particles [64], further underlying the importance of autophagy in HBV infection.

Similarly to HBV, the related Hepatitis C virus (HCV) is also able to affect autophagy to increase virus growth and survival in infected cells. Specifically, HCV induces autophagosome formation through a double mechanism: inducing a time-dependent dephosphorylation of ULK1 [65] and up-regulating transcriptional levels of Beclin 1 [66]. However, promotion of autophagosome formation occurs together with the inhibition of autophagosome maturation, mediated by over-expression of Rubicon (RUN domain and cysteine-rich domain containing, Beclin 1-interacting protein), a negative regulator of UVRAG which inactivates the UVRAG-PI3KC3 complex and Rab7 [67], and dislocation of V-ATPase [68]. Since HCV associates with lipid rafts inside autophagosome membranes to improve replication of its RNA [69] while autolysosomes would promote degradation of this replication complex, HCV manipulates host autophagy to improve its own replication. Moreover, autophagy manipulation also allows HCV release in the extracellular space, thus enabling mature HCV egression from infected cells and viral transmission [70].

3. Conclusions

Autophagy is a physiological pathway that mediates degradation of macromolecules and damaged structures, playing a key role in maintaining homeostasis, regulating crucial cellular functions, and mediating stress responses. Therefore, it is not surprising that a number of viruses evolved ingenious strategies to manipulate the host autophagic response in order to subvert cellular activities and to create a more conducive environment for viral replication [71]. Among them, a particularly important role for autophagy in mediating oncovirus infection, persistence, and tumorigenesis arose (reviewed in [72]). Recently, HPV16 has also emerged as an oncogenic virus with the ability of diverting cellular autophagy during its lifecycle, and significant evidence indicates that viral proteins dampen the autophagic response through many different mechanisms acting at each step of viral infection and tumorigenesis.

HPV16 impacts on autophagy both during the initial steps of infection (namely, adhesion and entry of viral particles into target cells) to avoid early clearance of viral particles by autophagic digestion (Figure 1), and later to help in promoting transformation of infected cells throughout the carcinogenic process (Figure 2, Table 1). Of note, the three different oncoviral proteins E5, E6, and E7 adopt different strategies to turn off host autophagy, further underlying how manipulation of the autophagic pathway is fundamental to finally promoting malignant transformation. Mechanistically, despite many different cellular pathways being affected by viral oncoproteins from diverse oncogenic viruses, a common outcome that seems to emerge from comparative analysis is the inhibition of the final autophagic step, namely autophagosome-lysosome fusion, as the main target of autophagy inhibition. EBV, HTLV-1, HBV, and HCV impair autolysosome formation, suggesting a deleterious role for autophagic degradation in oncovirus infection. Consistently, HPV16 also specifically dampens this step. Hence, given the importance of autolysosome formation on oncovirus tumorigenesis, and the potential of targeting this step for therapeutic purposes, more studies are necessary to better clarify the connection between HPV and autophagy, and to identify novel autophagy-based molecular targets for biomarker validation (Table 1) and/or drug development. To this end, the recently-characterized autophagic inducers able to increase autophagosome-lysosome fusion [73] could represent an important starting point to identify putatively useful molecules and to suggest relevant clinical treatment strategies against HPV infections.

In addition, even if most of the reported in vitro evidence has been obtained using HPV16 virions or oncoviral proteins, it could be very important to determine similarities and divergences in the host autophagy impact among different HR HPV genotypes, as well as to determine the role of autophagy during low-risk HPV infection. These results will highlight and strengthen the use of approaches aimed at modulating autophagy during the course of the HPV cycle.

It is evident that autophagy inhibition promotes the HPV lifecycle and tumor progression and that strategies aimed at restoring the autophagic response during HPV infection and carcinogenesis could be of importance to restrain HPV-mediated diseases.

Acknowledgments: Work in the S.C. laboratory related to the topics discussed in this review is supported by Associazione Italiana per la Ricerca sul Cancro (AIRC). D.M. is a recipient of the Fondazione Umberto Veronesi (FUV) fellowship.

References

1. De Martel, C.; Ferlay, J.; Franceschi, S.; Vignat, J.; Bray, F.; Forman, D.; Plummer, M. Global burden of cancers attributable to infections in 2008: A review and synthetic analysis. *Lancet Oncol.* **2012**, *13*, 607–615. [CrossRef]
2. Forman, D.; de Martel, C.; Lacey, C.J.; Soerjomataram, I.; Lortet-Tieulent, J.; Bruni, L.; Vignat, J.; Ferlay, J.; Bray, F.; Plummer, M.; et al. Global burden of human papillomavirus and related diseases. *Vaccine* **2012**, *30*, F12–F23. [CrossRef] [PubMed]
3. Zur Hausen, H. Papillomaviruses in the causation of human cancers—A brief historical account. *Virology* **2009**, *384*, 260–265. [CrossRef] [PubMed]
4. Giuliano, A.R.; Nyitray, A.G.; Kreimer, A.R.; Pierce Campbell, C.M.; Goodman, M.T.; Sudenga, S.L.; Monsonego, J.; Franceschi, S. EUROGIN 2014 roadmap: Differences in human papillomavirus infection natural history, transmission and human papillomavirus-related cancer incidence by gender and anatomic site of infection. *Int. J. Cancer* **2015**, *136*, 2752–2760. [CrossRef] [PubMed]
5. Song, S.; Tan, J.; Miao, Y.; Zhang, Q. Crosstalk of ER stress-mediated autophagy and ER-phagy: Involvement of UPR and the core autophagy machinery. *J. Cell. Physiol.* **2018**, *233*, 3867–3874. [CrossRef] [PubMed]
6. Hamacher-Brady, A.; Brady, N.R. Mitophagy programs: Mechanisms and physiological implications of mitochondrial targeting by autophagy. *Cell. Mol. Life Sci.* **2016**, *73*, 775–795. [CrossRef] [PubMed]
7. Hyttinen, J.M.; Amadio, M.; Viiri, J.; Pascale, A.; Salminen, A.; Kaarniranta, K. Clearance of misfolded and aggregated proteins by aggrephagy and implications for aggregation diseases. *Ageing Res. Rev.* **2014**, *18*, 16–28. [CrossRef] [PubMed]
8. Kimmey, J.M.; Stallings, C.L. Bacterial Pathogens versus Autophagy: Implications for Therapeutic Interventions. *Trends Mol. Med.* **2016**, *22*, 1060–1076. [CrossRef] [PubMed]
9. Kaushik, S.; Cuervo, A.M. The coming of age of chaperone-mediated autophagy. *Nat. Rev. Mol. Cell Biol.* **2018**, *19*, 365–381. [CrossRef] [PubMed]
10. Jacob, J.A.; Salmani, J.M.M.; Jiang, Z.; Feng, L.; Song, J.; Jia, X.; Chen, B. Autophagy: An overview and its roles in cancer and obesity. *Clin. Chim. Acta* **2017**, *468*, 85–89. [CrossRef] [PubMed]
11. Choi, A.M.; Ryter, S.W.; Levine, B. Autophagy in human health and disease. *N. Engl. J. Med.* **2013**, *368*, 1845–1846. [CrossRef] [PubMed]
12. Hamasaki, M.; Furuta, N.; Matsuda, A.; Nezu, A.; Yamamoto, A.; Fujita, N.; Oomori, H.; Noda, T.; Haraguchi, T.; Hiraoka, Y.; et al. Autophagosomes form at ER-mitochondria contact sites. *Nature* **2013**, *495*, 389–393. [CrossRef] [PubMed]
13. Ge, L.; Melville, D.; Zhang, M.; Schekman, R. The ER-Golgi intermediate compartment is a key membrane source for the LC3 lipidation step of autophagosome biogenesis. *eLife* **2013**, *2*, e00947. [CrossRef] [PubMed]
14. Kim, Y.C.; Guan, K.L. mTOR: A pharmacologic target for autophagy regulation. *J. Clin. Investig.* **2015**, *125*, 25–32. [CrossRef] [PubMed]
15. Stanley, R.E.; Ragusa, M.J.; Hurley, J.H. The beginning of the end: How scaffolds nucleate autophagosome biogenesis. *Trends Cell Biol.* **2014**, *24*, 73–81. [CrossRef] [PubMed]
16. Mizushima, N.; Yoshimori, T.; Levine, B. Methods in mammalian autophagy research. *Cell* **2010**, *140*, 313–326. [CrossRef] [PubMed]
17. Liang, C.; Lee, J.S.; Inn, K.S.; Gack, M.U.; Li, Q.; Roberts, E.A.; Vergne, I.; Deretic, V.; Feng, P.; Akazawa, C.; et al. Beclin1-binding UVRAG targets the class C Vps complex to coordinate autophagosome maturation and endocytic trafficking. *Nat. Cell Biol.* **2008**, *10*, 776–787. [CrossRef] [PubMed]

18. White, E. The role for autophagy in cancer. *J. Clin. Investig.* **2015**, *125*, 42–46. [CrossRef] [PubMed]
19. Jackson, W.T. Viruses and the autophagy pathway. *Virology* **2015**, *479–480*, 450–456. [CrossRef] [PubMed]
20. DiGiuseppe, S.; Bienkowska-Haba, M.; Sapp, M. Human Papillomavirus Entry: Hiding in a Bubble. *J. Virol.* **2016**, *90*, 8032–8035. [CrossRef] [PubMed]
21. Surviladze, Z.; Dziduszko, A.; Ozbun, M.A. Essential roles for soluble virion-associated heparan sulfonated proteoglycans and growth factors in human papillomavirus infections. *PLoS Pathog.* **2012**, *8*, e1002519. [CrossRef] [PubMed]
22. Surviladze, Z.; Sterk, R.T.; DeHaro, S.A.; Ozbun, M.A. Cellular entry of human papillomavirus type 16 involves activation of the phosphatidylinositol 3-kinase/Akt/mTOR pathway and inhibition of autophagy. *J. Virol.* **2013**, *87*, 2508–2517. [CrossRef] [PubMed]
23. Papinski, D.; Kraft, C. Regulation of Autophagy By Signaling Through the Atg1/ULK1 Complex. *J. Mol. Biol.* **2016**, *428 Pt A*, 1725–1741. [CrossRef] [PubMed]
24. Seglen, P.O.; Gordon, P.B. 3-Methyladenine: Specific inhibitor of autophagic/lysosomal protein degradation in isolated rat hepatocytes. *Proc. Natl. Acad. Sci. USA* **1982**, *79*, 1889–1892. [CrossRef] [PubMed]
25. Griffin, L.M.; Cicchini, L.; Pyeon, D. Human papillomavirus infection is inhibited by host autophagy in primary human keratinocytes. *Virology* **2013**, *437*, 12–19. [CrossRef] [PubMed]
26. Ishii, Y. Electron microscopic visualization of autophagosomes induced by infection of human papillomavirus pseudovirions. *Biochem. Biophys. Res. Commun.* **2013**, *433*, 385–389. [CrossRef] [PubMed]
27. Sapp, M.; Bienkowska-Haba, M. Viral entry mechanisms: Human papillomavirus and a long journey from extracellular matrix to the nucleus. *FEBS J.* **2009**, *276*, 7206–7216. [CrossRef] [PubMed]
28. Pinidis, P.; Tsikouras, P.; Iatrakis, G.; Zervoudis, S.; Koukouli, Z.; Bothou, A.; Galazios, G.; Vladareanu, S. Human Papilloma Virus' Life Cycle and Carcinogenesis. *Maedica* **2016**, *11*, 48–54. [PubMed]
29. Moody, C.A.; Laimins, L.A. Human papillomavirus oncoproteins: Pathways to transformation. *Nat. Rev. Cancer* **2010**, *10*, 550–560. [CrossRef] [PubMed]
30. Meisels, A.; Fortin, R. Condylomatous lesions of the cervix and vagina. I. Cytologic patterns. *Acta Cytol.* **1976**, *20*, 505–509. [CrossRef] [PubMed]
31. Venuti, A.; Paolini, F.; Nasir, L.; Corteggio, A.; Roperto, S.; Campo, M.S.; Borzacchiello, G. Papillomavirus E5: The smallest oncoprotein with many functions. *Mol. Cancer* **2011**, *10*, 140–158. [CrossRef] [PubMed]
32. Scheffner, M.; Werness, B.A.; Huibregtse, J.M.; Levine, A.J.; Howley, P.M. The E6 oncoprotein encoded by human papillomavirus types 16 and 18 promotes the degradation of p53. *Cell* **1990**, *63*, 1129–1136. [CrossRef]
33. Boyer, S.N.; Wazer, D.E.; Band, V. E7 protein of human papilloma virus-16 induces degradation of retinoblastoma protein through the ubiquitin-proteasome pathway. *Cancer Res.* **1996**, *56*, 4620–4624. [PubMed]
34. Darragh, T.M.; Colgan, T.J.; Thomas Cox, J.; Heller, D.S.; Henry, M.R.; Luff, R.D.; McCalmont, T.; Nayar, R.; Palefsky, J.M.; Stoler, M.H.; et al. The Lower Anogenital Squamous Terminology Standardization project for HPV-associated lesions: Background and consensus recommendations from the College of American Pathologists and the American Society for Colposcopy and Cervical Pathology. *Int. J. Gynecol. Pathol.* **2013**, *32*, 76–115. [CrossRef] [PubMed]
35. Belleudi, F.; Nanni, M.; Raffa, S.; Torrisi, M.R. HPV16 E5 deregulates the autophagic process in human keratinocytes. *Oncotarget* **2015**, *6*, 9370–9386. [CrossRef] [PubMed]
36. Mattoscio, D.; Casadio, C.; Miccolo, C.; Maffini, F.; Raimondi, A.; Tacchetti, C.; Gheit, T.; Tagliabue, M.; Galimberti, V.E.; De Lorenzi, F.; et al. Autophagy regulates UBC9 levels during viral-mediated tumorigenesis. *PLoS Pathog.* **2017**, *13*, e1006262. [CrossRef] [PubMed]
37. Zhou, X.; Munger, K. Expression of the human papillomavirus type 16 E7 oncoprotein induces an autophagy-related process and sensitizes normal human keratinocytes to cell death in response to growth factor deprivation. *Virology* **2009**, *385*, 192–197. [CrossRef] [PubMed]
38. Hanning, J.E.; Saini, H.K.; Murray, M.J.; Caffarel, M.M.; van Dongen, S.; Ward, D.; Barker, E.M.; Scarpini, C.G.; Groves, I.J.; Stanley, M.A.; et al. Depletion of HPV16 early genes induces autophagy and senescence in a cervical carcinogenesis model, regardless of viral physical state. *J. Pathol.* **2013**, *231*, 354–366. [CrossRef] [PubMed]
39. Carchman, E.H.; Matkowskyj, K.A.; Meske, L.; Lambert, P.F. Dysregulation of Autophagy Contributes to Anal Carcinogenesis. *PLoS ONE* **2016**, *11*, e0164273. [CrossRef] [PubMed]

40. Rademacher, B.L.; Matkowskyj, K.A.; Meske, L.M.; Romero, A.; Sleiman, H.; Carchman, E.H. The role of pharmacologic modulation of autophagy on anal cancer development in an HPV mouse model of carcinogenesis. *Virology* **2017**, *507*, 82–88. [CrossRef] [PubMed]
41. Cheng, H.Y.; Zhang, Y.N.; Wu, Q.L.; Sun, X.M.; Sun, J.R.; Huang, X. Expression of beclin 1, an autophagy-related protein, in human cervical carcinoma and its clinical significance. *Eur. J. Gynaecol. Oncol.* **2012**, *33*, 15–20. [PubMed]
42. Wang, Z.H.; Xu, L.; Wang, Y.; Cao, M.Q.; Li, L.; Bai, T. Clinicopathologic correlations between human papillomavirus 16 infection and Beclin 1 expression in human cervical cancer. *Int. J. Gynecol. Pathol.* **2011**, *30*, 400–406. [CrossRef] [PubMed]
43. Fang, W.; Shu, S.; Yongmei, L.; Endong, Z.; Lirong, Y.; Bei, S. miR-224-3p inhibits autophagy in cervical cancer cells by targeting FIP200. *Sci. Rep.* **2016**, *6*, 33229. [CrossRef] [PubMed]
44. Wang, H.Y.; Yang, G.F.; Huang, Y.H.; Huang, Q.W.; Gao, J.; Zhao, X.D.; Huang, L.M.; Chen, H.L. Reduced expression of autophagy markers correlates with high-risk human papillomavirus infection in human cervical squamous cell carcinoma. *Oncol. Lett.* **2014**, *8*, 1492–1498. [CrossRef] [PubMed]
45. Chen, T.C.; Hung, Y.C.; Lin, T.Y.; Chang, H.W.; Chiang, I.P.; Chen, Y.Y.; Chow, K.C. Human papillomavirus infection and expression of ATPase family AAA domain containing 3A, a novel anti-autophagy factor, in uterine cervical cancer. *Int. J. Mol. Med.* **2011**, *28*, 689–696. [PubMed]
46. Wang, F.; Shan, S.; Huo, Y.; Xie, Z.; Fang, Y.; Qi, Z.; Chen, F.; Li, Y.; Sun, B. MiR-155-5p inhibits PDK1 and promotes autophagy via the mTOR pathway in cervical cancer. *Int. J. Biochem. Cell Biol.* **2018**, *99*, 91–99. [CrossRef] [PubMed]
47. Hara, T.; Takamura, A.; Kishi, C.; Iemura, S.; Natsume, T.; Guan, J.L.; Mizushima, N. FIP200, a ULK-interacting protein, is required for autophagosome formation in mammalian cells. *J. Cell Biol.* **2008**, *181*, 497–510. [CrossRef] [PubMed]
48. De Leo, A.; Colavita, F.; Ciccosanti, F.; Fimia, G.M.; Lieberman, P.M.; Mattia, E. Inhibition of autophagy in EBV-positive Burkitt's lymphoma cells enhances EBV lytic genes expression and replication. *Cell Death Dis.* **2015**, *6*, e1876. [CrossRef] [PubMed]
49. Granato, M.; Santarelli, R.; Farina, A.; Gonnella, R.; Lotti, L.V.; Faggioni, A.; Cirone, M. Epstein-barr virus blocks the autophagic flux and appropriates the autophagic machinery to enhance viral replication. *J. Virol.* **2014**, *88*, 12715–12726. [CrossRef] [PubMed]
50. Chan, C.P.; Kok, K.H.; Jin, D.Y. Human T-Cell Leukemia Virus Type 1 Infection and Adult T-Cell Leukemia. *Adv. Exp. Med. Biol.* **2017**, *1018*, 147–166. [PubMed]
51. Tang, S.W.; Chen, C.Y.; Klase, Z.; Zane, L.; Jeang, K.T. The cellular autophagy pathway modulates human T-cell leukemia virus type 1 replication. *J. Virol.* **2013**, *87*, 1699–1707. [CrossRef] [PubMed]
52. Ren, T.; Takahashi, Y.; Liu, X.; Loughran, T.P.; Sun, S.C.; Wang, H.G.; Cheng, H. HTLV-1 Tax deregulates autophagy by recruiting autophagic molecules into lipid raft microdomains. *Oncogene* **2015**, *34*, 334–345. [CrossRef] [PubMed]
53. Wang, W.; Zhou, J.; Shi, J.; Zhang, Y.; Liu, S.; Liu, Y.; Zheng, D. Human T-cell leukemia virus type 1 Tax-deregulated autophagy pathway and c-FLIP expression contribute to resistance against death receptor-mediated apoptosis. *J. Virol.* **2014**, *88*, 2786–2798. [CrossRef] [PubMed]
54. Goncalves, P.H.; Ziegelbauer, J.; Uldrick, T.S.; Yarchoan, R. Kaposi sarcoma herpesvirus-associated cancers and related diseases. *Curr. Opin. HIV AIDS* **2017**, *12*, 47–56. [CrossRef] [PubMed]
55. Leidal, A.M.; Lee, P.W.; McCormick, C. Viral subversion of autophagy impairs oncogene-induced senescence. *Autophagy* **2012**, *8*, 1138–1140. [CrossRef] [PubMed]
56. Leidal, A.M.; Cyr, D.P.; Hill, R.J.; Lee, P.W.; McCormick, C. Subversion of autophagy by Kaposi's sarcoma-associated herpesvirus impairs oncogene-induced senescence. *Cell Host Microbe* **2012**, *11*, 167–180. [CrossRef] [PubMed]
57. Wen, H.J.; Yang, Z.; Zhou, Y.; Wood, C. Enhancement of autophagy during lytic replication by the Kaposi's sarcoma-associated herpesvirus replication and transcription activator. *J. Virol.* **2010**, *84*, 7448–7458. [CrossRef] [PubMed]
58. Michielsen, P.; Ho, E. Viral hepatitis B and hepatocellular carcinoma. *Acta Gastro-Enterol. Belg.* **2011**, *74*, 4–8.
59. Sir, D.; Tian, Y.; Chen, W.L.; Ann, D.K.; Yen, T.S.; Ou, J.H. The early autophagic pathway is activated by hepatitis B virus and required for viral DNA replication. *Proc. Natl. Acad. Sci. USA* **2010**, *107*, 4383–4388. [CrossRef] [PubMed]

60. Tian, Y.; Sir, D.; Kuo, C.F.; Ann, D.K.; Ou, J.H. Autophagy required for hepatitis B virus replication in transgenic mice. *J. Virol.* **2011**, *85*, 13453–13456. [CrossRef] [PubMed]
61. Zhang, H.T.; Chen, G.G.; Hu, B.G.; Zhang, Z.Y.; Yun, J.P.; He, M.L.; Lai, P.B. Hepatitis B virus x protein induces autophagy via activating death-associated protein kinase. *J. Viral Hepat.* **2014**, *21*, 642–649. [CrossRef] [PubMed]
62. Zhou, T.; Jin, M.; Ding, Y.; Zhang, Y.; Sun, Y.; Huang, S.; Xie, Q.; Xu, C.; Cai, W. Hepatitis B virus dampens autophagy maturation via negative regulation of Rab7 expression. *Biosci. Trends* **2016**, *10*, 244–250. [CrossRef] [PubMed]
63. Liu, B.; Fang, M.; Hu, Y.; Huang, B.; Li, N.; Chang, C.; Huang, R.; Xu, X.; Yang, Z.; Chen, Z.; et al. Hepatitis B virus X protein inhibits autophagic degradation by impairing lysosomal maturation. *Autophagy* **2014**, *10*, 416–430. [CrossRef] [PubMed]
64. Doring, T.; Zeyen, L.; Bartusch, C.; Prange, R. Hepatitis B Virus Subverts the Autophagy Elongation Complex Atg5-12/16L1 and Does Not Require Atg8/LC3 Lipidation for Viral Maturation. *J. Virol.* **2018**, 92. [CrossRef] [PubMed]
65. Hansen, M.D.; Johnsen, I.B.; Stiberg, K.A.; Sherstova, T.; Wakita, T.; Richard, G.M.; Kandasamy, R.K.; Meurs, E.F.; Anthonsen, M.W. Hepatitis C virus triggers Golgi fragmentation and autophagy through the immunity-related GTPase M. *Proc. Natl. Acad. Sci. USA* **2017**, *114*, E3462–E3471. [CrossRef] [PubMed]
66. Shrivastava, S.; Bhanja Chowdhury, J.; Steele, R.; Ray, R.; Ray, R.B. Hepatitis C virus upregulates Beclin1 for induction of autophagy and activates mTOR signaling. *J. Virol.* **2012**, *86*, 8705–8712. [CrossRef] [PubMed]
67. Wang, L.; Tian, Y.; Ou, J.H. HCV induces the expression of Rubicon and UVRAG to temporally regulate the maturation of autophagosomes and viral replication. *PLoS Pathog.* **2015**, *11*, e1004764. [CrossRef] [PubMed]
68. Taguwa, S.; Kambara, H.; Fujita, N.; Noda, T.; Yoshimori, T.; Koike, K.; Moriishi, K.; Matsuura, Y. Dysfunction of autophagy participates in vacuole formation and cell death in cells replicating hepatitis C virus. *J. Virol.* **2011**, *85*, 13185–13194. [CrossRef] [PubMed]
69. Kim, J.Y.; Wang, L.; Lee, J.; Ou, J.J. Hepatitis C Virus Induces the Localization of Lipid Rafts to Autophagosomes for Its RNA Replication. *J. Virol.* **2017**, *91*. [CrossRef] [PubMed]
70. Shrivastava, S.; Devhare, P.; Sujijantarat, N.; Steele, R.; Kwon, Y.C.; Ray, R.; Ray, R.B. Knockdown of Autophagy Inhibits Infectious Hepatitis C Virus Release by the Exosomal Pathway. *J. Virol.* **2016**, *90*, 1387–1396. [CrossRef] [PubMed]
71. Kudchodkar, S.B.; Levine, B. Viruses and autophagy. *Rev. Med. Virol.* **2009**, *19*, 359–378. [CrossRef] [PubMed]
72. Silva, L.M.; Jung, J.U. Modulation of the autophagy pathway by human tumor viruses. *Semin. Cancer Biol.* **2013**, *23*, 323–328. [CrossRef] [PubMed]
73. Hundeshagen, P.; Hamacher-Brady, A.; Eils, R.; Brady, N.R. Concurrent detection of autolysosome formation and lysosomal degradation by flow cytometry in a high-content screen for inducers of autophagy. *BMC Biol.* **2011**, *9*, 38. [CrossRef] [PubMed]

8

Cigarette Smoking Promotes Infection of Cervical Cells by High-Risk Human Papillomaviruses, but not Subsequent E7 Oncoprotein Expression

Kimon Chatzistamatiou [1,*], Theodoros Moysiadis [2], Dimos Vryzas [3], Ekaterini Chatzaki [4], Andreas M. Kaufmann [5], Isabel Koch [6], Erwin Soutschek [6], Oliver Boecher [6], Athena Tsertanidou [7], Nikolaos Maglaveras [8], Pidder Jansen-Duerr [9] and Theodoros Agorastos [7]

1 2nd Department of Obstetrics and Gynecology, Aristotle University of Thessaloniki, Hippokratio General Hospital, 54642 Thessaloniki, Greece
2 Institute of Applied Biosciences, Centre for Research & Technology-Hellas, 57001 Thessaloniki, Greece; moysiadis.theodoros@gmail.com
3 Medical School, Democritus University of Thrace, 68100 Alexandroupolis, Greece; d.vrizas@gmail.com
4 Laboratory of Pharmacology, Medical School, Democritus University of Thrace, 68100 Alexandroupolis, Greece; achatzak@med.duth.gr
5 Department of Gynecology, Charité-Universitaetsmedizin Berlin, Campus Benjamin Franklin, 10117 Berlin, Germany; andreas.kaufmann@charite.de
6 Mikrogen GmbH, 82061 Neuried, Germany; koch@mikrogen.de (I.K.); soutschek@mikrogen.de (E.S.); boecher@mikrogen.de (O.B.)
7 4th Department of Obstetrics and Gynecology, Aristotle University of Thessaloniki, Hippokratio General Hospital, 54642 Thessaloniki, Greece; athetse@gmail.com (A.T.); agorast@auth.gr (T.A.)
8 Lab of Computing and Medical Informatics, Department of Medicine, Aristotle University of Thessaloniki, 54124 Thessaloniki, Greece; nicmag@med.auth.gr
9 Research Institute for Biomedical Aging Research, University of Innsbruck, A-6020 Innsbruck, Austria; pidder.jansen-duerr@uibk.ac.at
* Correspondence: kimon.chatzistamatiou@gmail.com

Abstract: Persistent cervical infection with high-risk human papillomaviruses (hrHPVs) is a necessary, but not sufficient, condition for the development of cervical cancer. Therefore, there are other co-factors facilitating the hrHPV carcinogenic process, one of which is smoking. To assess the effect of smoking on high-risk (hr) HPV DNA positivity and on the expression of HPV E7 oncoprotein, as a surrogate of persistent hrHPV infection, we used data from women recruited for the PIPAVIR project, which examined the role of E7 protein detection in cervical cancer screening. Women were tested for hrHPV DNA, using Multiplex Genotyping (MPG), and E7 protein, using a novel sandwich ELISA method, and gave information on their smoking habits. Among 1473 women, hrHPV prevalence was 19.1%. The odds ratio (OR) for hrHPV positivity of smokers compared to non-smokers was 1.785 (95% confidence intervals (CI): 1.365–2.332, $p < 0.001$). The ORs for E7 positivity, concerning hrHPV positive women, ranged from 0.720 to 1.360 depending on the E7 detection assay used, but this was not statistically significant. Smoking increases the probability of hrHPV infection, and smoking intensity is positively associated to this increase. Smoking is not related to an increased probability of E7 protein positivity for hrHPV positive women.

Keywords: high-risk HPV infection; E7 oncoprotein; cigarette smoking; HPV carcinogenesis; cervical cancer

1. Introduction

Persistent cervical infection with high-risk human papillomaviruses (hrHPVs) is a necessary condition for the development of cervical cancer since it has been shown that virtually all cases of cervical cancer are related to HPV DNA detection [1]. However, other co-factors seem to play a role in cervical cancer development, either by facilitating hrHPV transmission and infection or by accelerating the carcinogenic process leading to cervical cancer [2]. The most important factors identified as co-factors to cervical carcinogenesis in hrHPV positive women are high parity, long-term oral contraceptive (OC) use, smoking, and co-infection with other sexually transmitted agents [3]. Particularly, parity and smoking seem to be more consistently regarded as factors modulating the risk of progression from HPV infection to cervical precancer and cancer [4].

Smoking has been shown to be a factor associated to a higher probability of hrHPV infection by studies based on hrHPV DNA detection [5–7]. However, persistent hrHPV infection, characterized by the expression of E6 and E7 oncoproteins, which play a crucial role in HPV-related carcinogenesis [8], has not yet been studied thoroughly concerning probable associations to smoking.

The present analysis aims to assess the effect of smoking on the expression of HPV E7 oncoprotein, as a surrogate of persistent hrHPV infection, and therefore to investigate whether smoking is a factor involved in the progression of hrHPV infection to cervical precancer and cancer in the clinical setting of the PIPAVIR (detection of persistent infections by human papillomaviruses) study, a study designed for the development and initial clinical assessment of a novel HPV E7 detection method for cervical cancer screening. According to the PIPAVIR study, hrHPV cervical infection was assessed using a PCR (polymerace chain reaction)-based HPV DNA detection method and five different sandwich ELISA assays identifying the E7 protein of different combinations of hrHPV types.

2. Results

2.1. Demographic Data

Demographic characteristics of the study population have been described previously [9]. Briefly, the current analysis was conducted on 1473 women aged 30–60 years old who were enrolled for the PIPAVIR project and for whom there was a valid result for hrHPV DNA genotyping and E7 protein detection. Among these women, 776 (52.7%) were non-smokers, 605 (41.1%) were active smokers and there was a smaller group of 92 women (6.2%) who were ex-smokers as defined in the methods section. Among smokers, 181 women had a SII (Smoking intensity Index) $\leq$ 50, 161 women had 50 < SII $\leq$ 100, and 251 women had a SII > 100. High-risk HPV prevalence for the study population was 19.1% (282 women found positive for hrHPV out of 1473) (Table 1).

Table 1. Demographic characteristics of the study population.

Demographic Characteristic	*n* = 1473
Age (years)	
Mean (sd)	43.52 (8.05)
Median (range)	44 (30–60)
Children	
No	314 (21.3%)
Yes	1159 (78.7%)
Smoking History	
No	776 (52.7%)
Yes	605 (41.1%)
Ex-Smoker	92 (6.2%)

Table 1. *Cont.*

Demographic Characteristic	$n = 1473$
SII	1369
=0	776
<50	181
50–100	161
>100	251
Pap Test (at least once)	
No	28 (1.62%)
Yes	1695 (98.37%)
hrHPV	
Negative	1191 (80.9%)
Positive	282 (19.1%)

hrHPV: high-risk human papillomavirus; SII: smoking intensity index (referring to the product of the number of cigarettes smoked per day by the years of smoking by 365 divided by 1000).

2.2. Sample Stratification According to Smoking and Age

We considered different groups of women regarding their smoking status (smokers, ex-smokers and non-smokers) and performed the analysis according to three distinct scenarios: (a) all three categories are independent; (b) the ex-smokers and smoker groups are merged; and (c) the ex-smokers and non-smoker groups are merged.

SII was used both as a continuous variable and as a categorical one. In particular, women were assigned to four groups, with group "1" consisting of non-smokers, while smokers were distributed into three groups (2, 3, and 4) with SII values less than 50, between 50 and 100, and over 100, respectively. We assessed separate age categories, but the analysis did not yield significant findings and differences between these categories.

2.3. Smoking and hrHPV DNA Positivity

The analysis was initially focused on the high-risk HPV DNA detection, resulting to odds ratios suggesting that smoking is, indeed, significantly related to testing positive for hrHPV. As shown in Table 2, in Scenario (a), the odds ratio of smokers compared to the reference category (non-smokers) was 1.785 (95% confidence intervals (CI): 1.365–2.332, $p < 0.001$) indicating that the odds of being hrHPV positive were 1.785 times higher for smokers compared to non-smokers. The same has been confirmed in Scenarios (b) and (c), for which the incorporation of ex-smokers to the smoking and non-smoking group has resulted in OR > 1, 1.626 (95% CI: 1.251–2.113, $p < 0.001$) and 1.838 (95% CI: 1.415–2.388, $p < 0.001$), respectively, indicating statistically significantly higher odds of being hrHPV positive in the smoking group for both scenarios. The comparison of ex-smokers to non-smokers yielded a not statistically significant difference (OR = 0.735, 95% CI: 0.380–1.421, $p = 0.360$) in Scenario (a). The ORs at subsequent analysis were also adjusted for the number of pregnancies and the use of oral contraceptives yielding non-significant differences compared to the unadjusted ORs.

Table 2. Association between smoking and high-risk human papillomavirus status. The cross-tabulation tables are displayed per horizontal panel along with the odds ratios (95% CI) and the corresponding *p*-values (binary logistic regression).

	hrHPV (−)	hrHPV (+)	Total	OR (95% CI)	*p*-Value	Adjusted OR * (95% CI)	*p*-Value
Scenario (a)							
Non-Smokers (reference)	655 (84.4%)	121 (15.6%)	776 (100%)				
Smokers	455 (75.2%)	150 (24.8%)	605 (100%)	1.785 (1.365–2.332)	<0.001	1.788 (1.367–2.340)	<0.001
Ex-Smokers	81 (88%)	11 (12%)	92 (100%)	0.735 (0.380–1.421)	0.360	0.682 (0.350–1.329)	0.261
Total	1191 (80.9%)	282 (19.1%)	1473 (100%)				
Scenario (b)							
Non-Smokers (reference)	655 (84.4%)	121 (15.6%)	776 (100%)				
Smokers and ex-smokers	536 (76.9%)	161 (23.1%)	697 (100%)	1.626 (1.251–2.113)	<0.001	1.618 (1.244–2.106)	<0.001
Total	1191 (80.9%)	282 (19.1%)	1473 (100%)				
Scenario (c)							
Non- and ex-smokers (reference)	736 (84.8%)	132 (15.2%)	868 (100%)				
Smokers	455 (75.2%)	150 (24.8%)	605 (100%)	1.838 (1.415–2.388)	<0.001	1.856 (1.427–2.413)	<0.001
Total	1191 (80.9%)	282 (19.1%)	1473 (100%)				
Smoking Intensity Index							
SII = 0 (reference)	655 (84.4%)	121 (15.6%)	776 (100%)				
0 < SII < 50	133 (73.5%)	48 (26.5%)	181 (100%)	1.954 (1.332–2.865)	0.001	1.828 (1.242–2.690)	0.002
50 < SII < 100	112 (69.6%)	49 (30.4%)	161 (100%)	2.368 (1.607–3.490)	<0.001	2.272 (1.538–3.356)	<0.001
SII > 100	201 (80.1%)	50 (19.9%)	251 (100%)	1.347 (0.934–1.941)	0.111	1.471 (1.015–2.131)	0.041
Total	1101 (80.4%)	268 (19.6%)	1369 (100%)				

hrHPV: High-Risk Human Papillomavirus; SII: Smoking Intensity Index; OR: Odds Ratio; * adjustment was performed for the number of pregnancies and the use of oral contraceptives.

Concerning SII, it was shown that women who were smokers had higher odds to be positive for hrHPV compared to non-smokers, regardless of them being light, medium or heavy smokers. In detail, women with SII below 50 presented OR of 1.954 (95% CI: 1.332–2.865, $p = 0.001$), and women with SII between 50 and 100 presented higher OR (2.368 (95% CI: 1.607–3.490, $p < 0.001$)), which means that women who had smoked more were more probable to be hrHPV positive compared to non-smokers than those who were light smokers. However, interestingly, the same result was not observed concerning heavy smokers, since they presented OR (unadjusted) of 1.347 (95% CI: 0.934–1.941, $p = 0.111$) of being hrHPV positive, a value which did not reveal a statistically significant difference compared to non-smokers as one would expect (Table 2). However, statistical significance was obtained after adjustment, which yielded a OR of 1.471 (95% CI: 1.015–2.131, $p = 0.041$).

2.4. Smoking and E7 Detection

Regarding the results for the five different E7-testing assays, labeled recomWell HPV 16/18/45 KJ_{high}, recomWell HPV 39/51/56/59, recomWell HPV 16/31/33/35/52/58, recomWell HPV HR screen and recomWell HPV 16/18/45 KJ_{low}, only women who had previously given a hrHPV positive result were selected, namely 282 women (19.1%). Among them, 121 (42.9%) were non-smokers, 150 (53.2%) were current smokers and 11 (3.9%) were ex-smokers (Table 3).

Table 3. Frequency table regarding smoking status for hrHPV positive women.

	Frequency (*n*)	**Percentage (%)**
Non-Smokers	121	42.9
Smokers	150	53.2
Ex-Smokers	11	3.9
Total	282	100.0
Smoking Intensity Index		
SII = 0	121	45.1
0 < SII < 50	48	17.9
50 < SII < 100	49	18.3
SII > 100	50	18.7

SII: Smoking Intensity Index.

This analysis involved the three scenarios as previously described and yielded no statistically significant differences concerning E7 positivity for hrHPV positive women who were smokers compared to non-smokers in either of the three (Table 4). The most important finding was the higher than 1 ORs, presented by recomWell HPV 39/51/56/59 assay, in all three scenarios examined, meaning that smokers who were hrHPV positive tended to have an increased possibility for positive E7 testing compared to non-smokers, however none of these results were statistically significant and, therefore, this correlation could not be established. Adjustment did not reveal statistically significant differences for any of the E7 tests examined and the ORs were only changed slightly.

Table 4. Association between smoking/smoking intensity and E7 for hrHPV positive women. The odds ratios (95% CI) and the corresponding *p*-values (binary logistic regression) are displayed per horizontal panel. Cross-tabulation tables are not displayed due to space limitation.

E7 Detection Assays	recomWell HPV 16/18/45 KJ_{high}	recomWell HPV 39/51/56/59	recomWell HPV 16/31/33/35/52/58	recomWell HPV HR Screen	recomWell HPV 16/18/45 KJ_{low}
	OR (*p*-Value) 95% CI	OR (*p*-Value) 95% CI	OR (*p*-Value) 95% CI	OR (*p*-Value) 95% CI	OR (*p*-Value) 95% CI
Scenario (a)					
Non-Smokers (reference)					
Smokers	0.720 (0.182) 0.444–1.167	1.360 (0.233) 0.820–2.256	0.907 (0.707) 0.545–1.510	0.966 (0.889) 0.591–1.578	0.838 (0.492) 0.506–1.388
Ex-Smokers	1.971 (0.334) 0.498–7.798	1.798 (0.357) 0.516–6.261	1.101 (0.883) 0.305–3.980	0.773 (0.684) 0.223–2.676	0.987 (0.984) 0.274–3.562
Scenario (b)					
Non-Smokers (reference)					
Smokers and ex-smokers	0.767 (0.275) 0.477–1.235	1.387 (0.198) 0.843–2.283	0.919 (0.742) 0.557–1.518	0.951 (0.838) 0.587–1.542	0.848 (0.514) 0.516–1.392
Scenario (c)					
Non- and ex-Smokers (reference)					
Smokers	0.683 (0.114) 0.426–1.096	1.290 (0.309) 0.790–2.108	0.899 (0.676) 0.547–1.479	0.987 (0.958) 0.612–1.594	0.839 (0.485) 0.513–1.373
Smoking Intensity Index (SII)	0.996 (0.093) 0.992–1.001	1.000 (0.902) 0.995–1.004	0.999 (0.588) 0.994–1.003	0.997 (0.135) 0.992–1.001	0.999 (0.688) 0.995–1.004
SII = 0 (reference)					
0 < SII < 50	1.035 (0.921) 0.525–2.039	1.414 (0.329) 0.706–2.832	0.963 (0.918) 0.474–1.957	0.901 (0.879) 0.456–1.781	0.947 (0.879) 0.471–1.904
%cmidrule1-6 50 < SII < 100	0.710 (0.313) 0.364–1.382	1.758 (0.105) 0.889–3.477	1.119 (0.715) 0.560–2.236	1212 (0.993) 0.606–2.424	1.003 (0.993) 0.503–1.998
SII > 100	0.581 (0.109) 0.299–1.130	0.925 (0.831) 0.452–1.894	0.749 (0.434) 0.363–1.545	0.819 (0.182) 0.420–1.598	0.607 (0.182) 0.292–1.263

SII: Smoking Intensity Index; OR: Odds Ratio.

3. Discussion

According to the International Agency for Research on Cancer (IARC) certain types of human papillomaviruses are considered carcinogenic to humans. These types are HPV 16, the most potent carcinogen, and others (HPVs 18, 31, 33, 35, 39, 45, 51, 52, 56, 58, and 59) for which there is sufficient evidence concerning carcinogenicity. Other types are also thought to have a carcinogenic potential for which there is limited evidence [10]. The natural history of hrHPV infection, however, dictates that only a minority of cervical hrHPV infections will eventually lead to cervical cancer [11]. On the other hand, virtually all cases of cervical cancer are related to a hrHPV infection [1]. Therefore, according to these facts, hrHPV infection is considered a necessary but not sufficient condition for cervical carcinogenesis, which is facilitated by certain known or unknown co-factors [4].

Smoking is a known co-factor for hrHPV related carcinogenesis. In general, women who smoke are more frequently found hrHPV positive compared to non-smokers according to a study conducted in Greece [5], the main site of the presented study too, as well as according to other studies in different countries [6,12–15]. Moreover, the intensity of smoking seems to play a role in the risk of being hrHPV positive, as has been shown by a pooled IARC analysis [16], and by the Greek study too [5]. The former analysis considered women who smoked at least one cigarette per day for at least one year as smokers and the latter used the SII, a variable introduced by our group then, and used in the current analysis too, which is strongly influenced by the time during which a woman has been a smoker. This is an advantage of that index which may allow for more robust results. The current analysis has shown a higher probability for women who currently smoke compared to non-smokers and an even higher probability for women who smoke more. However, as already noted, women who have an SII > 100 (heavy smokers) do not differ statistically significantly compared to non-smokers in the unadjusted model, but they do differ statistically significantly in the adjusted model. Although this is a rather unexpected result, it could be justified, since the effect of smoking on HPV testing does not necessarily exhibit a monotone relation, and, moreover, the thresholds 50 and 100 used to define the categories are arbitrary.

A third group of women, comprised of ex-smokers, was also identified which involved women who had stopped smoking at least a year ago, according to previous research practices [16]. These women were only 92/1473 (6.2%) in the current analysis, however, since they are a special group, they were analyzed as such (Scenario (a)) and also as being either smokers (Scenario (b)) or non-smokers (Scenario (c)). Ex-smokers and non-smokers did not exhibit any statistically significant differences in the odds for being hrHPV positive. Furthermore, when ex-smokers were grouped with smokers, the OR of being hrHPV positive was decreased compared to the one deriving from them being grouped with non-smokers (1.626 vs. 1.838). These results might suggest the rapid effect of the protective role of smoking cessation concerning hrHPV infection.

The molecular mechanism of hrHPV induced carcinogenesis is based on the crucial role of E6 and E7 viral oncoproteins which are involved in the degradation of p53 [17–19] and reduced activity of pRB (Retinoblastoma protein) [20,21], respectively. The current analysis, apart from examining the role of smoking on the probability of hrHPV cervical infection, mainly aimed to assess its role on the progression of hrHPV infection from a transient to a productive one, identified by the detection of E7 protein, the most potent factor in HPV related carcinogenesis. To our knowledge, this is being investigated for the first time, and, interestingly, it was shown that smoking did not increase the probability of E7 positivity, as was determined using the novel E7 assays developed for clinical use during the PIPAVIR project. The target group for this analysis had been women of the PIPAVIR cohort, tested positive for hrHPV DNA using a PCR-based method, which implies the detection of hrHPV DNA but does not give information on the occurrence of transient or transforming hrHPV infection.

The modifying role of smoking in the natural history of hrHPV infection is not yet clear and epidemiologic studies often reach controversial results, probably due to methodologic issues which produce bias, since elimination of confounding is not always possible [22]. It seems that smoking interferes with the immune system by suppressing it [23] and this mechanism has been proposed as the main link between smoking and hrHPV infection. Recently, the first relevant study to investigate by complex modeling indirect (antibody-mediated) and direct (antibody-independent) effects of smoking on HPV infection showed that current smokers had 29% increased odds for HPV16 infection by the indirect mechanism, an effect which was statistically significant, whereas this was not shown for the direct effect of smoking [24]. Moreover, the indirect effect of smoking was more intense (61% increased odds) in women who smoked more per day, but a significant increase was not observed concerning smoking duration. Former smokers were also not reported to express a significant antibody-mediated effect on HPV16 infection. These findings describe a mechanism which could explain the results of the current analysis, namely the higher probability, and the insignificantly different probability of hrHPV infection in current and former smokers, respectively, compared to never-smokers. An immune mechanism behind the effect of smoking on hrHPV infection would most probably imply an effect early in the natural history of the infection, by preventing the virus–host cell membrane interaction, and not on later stages characterized by expression of E6/E7 viral proteins. In our sub-analysis, the non-significant effect of smoking on the detection of E7 protein, an indicator of persistent hrHPV infection in vivo, supports further the suggestion, that cigarette smoking has a more prominent role in earlier stages of HPV-related carcinogenesis. However, more research is needed to clarify the molecular mechanisms by which smoking interacts with the natural history of hrHPV infection.

As a conclusion, smoking seems to increase the probability of hrHPV infection, and smoking intensity is positively associated to this increase. However, smoking is not related to an increased probability of E7 protein positivity concerning hrHPV positive women.

4. Materials and Methods

4.1. Study Design and Sampling

The PIPAVIR project was a study conducted between August 2012 and August 2015, aiming at the development and initial clinical assessment of a novel ELISA hrHPVE7 oncoprotein detection method for cervical cancer screening. The study design has been described in detail previously [9]. Briefly, participants were non-pregnant women aged 30–60, without a history of cervical intraepithelial neoplasia (CIN), who visited the Family Planning Centre, Hippokratio Hospital of Thessaloniki, Greece and the Department of Gynecology and Obstetrics in Im Mare Klinikum, Kiel, Germany and gave their written informed consent to participate. Subsequently, a cervicovaginal sample, used for Thinprep cytology, HPV DNA genotyping and hrHPVE7 protein detection was taken and women with a positive cytology or hrHPV DNA result were referred to colposcopy followed by biopsy and/or endocervical curettage (ECC), when needed. Sampling was performed using the CervexBrush (Rovers Medical Devices, Lekstraat 10, NL-5347, KV Oss, The Netherlands) and the Cytobrush (CooperSurgical, Inc., 95 Corporate Drive, Trumbull, CT, USA) according to the manufacturer's instructions. After sampling both brushes were immersed in a vial containing collection fluid (PreservCyt, Hologic, Bedford, MA, USA).

4.2. Smoking Habits Assessment

A personal information sheet was filled in for each woman enrolled in the study including demographic characteristics. Concerning smoking, each woman was asked to identify herself as a current smoker, a non-smoker (never smoker), or an ex-smoker, meaning that she had quit smoking at least one year prior to recruitment. Smokers were also asked for more specific information about their smoking habits, in particular, about the number of cigarettes smoked per day as well as their smoking years. This information allowed a more quantitative approach through Smoking Intensity Index (SII), a new variable created by the number of cigarettes smoked per day multiplied by 365 and by years of smoking and divided by 1000 (n (cigarette/day) $\times$ 365 $\times$ n (years) $\div$ 1000) [5]. In this way, women were divided into light smokers (SII < 50), medium smokers (SII 50–100) and heavy smokers (SII > 100).

4.3. HPV DNA Genotyping

HPV genotyping by Multiplexed Genotyping (MPG), consisting of a consensus broad-spectrum GP5+/GP6+ primer multiplex PCR and a type-specific probe read out by Luminex technology was performed at the CHARITE, Gynecologic Tumor Immunology Laboratory, Berlin, Germany [25,26]. MPG, targets the L1 gene sequence, and identifies hrHPV types 16, 18, 26, 31, 33, 35, 39, 45, 51, 52, 53, 56, 58, 59, 66, 68, 73, and 82, and low-risk HPVs 6, 11, 42, 54, 70, 72 and 90.

4.4. hrHPVE7 Testing

A sandwich ELISA method was developed during the PIPAVIR project to detect hrHPVE7 protein, as a result of a collaboration between the laboratories of Innsbruck University and Mikrogen GmbH, Neuried, Germany [9]. The detection strategy for hrHPVE7 was based on five different assays: "recomWell HPV 16/18/45 KJhigh", "recomWell HPV 16/18/45 KJlow", "recomWell HPV 39/51/56/59", "recomWell HPV 16/31/33/35/52/58" and "recomWell HPV HR screen" (for 16, 18, 31, 33, 35, 39, 45, 51, 52, 56, 58 and 59 E7). Each assay refers to different combinations of hrHPVE7 or different concentrations of the relevant ELISA antibodies and yields a measurement of the optical density, a continuous variable.

4.5. Statistical Analysis

The main purpose of the analysis was to validate the association between HPV infection, smoking and smoking intensity, and to assess for the first time the association between E7 protein detection, smoking and smoking intensity for hrHPV positive women. The result of the HPV test was a categorical binary variable, whereas the optical density value derived from the measurement of the E7 oncoprotein was a continuous numerical variable. For this analysis, this continuous variable was transformed into a binary one, by a cut-off value calculated using receiver operating characteristic curve analysis. The impact of smoking and smoking intensity on hrHPV infection was evaluated empirically by cross-tabulation matrices (in the case of E7, cross-tabulation matrices are not displayed due to space limitation). The binary logistic regression was selected to statistically evaluate the association between HPV/E7 and smoking/smoking intensity, since it was of interest to investigate three categories in terms of smoking (smokers, ex-smokers and non-smokers) and four categories in terms of smoking intensity (SII = 0, $0 < \text{SII} \leq 50$, $50 < \text{SII} \leq 100$ and SII > 100), and their binary correlations (one category was selected each time as a reference category). The selected regression model was applied unadjusted to assess smoking/smoking intensity, and adjusted for the number of pregnancies and the combined oral contraceptive use (COC). The odds ratios (ORs) and adjusted ORs along with their 95% confidence intervals (CI) and the corresponding p-values were computed. The significance level was set to 0.05. p-value < 0.05 indicated statistically significantly different ORs compared to one. All analyses were performed using the statistical software SPSS Statistics 22.0.

4.6. Ethical Approval

Allproceduresperformed in studies involving human participants were in accordance with the ethical standards of the institutional and/or national research committee and with the 1964 Helsinki declaration and its later amendments or comparable ethical standards. The study was performed subject to the: Bioethics Committee of the Medical School of Aristotle University of Thessaloniki/Greece (33-31/8/2012) and the Ethikkommission der CHARITE-Universitätsmedizin Berlin/Germany (Available online: http://www.charite.de/fakultaet/kommissionen/ethikkommission.html). Project identification date: 31 August 2012.

4.7. Informed Consent

Informed consent was obtained from all individual participants included in the study.

Acknowledgments: The authors acknowledge the contribution of the following colleagues and collaborators: Achim Schneider, Lutz Gissmann, and Silvia de Sanjose for their invaluable advice during the project; Ursula Schiller for expert technical assistance with Multiplex Genotyping; Garifallia Michalaki and Georgia Kiriafini for nursing assistance during sampling; Thomas Theodoridis † and Fausto Carcea for clinical supervision during the sampling procedure; Anastasia Kitsou for administrative work; and The Mikrogen team (Mira Kellner, Stefanie Fehrmann, Mandy Fleischhauer, Steven McNamara, Melanie Thiessen, and Sophie Vetter) for technical assistance with E7 ELISA test development and measurements of the clinical samples. The trial was funded by the EU 7th framework programme, FP7-HEALTH-304927. (†: deceased).

Author Contributions: Theodoros Agorastos and Kimon Chatzistamatiou conceived and designed the study, wrote the manuscript and supervised statistical analysis; Theodoros Moysiadis and Dimos Vryzas performed the statistical analysis and wrote parts of the manuscript; Athena Tsertanidou handled the database and drafted parts of the manuscript; Ekaterini Chatzaki supervised statistical analysis and reviewed the manuscript; Andreas M. Kaufmann performed and supervised all hrHPV DNA testing and wrote parts of the manuscript; Isabel Koch, Erwin Soutschek and Oliver Boecher participated in the E7 testing development, performed and supervised all E7 testing and wrote parts of the manuscript; Nikolaos Maglaveras supervised software design for the data capturing system used for the study and reviewed the manuscript; and Pidder Jansen-Duerr developed the E7 testing method and reviewed the manuscript.

Abbreviations

hrHPV	High-Risk Human Papillomavirus
DNA	Directory Deoxyribonucleic Acid
ELISA	Enzyme-Linked Immune Sorbent Assay
IARC	International Agency for the Research on Cancer
OR	Odds Ratio
pRB	Retinoblastoma Protein
PCR	Polymerace Chain Reaction
MPG	Multiplex Genotyping
CIN	Cervical Intraepithelial Neoplasia

References

1. Walboomers, J.M.M.; Jacobs, M.V.; Manos, M.M.; Bosch, F.X.; Kummer, J.A.; Shah, K.V.; Snijders, P.J.F.; Peto, J.; Meijer, C.J.L.M.; Munoz, N. Human papillomavirus is a necessary cause of invasive cervical cancer worldwide. *J. Pathol.* **1999**, *189*, 12–19. [CrossRef]
2. Arends, M.J.; Buckley, C.H.; Wells, M. Aetiology, pathogenesis, and pathology of cervical neoplasia. *J. Clin. Pathol.* **1998**, *51*, 96–103. [CrossRef] [PubMed]
3. Castellsague, X.; Bosch, F.X.; Munoz, N. Environmental co-factors in HPV carcinogenesis. *Virus. Res.* **2002**, *89*, 191–199. [CrossRef]
4. Castellsague, X.; Munoz, N. Chapter 3: Cofactors in human papillomavirus carcinogenesis—Role of parity, oral contraceptives, and tobacco smoking. *J. Natl. Cancer Inst. Monogr.* **2003**, *2003*, 20–28. [CrossRef]
5. Chatzistamatiou, K.; Katsamagas, T.; Zafrakas, M.; Zachou, K.; Orologa, A.; Fitsiou, F.; Theodoridis, T.; Konstantinidis, T.; Konstantinidis, T.C.; Agorastos, T. Smoking and genital human papilloma virus infection in women attending cervical cancer screening in greece. *World J. Obstet. Gynecol.* **2013**, *2*, 53. [CrossRef]
6. Pista, A.; de Oliveira, C.F.; Cunha, M.J.; Paixao, M.T.; Real, O.; Group, C.P.S. Risk factors for human papillomavirus infection among women in portugal: The cleopatre portugal study. *Int. J. Gynaecol. Obstet.* **2012**, *118*, 112–116. [CrossRef] [PubMed]
7. Vaccarella, S.; Franceschi, S.; Herrero, R.; Munoz, N.; Snijders, P.J.; Clifford, G.M.; Smith, J.S.; Lazcano-Ponce, E.; Sukvirach, S.; Shin, H.R.; et al. Sexual behavior, condom use, and human papillomavirus: Pooled analysis of the iarc human papillomavirus prevalence surveys. *Cancer Epidemiol. Biomark. Prev.* **2006**, *15*, 326–333. [CrossRef] [PubMed]

8. Narisawa-Saito, M.; Kiyono, T. Basic mechanisms of high-risk human papillomavirus-induced carcinogenesis: Roles of E6 and E7 proteins. *Cancer Sci.* **2007**, *98*, 1505–1511. [CrossRef] [PubMed]
9. Agorastos, T.; Chatzistamatiou, K.; Moysiadis, T.; Kaufmann, A.M.; Skenderi, A.; Lekka, I.; Koch, I.; Soutschek, E.; Boecher, O.; Kilintzis, V.; et al. Human papillomavirus E7 protein detection as a method of triage to colposcopy of hpv positive women, in comparison to genotyping and cytology. Final results of the pipavir study. *Int. J. Cancer* **2017**, *141*, 519–530. [CrossRef] [PubMed]
10. Bouvard, V.; Baan, R.; Straif, K.; Grosse, Y.; Secretan, B.; El Ghissassi, F.; Benbrahim-Tallaa, L.; Guha, N.; Freeman, C.; Galichet, L.; et al. A review of human carcinogens—Part b: Biological agents. *Lancet Oncol.* **2009**, *10*, 321–322. [CrossRef]
11. Schiffman, M.; Castle, P.E.; Jeronimo, J.; Rodriguez, A.C.; Wacholder, S. Human papillomavirus and cervical cancer. *Lancet* **2007**, *370*, 890–907. [CrossRef]
12. Herrero, R.; Castle, P.E.; Schiffman, M.; Bratti, M.C.; Hildesheim, A.; Morales, J.; Alfaro, M.; Sherman, M.E.; Wacholder, S.; Chen, S.; et al. Epidemiologic profile of type-specific human papillomavirus infection and cervical neoplasia in guanacaste, costa rica. *J. Infect. Dis.* **2005**, *191*, 1796–1807. [CrossRef] [PubMed]
13. Remschmidt, C.; Kaufmann, A.M.; Hagemann, I.; Vartazarova, E.; Wichmann, O.; Delere, Y. Risk factors for cervical human papillomavirus infection and high-grade intraepithelial lesion in women aged 20 to 31 years in germany. *Int. J. Gynecol. Cancer* **2013**, *23*, 519–526. [CrossRef] [PubMed]
14. Roteli-Martins, C.M.; Panetta, K.; Ferreira Alves, V.A.; Coelho Siqueira, S.A.; Syrjänen, K.J.; Mauricette Derchain, S.F.C. Cigarette smoking and high-risk HPV DNA as predisposing factors for high-grade cervical intraepithelial neoplasia (cin) in young brazilian women. *Acta Obstet. Gynecol. Scand.* **1998**, *77*, 678–682. [CrossRef] [PubMed]
15. Sellors, J.W.; Mahony, J.B.; Kaczorowski, J.; Lytwyn, A.; Bangura, H.; Chong, S.; Lorincz, A.; Dalby, D.M.; Janjusevic, V.; Keller, J.L. Prevalence and predictors of human papillomavirus infection in women in ontario, canada. Survey of HPV in ontario women (show) group. *CMAJ* **2000**, *163*, 503–508. [PubMed]
16. Vaccarella, S.; Herrero, R.; Snijders, P.J.; Dai, M.; Thomas, J.O.; Hieu, N.T.; Ferreccio, C.; Matos, E.; Posso, H.; de Sanjose, S.; et al. Smoking and human papillomavirus infection: Pooled analysis of the international agency for research on cancer HPV prevalence surveys. *Int. J. Epidemiol.* **2008**, *37*, 536–546. [CrossRef] [PubMed]
17. Hiller, T.; Poppelreuther, S.; Stubenrauch, F.; Iftner, T. Comparative analysis of 19 genital human papillomavirus types with regard to p53 degradation, immortalization, phylogeny, and epidemiologic risk classification. *Cancer Epidemiol. Biomark. Prev.* **2006**, *15*, 1262–1267. [CrossRef] [PubMed]
18. Mesplede, T.; Gagnon, D.; Bergeron-Labrecque, F.; Azar, I.; Senechal, H.; Coutlee, F.; Archambault, J. P53 degradation activity, expression, and subcellular localization of E6 proteins from 29 human papillomavirus genotypes. *J. Virol.* **2012**, *86*, 94–107. [CrossRef] [PubMed]
19. Zimmermann, H.; Degenkolbe, R.; Bernard, H.U.; O'Connor, M.J. The human papillomavirus type 16 e6 oncoprotein can down-regulate p53 activity by targeting the transcriptional coactivator cbp/p300. *J. Virol.* **1999**, *73*, 6209–6219. [PubMed]
20. Fiedler, M.; Muller-Holzner, E.; Viertler, H.P.; Widschwendter, A.; Laich, A.; Pfister, G.; Spoden, G.A.; Jansen-Durr, P.; Zwerschke, W. High level HPV-16 E7 oncoprotein expression correlates with reduced prb-levels in cervical biopsies. *FASEB J.* **2004**, *18*, 1120–1122. [CrossRef] [PubMed]
21. Frolov, M.V.; Dyson, N.J. Molecular mechanisms of e2f-dependent activation and prb-mediated repression. *J. Cell Sci.* **2004**, *117*, 2173–2181. [CrossRef] [PubMed]
22. Franco, E.L.; Spence, A.R. Commentary: Smoking and human papillomavirus infection: The pursuit of credibility for an epidemiologic association. *Int. J. Epidemiol.* **2008**, *37*, 547–548. [CrossRef] [PubMed]
23. Sopori, M. Effects of cigarette smoke on the immune system. *Nat. Rev. Immunol.* **2002**, *2*, 372–377. [CrossRef] [PubMed]
24. Eldridge, R.C.; Pawlita, M.; Wilson, L.; Castle, P.E.; Waterboer, T.; Gravitt, P.E.; Schiffman, M.; Wentzensen, N. Smoking and subsequent human papillomavirus infection: A mediation analysis. *Ann. Epidemiol.* **2017**, *27*, 724–730.e721. [CrossRef] [PubMed]
25. Schmitt, M.; Bravo, I.G.; Snijders, P.J.; Gissmann, L.; Pawlita, M.; Waterboer, T. Bead-based multiplex genotyping of human papillomaviruses. *J. Clin. Microbiol.* **2006**, *44*, 504–512. [CrossRef] [PubMed]
26. Schmitt, M.; Dondog, B.; Waterboer, T.; Pawlita, M. Homogeneous amplification of genital human alpha papillomaviruses by pcr using novel broad-spectrum GP5+ and GP6+ primers. *J. Clin. Microbiol.* **2008**, *46*, 1050–1059. [CrossRef] [PubMed]

9

A Role of Sp1 Binding Motifs in Basal and Large T-Antigen-Induced Promoter Activities of Human Polyomavirus HPyV9 and its Variant UF-1

Ugo Moens [1,*], Xiaobo Song [2], Marijke Van Ghelue [3], John A. Lednicky [4] and Bernhard Ehlers [5]

1 Molecular Inflammation Research Group, Department of Medical Biology, Faculty of Health Sciences, University of Tromsø, 9037 Tromsø, Norway
2 Host Microbe Interaction Research Group, Department of Medical Biology, Faculty of Health Sciences, University of Tromsø, 9037 Tromsø, Norway; xiaobo.song@uit.no
3 Department of Medical Genetics, University Hospital Northern-Norway, 9038 Tromsø, Norway; marijke.van.ghelue@unn.no
4 Department of Environmental and Global Health, College of Public Health and Health Professions, University of Florida, Gainesville , FL 32603, USA; jlednicky@phhp.ufl.edu
5 Division 12, Measles, Mumps, Rubella and Viruses Affecting Immunocompromised Patients, Robert Koch Institute, 13353 Berlin, Germany; EhlersB@rki.de
* Correspondence: ugo.moens@uit.no

Abstract: Human polyomavirus 9 (HPyV9) was originally detected in the serum of a renal transplant patient. Seroepidemiological studies showed that ~20–50% of the human population have antibodies against this virus. HPyV9 has not yet been associated with any disease and little is known about the route of infection, transmission, host cell tropism, and genomic variability in circulating strains. Recently, the HPyV9 variant UF-1 with an eight base-pair deletion, a thirteen base-pair insertion and with point mutations, creating three putative Sp1 binding sites in the late promoter was isolated from an AIDS patient. Transient transfection studies with a luciferase reporter plasmid driven by HPyV9 or UF1 promoter demonstrated that UF1 early and late promoters were stronger than HPyV9 promoters in most cell lines, and that the UF1 late promoter was more potently activated by HPyV9 large T-antigen (LTAg). Mutation of two Sp1 motifs strongly reduced trans-activation of the late UF1 promoter by HPyV9 LTAg in HeLa cells. In conclusion, the mutations in the UF1 late promoter seem to strengthen its activity and its response to stimulation by HPyV9 LTAg in certain cells. It remains to be investigated whether these promoter changes have an influence on virus replication and affect the possible pathogenic properties of the virus.

Keywords: large T antigen; luciferase; mutation; non-coding control region; Sp1

---

1. Introduction

Polyomaviruses, a family of small, double-stranded DNA viruses that infects mammals, birds, and fish, while polyomavirus-like sequences have also been found in reptiles and invertebrates [1–3]. In 1971, the first two human polyomaviruses BK polyomavirus (BKPyV) and JC polyomavirus (JCPyV) were described and named after the initials of the patients from which they were isolated [4,5]. It was not until 2007 that a third human polyomavirus (Karolinska Institute polyomavirus, KIPyV) was isolated [6], and since then, another 11 not previously reported polyomaviruses have been detected in human samples. These are: Washington University polyomavirus (WUPyV), Merkel cell polyomavirus (MCPyV), HPyV6, HPyV7, *Trichodysplasia spinulosa* polyomavirus (TSPyV), HPyV9, Malawi polyomavirus (MWPyV), Saint Louis polyomavirus (STLPyV), HPyV12, New Jersey polyomavirus (NJPyV), and Lyon IARC polyomavirus (LiPyV) [1,7]. All human polyomaviruses share high sequence similarity in their coding region, whereas the non-coding control region (NCCR) that encompassed the origin of replication and the transcription control sequences for the early and late genes is highly variable [1]. Also, between isolates of a human polyomavirus species, hypervariability in the NCCR is observed. The changes in the NCCR of specific human polyomavirus affect the life cycle of the virus and may be associated with the pathogenic properties of the virus, as was convincingly shown for BKPyV and JCPyV [8,9]. Major mutations and rearrangements in the NCCR have been reported in other human polyomaviruses [10], but the effects on the properties of the virus have been scarcely examined [11,12].

HPyV9 was first detected in the serum from a renal transplant patient under immunosuppressive treatment [13], and shortly after from the facial surface of a Merkel cell carcinoma patient [14]. Seroprevalence of HPyV9 varies between ~6–39% in children (<1 month-11 years), ~18–47% in healthy adults (>16 years of age), and reaches 70% in individuals aged 80 years and older [15–25]. Whether different HPyV9 variants are circulating in the human population has scarcely been investigated. The genomes of both isolates differed by just two nucleotides. Recently, a third HPyV9 variant was isolated from peripheral blood monocytes of an AIDS patient [26]. This variant, designated HPyV9 UF-1 (hereafter referred to as UF-1), differs by two amino acids in the VP2/3 proteins relative to the originally isolated HPyV9 genomes [13,14], and has several changes in its non-coding control region (NCCR). The biological implications of these mutations on the viral life cycle are not known as permissive cell systems for virus propagation are lacking [25,26].

In an effort to determine the biological importance of the differences in NCCR between HPyV9 and UF-1, we examined their relative early and late promoter strength in seven different human cell lines and studied the effect of HPyV9 LTAg on the early and late promoter activities. The possible role of additional Sp1 sites in UF-1 NCCR when compared to the HPyV9 NCCR in promoter activity was also investigated.

2. Results

2.1. The Basal UF-1 Early and Late Promoter Activity is Higher than that of HPyV9 in Most Cell Lines Tested

The NCCR of the originally described HPyV9 isolates [13,14] is identical, while the UF-1 variant contains several point mutations, a 13-base pair insertion and an 8-base pair deletion as compared to HPyV9 (Figure 1).

HPyV9 DNA can be found in urine, throat swabs, skin, and blood from healthy individuals [13,27–31], while viremia was observed in renal recipients and in patients with both kidney and pancreas transplantation [20]. However, little is known about the route of infection, cell tropism, transmission, and life cycle of HPyV9 and a permissive cell culture system is lacking. The effect of mutations in the NCCR on the properties of HPyV9 has not been investigated. In an effort to identify cell lines that may sustain the viral life cycle and to assess the effect of these differences on promoter activity, we compared the relative activities of the early and late HPyV9 promoter (hereafter referred to as H9-E and H9-L) with those of the early and late UF-1 promoter (UF1-E and UF1-L) in seven different human cell lines. The cell tropism of HPyV9 is not known, but because our previous study had shown that the HPyV9 promoter is relatively strong in BEL7402, HEK293, SK-N-BE, and SW480 cells, we selected these cells [12]. In addition, we chose one osteosarcoma cell line (U2OS), one human papillomavirus (HPV)-negative cervical cancer cell line (C33A), and the HPV18-positive cervical cancer cell line HeLa. The luciferase values that were obtained for all four promoters in HEK293 cells were 5–10× higher than the values for the corresponding promoters in C33A cells. The promoter activities in HeLa and U2OS were comparable, but up to 20-fold lower than in C33A cells. All four promoters had lowest activity in SK-N-BE, SW480, and BEL7402 cells (see Supplementary Materials). Differences in promoter strength may be partially explained by differences in transfection efficiency. Using an enhanced green fluorescent protein expression vector, it was estimated that all cells had a transfection efficiency of more than 75%, except for the BEL7402 (~50%) and C33A cells (~30%). For both HPyV9 and UF1, their late promoter was stronger (2- to 7-fold) than their early promoter in most of the cell lines that were tested. In HeLa and U2OS cells, the H9-E and H9-L promoters possessed similar activities and the same was true for the UF1-E and UF1-L promoters in BEL7402. In C33A cells, UF1-E was three-fold stronger than UF1-L (Appendix A). To compare the activity of all four promoters, we set the H9-E promoter activity arbitrarily as 100%. The UF1-E promoter was stronger as compared with the H9-E promoter in all cell lines, except in HeLa and U2OS cells where the basal HPyV9 promoter activity was, respectively, ~two- to four-fold higher (Figure 2). The H9-L promoter activity was higher than the H9-E promoter activity, except in U2OS (Figure 2) where there was no significant difference. For the late promoter activity, there was a tendency that the H9-L and UF1-L were comparable, except for C33A and HeLa cells, in which the H9-L promoter was on average approximately four-fold stronger, while UF1-L was stronger in BEL7402 cells (Figure 2). The UF1-E promoter was stronger than the UF1-L promoter in C33A, HeLa, and SK-N-BE cells, but weaker in HEK293, SW480, and U2OS cells (Figure 2). UF1-E and UF1-L promoter activity was comparable in BEL7402 cells (Figure 2).

```
HPyV9-UF1      GGCTCTGCAAAAAGTAAAATAAGTCTTACTACCTGAGAATCAAGTTAATTAAGTTTCAAA
HPyV9          GGCTCTGCAAAAAGTAAAATAAGTCTTACTACCTGAGAATCAAGTTAATTAAGTTTCAAA
IPPyV          GGCTCTGCAAAAAGTAAAATAAGTCTTACTACCTGAGAATCAAGTTAATTAAGTTTCAAA
               ************************************************************

HPyV9-UF1      TAGGTGGAGAACTTCTTACCTTAATGAGTTTTGGAAAAGCCTCCAAAGCCTCTCTCTTTG
HPyV9          TAGGTGGAGAACTTCTTACCTTAATGAGTTTTGGAAAAGCCTCCAAAGCCTCTCTCTTTG
IPPyV          TAGGTGGAGAACTTCTTACCTTAATGAGTTTTGGAAAAGCCTCCAAAGCCTCTCTCTTTG
               ************************************************************

HPyV9-UF1      TTGCAACAAGAGAAGGAGGCAAGGGGCCCCTGGCCTCTTATACTATCAAAAAAAAACCTG
HPyV9          TTGCAACAAGAGAAGGAGGCAAGGGGCCCCTGGCCTCTTATACTATCAAAAAAAAACCTG
IPPyV          TTGCAACAAGAGAAGGAGGCAAGGGGCCCCTGGCCTCTTATACTATCAAAAAAAAACCTG
               ************************************************************

HPyV9-UF1      TGTTGCCATAGTGATTTTGAGTGACAGGGAGATAACTTTGCAAATTGAGGAACTAAATGA
HPyV9          TGTTGCCATAGTGATTTTGAGTGACAGGGAGATAACTTTGCAAATTGAGGAACTAAATGA
IPPyV          TGTTGCCATAGTGATTTTGAGTGACAGGGAGATAACTTTGCAAATTGAGGAACTAAATGA
               ************************************************************
                                                  AML-1a    AML-1a
HPyV9-UF1      CAGGTTATTTTTGCAGAATCAACTCTAGAGGAAG--------ACTGTGGTATCTACCTGC
HPyV9          CAGGTTATTTTTGCAGAATCAACTCTAGAGGAAACTGTGGTATCTGTGGTATCTACCTGC
IPPyV          CAGGTTATTTTTGCAGAATCAACTCTAGAGGAAACTGTGGTATCTGTGGTATCTACCTGC
               *********************************          *****************
                         Sp1
HPyV9-UF1      TCCACTTGACCGCCCGGGGACTTTTTGGTGACATTCCGGTGTGTGACAAACAAGTGCATT
HPyV9          TCCACTTGACTGCCTGGGGACTTTTTGGTGACATTCCGGTGTGTGACAAACAAGTGCATT
IPPyV          TCCACTTGACTGCCTGGGGACTTTTTGGTGACATTCCGGTGTGTGACAAACAAGTGCATT
               ********** *** *********************************************

HPyV9-UF1      GCGGTTTCCGTTGCTAGGTGGCGCTTAGCAACCCCTCTAGATACCAGTAGCTAAGGGGGA
HPyV9          GCGGTTTCCGTTGCTAGGTGGCGCTTAGCAACCCCTCTAGATACCAGTAGCTAAGGGGGA
IPPyV          GCGGTTTCCGTTGCTAGGTGGCGCTTAGCAACCCCTCTAGATACCAGTAGCTAAGGGGGA
               ************************************************************
                                                         Sp1
HPyV9-UF1      AGTGAAATCATAGCAACCTAGAGGAAGCTCATTTCATCAGCGCCGCCCAGAAGCCCGCCT
HPyV9          AGTGAAATCATTTTGCCATCTAAGG-------------TCAACAGAGAAGAAGCCCGCCT
IPPyV          AGTGAAATCATTTTGCCATCTAAGG-------------TCAACAGAGAAGAAGCCCGCCT
               ***********     * *  * *                  * *   ************
                        Sp1
HPyV9-UF1      TTTTTAAACCCGCCCATTTTGAACTTGGGTAAAGT
HPyV9          AAACTAAACGCGCCAATTTTGAACTTGGGTAAGAT
IPPyV          AAACTAAACGCGCCAATTTTGAACTTGGGTAAGAT
                   ***** **** ********************
```

Figure 1. Alignment of the non-coding control region (NCCR) of human polyomavirus 9 (HPyV9), IPPyV and UF-1. IPPyV was the name given to the HPyV9 variant described by Sauvage et al. [14]. Indels are indicated by dashes, asterisk show identical nucleotides. The putative Sp1 and AML-1a motifs are shown in a frame, while the putative HPyV9 LTAg-binding motifs are underlined. The NCCR is depicted in an early-to-late direction.

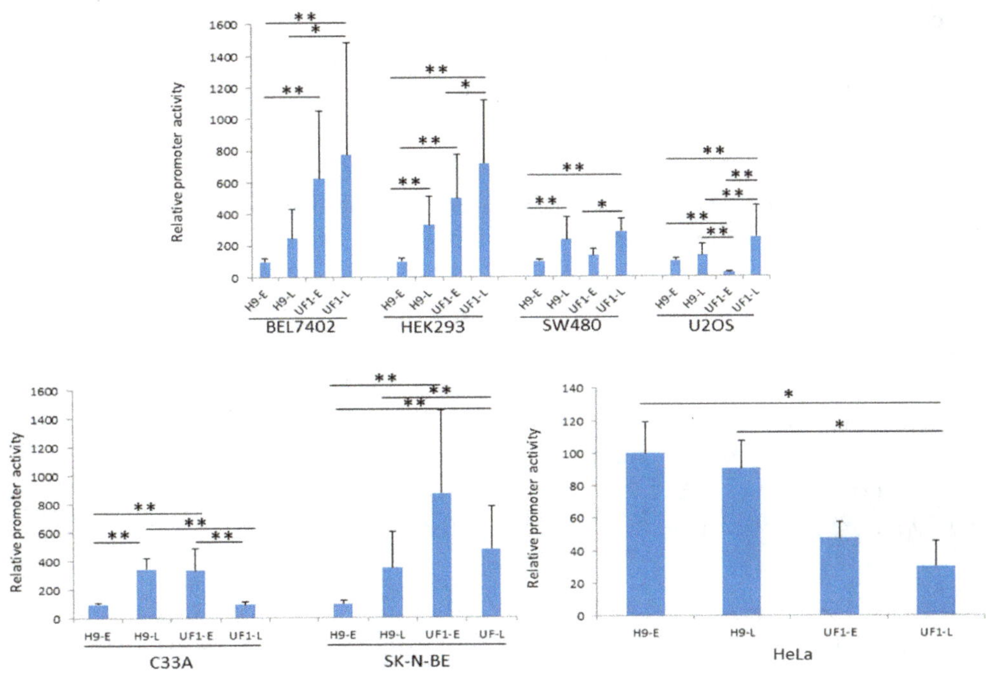

Figure 2. Comparison of the early and late promoter activity of HPyV9 and UF-1 in different human cell lines. Cells that were ~70% confluent were transfected with 400 ng of a reporter plasmid containing the *luciferase* gene driven by either the early or the late promoter of HPyV9 (respectively UF-1). Cells were harvest the next day and luciferase activity was measured and corrected for protein concentration in the sample. Each experiment was repeated three to five times, each time with three independent parallels and the results are the average of 9–15 values ± standard deviation. The activity of the HPyV9 early promoter (H9-E) was arbitrary set as 100. * $p < 0.05$; ** $p < 0.01$.

2.2. The UF1 Late Promoter is more Potently Activated by HPyV9 LTAg than the HPyV9 Late Promoter

Next, we examined the effect of HPyV9 LTAg on the early and late promoter activities of HPyV9 and UF-1. We decided to test in BEL7402, HEK293, and HeLa cells because some of the most obvious differences in early and/or late promoter activities between HPyV9 and UF-1 were observed in these cells. Cells were transfected with HPyV9 LTAg expression plasmid (the UF-1 LTAg amino acid sequence is completely identical to HPyV9) or with empty vector. Expression of HPyV9 LTAg in the three cell lines was confirmed by western blotting (Figure 3).

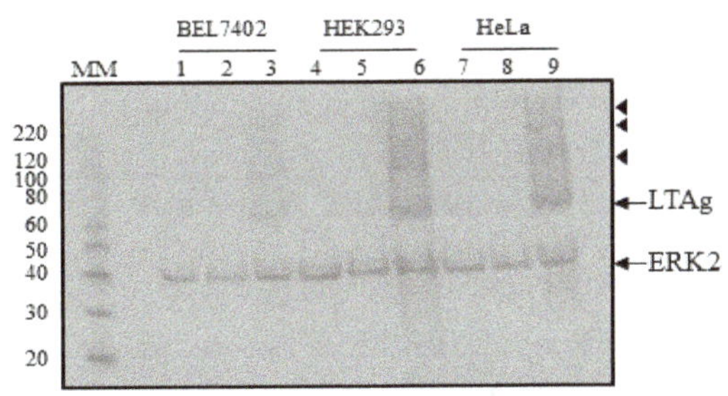

Figure 3. Detection of ectopic expressed HPyV9 LTAg in BEL7402, HEK293, and HeLa cells. Lanes 1–3: lysates of BEL7402 cells, lanes 4–6: lysates of HEK293 cells; lanes 7–9: lysates of HeLa cells. Lanes 1, 4, and 7: mock transfected cells; lanes 2, 5, and 8: cells transfected with empty vector pcDNA3.1(+); lanes 3, 6, and 9: cells transfected with expression plasmid for hemagglutinin (HA) tagged LTAg of HPyV9. Western blots were performed with anti-HA and anti-ERK2 antibodies. The molecular mass (in kDa) of the markers (lane MM) is indicated. The arrows indicate the bands corresponding to HPyV9, LTAg, and ERK2, respectively, while the bands marked with arrowheads may represent post-translationally modified HPyV9 LTAg.

Cells were co-transfected with (i) the reporter plasmid containing the luciferase gene under control of the early or late promoter of, respectively, HPyV9 or UF-1, and (ii) empty expression vector pcDNA3.1(+) or HPyV9 LTAg vector. Dose-dependent transfection studies with increasing amounts of HPyV9 LTAg (0–1200 ng) or corresponding empty vector showed that co-transfection with 400 ng plasmid DNA/10^5 cells was optimal (Ugo Moens, University of Tromsø, Norway, unpublished results, 2017). Higher concentrations hampered the HPyV9 promoters probably because the empty and HPyV9 LTAg expression vectors contain the strong cytomegalovirus promoter that competes out transcription factors required for the HPyV9 promoter. HPyV9 LTAg stimulated both H9-E and UF1-E promoters in all cell lines tested, although the luciferase expression levels differed (Figure 4A–C). H9-L and UF1-L promoters were also activated by HPyV9 LTAg, except in HEK293 cells were a 50% reduction was observed for H9-L (Figure 4B). Strongest activation was measured for the UF1-L promoter in HeLa cells (~36-fold increase; Figure 4C), while a 1.3- to 6.5-fold HPyV9 LTAg-mediated increase was observed for the HPyV9 and UF-1 promoters in the other experiments. No significant HPyV9 LTAg-mediated induction was observed for H9-L in BEL7402 cells (Figure 4A), while for all other promoters in all cell lines tested HPyV9 LTAg significantly affected promoter activity.

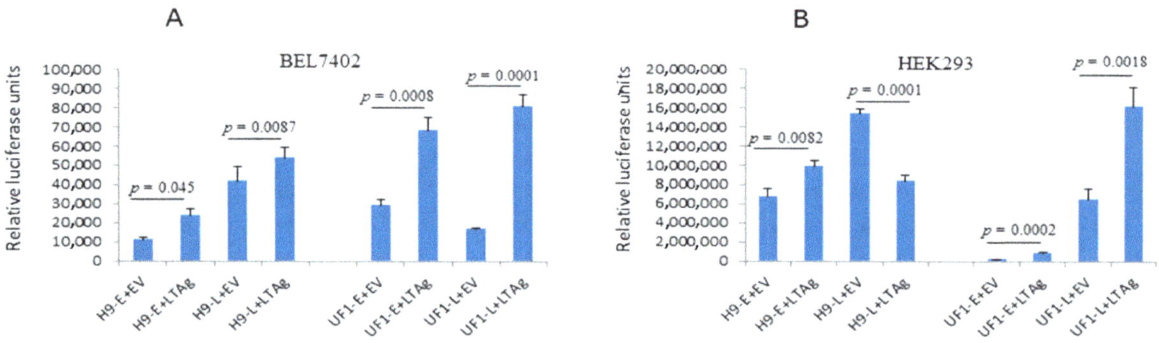

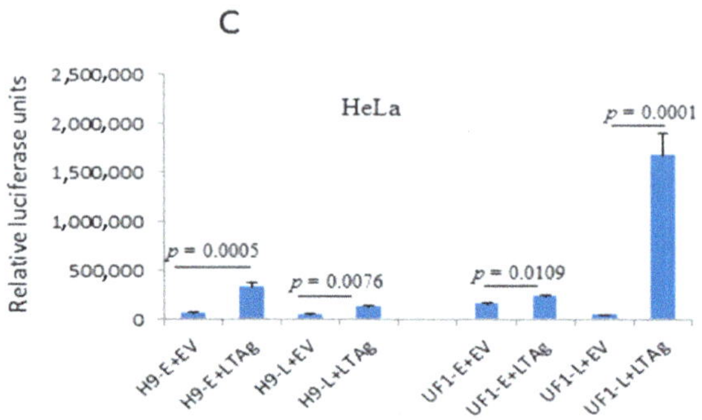

D

promoter	BEL7402	HEK293	HeLa
H9-E	2.7 (2.1-6.2)	1.9 (1.2-3.1)	6.5 (4.3-9.0)
UF1-E	1.6 (0.8-2.3)	4.5 (2.3-7.1)	1.4 (1.0-1.8)
H9-L	1.3 (1.1-1.4)	0.5 (0.1-0.8)	2.5 (1.7-2.9)
UF1-L	5.2 (4.5-6.2)	1.6 (1.0-2.5)	35.7 (22.9-50.9)

Figure 4. Effect of HPyV9 LTAg on the early and late promoters of HPyV9 and UF1. BEL7402 (panel **A**), HEK293 (panel **B**), and HeLa cells (panel **C**) were co-transfected with the luciferase reporter plasmid containing the H9-E, H9-L, UF1-E, or UF1-L promoter, and empty vector pcDNA3.1(+) or expression plasmid for HPyV9 LTAg. Luciferase values were measured and corrected for the protein concentration. Each experiment was repeated three to five times (Appendix A). A representative experiment for each promoter and each cell line is shown. Each bar represents the average of three independent parallels ± standard deviation. (Panel **D**) Fold induction of the promoter by HPyV9 LTAg. The average (range) of three to four independent experiments is given.

2.3. *Sp1 Sites in the Late UF1 Promoter and HPyV9 LTAg*

One of the differences between the UF1 and HPyV9 promoter is the presence of three Sp1 sites in the UF1 promoter, two of which are in the late promoter region (Figure 1). Previous studies had shown that SV40 LTAg can interact with Sp1 and stimulate transcription from Sp1 site-containing promoters [32,33]. Because Sp1 is constitutively expressed in HeLa cells [34] and the UF1-L promoter activity was strongest (on average 35-fold) stimulated by HPyV9 LTAg in these cells (Figure 4C,D), we examined the effect of HPyV9 LTAg on the activity of the UF1-E promoter and the UF1-L promoter in which the two distal Sp1 sites were mutated in HeLa cells. Destroying these two Sp1 motifs did not have an effect on the early promoter activity in the absence of HPyV9 LTAg, because the Sp1-mutated early promoter's activity did not change (1.1 ± 0.4-fold; $n = 4$) when compared with that of the non-mutated promoter. However, the activity of the mutated late promoter was 2-fold (±0.3; $n = 4$) higher than the non-mutated promoter. Disrupting the two Sp1 sites reduced HPyV9 LTAg activation of the mutated late promoter 7.3-fold ± 1.5 ($n = 4$) as compared to the wild-type UF1 late promoter, while transactivation of the mutated UF1-E promoter was only slightly (1.1-fold ± 0.7) reduced (Figure 5).

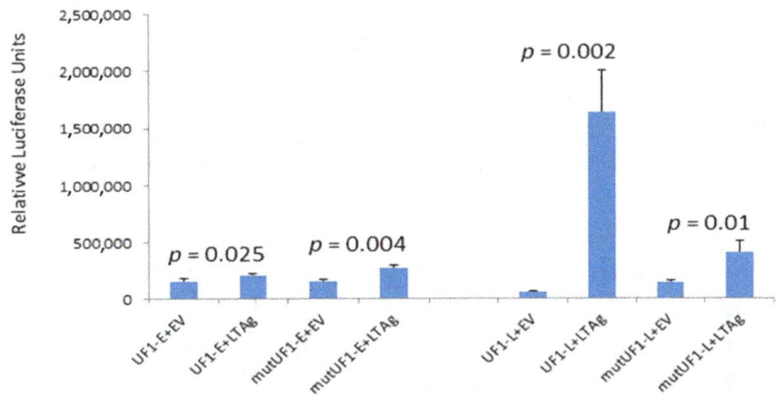

plasmid	exp1	exp2	exp3	exp4
UF1-E	2.9	1.3	1.0	1.0
mutUF1-E	2.7 (1.1xreduced)	1.8 (0.7xreduced)	1.9 (1.9xenhanced)	0.5 (2xreduced)
UF1-L	44	26	29	20
mutUF1-L	6.1 (7.2xreduced)	2.8 (9.3xreduced)	5.2 (5.6xreduced)	2.8 (7.1xreduced)

Figure 5. Effect of mutation of putative Sp1 motifs on HPyV9 LTAg-induced UF1-E and UF1-L promoter activity. HeLa cells were co-transfected with the luciferase reporter plasmid with wild-type UF1-E (respectively UF1-L) promoter or with mutated UF1-E (respectively mutated UF1-L) promoter and empty expression vector pcDNA3.1(+) (EV) or HPyV9 LTAg expression vector. Luciferase activity was measured and corrected for protein concentration. Top panel: a representative experiment is shown. The bars show the average of three parallels ± SD. Each experiment was performed 4 times (Appendix A). The table summarizes the promoter activity in the presence of HPyV9 LTAg (shown as fold induction) of the four experiments.

3. Discussions

HPyV9 is one of the 14 polyomaviruses that have been detected in humans so far, but little is known about the biology of this virus. Antibodies against HPyV9 can be detected in ~20–50% of the adult population, while viral DNA is found with low frequency in urine, blood, skin, and throat swabs (reviewed in [31]). Attempts to propagate HPyV9 in cell culture have been unsuccessful [25,26]. In this study, we compared the early and late promoter activity of the originally described HPyV9 variant with that of the UF-1 isolate in seven different human cell lines. We had previously monitored the HPyV9 early and late promoter in ten different cell lines [12], and selected BEL7402, HEK293, SK-N-BE, and SW480 because their promoter activity was highest in these cells. In addition, we have also tested the human cervical cancer cell lines C33A and HeLa, and human osteosarcoma U2OS cells. The UF-1 early and late promoters were stronger than the HPyV9 promoters in most cell lines, except for HeLa cells where both UF-1 early and late promoter activity was lower than those of HPyV9, while the UF-1 early promoter was weaker in U2OS and the UF-1 late was weaker in C33A and SK-N-BE cells. The molecular basis for the difference in activity between the HPyV9 and UF-1 promoters in a specific cell line and between the different cells remains elusive, but the most obvious reason is the difference in the repertoire of transcription factors in each cell line.

Our study is the first to show that HPyV9 LTAg, in agreement with LTAg of other human polyomaviruses, influences early and late promoter activities. The amplitude of promoter activity stimulation by HPyV9 LTAg was different, with the late UF-1 promoter more inducible by HPyV9 LTAg, despite conservation of the putative LTAg-binding sites in the HPyV9 and UF-1 NCCR (Figure 1). The LTAg of polyomaviruses not only affects transcription by direct binding to these motifs in the promoter, but also through interaction with cellular transcription factors [1]. Hence, the differences in HPyV9 LTAg-induced promoter activity can be explained by distinct transcription factor binding sites. The UF-1 promoter possesses three putative Sp1 sites (5′-CCGCCC-3′), two of which are located in its late promoter region (Figure 1). The 5′-AGCTCATTTCATC-3′ deletion in HPyV9 does not seem to encompass putative binding sites using the PROMO transcription factor binding site prediction algorithm [35,36], while nucleotides spanning this region contain mutations that create putative binding motifs for STAT5A and Elk1. The possible contribution of these transcription factors in the regulation of HPyV9 promoter activity was not tested. Sp1 is known to be ubiquitously expressed at medium to high levels in all of the mammalian cells and tissues that have been tested [37,38]. LTAg of SV40 can interact with Sp1 and stimulate transcription of Sp1-containing promoters [32,33]. Accordingly, HPyV9 LTAg may more potently stimulate the UF1 promoter than the HPyV9 promoter through the additional Sp1 sites. However, the Sp1 family member Sp3 binds to the Sp1 motif with an affinity and specificity that are comparable with Sp1. Like Sp1, Sp3 is ubiquitously expressed, but in contrast to Sp1, Sp3 acts as a repressor [39,40]. Mutation of the two distal Sp1 sites did not affect the activity of the UF1-E promoter, but stimulated the UF1-L promoter in the absence of HPyV9 LTAg. However, in the presence of HPyV9 LTAg, induction of UF1-L was severely impaired. SV40 LTAg physically interacts with Sp1 and stimulates transcription of Sp1-containing promoters [32,33]. It is not known whether Sp3 can bind SV40 LTAg, but Sp1 and Sp3 share only ~40% amino acid identity. Whether HPyV9 LTAg binds Sp1 has to be proven, but supposing that HPyV9 LTAg does, the following scenario can be imagined. In the absence of HPyV9 LTAg, Sp1, and Sp3 may compete for the same motifs, and, depending on which of the proteins binds, stimulate (Sp1) or repress (Sp3) the promoter. HPyV9 LTAg, through its association with Sp1 may help to recruit Sp1 to the promoter, but not Sp3 assuming that Sp3 cannot interact with HPyV9 LTAg. Thus, intact Sp1 sites may bind Sp1 or Sp3, while HPyV9 LTAg may help to recruit Sp1 but not Sp3 to these sites. The deletion of the Sp1 site prevents Sp1 and Sp3 binding, but also LTAg-mediated recruitment of Sp1. The fact that the early and late promoter of HPyV9 was also induced by HPyV9 LTAg in HeLa cells suggests that other, Sp1-independent pathways, are operational. The UF1 promoter lacks a putative AML-1a (=RUNX1) site. This transcription factor is highly expressed in cells from myeloid and lymphoid tissue, while low transcript levels are found in HEK293, HeLa, U2OS, colon cancer, and neuroblastoma cells [37,38]. Because our study did not include myeloid and lymphoid cells, but rather cells with no or low AMP-1a expression levels, we did not investigate the role of this transcription factor. However, we cannot rule out a role of this transcription factor in HPyV9 promoter activity. Finally, LTAg stimulated the HPyV9 and UF-1 early and late promoters in all cell lines, except in HEK293 cells where the H9-L promoter was inhibited in the presence of LTAg. Further studies are required to unveil the mechanism of LTAg on the HPyV9 and UF-1 promoters in the different cell lines.

The natural host cells for HPyV9 remain unknown. In a previous study, we measured VP1 mRNA levels in Vero, RH, COS-7, HEK293 (all kidney-derived), BJAB (mouse hybridoma), HeLa, Huh-7 (hepatocellular carcinoma), and NIH3T3 (mouse fibroblast) cells transfected with complete HPyV9 genome [25]. VP1 mRNA levels increased over time, with highest copy number quantified in Vero cells and minimal increase in NIH3T3 and HeLa cells. Despite early (luciferase reporter assays in this study) and late viral promoter activity (this study and [25]), none of the cell lines were able to support a complete viral life cycle. These findings indicate that the presence of actively transcribed HPyV9 genome is not sufficient to generate infectious virus particles.

Western blot detection of ectopic expression of HPyV9 LTAg resulted in several bands corresponding to a molecular mass higher than the predicted 78 kDa. Toptan and colleagues, using a monoclonal antibody cross-reacting with the J-domain common in human polyomavirus LTAg and small T-antigen [41], detected bands corresponding to >78 kDa in HEK293 cells transfected with a HPyV9 LTAg expression plasmid [42]. These bands may represent post-translationally modified HPyV9 LTAg, as was shown for LTAg of other polyomaviruses [43–45].

In conclusion, the basal promoter activity of the HPyV9 variant UF-1, which was isolated from an AIDS patient, is stronger and more potently induced by LTAg in most of the cell lines we examined when compared with the original HPyV9 isolate. This suggests that the mutations in the UF1 NCCR may affect the life cycle of the virus. Further screening of specimens from healthy and diseased individuals is necessary to determine the host cell(s) for HPyV9, which may allow for establishing permissive cell cultures for virus propagation and unveiling a possible association with diseases.

4. Materials and Methods

4.1. Cell Lines

The human cell lines C33A (human papilloma virus negative cervical cancer), HEK293 (human embryonal kidney cells), HeLa (human papilloma virus positive cervical cancer), SK-N-BE (neuroblastoma), SW480 (colorectal adenocarcinoma), and U2OS (osteosarcoma) were maintained in Dulbecco's Modified Eagle's Medium (Sigma D5796, St. Louis, MO, USA), country) with 10% foetal bovine serum (Life Technologies, cat. No. 10500-064, Carlsbad, CA, USA). BEL-7402 (human hepatocellular carcinoma) was grown in DMEM with low glucose (Sigma D5546, St. Louis, MO, USA) [12]. All of the cells were kept at 37 °C in a humidified CO_2 incubator.

4.2. Plasmids

The luciferase reporter plasmids containing the HPyV9 early or late promoter have been previously described [12]. The corresponding plasmids with the HPyV9 UF-1 early or late promoter were synthesized by GenScript as follows (Piscataway, NJ, USA). The complete NCCR was synthesized with cohesive *Hind*III at both ends and subsequently cloned in the *Hind*III site of pUC58. The NCCR was then excised and cloned in either orientation in the *Hind*III site of pGL3-basic. The empty expression vector pcDNA3.1(+) was purchased from Thermo Fisher Scientific (Waltham, MA, USA). The HPyV9 large T antigen expression vector was made by GenScript by cloning the synthesized cDNA into the *Kpn*I-*Apa*I sites of their pcDNA3.1+C-HA vector. Expressed HPyV9 LTAg has a C-terminal linked HA tag. The luciferase reporter plasmids with the UF1-E or UF1-L promoter with mutated Sp1 sites were generated by PCR-based site directed mutagenesis using primers mutSp1_UF1_fwd: 5′-GCTCATTTCATCAGCGCAGAGAAGAAGCCCGCCTTTTTAAACGCGCCAATTTTGAACTTGG-3′ and mutSp1_UF1_rev: 5′-CCAAGTTCAAAATTGGCGCGTTTAAAAAG GCGGGCTTCTTCTC TGCGCTGATGAAATGAGC-3′. These primers result in the mutation of the two Sp1 sites proximal to the late coding region (Figure 1). Mutations were confirmed by sequencing.

4.3. Transfection

Cells were seeded into twelve-well cell culture plates (Falcon cat. No. 353072, Corning Incorporated, Corning, NY, USA) and transfected the following day. Cells were approximately 70% confluent on the day of transfection. JetPrime (Polyplus-transfection, Illkirch, France) was used as transfection reagent. Cells were transfected with 400 ng luciferase plasmid according to the manufacturer's instructions. In the experiments where the effect of HPyV9 LTAg on promoter activity was monitored, co-transfections were performed using 400 ng empty vector (pcDNA3.1(+)) or 400 ng HPyV9 LTAg expression plasmid.

4.4. Luciferase Assay

Cells were lysed 24 h post-transfection in 100 μL Luciferase Assay Tropix Lysis solution (Applied Biosystems, Foster City, CA, USA) with DTT added to a final concentration of 0.5 mM (Sigma, St. Louis, MO, USA) freshly added. Cells were scraped and transferred to Eppendorf tubes, followed by 3 min centrifugation at 12,000× *g* at room temperature in a Microfuge 22R centrifuge (Beckman Coulter, Bea, CA, USA). A 20 μL aliquot of each supernatant was subsequently transferred to 96-well microtiter plate and 50 μL luciferase buffer (Promega, Madison, WI, USA) was added. Light units were measured using CLARIOstar monochromator (520–620 nm) microplate reader (BMG Labtech GmbH, Ortenberg, Germany).

4.5. Protein Concentration Assay

The luciferase value of each sample was corrected relative to its protein concentration. Protein concentrations were determined using the Protein Quantification Assay from Macherey-Nagel (Düren, Germany) according to the manufacturer's instructions with OD570 measured using the CLARIOstar monochromator microplate reader.

4.6. Western Blotting

Cells were plated out in 6-well plates and the next day transfected with 2 μg DNA (either pcDNA3.1 or pcDNA3.1-LTAg) using JetPrime. Twenty-four hours after transfection, the cells were washed briefly with phosphate-buffered saline (PBS, Biochrom GmbH) and harvested in NuPage LDS sample buffer (Invitrogen, Carlsbad, CA, USA) with 100 mM DTT. The samples were sonicated and heated for 10 min at 70 °C. Proteins were separated on NuPAGE™ Novex™ 4–12% Bis-Tris Protein Gels (Thermo Fisher Scientific, Waltham, MA, USA) and transferred onto a 0.45 μm PVDF Membrane (Merck Life Science AS, Darmstadt, Germany). The membrane was blocked in TBST (Tris-buffered saline with 0.1% Tween-20; Sigma, St. Louis, MO, USA) containing 5% (w/v) skim milk powder. Incubation with the primary anti-HA antibody (SC-805 Santa Cruz Biotechnology, Dallas TX, USA) or anti-ERK2 antibody (SC-154) was overnight at 4 °C in blocking buffer. Following three washes in TBST, the membrane was incubated with the polyclonal swine anti-rabbit secondary antibody that was conjugated with alkaline phosphatase (D0306, Dako, Santa Clara, CA, USA) solution for 1 h at room temperature. After four washes, detection and visualization were performed using CDP-Star chemiluminescent (C0712, Sigma, St. Louis, MO, USA) and the ImageQuant LAS 4000 imager (GE Healthcare, Pittsburgh, PA, USA). MagicMark™ XP Western Protein Standard (Thermo Fisher Scientific, Waltham, MA, USA) was used to estimate the molecular mass of the detected proteins.

4.7. Statistical Analysis

The *t*-test was employed to determine statistical differences between the promoter activities.

Acknowledgments: This work was supported by grants from Raagholt Stiftelse and Olav and Erna Aakre foundation against cancer (A65242). The University of Tromsø covered the costs to publish in open access.

Author Contributions: Ugo Moens, John A. Lednicky and Bernhard Ehlers conceived the study. Ugo Moens, Xiaobo Song, Marijke Van Ghelue and Bernhard Ehlers designed the experiments. Ugo Moens, Xiaobo Song and Marijke Van Ghelue performed the experiments. Ugo Moens, Xiaobo Song, Marijke Van Ghelue, John A. Lednicky and Bernhard Ehlers analyzed the data. Ugo Moens, Xiaobo Song, Marijke Van Ghelue, John A. Lednicky and Bernhard Ehlers wrote the paper.

Abbreviations

EV	empty vector
HPyV9	human polyomavirus 9
H9-E	early promoter HPyV9
H9-L	late promoter HPyV9
LTAg	large T-antigen
NCCR	non-coding control region
PCR	Polymerase Chain Reaction
UF1-E	early promoter UF1
UF1-L	late promoter UF1

Appendix A

Transfection results in seven human cell lines; effect of LTAg on HPyV9 and UF1 promoter in BEL7402, HEK293 and HeLa cells; effect of Sp1 mutation on LTAg-induced promoter activity.

References

1. Moens, U.; Krumbholz, A.; Ehlers, B.; Zell, R.; Johne, R.; Calvignac-Spencer, S.; Lauber, C. Biology, evolution, and medical importance of polyomaviruses: An update. *Infect. Genet. Evol.* **2017**, *54*, 18–38. [CrossRef] [PubMed]
2. Buck, C.B.; Van Doorslaer, K.; Peretti, A.; Geoghegan, E.M.; Tisza, M.J.; An, P.; Katz, J.P.; Pipas, J.M.; McBride, A.A.; Camus, A.C.; et al. The Ancient Evolutionary History of Polyomaviruses. *PLoS Pathog.* **2016**, *12*, e1005574. [CrossRef] [PubMed]
3. Moens, U.; Calvignac-Spencer, S.; Lauber, C.; Ramqvist, T.; Feltkamp, M.C.W.; Daugherty, M.D.; Verschoor, E.J.; Ehlers, B. ICTV Report Consortium. ICTV Virus Taxonomy Profile: Polyomaviridae. *J. Gen. Virol.* **2017**, *98*, 1159–1160. [PubMed]
4. Gardner, S.D.; Field, A.M.; Coleman, D.V.; Hulme, B. New human papovavirus (B.K.) isolated from urine after renal transplantation. *Lancet* **1971**, *1*, 1253–1257. [CrossRef]
5. Padgett, B.L.; Walker, D.L.; ZuRhein, G.M.; Eckroade, R.J.; Dessel, B.H. Cultivation of papova-like virus from human brain with progressive multifocal leucoencephalopathy. *Lancet* **1971**, *1*, 1257–1260. [CrossRef]
6. Allander, T.; Andreasson, K.; Gupta, S.; Bjerkner, A.; Bogdanovic, G.; Persson, M.A.; Dalianis, T.; Ramqvist, T.; Andersson, B. Identification of a third human polyomavirus. *J. Virol.* **2007**, *81*, 4130–4136. [CrossRef] [PubMed]
7. Gheit, T.; Dutta, S.; Oliver, J.; Robitaille, A.; Hampras, S.; Combes, J.D.; McKay-Chopin, S.; Le Calvez-Kelm, F.; Fenske, N.; Cherpelis, B.; et al. Isolation and characterization of a novel putative human polyomavirus. *Virology* **2017**, *506*, 45–54. [CrossRef] [PubMed]
8. Gosert, R.; Rinaldo, C.H.; Funk, G.A.; Egli, A.; Ramos, E.; Drachenberg, C.B.; Hirsch, H.H. Polyomavirus BK with rearranged noncoding control region emerge in vivo in renal transplant patients and increase viral replication and cytopathology. *J. Exp. Med.* **2008**, *205*, 841–852. [CrossRef] [PubMed]
9. Gosert, R.; Kardas, P.; Major, E.O.; Hirsch, H.H. Rearranged JC virus noncoding control regions found in progressive multifocal leukoencephalopathy patient samples increase virus early gene expression and replication rate. *J. Virol.* **2010**, *84*, 10448–10456. [CrossRef] [PubMed]
10. NCBI Resource Coordinators. Database Resources of the National Center for Biotechnology Information. *Nucleic Acids Res.* **2017**, *45*, D12–D17.
11. Song, X.; Van Ghelue, M.; Ludvigsen, M.; Nordbø, S.A.; Ehlers, B.; Moens, U. Characterization of the non-coding control region of polyomavirus KI isolated from nasopharyngeal samples from patients with respiratory symptoms or infection and from blood from healthy blood donors in Norway. *J. Gen. Virol.* **2016**, *97*, 1647–1657. [CrossRef] [PubMed]

12. Moens, U.; Van Ghelue, M.; Ludvigsen, M.; Korup-Schulz, S.; Ehlers, B. Early and late promoters of BKPyV, MCPyV, TSPyV, and HPyV12 are among the strongest of all known human polyomaviruses in 10 different cell lines. *J. Gen. Virol.* **2015**, *96*, 2293–2303. [CrossRef] [PubMed]
13. Scuda, N.; Hofmann, J.; Calvignac-Spencer, S.; Ruprecht, K.; Liman, P.; Kuhn, J.; Hengel, H.; Ehlers, B. A novel human polyomavirus closely related to the african green monkey-derived lymphotropic polyomavirus. *J. Virol.* **2011**, *85*, 4586–4590. [CrossRef] [PubMed]
14. Sauvage, V.; Foulongne, V.; Cheval, J.; Ar Gouilh, M.; Pariente, K.; Dereure, O.; Manuguerra, J.C.; Richardson, J.; Lecuit, M.; Burguière, A.; et al. Human polyomavirus related to African green monkey lymphotropic polyomavirus. *Emerg. Infect. Dis.* **2011**, *17*, 1364–1370. [CrossRef] [PubMed]
15. Trusch, F.; Klein, M.; Finsterbusch, T.; Kuhn, J.; Hofmann, J.; Ehlers, B. Seroprevalence of human polyomavirus 9 and cross-reactivity to African green monkey-derived lymphotropic polyomavirus. *J. Gen. Virol.* **2012**, *93*, 698–705. [CrossRef] [PubMed]
16. Karachaliou, M.; Chatzi, L.; Roumeliotaki, T.; Kampouri, M.; Kyriklaki, A.; Koutra, K.; Chalkiadaki, G.; Michel, A.; Stiaki, E.; Kogevinas, M.; et al. Common infections with polyomaviruses and herpesviruses and neuropsychological development at 4 years of age, the Rhea birth cohort in Crete, Greece. *J. Child Psychol. Psychiatry* **2016**, *57*, 1268–1276. [CrossRef] [PubMed]
17. Karachaliou, M.; Waterboer, T.; Casabonne, D.; Chalkiadaki, G.; Roumeliotaki, T.; Michel, A.; Stiakaki, E.; Chatzi, L.; Pawlita, M.; Kogevinas, M.; et al. The Natural History of Human Polyomaviruses and Herpesviruses in Early Life—The Rhea Birth Cohort in Greece. *Am. J. Epidemiol.* **2016**, *183*, 671–679. [CrossRef] [PubMed]
18. Gossai, A.; Waterboer, T.; Nelson, H.H.; Michel, A.; Willhauck-Fleckenstein, M.; Farzan, S.F.; Hoen, A.G.; Christensen, B.C.; Kelsey, K.T.; Marsit, C.J.; et al. Seroepidemiology of Human Polyomaviruses in a US Population. *Am. J. Epidemiol.* **2016**, *83*, 61–69. [CrossRef] [PubMed]
19. Sroller, V.; Hamsikova, E.; Ludvikova, V.; Musil, J.; Nemeckova, S.; Salakova, M. Seroprevalence rates of HPyV6, HPyV7, TSPyV, HPyV9, MWPyV and KIPyV polyomaviruses among the healthy blood donors. *J. Med. Virol.* **2016**, *88*, 1254–1261. [CrossRef] [PubMed]
20. Van der Meijden, E.; Wunderink, H.F.; van der Blij-de Brouwer, C.S.; Zaaijer, H.L.; Rotmans, J.I.; Bavinck, J.N.; Feltkamp, M.C. Human polyomavirus 9 infection in kidney transplant patients. *Emerg. Infect. Dis.* **2014**, *20*, 991–999. [CrossRef] [PubMed]
21. Van der Meijden, E.; Bialasiewicz, S.; Rockett, R.J.; Tozer, S.J.; Sloots, T.P.; Feltkamp, M.C. Different serologic behavior of MCPyV, TSPyV, HPyV6, HPyV7 and HPyV9 polyomaviruses found on the skin. *PLoS ONE* **2013**, *8*, e81078. [CrossRef] [PubMed]
22. Nicol, J.T.; Robinot, R.; Carpentier, A.; Carandina, G.; Mazzoni, E.; Tognon, M.; Touzé, A.; Coursaget, P. Age-specific seroprevalences of merkel cell polyomavirus, human polyomaviruses 6, 7, and 9, and trichodysplasia spinulosa-associated polyomavirus. *Clin. Vaccine Immunol.* **2013**, *20*, 363–368. [CrossRef] [PubMed]
23. Nicol, J.T.; Touzé, A.; Robinot, R.; Arnold, F.; Mazzoni, E.; Tognon, M.; Coursaget, P. Seroprevalence and cross-reactivity of human polyomavirus 9. *Emerg. Infect. Dis.* **2012**, *18*, 1329–1332. [CrossRef] [PubMed]
24. Robles, C.; Casabonne, D.; Benavente, Y.; Costas, L.; Gonzalez-Barca, E.; Aymerich, M.; Campo, E.; Tardon, A.; Jiménez-Moleón, J.J.; Castano-Vinyals, G.; et al. Seroreactivity against Merkel cell polyomavirus and other polyomaviruses in chronic lymphocytic leukaemia, the MCC-Spain study. *J. Gen. Virol.* **2015**, *96*, 2286–2292. [CrossRef] [PubMed]
25. Korup-Schulz, S.V.; Lucke, C.; Schmuck, R.; Moens, U.; Ehlers, B. Large T antigen variants of human polyomaviruses 9 and 12 and seroreactivity against their N-terminus. *J. Gen. Virol.* **2017**, *98*, 704–714. [CrossRef] [PubMed]
26. Lednicky, J.A.; Butel, J.S.; Luetke, M.C.; Loeb, J.C. Complete genomic sequence of a new Human polyomavirus 9 strain with an altered noncoding control region. *Virus Genes* **2014**, *49*, 490–492. [CrossRef] [PubMed]
27. Siebrasse, E.A.; Bauer, I.; Holtz, L.R.; Le, B.M.; Lassa-Claxton, S.; Canter, C.; Hmiel, P.; Shenoy, S.; Sweet, S.; Turmelle, Y.; et al. Human polyomaviruses in children undergoing transplantation, United States, 2008–2010. *Emerg. Infect. Dis.* **2012**, *18*, 1676–1679. [CrossRef] [PubMed]
28. Csoma, E.; Sapy, T.; Meszaros, B.; Gergely, L. Novel human polyomaviruses in pregnancy: Higher prevalence of BKPyV, but no WUPyV, KIPyV and HPyV9. *J. Clin. Virol.* **2012**, *55*, 262–265. [CrossRef] [PubMed]

29. Hampras, S.S.; Giuliano, A.R.; Lin, H.Y.; Fisher, K.J.; Abrahamsen, M.E.; McKay-Chopin, S.; Gheit, T.; Tommasino, M.; Rollison, D.E. Natural History of Polyomaviruses in Men: The HPV Infection in Men (HIM) Study. *J. Infect. Dis.* **2015**, *211*, 1437–1446. [CrossRef] [PubMed]
30. Foulongne, V.; Sauvage, V.; Hebert, C.; Dereure, O.; Cheval, J.; Gouilh, M.A.; Pariente, K.; Segondy, M.; Burguière, A.; Manuguerra, J.C.; et al. Human skin microbiota: High diversity of DNA viruses identified on the human skin by high throughput sequencing. *PLoS ONE* **2012**, *7*, e38499. [CrossRef] [PubMed]
31. Ehlers, B.; Wieland, U. The novel human polyomaviruses HPyV6, 7, 9 and beyond. *APMIS* **2013**, *121*, 783–795. [CrossRef] [PubMed]
32. Johnston, S.D.; Yu, X.M.; Mertz, J.E. The major transcriptional transactivation domain of simian virus 40 large T antigen associates nonconcurrently with multiple components of the transcriptional preinitiation complex. *J. Virol.* **1996**, *70*, 1191–1202. [PubMed]
33. Gilinger, G.; Alwine, J.C. Transcriptional activation by simian virus 40 large T antigen: Requirements for simple promoter structures containing either TATA or initiator elements with variable upstream factor binding sites. *J. Virol.* **1993**, *67*, 6682–6688. [PubMed]
34. Briggs, M.R.; Kadonaga, J.T.; Bell, S.P.; Tjian, R. Purification and biochemical characterization of the promoter-specific transcription factor, Sp1. *Science* **1986**, *234*, 47–52. [CrossRef] [PubMed]
35. Messeguer, X.; Escudero, R.; Farre, D.; Nunez, O.; Martinez, J.; Alba, M.M. PROMO: Detection of known transcription regulatory elements using species-tailored searches. *Bioinformatics* **2002**, *18*, 333–334. [CrossRef] [PubMed]
36. Farre, D.; Roset, R.; Huerta, M.; Adsuara, J.E.; Rosello, L.; Alba, M.M.; Messeguer, X. Identification of patterns in biological sequences at the ALGGEN server: PROMO and MALGEN. *Nucleic Acids Res.* **2003**, *31*, 3651–3653. [CrossRef] [PubMed]
37. Uhlen, M.; Fagerberg, L.; Hallstrom, B.M.; Lindskog, C.; Oksvold, P.; Mardinoglu, A.; Sivertsson, Å.; Kampf, C.; Sjöstedt, E.; Asplund, A.; et al. Proteomics. Tissue-based map of the human proteome. *Science* **2015**, *347*, 1260419. [CrossRef] [PubMed]
38. Thul, P.J.; Akesson, L.; Wiking, M.; Mahdessian, D.; Geladaki, A.; Ait Blal, H.; Alm, T.; Asplund, A.; Björk, L.; Breckels, L.M.; et al. A subcellular map of the human proteome. *Science* **2017**, *356*, eaal3321. [CrossRef] [PubMed]
39. De Luca, P.; Majello, B.; Lania, L. Sp3 represses transcription when tethered to promoter DNA or targeted to promoter proximal RNA. *J. Biol. Chem.* **1996**, *271*, 8533–8536. [CrossRef] [PubMed]
40. Lania, L.; Majello, B.; De Luca, P. Transcriptional regulation by the Sp family proteins. *Int. J. Biochem. Cell Biol.* **1997**, *29*, 1313–1323. [CrossRef]
41. Gupta, T.; Robles, M.T.; Schowalter, R.M.; Buck, C.B.; Pipas, J.M. Expression of the small T antigen of Lymphotropic Papovavirus is sufficient to transform primary mouse embryo fibroblasts. *Virology* **2016**, *487*, 112–120. [CrossRef] [PubMed]
42. Toptan, T.; Yousem, S.A.; Ho, J.; Matsushima, Y.; Stabile, L.P.; Fernandez-Figueras, M.T.; Bhargava, R.; Ryo, A.; Moore, P.S.; Chang, Y. Survey for human polyomaviruses in cancer. *JCI Insight* **2016**, *1*, e85562. [CrossRef] [PubMed]
43. Schmitt, M.K.; Mann, K. Glycosylation of simian virus 40 T antigen and localization of glycosylated T antigen in the nuclear matrix. *Virology* **1987**, *156*, 268–281. [CrossRef]
44. Klockmann, U.; Deppert, W. Acylation: A new post-translational modification specific for plasma membrane-associated simian virus 40 large T-antigen. *FEBS Lett.* **1983**, *151*, 257–259. [CrossRef]
45. Grasser, F.A.; Scheidtmann, K.H.; Tuazon, P.T.; Traugh, J.A.; Walter, G. In vitro phosphorylation of SV40 large T antigen. *Virology* **1988**, *165*, 13–22. [CrossRef]

10

Detection of HPV16 in Esophageal Cancer in a High-Incidence Region of Malawi

Anja Lidwina Geßner [1,2], Angelika Borkowetz [3], Michael Baier [4], Angela Göhlert [5], Torsten J. Wilhelm [6], Alexander Thumbs [7], Eric Borgstein [7], Lars Jansen [2], Katrin Beer [2], Henning Mothes [1,*,†] and Matthias Dürst [2,*,†]

1 Department of General, Visceral and Vascular Surgery, Jena University Hospital—Friedrich-Schiller-University; 07747 Jena, Germany; anja.l.gessner@gmail.com
2 Department of Gynecology, Jena University Hospital—Friedrich-Schiller-University, 07747 Jena, Germany; Lars.Jansen@med.uni-jena.de (L.J.); Katrin.Beer@med.uni-jena.de (K.B.)
3 Department of Urology, Technische Universität Dresden; 01307 Dresden, Germany; angelika@borkowetz.de
4 Institute for Medical Microbiology, Jena University Hospital—Friedrich-Schiller-University, 07747 Jena, Germany; Michael.Baier@med.uni-jena.de
5 Institute for Pathology, Jena University Hospital—Friedrich-Schiller-University, 07743 Jena, Germany; pathologiejena@gmail.com
6 Department of Surgery, University Medical Centre Mannheim, 68167 Mannheim, Germany; tjwilhelm@web.de
7 Department of Surgery, Queen Elizabeth Central Hospital—College of Medicine, Blantyre 3, Malawi; alex_thumbs@hotmail.com (A.T.); eborg@me.com (E.B.)
* Correspondence: Henning.Mothes@med.uni-jena.de (H.M.); matthias.duerst@med.uni-jena.de (M.D.);

† These authors contributed equally to this work.

Abstract: This study was designed to explore the role of human papillomavirus (HPV) in esophageal squamous cell carcinoma (ESCC). Fifty-five patients receiving diagnostic upper gastrointestinal endoscopy at Zomba Central Hospital or Queen Elizabeth Hospital in Blantyre (Malawi) in 2010, were included in our study. Formalin-fixed paraffin-embedded biopsies were collected for histopathological diagnosis. HPV DNA was detected using multiplex Quantitative PCR (qPCR) and in situ hybridization (ISH). p16^{INK4a} staining served as a surrogate marker for HPV oncogene activity. Cell proliferation was determined by Ki-67 staining. Human immunodeficiency virus (HIV) status was evaluated by serology. Data on the consumption of alcohol and tobacco, and history of tuberculosis (TBC), oral thrush, and Herpes zoster, were obtained by questionnaire. Forty patients displayed ESCC, three displayed dysplastic epithelium, and 12 displayed normal epithelium. HPV16 was detected in six ESCC specimens and in one dysplastic lesion. Among HPV-positive patients, viral load varied from 0.001 to 2.5 copies per tumor cell. HPV DNA presence could not be confirmed by ISH. p16^{INK4a} positivity correlated with the presence of HPV DNA ($p = 0.03$). Of particular note is that the Ki-67 proliferation index, in areas with diffuse nuclear or cytoplasmatic p16^{INK4a} staining $\geq 50\%$, was significantly higher in HPV-positive tumors compared to the corresponding p16^{INK4a} stained areas of HPV-negative tumors ($p = 0.004$). HPV infection in ESCC was not associated with the consumption of tobacco or alcohol, but there were significantly more patients drinking locally brewed alcohol among HPV-positive tumor patients compared to non-tumor patients ($p = 0.02$) and compared to HPV-negative tumor patients ($p = 0.047$). There was no association between HIV infection, history of TBC, Herpes zoster, oral thrush, or HPV infection, in ESCC patients. Our indirect evidence for viral oncogene activity is restricted to single tumor cell areas, indicative of the role of HPV16 in the development of ESCC. The inhomogeneous presence of the virus within the tumor is reminiscent of the "hit and run" mechanism discussed for β-HPV types, such as HPV38.

Keywords: human Papillomavirus; esophageal squamous cell carcinoma; p16^{INK4a} immunohistochemistry; Ki-67 proliferation index; polymerase chain reaction; in situ hybridization; alcohol; smoking

1. Introduction

Esophageal cancer (EC) is the eighth most common cancer diagnosis, with an incidence of 456,000 cases worldwide in 2012. More prevalent than other tumors, it shows high geographical variation, with high occurrence in a variety of regions, including the Asian esophageal cancer belt in Southcentral Asia, Turkmenistan, and parts of Southern Africa. Malawi belongs to the high incidence area of EC in Africa [1,2]. Although an epidemiological shift is taking place, with rising numbers of histologically proven adenocarcinoma of the esophagus in Western countries, the esophageal squamous cell carcinoma (ESCC) is still the predominant histopathological type worldwide [2]. This indicates that the incidence of EC is influenced by lifestyle and other exogenous factors. The impact of non-infectious agents, such as tobacco and alcohol, as well as infectious causes such as human papillomavirus (HPV), on the pathogenesis of ESCC is discussed in the literature [3,4].

In Malawi, ESCC is endemic, with an incidence rate of 24.2 per 100,000 residents, compared to 3.2 per 100,000 in the United States or 6.6 per 100,000 in the U.K. [1]. HPV prevalence in Eastern Africa is one of the highest in the world and exceeds 30% among women, even with normal cervical cytology [5]. HPV-related cervical cancer is the most frequent cancer type in Malawi with an incidence rate of 75.9 per 100,000 women [1].

Since the discovery of HPV type 16 in 1984, the role of HPV in human malignancies has become well established [6,7]. HPV is responsible for 2% of the cancer burden in developed regions and 7% of the cancer burden in developing regions [5]. High-risk HPV (hr-HPV) is known to play a causal role in the development of cervical, anal, head and neck cancer, and squamous cell carcinoma of the vulva [8,9]. HPV16 is the most prevalent representative in human cancers [10]. HPV is known for its strict tropism and commonly infects basal cells of stratified squamous epithelia [11]. HPV infection of the cervical epithelium is either productive or transforming in nature. Productive infections are characterized by koilocytes, typically seen in cervical intraepithelial neoplasia (CIN) 1 lesions. In productive infections, replication and expression of viral proteins are tightly regulated and are linked to the process of epithelial differentiation. Transforming viral infections are the result of virus persistence and the constitutive activity of viral oncogenes E6 and E7, which promote genetic instability. Accumulating genetic changes are a prerequisite for immortalization and tumorigenicity. Transforming infections usually result in CIN2/3 lesions, characterized by deregulated viral E6/E7 expression in the basal and parabasal layers, which distinguishes them from virus productive lesions [12]. Since the first report of HPV in esophageal tumors in 1982 [13], many studies have been conducted exploring the role of HPV in EC, but results remained inconclusive [14,15].

Tests for the detection of persistent HPV infections in tissue samples are manifold, and include in situ hybridization and target nucleic acid amplification. In particular, polymerase chain reaction (PCR) is used for HPV detection, genotyping, and viral load determination. In situ hybridization detects the viral genome within the histopathological context of a lesion [16]. Determination of the proliferation index, oncogene activity, and serological response provides further information on the role of HPV in ESCC. Ki-67 is a proliferation marker detected by immunohistochemistry and is commonly used to determine cell proliferation within tumor areas [17,18]. Immunohistochemical staining of p16^{INK4a} reflects the oncogene activity of the deregulated E7 oncoprotein, which inhibits retinoblastoma (Rb) [18,19].

To further explore the role of HPV in ESCC we applied different methods to investigate HPV infection in the tissue samples of patients with ESCC. Our study is the first to combine real-time qPCR, in situ hybridization, and the immunohistochemical staining of Ki-67 and p16^{INK4a} to determine HPV load, HPV prevalence within the histopathological context, the proliferation index, and oncogene activity, respectively. Moreover, further risk factors such as smoking, alcohol consumption, and human immunodeficiency virus (HIV) infection, were also considered.

2. Results

A total of 55 patients attending upper gastrointestinal endoscopy in 2010 were included in this study. The tumor group consisted of 40 histologically proven ESCC patients. Three patients showed a dysplastic epithelium in their histopathology. For 12 patients, no histopathological signs of ESCC were found in their biopsies, therefore these patients were defined as non-tumor patients.

2.1. Multiplex Real-Time PCR Assay

All 55 samples were examined using a multiplex real-time PCR assay for the detection of HPV16, 18, 31, and 45, in two independent runs. Since we could only detect HPV16 DNA, three further independent duplex real-time PCRs for the detection of HPV16 and β-globin were performed for most of the samples. For five samples, HPV16-positivity was reproducible in all runs, and for two samples it was reproducible in 4 of 5 and 3 of 5 runs, respectively. Thus, HPV16 DNA was detected in seven (12.7%) samples at least three times. These seven samples were considered HPV16-positive (Table 1). One of the 7 HPV-positive samples showed dysplastic epithelium. Six of these 7 HPV-positive biopsies stemmed from tumor patients. Therefore, HPV-positivity among patients with histopathologically confirmed ESCC was 15%. Of 15 samples, sufficient material was left for renewed DNA extraction and further HPV genotyping. For these samples, multiplex real-time PCR was performed for all 7 hrHPV types (HPV16, 18, 31, 33, 45, 52, and 58). HPV16-positivity could be confirmed in 2 of 3 cases. No other HPV genotypes could be detected. Reproducibility of the PCR results for HPV16 showed high overall agreement (Fleiss's kappa = 0.6868, SE = 0.0348). Viral load, estimated by ΔC_T in a multiplex real-time PCR assay, in HPV-positive samples, varied from 0.001 to 1.01 HPV copies per cell. When normalizing for the proportion of tumor cells within the sections, viral load varied from 0.001 to 2.5 HPV copies per tumor cell. In four samples, viral load ranged from 1.3 to 2.5 HPV copies per tumor cell.

Table 1. Human papillomavirus (HPV), human immunodeficiency virus (HIV), p16, and Ki-67 status in esophageal squamous cell carcinoma (ESCC), dysplasia and histopathologically normal biopsies.

Study Number	Hist. Diagn. [1]	Grading	Ki-67 Index Overall [2]	$p16^{INK4a}$ Staining Pattern [3]	$p16^{INK4a}$ Stained Tumor Cell Areas [4]	% of Ki-67 Stained Cells within $p16^{INK4a}$ Stained Areas	HPV qPCR [5]	HPV ISH [6]	HIV Status [7]
HV090610-44	1	1	3	2	1	71	1	0	0
DV280610-68	1	1	3	1	2	88	1	0	0
EC070710-73	2	-	3	1	2	51	1	0	0
FN280710-105	1	1	3	1	2	67	1	0	-
AM240510-22	1	2	3	1	2	61	1	0	0
JS070710-72	1	1	3	1	2	72	1	0	0
MRD160610-57	1	2	-	-	-	-	1	-	0
PD120510-1	1	2	3	0	0	-	0	0	0
CC050710-70	2	-	3	0	0	-	0	0	1
AM14071087	1	2	2	2	2	26	0	0	0
EK070610-40	1	2	3	0	0	-	0	-	0
SN210710-101	1	2	3	0	0	-	0	-	-
SN210710-96	1	2	3	0	0	-	0	-	-
HM070710-74	1	3	-	-	-	-	0	-	0
LI190710-91	1	2	-	-	-	-	0	-	-
DM090610-41	1	2	3	1	2	60	0	-	0
AOM160610-55	1	2	3	1	2	53	0	0	0
MG260510-29	1	2	3	1	2	55	0	-	0
EO160610-59	1	3	3	1	1	60	0	-	0
LC150610-97	1	2	2	1	2	25	0	-	-
LM180510-10	1	2	3	0	0	-	0	-	1
PL130710-81	1	2	2	1	2	46	0	0	0
RC170510-7	1	2	2	1	2	45	0	-	0
EM280610-69	2	-	3	1	2	67	0	-	0
LN180510-12	1	3	2	0	0	-	0	-	-
ESL190510-17	1	3	2	1	2	48	0	-	0
JC260710-102	1	2	3	0	0	-	0	-	-
DL160610-56	1	3	2	1	2	49	0	-	1
JC120510-2	1	2	3	0	0	-	0	-	0
JN140610-54	1	2	0	1	1	0	0	0	0
RM170510-8	1	2	2	1	2	47	0	-	0
ZV160610-58	1	2	3	1	1	62	0	-	0
SF270710-103	1	-	2	0	0	-	0	-	-
FM140710-86	1	3	3	1	2	54	0	-	0
OD120510-3	1	2	2	1	2	45	0	-	0
DK280610-66	1	3	3	1	1	78	0	-	0
MJ020610-34	1	3	2	0	0	-	0	-	1
JK140710-89	1	2	3	1	2	65	0	-	0
RM210710-93	1	2	3	0	0	-	0	-	-
AM140710-84	1	3	3	0	0	-	0	-	0
BZ020610-35	1	2	0	1	2	8	0	0	0
GP140710-90	1	2	2	1	1	41	0	-	-

Table 1. *Cont.*

Study Number	Hist. Diagn. [1]	Grading	Ki-67 Index Overall [2]	$p16^{INK4a}$ Staining Pattern [3]	$p16^{INK4a}$ Stained Tumor Cell Areas [4]	% of Ki-67 Stained Cells within $p16^{INK4a}$ Stained Areas	HPV qPCR [5]	HPV ISH [6]	HIV Status [7]
JC070710-77	1	2	3	2	2	28	0	-	0
RM150610-98	0	-	2	0	0	-	0	-	0
MJ240510-23	0	-	3	0	0	-	0	-	1
EC070710-73	0	-	2	0	0	-	0	-	-
DD170510-9	0	-	3	0	0	-	0	-	0
FN280610-62	0	-	2	0	0	-	0	-	0
AM140610-50	0	-	2	0	0	-	0	-	1
KG170510-4	0	-	1	0	0	-	0	-	1
AD130710-83	0	-	2	0	0	-	0	-	0
AT090610-45	0	-	1	0	0	-	0	-	0
GN280710-107	0	-	2	0	0	-	0	-	-
MB140610-51	0	-	3	0	0	-	0	-	0
JM130710-82	0	-	0	1	1	6	0	-	-

[1] Histological diagnosis: 0 = tumor-free epithelium, 1 = ESCC, 2 = dysplastic epithelium; [2] Ki-67 index: $0 \leq 10\%$: negative, 1 = 10–20%: weakly positive, 2 = 20–50%: positive, $3 \geq 50\%$: strongly positive; [3] $p16^{INK4a}$ staining pattern: 0 = focal staining, 1 = diffuse cytoplasmatic positive, 2 = diffuse nuclear positive; [4] $p16^{INK4a}$ staining in tumor tissue: 0 = $p16^{INK4a}$ negative, 1 = 20–50% positive, $2 \geq 50\%$ strong positive; [5] HPV in PCR: 0 = negative, 1 = positive; [6] In situ hybridization: 0 = negative; [7] HIV status: 0 = negative, 1 = positive; - not done/not evaluable.

2.2. Histopathological Analysis, p16^{INK4a} Status, and the Ki-67 Proliferation Index

Six of the 7 HPV16-positive cases were tumor patients with histopathologically confirmed ESCCs. One HPV16-positive case showed dysplastic epithelium. The difference in HPV prevalence between tumor patients and non-tumor patients was not statistically significant ($p = 0.3$). Information about grading was available for 39 tumor patients. Grading of ESCC samples among HPV-positive tumor samples was significantly lower compared to HPV-negative tumor samples ($p < 0.001$).

Evaluation of p16^{INK4a} and Ki-67 status was not possible for two HPV-negative and one HPV-positive sample, as there were no tumor cell areas visible in the stained section.

Three tumor samples showed a diffuse nuclear p16^{INK4a} staining pattern (overall ≥20% of tumor cell areas), of which one sample was HPV-positive (Figure 1, Table 1). Diffuse cytoplasmatic p16^{INK4a} positive staining (overall ≥20% of tumor cell areas) was found in 25 samples, of which 22 were tumor samples.

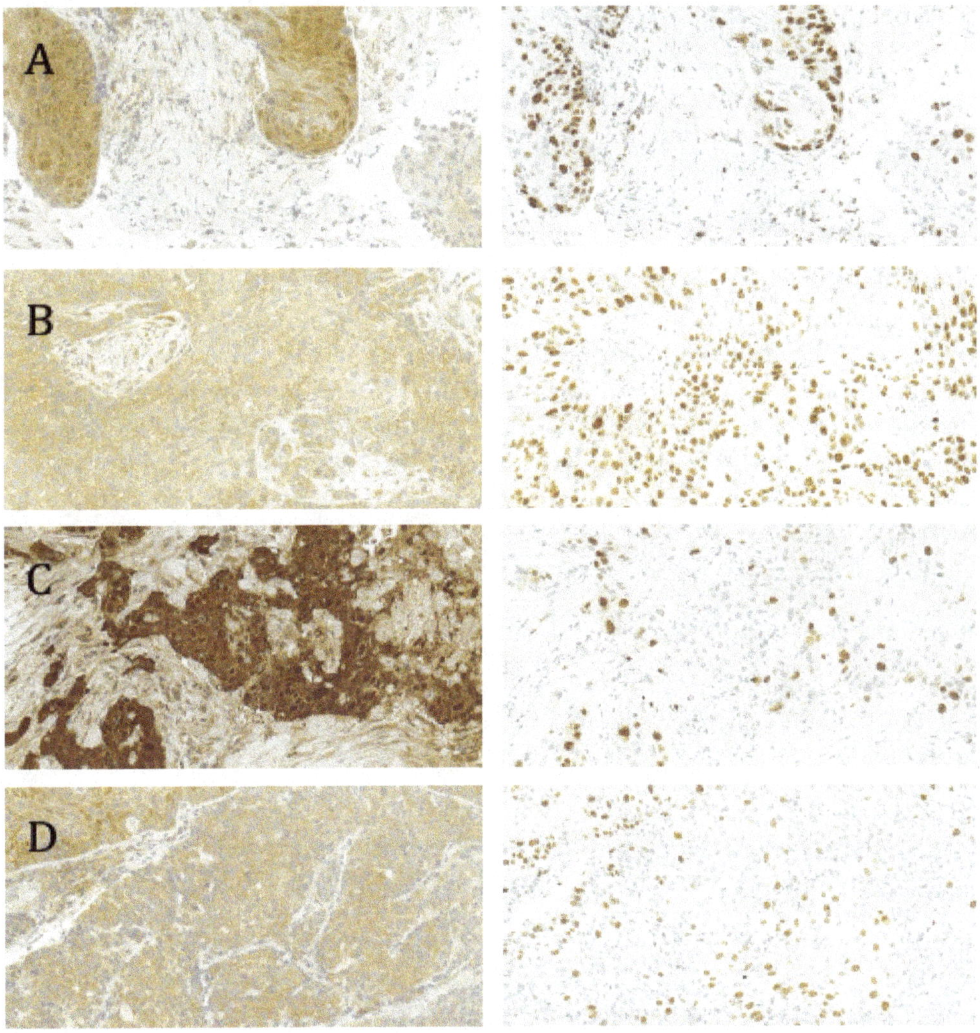

Figure 1. p16^{INK4a} and Ki-67 staining (left and right panels, respectively) in consecutive sections of human papillomavirus (HPV)-positive and HPV-negative tumor biopsies. (**A**) HPV-positive, diffuse nuclear p16^{INK4a} staining; (**B**) HPV-positive, diffuse cytoplasmatic p16^{INK4a} staining; (**C**) HPV-negative, diffuse nuclear p16^{INK4a} staining. (**D**) HPV-negative, diffuse cytoplasmatic p16^{INK4a} staining. The proliferation index (Ki-67 staining) is significantly higher in HPV-positive tumors. (**A**–**D**) were obtained with 20× magnification.

Biopsies showing diffuse nuclear or diffuse cytoplasmic p16^{INK4a} staining <20%, as well as biopsies with only focally stained cells, were defined as p16^{INK4a}-negative. All HPV-positive tumor samples showed either diffuse nuclear or diffuse cytoplasmatic p16^{INK4a} staining $\geq$20%. Positive p16^{INK4a} staining was significantly associated with HPV status (p = 0.03). This association was still significant when positive nuclear or cytoplasmatic p16^{INK4a} staining $\geq$50% was taken as the threshold level for p16^{INK4a}-positivity (p = 0.03).

All HPV positive samples, overall, showed a high Ki-67 proliferation index. HPV-positive samples (median: 69.0%, interquartile range: 17.5%) had a higher proliferation index compared to HPV-negative tumors, including dysplastic samples (median: 52.0%, interquartile range: 25.8%; p = 0.03). Additionally, we determined the Ki-67 proliferation index exclusively in tumor areas with diffuse nuclear or cytoplasmatic p16^{INK4a} staining (Figure 1, Table 1). Interestingly, the Ki-67 proliferation index of HPV-positive samples (median: 69.5%, interquartile range: 27.0%) with nuclear or cytoplasmatic p16^{INK4a} staining $\geq$50% was significantly higher compared to the corresponding areas of HPV-negative tumor samples (median: 47.0%, interquartile range: 26.0%; p = 0.004). This association was also significant when the Ki-67 proliferation index of HPV-positive samples (median: 71.0%, interquartile range: 16.0%) with nuclear or cytoplasmatic p16^{INK4a} staining $\geq$20% was compared to HPV-negative tumor samples (median: 47.5%, interquartile range: 28.0%; p = 0.003).

2.3. *In Situ Hybridization*

In situ hybridization was performed for all HPV-positive biopsies, except for one tumor, which lacked tumor cell areas in subsequent sectioning performed after histopathological diagnosis and HPV-genotyping. Moreover, HPV-negative tumors showing diffuse nuclear and cytoplasmic p16^{INK4A} staining, as well as tumors with only focal p16^{INK4a} staining, were also hybridized (Table 1). Only unspecific background in situ hybridization signals were seen, and no difference in signal patterns between HPV-positive and -negative tumor samples was observed. HPV16-positive cervical carcinomas and precancers were used as positive controls and showed the expected HPV DNA signal pattern (Figure S1, supplementary materials).

2.4. *Association between HPV Status, Clinical Data and Risk Factors*

Clinical data regarding history of TBC, oral thrush, and Herpes zoster, as well as tobacco and alcohol consumption, were collected. These results are summarized in Table 2. HIV status was determined serologically.

HPV-positivity among ESCC patients was not significantly associated with positive HIV status, when HPV-positive tumor patients were compared to HPV-negative tumor patients (0.0% vs. 12.9%; p = 1.0) and non-tumor patients (0.0% vs. 20.0%; p = 0.5)

HPV-positivity among ESCC patients was not significantly associated with smoking, when HPV-positive tumor patients were compared to HPV-negative tumor patients (80.0% vs. 46.4%; p = 0.3) and non-tumor patients (80.0% vs. 40.0%; p = 0.3).

HPV-positivity in ESCC patients was not significantly associated with the consumption of alcohol, when HPV-positive tumor patients were compared to HPV-negative tumor patients (80.0% vs. 35.7%; p = 0.1) and non-tumor patients (80.0% vs. 20.0%; p = 0.09). There were significantly more patients who drank locally brewed alcohol among HPV-positive tumor patients compared to HPV-negative tumor patients (80.0% vs. 28.6%; p = 0.047) and non-tumor patients (80.0% vs. 10.0%; p = 0.02).

Table 2. Clinico-pathological characteristics of patients.

Patients' Characteristics	HPV-Positive Tumor Patients (n = 6)	HPV-Negative Tumor Patients (n = 34)	P Value *	Non-Tumor Patients (n = 12)	P Value **
General					
Age (years)	50.0 (13.0)	50.5 (28.0)	0.7	43.5 (38.0)	0.7
Male gender	5 (83.3%)	22 (64.7%)	0.6	9 (81.8%)	1.0
Risk factors					
Oral thrush in history	1 (20.0%)	4 (14.3%)	1.0	1 (10.0%)	1.0
Tuberculosis in history	0 (0.0%)	4 (14.3%)	1.0	0 (0.0%)	a
Herpes zoster in history	0 (0.0%)	2 (7.1%)	1.0	0 (0.0%)	a
HIV-positive	0 (0.0%)	4 (12.9%)	1.0	2 (20.0%)	0.5
Smoker	4 (80.0%)	13 (46.4%)	0.3	4 (40.0%)	0.3
Pack years	3.0 (12.0)	0.0 (4.0)	0.1	0.0 (1.0)	0.05
Duration of smoking (years)	15.0 (30.0)	0.0 (9.0)	0.08	0.0 (4.0)	0.04
Number of cigarettes	6.0 (7.0)	0.0 (6.0)	0.2	0.0 (5.0)	0.05
Alcohol	4 (80.0%)	10 (35.7%)	0.1	2 (20.0%)	0.09
Locally brewed alcohol	4 (80.0%)	8 (28.6%)	0.047	1 (10.0%)	0.02
Duration of drinking alcohol (years)	17.0 (21.0)	0.0 (5.0)	0.1	0.0 (1.0)	0.02

Data are presented as absolute numbers with percentages in brackets or medians with interquartile ranges in brackets. * Comparison of HPV-positive tumor patients with HPV-negative tumor patients. ** Comparison of HPV-positive tumor patients with non-tumor patients. (a) An exact Fisher test was not performed as the variable is a constant.

3. Discussion

This study is, according to our knowledge, the first study combining PCR, in situ hybridization, Ki-67 index, and p16^{INK4a} immunohistochemistry, to investigate the role of HPV in ESCC carcinogenesis and the first study investigating HPV in ESCC in Malawi, a high-risk area for ESCC.

The prevalence rate of HPV in both ESCC and benign esophageal papillomas, varies greatly all over the world [4,20] and even within the same high-risk area [21,22]. Such great geographic differences have not been reported for other HPV-associated malignancies, such as cervical cancer [9]. In our study, HPV prevalence was 15% in ESCC and HPV was found only in ESCC and dysplasia. No other HPV-type, except HPV16, was found. Similar to another local study from Zambia published in 2015, HPV was prevalent in a limited number of samples [23]. In Zambia, esophageal cancer is the fifth most common diagnosed cancer [1]. HPV was detected only in 2 out of 44 samples from ESCC patients using PCR [23]. A recent meta-analysis reported an overall prevalence of HPV in ESCC of 19.8% in Africa and 27.7% worldwide, with HPV16 being the most prevalent HPV type [20]. There was plausible variation across different detection methods and most of the studies used PCR for the detection of HPV [20,24]. PCR is known to be highly sensitive and, therefore, is not only susceptible to cross contamination, but also to substances inhibiting the reaction. This can lead to false-positive as well as false-negative HPV findings in tissue samples [16]. We have taken all possible precautionary measures to prevent the contamination of DNA during extraction and to ensure high quality DNA for PCR analyses. Despite over-fixation in buffered formalin, false-negative results are unlikely, as shown by Ct-values $\leq$35 for amplified β-globin, which was used as an internal control. Reproducibility of the PCR results for HPV16 on different days was tested and showed high overall agreement (Fleiss's Kappa = 0.6868, SE = 0.0348).

To define the prevalence of HPV within the histopathological context of the tumor samples and to determine viral load, quantitative PCR and HPV-DNA in situ hybridization was carried out. In cervical cancer, the HPV genome is usually present and transcriptionally active in every tumor cell, and a viral load of several hundred copies per cell has been reported [9,25]. Additionally, a distinct staining pattern, characteristic for the severity of neoplastic cervical lesions, is evident by in situ hybridization [26,27]. In our study, the average Ct value for HPV16 in qPCR showed that HPV was not present in every cell and, if present at all, only a few viral genomes per cell persisted. Viral load exceeded 1.0 copy per cell in four of the tested HPV-positive samples. The maximum number of copies per cell was 2.5. This is consistent with results from another study examining viral load in ESCC with a multiplex real-time PCR assay. Si and colleagues found less than 10 copies per genome equivalent in 65% of the HPV16-positive esophageal cancer samples [28]. Reasons for the low viral load in ESCC samples may be the unequal distribution of HPV in the tumor tissue, or even loss of HPV during cancer development. Moreover, there is some evidence that low HPV copy numbers may be related to the physical state of the viral genome. Si and colleagues showed that the majority of ESCC samples harboured viral DNA in an integrated form [29]. The weakness of that study, however, was that the integration data was based on E2/E6 ratios, which provide only indirect evidence for HPV-integration. Thus far, sequencing of viral–cellular DNA junctions, and validation by PCR, has not been reported for ESCC. Integration of the viral genome may promote cellular growth and increase viral genome expression, compared to the extrachromosomal status of the viral genome [30].

In situ hybridization was carried out on 13 samples. In situ hybridization staining patterns among HPV-positive tumor samples did not differ from HPV-negative tumor samples (Table 1). This is in line with the determined low HPV-genome copy number per cell, which is below the detection limit of most in situ hybridization assays and excludes the presence of interspersed single cells with high copy numbers. Possibly, storage of the tissue in formaldehyde for several weeks before embedding in paraffin may have lowered sensitivity and worsened the background. This assumption seems likely since the control tissues, comprising cervical pre-cancer and cancer fixed in formaldehyde overnight, showed prominent and distinct hybridization signals (Figure S1, supplementary materials). Malik and colleagues tested p16-positive ESCC samples with in situ hybridization. None of the $p16^{INK4a}$-positive samples showed high-risk HPV by in situ hybridization [31]. In another study, less than 10% of the cells showed positive weak staining in samples that tested positive by in situ hybridization [32]. Cooper and colleagues found HPV in 52% of ESCC samples by in situ hybridization. The virus was detected in isolated cells only, or aggregates of tumor cells, and the signals were punctuated in all HPV-positive tumors [33]. A punctuated signal pattern is reported for HPV-positive tumor cells with an integrated HPV genome [34].

To assess viral E7 oncogene activity in tumor samples, we performed $p16^{INK4a}$ immunohistochemistry. Transforming viral infections are characterized by an upregulation of $p16^{INK4a}$, due to the inactivation of Rb as a result of deregulated E7 gene expression. Strong, homogeneous and diffuse staining of $p16^{INK4a}$ throughout the whole cancer tissue is a well-defined surrogate marker for HPV-induced cervical cancer [35]. In the literature, there is consent that, in contrast to cervical cancer, diffuse nuclear and/or cytoplasmic $p16^{INK4a}$ staining of the entire lesion is not characteristic for ESCC. Unfortunately, the definition of $p16^{INK4a}$-positivity in ESCC varies greatly and, thus, complicates direct comparison between studies. In our study, we showed that diffuse $p16^{INK4a}$ staining of $\geq$20% of tumor cell areas was significantly associated with HPV positivity ($p = 0.03$). This correlation remained significant when raising the threshold level for $p16^{INK4a}$-positivity to $\geq$50% of tumor cell areas ($p = 0.03$). Our results are consistent with another study investigating $p16^{INK4a}$ expression in ESCC. Cao and colleagues found a correlation between HPV and $p16^{INK4a}$ (86.2% $p16^{INK4a}$-positive samples among HPV-positive tumors). Strong nuclear and/or cytoplasmatic staining over 70% was defined as positive in that study [36]. $p16^{INK4a}$ staining of only parts of the tumor tissue is also described for tumors of other sites, such as vulvar neoplasias [37], carcinomas of the head and neck [38], anal cancers [39], and colorectal carcinomas [40]. Moreover, expression of $p16^{INK4a}$ in tumor cells can be lost in the cells due to molecular events, such as promotor methylation or loss of heterozygosity [41].

However, there are several studies in which no correlation between $p16^{INK4a}$-positivity and HPV status in ESCC was evident [15,31,32,42,43]. In those studies, the definition of $p16^{INK4a}$-positivity ranged from $\geq$10% positive cells to 100%. Clearly, focal staining for $p16^{INK4a}$, or even patchy staining, most likely reflects numerous physiological processes unrelated to HPV E7 expression [44]. However, it is of particular note that, in our study, the Ki-67 proliferation index in areas with diffuse nuclear or cytoplasmic $p16^{INK4a}$ staining in at least 20% of tumor cell areas, or even in over 50% of tumor cell areas, was significantly higher in HPV-positive tumors than in the corresponding $p16^{INK4a}$ stained areas of HPV-negative tumors ($p = 0.003$ and $p = 0.004$, respectively). It may be speculated that this increase in the Ki-67 proliferation index is due to the functional inactivation of Rb by the HPV E7 oncoprotein, which would provide support for a role of HPV in ESCC. No other published studies have compared the Ki-67 proliferation index in p16^{INK4a}-positive tumor cell areas in relation to HPV status. Clearly, elevated levels of p16^{INK4a} expression are not exclusive for HPV-induced cancer. As reviewed by Witkiewicz and colleagues, there are two complementary models for the occurrence of tumors with high expression of p16^{INK4a}. First, oncogenic stresses induce p16^{INK4a}, thereby limiting tumorigenic progression. This is followed by the inactivation of Rb, which facilitates disease progression. Second, loss of Rb yields oncogenic stress, which induces p16^{INK4a} [45]. Since the Ki-67 proliferation index in p16^{INK4a} stained areas differed significantly between HPV-positive and HPV-negative tumors, the first model most likely applies for HPV-negative ESCC. This would align with the observations of Bai and colleagues, who showed that expression of p16^{INK4a} increases during esophageal squamous cell cancer progression [46].

The risk of developing cervical intraepithelial neoplasias increases in patients with an impaired immune system [9]. However, we found no association between HPV infection and HIV infection in patients with ESCC in Malawi. Significantly more HPV-positive tumor patients consumed locally brewed alcohol than non-tumor patients ($p = 0.02$) and HPV-negative tumor patients ($p = 0.047$). Locally brewed alcohol may act synergistically with the HPV infection in the pathogenesis of ESCC. The carcinogenic effect of locally brewed alcohol products could be mediated by acetaldehyde, the main metabolite of ethanol [47], but also by contaminants, such as fumonisins [48].

Limitations of our study include the low number of patients, especially the low number of HPV-positive tumor patients in the statistical analysis of further risk factors, and the long storage of the biopsies in formalin before embedding in paraffin. A self-generated questionnaire was used for the assessment of further risk factors, as no validated questionnaire for these factors was available in 2010. Not all questions could be answered by all patients. Furthermore, in Africa, patients with ESCC usually present themselves at a late stage of disease [49]. Therefore, only a few patients with early stages and dysplasias were included in this study.

Several observations in our study, such as the presence of HPV16 DNA in cancers and dysplasias only, and the increased proliferation rate in p16^{INK4a}-overexpressed tumor areas in HPV-positive cancers, suggest that HPV infection might play a role in a subset of ESCC patients. However, since the HPV genome does not seem to be present in every tumor cell, the theory of "hit and run" [50], in analogy to some β-HPV types, should also be considered for HPV-associated carcinogenesis in ESCC. This theory is discussed for HPV-initiated cutaneous squamous cell carcinomas and bovine papillomavirus type 4-initiated ESCC in cattle [51]. β-HPVs are found in epidermodysplasia verruciformis-associated cutaneous squamous cell carcinomas at UV-exposed skin sites. Furthermore, there is evidence that β-HPVs can act as co-factors in the development of UV-induced keratinocyte carcinomas in immunosuppressed individuals. A synergism between β-HPVs and DNA damage from UV light is suggested. β-HPVs in keratinocyte carcinoma are suggested to play a role during the early stages of carcinogenesis, by promoting the accumulation of UV-induced mutations and oncogenic transformation. Their persistence is not mandatory for the maintenance of the malignant phenotype and tumor viruses may be lost after initiation [51–54]. Studies investigating human skin biopsies have shown low viral loads in squamous cell carcinomas examined by qPCR assays and confirmed by in situ hybridization, showing only few positive nuclei per section [55]. In β-HPV-associated skin lesions, HPV status and p16^{INK4a} expression do not correlate [56]. In cervical cancer, the continued viral oncogene activity of high risk HPV types of the α genus is characteristic. However, this mechanism may not be exclusive, in particular, when considering the carcinogenic role of HPV16 in other epithelial sites such as the oesophagus. Indeed, 47.4% of ESCC display somatic TP53 mutations [57], which would be in line with a "hit and run" mechanism.

In conclusion, the role of human papillomaviruses in the carcinogenesis of esophageal squamous cell carcinomas, still remains elusive. To further investigate the role of HPV in ESCC, more studies, with a stronger focus on molecular events in the context of intratumoral heterogeneity and other risk factors, need to be conducted.

4. Material and Methods

Fifty-five patients receiving diagnostic upper gastrointestinal endoscopy at Zomba Central Hospital or Queen Elizabeth Hospital in Blantyre in 2010, were included in our study. Biopsies of all 55 patients were taken and immediately fixed in buffered formalin. Sera were collected from 42 patients. Further investigation was done at Jena University Hospital, where the biopsies were embedded in paraffin. Formalin-fixed paraffin-embedded (FFPE) tissue blocks were cut as consecutive sections. The first and the last sections (3 μm) were stained with hematoxylin and eosin for histopathological diagnosis. The sections in-between (15 μm) were used for DNA extraction and subsequent qPCR. Renewed sectioning was performed for in situ hybridization and immunostaining for $p16^{INK4a}$ and Ki-67.

All patients gave their written consent prior to endoscopy (signature or finger print). The study was approved by the College of Medicine Research and Ethics Committee (COMREC) (approved 21 May 2010, Nr P.04/10/930).

4.1. DNA Extraction

Consecutive serial tissue sections were prepared for DNA extraction. The first and last sections were hematoxylin-eosin-stained for histopathological diagnosis and to estimate the fraction of tumor cells in the tissue. Ten 15 μm tissue sections were deparaffinized with xylene, re-hydrated through a graded ethanol series, and dissolved in a 500 μL digestion buffer (100 mM NaCl, 10 mM Tris-HCl pH8.0, 25 mM EDTA (Ethylenediaminetetraacetic acid), and 0.5% SDS (Sodium dodecyl sulfate)). Proteinase K (100 μg/mL) digestion was done overnight at 56 °C. For DNA extraction, an equal volume of phenol/chloroform/isoamylalcohol was added and vortexed intermittently for 30 s. After centrifugation, the DNA in the aqueous phase was precipitated by adding 1/10th volume of 7.5 M ammonium acetate and 2 volumes of ice-cold ethanol, followed by another centrifugation step. The pellet was dissolved in 25 μL 10 mM Tris/HCl pH = 8.0. DNA concentration and quality were determined by spectrophotometry (NanoDrop ND-1000, Thermo Fisher Scientific GmbH, Dreieich, Germany).

4.2. Multiplex Real-Time PCR Assay (TaqMan Format)

A multiplex real-time PCR assay (TaqMan format [58]), designed to detect and quantify the 7 most frequent high-risk HPV types (HPV16, 18, 31, 33, 45, 52, and 58) in cervical cancer, was used. It was conducted as described previously [59]. Reactions were run in an ABI 7300 cycler (Applied Biosystems, Darmstadt, Germany), which is able to detect 4 different fluorescent dyes simultaneously. Reaction mix 1 detects HPV16, 18, 31, and 45, and reaction mix 2 detects HPV33, 52, 58, and β-globin. The PCR primers targeted the LCR/E6/E7 regions of the HPV genomes and the PCR products were detected by the corresponding TaqMan probes. DNA quality and the relative viral copy number were determined by amplifying the housekeeping gene *β-globin* in one of the multiplex reactions. PCR was performed in an end volume of 25 μL, containing 9.5 μL DNA (total of 50 ng), 12.5 μL Platinum Quantitative PCR SuperMix-UDG (Invitrogen, Carlsbad, CA, USA) to prevent amplification of carry-over products [60], 10 pmol of each primer, and 1–5 pmol of each probe. Forty-five PCR cycles were run at 94 °C for 15 s, 50 °C for 20 s, and 60 °C for 40 s, each [59].

Reproducibility of the PCR results was assessed by repeating the assay on different days. All samples were tested for HPV16-positivity at least three times. All samples were tested with a primer mixture of reaction 1 twice, and once with a primer mixture containing primers for HPV16 and β-globin only (duplex PCR). For 30 samples, there was enough DNA left to repeat the duplex PCR assay twice (Table 3).

Ct values of the HPV16 genome and β-globin were used to calculate the viral load of HPV16, as copies per cell in 6 of 7 HPV-positive samples, according to the following formula:

$$\left(\frac{1}{2^{\Delta Ct}}\right) \div 2$$

To assess validity, viral load was determined the same way in HPV16-positive cell lines with a known viral load: SiHa (Ct HPV16: 21.8, Ct β-globin: 23.93, viral load: 2.2 copies per cell), HPKII (Ct HPV16: 21.48, Ct β-globin: 24.77, viral load: 5.0 copies per cell), and MRI (Ct HPV16: 19.07, Ct β-globin: 24.96, viral load: 29.7 copies per cell).

Fifteen samples were of sufficient size for renewed sectioning and DNA extraction. These samples were analysed for all 7 hrHPV types (Table 3).

Table 3. HPV DNA detection by multiplex PCR.

PCR Primer Mixture		Run	Number of Tested Samples
Mixture 1	HPV16, 18, 31, 45	1 2	55 55
Duplex only	HPV16 and β-globin	1 2 3	55 30 30
Mixture 1 and 2	HPV16, 18, 31, 45 and HPV33, 52, 58 and β-globin	1	15

4.3. *In Situ Hybridization*

The FFPE sections were deparaffinized with xylene, re-hydrated through a graded ethanol series, and washed in phosphate buffered saline (PBS). In situ hybridization was conducted using Dako GenPoint HPV Biotinylated DNA probe according to the manufacturer's protocol (GenPoint/Dako Cytomation, Agilent, Waldbronn, Germany). GenPoint HPV Probe contains HPV genomic clones in the form of double-stranded fragments of 500 base pairs or less and multiple biotinylated oligonucleotides from 25 to 40 bases in length. Briefly, the deparaffinized tissue sections were incubated with 0.8% pepsin at 37 °C for 5–10 min, rinsed in deionized water, immersed in 0.3% H_2O_2 in methanol for 20 min, and rinsed in deionized water again. A drop of GenPoint HPV Biotinylated DNA probe was added to the air-dried sections. After heat denaturation at 90 °C, hybridization was done in a pre-warmed humid chamber at 37 °C for 16 to 18 h. The slides were washed in Tris-buffered saline with Tween20 (TBST). Hybridized probes were detected by streptavidin-HRP.

4.4. *$p16^{INK4a}$*

The expression of p16 was analysed by immunohistochemistry. A CINtec Histology Kit (biogen/Roche, Tucson, AZ, USA) containing the monoclonal antibody anti-$p16^{INK4a}$ (E6H4) was used according to the manufacturer's protocol for qualitative detection of $p16^{INK4a}$ proteins in FFPE tissue sections. After adding the substrate-chromogen solution, slides were washed with distilled water and stained with Harris Hematoxylin, followed by washing with tap water. Finally, the sections were dehydrated and mounted with coverslips. Similar to cervical precancers (CIN2/3) and cancer, samples showing a diffuse nuclear $p16^{INK4a}$ staining pattern were considered to be positive. Furthermore, samples showing a diffuse cytoplasmic $p16^{INK4a}$ staining pattern were also defined as $p16^{INK4a}$-positive. However, in contrast to cervical cancer, $p16^{INK4a}$ staining showed considerable intratumoral heterogeneity. To account for this, $p16^{INK4a}$-positivity was scored in the context of all tumor cell areas, i.e., biopsies in which 20–50% and >50% of tumor cell areas showed diffuse staining were scored as positive and strongly positive, respectively. Biopsies with focal staining or with less than 20% diffuse staining were defined as $p16^{INK4a}$-negative.

4.5. *Ki-67*

To assess the proliferation activity of tumor cells, the Ki-67 index was determined. Ki-67-positive cells were counted in the area with the highest Ki-67 staining. In addition to this overall Ki-67 index, we determined the Ki-67 index in p16^{INK4a}-positive areas. At least 200–300 cells were counted. Ki-67 staining indices were grouped into negative (<10%), weakly positive (10–20%), positive (20–50%), and strongly positive (>50%).

4.6. *Serological Analysis (ELISA)*

The HIV status of each patient was determined using a commercially available ELISA system (Enzygnost® HIV Integral II, Siemens, Erlangen, Germany) according to the manufacturer's protocol. The tests were performed in a BEPIII system (Siemens, Germany). Both HIV specific antibodies and the HIV p24 antigen, were detected, ensuring very high diagnostic sensitivity.

4.7. *Data on Alcohol Consumption and Smoking, and Clinical Data*

Data on the consumption of alcohol and tobacco, as well as clinical data such as history of Herpes zoster, tuberculosis (TBC), and oral thrush, were collected using a self-generated questionnaire. Not all questions could be answered by all patients.

4.8. *Pathological Analysis*

Histopathological diagnosis, grading and analysis of p16^{INK4a}, and Ki-67 immunohistochemistry were done by ALG and an experienced pathologist (AG) at Jena University Hospital.

4.9. *Statistical Analysis*

To assess the relationship between the epidemiological data and clinical data, an exact Fisher test was used for nominal data and a Mann–Whitney U test was used for metric data. Therefore, median and interquartile ranges were determined to describe the variance of the metric variables. Results were considered statistically significant when $p < 0.05$. To test the reproducibility of the PCR results, Fleiss's Kappa was assessed. Statistical analysis was performed with IBM SPSS Statistics 22.

Author Contributions: Michael Baier, Torsten J. Wilhelm, Alexander Thumbs, Eric Borgstein, Henning Mothes, and Matthias Dürst, conceived and designed the study. Anja Lidwina Geßner, Angelika Borkowetz, Torsten J. Wilhelm, Alexander Thumbs, Eric Borgstein, and Henning Mothes, were responsible for the biopsies and patient management. Anja Lidwina Geßner, Lars Jansen, and Katrin Beer performed the experimental work. Anja Lidwina Geßner, Angela Göhlert, Henning Mothes, and Matthias Dürst, analyzed the data. Anja Lidwina Geßner, Henning Mothes, and Matthias Dürst wrote the paper. All authors approved the final manuscript version.

Abbreviations

HPV	Human papillomavirus
ESCC	Esophageal squamous cell carcinoma
Rb	Retinoblastoma
qPCR	Quantitative PCR
TBC	Tuberculosis
HIV	Human immunodeficiency virus

References

1. The International Agency for Research on Cancer (IARC). GLOBOCAN 2012: Estimated Cancer Incidence, Mortality and Prevalence Worldwide in 2012. Available online: http://globocan.iarc.fr/Default.aspx (accessed on 27 January 2018).

2. Arnold, M.; Soerjomataram, I.; Ferlay, J.; Forman, D. Global incidence of oesophageal cancer by histological subtype in 2012. *Gut* **2015**, *64*, 381–387. [CrossRef] [PubMed]
3. Napier, K.J.; Scheerer, M.; Misra, S. Esophageal cancer: A Review of epidemiology, pathogenesis, staging workup and treatment modalities. *World J. Gastrointest. Oncol.* **2014**, *6*, 112–120. [CrossRef] [PubMed]
4. Syrjanen, K.J. HPV infections and oesophageal cancer. *J. Clin. Pathol.* **2002**, *55*, 721–728. [CrossRef] [PubMed]
5. Forman, D.; de Martel, C.; Lacey, C.J.; Soerjomataram, I.; Lortet-Tieulent, J.; Bruni, L.; Vignat, J.; Ferlay, J.; Bray, F.; Plummer, M.; et al. Global burden of human papillomavirus and related diseases. *Vaccine* **2012**, *30* (Suppl. 5), F12–F23. [CrossRef] [PubMed]
6. Durst, M.; Gissmann, L.; Ikenberg, H.; zur Hausen, H. A papillomavirus DNA from a cervical carcinoma and its prevalence in cancer biopsy samples from different geographic regions. *Proc. Natl. Acad. Sci. USA* **1983**, *80*, 3812–3815. [CrossRef] [PubMed]
7. Zur Hausen, H. Papillomaviruses and cancer: From basic studies to clinical application. *Nat. Rev. Cancer* **2002**, *2*, 342–350. [CrossRef] [PubMed]
8. Gillison, M.L.; Shah, K.V. Human papillomavirus-associated head and neck squamous cell carcinoma: Mounting evidence for an etiologic role for human papillomavirus in a subset of head and neck cancers. *Curr. Opin. Oncol.* **2001**, *13*, 183–188. [CrossRef] [PubMed]
9. Gillison, M.L.; Shah, K.V. Chapter 9: Role of mucosal human papillomavirus in nongenital cancers. *J. Natl. Cancer Inst. Monogr.* **2003**, *2003*, 57–65. [CrossRef]
10. Zur Hausen, H. Papillomaviruses causing cancer: Evasion from host-cell control in early events in carcinogenesis. *J. Natl. Cancer Inst.* **2000**, *92*, 690–698. [CrossRef] [PubMed]
11. Durst, M.; Glitz, D.; Schneider, A.; zur Hausen, H. Human papillomavirus type 16 (HPV 16) gene expression and DNA replication in cervical neoplasia: Analysis by in situ hybridization. *Virology* **1992**, *189*, 132–140. [CrossRef]
12. Snijders, P.J.; Steenbergen, R.D.; Heideman, D.A.; Meijer, C.J. HPV-mediated cervical carcinogenesis: Concepts and clinical implications. *J. Pathol.* **2006**, *208*, 152–164. [CrossRef] [PubMed]
13. Syrjanen, K.J. Histological changes identical to those of condylomatous lesions found in esophageal squamous cell carcinomas. *Arch. Geschwulstforsch.* **1982**, *52*, 283–292. [PubMed]
14. The International Agency for Research on Cancer (IARC). *Monographs on the Evaluation of Carcinogenic Risks to Humans*; IARC: Lyon, France, 2012; Volume 100B.
15. Halec, G.; Schmitt, M.; Egger, S.; Abnet, C.C.; Babb, C.; Dawsey, S.M.; Flechtenmacher, C.; Gheit, T.; Hale, M.; Holzinger, D.; et al. Mucosal α-papillomaviruses are not associated with esophageal squamous cell carcinomas: Lack of mechanistic evidence from South Africa, China and Iran and from a world-wide meta-analysis. *Int. J. Cancer* **2016**, *139*, 85–98. [CrossRef] [PubMed]
16. Hubbard, R.A. Human papillomavirus testing methods. *Arch. Pathol. Lab. Med.* **2003**, *127*, 940–945. [PubMed]
17. Gerdes, J.; Lemke, H.; Baisch, H.; Wacker, H.H.; Schwab, U.; Stein, H. Cell cycle analysis of a cell proliferation-associated human nuclear antigen defined by the monoclonal antibody Ki-67. *J. Immunol.* **1984**, *133*, 1710–1715. [PubMed]
18. Keating, J.T.; Cviko, A.; Riethdorf, S.; Riethdorf, L.; Quade, B.J.; Sun, D.; Duensing, S.; Sheets, E.E.; Munger, K.; Crum, C.P. Ki-67, cyclin E, and p16INK4 are complimentary surrogate biomarkers for human papilloma virus-related cervical neoplasia. *Am. J. Surg. Pathol.* **2001**, *25*, 884–891. [CrossRef] [PubMed]
19. Dyson, N.; Howley, P.M.; Munger, K.; Harlow, E. The human papilloma virus-16 E7 oncoprotein is able to bind to the retinoblastoma gene product. *Science* **1989**, *243*, 934–937. [CrossRef] [PubMed]
20. Petrick, J.L.; Wyss, A.B.; Butler, A.M.; Cummings, C.; Sun, X.; Poole, C.; Smith, J.S.; Olshan, A.F. Prevalence of human papillomavirus among oesophageal squamous cell carcinoma cases: Systematic review and meta-analysis. *Br. J. Cancer* **2014**, *110*, 2369–2377. [CrossRef] [PubMed]
21. Guo, F.; Liu, Y.; Wang, X.; He, Z.; Weiss, N.S.; Madeleine, M.M.; Liu, F.; Tian, X.; Song, Y.; Pan, Y.; et al. Human papillomavirus infection and esophageal squamous cell carcinoma: A case-control study. *Cancer Epidemiol. Biomark. Prev.* **2012**, *21*, 780–785. [CrossRef] [PubMed]
22. Zhou, X.B.; Guo, M.; Quan, L.P.; Zhang, W.; Lu, Z.M.; Wang, Q.H.; Ke, Y.; Xu, N.Z. Detection of human papillomavirus in Chinese esophageal squamous cell carcinoma and its adjacent normal epithelium. *World J. Gastroenterol.* **2003**, *9*, 1170–1173. [CrossRef] [PubMed]

23. Kayamba, V.; Bateman, A.C.; Asombang, A.W.; Shibemba, A.; Zyambo, K.; Banda, T.; Soko, R.; Kelly, P. HIV infection and domestic smoke exposure, but not human papillomavirus, are risk factors for esophageal squamous cell carcinoma in Zambia: A case-control study. *Cancer Med.* **2015**, *4*, 588–595. [CrossRef] [PubMed]
24. Liyanage, S.S.; Rahman, B.; Ridda, I.; Newall, A.T.; Tabrizi, S.N.; Garland, S.M.; Segelov, E.; Seale, H.; Crowe, P.J.; Moa, A.; et al. The aetiological role of human papillomavirus in oesophageal squamous cell carcinoma: A meta-analysis. *PLoS ONE* **2013**, *8*, e69238. [CrossRef] [PubMed]
25. Zhang, D.; Zhang, Q.; Zhou, L.; Huo, L.; Zhang, Y.; Shen, Z.; Zhu, Y. Comparison of prevalence, viral load, physical status and expression of human papillomavirus-16, -18 and -58 in esophageal and cervical cancer: A case-control study. *BMC Cancer* **2010**, *10*, 650. [CrossRef] [PubMed]
26. Schneider, A.; Oltersdorf, T.; Schneider, V.; Gissmann, L. Distribution pattern of human papilloma virus 16 genome in cervical neoplasia by molecular in situ hybridization of tissue sections. *Int. J. Cancer* **1987**, *39*, 717–721. [CrossRef] [PubMed]
27. Evans, M.F.; Mount, S.L.; Beatty, B.G.; Cooper, K. Biotinyl-tyramide-based in situ hybridization signal patterns distinguish human papillomavirus type and grade of cervical intraepithelial neoplasia. *Mod. Pathol.* **2002**, *15*, 1339–1347. [CrossRef] [PubMed]
28. Si, H.X.; Tsao, S.W.; Poon, C.S.; Wang, L.D.; Wong, Y.C.; Cheung, A.L. Viral load of HPV in esophageal squamous cell carcinoma. *Int. J. Cancer* **2003**, *103*, 496–500. [CrossRef] [PubMed]
29. Si, H.X.; Tsao, S.W.; Poon, C.S.; Wong, Y.C.; Cheung, A.L. Physical status of HPV-16 in esophageal squamous cell carcinoma. *J. Clin. Virol.* **2005**, *32*, 19–23. [CrossRef] [PubMed]
30. Jeon, S.; Allen-Hoffmann, B.L.; Lambert, P.F. Integration of human papillomavirus type 16 into the human genome correlates with a selective growth advantage of cells. *J. Virol.* **1995**, *69*, 2989–2997. [PubMed]
31. Malik, S.M.; Nevin, D.T.; Cohen, S.; Hunt, J.L.; Palazzo, J.P. Assessment of immunohistochemistry for p16INK4 and high-risk HPV DNA by in situ hybridization in esophageal squamous cell carcinoma. *Int. J. Surg. Pathol.* **2011**, *19*, 31–34. [CrossRef] [PubMed]
32. Herbster, S.; Ferraro, C.T.; Koff, N.K.; Rossini, A.; Kruel, C.D.; Andreollo, N.A.; Rapozo, D.C.; Blanco, T.C.; Faria, P.A.; Santos, P.T.; et al. HPV infection in Brazilian patients with esophageal squamous cell carcinoma: Interpopulational differences, lack of correlation with surrogate markers and clinicopathological parameters. *Cancer Lett.* **2012**, *326*, 52–58. [CrossRef] [PubMed]
33. Cooper, K.; Taylor, L.; Govind, S. Human papillomavirus DNA in oesophageal carcinomas in South Africa. *J. Pathol.* **1995**, *175*, 273–277. [CrossRef] [PubMed]
34. Cooper, K.; Herrington, C.S.; Stickland, J.E.; Evans, M.F.; McGee, J.O. Episomal and integrated human papillomavirus in cervical neoplasia shown by non-isotopic in situ hybridisation. *J. Clin. Pathol.* **1991**, *44*, 990–996. [CrossRef] [PubMed]
35. Sano, T.; Oyama, T.; Kashiwabara, K.; Fukuda, T.; Nakajima, T. Immunohistochemical overexpression of p16 protein associated with intact retinoblastoma protein expression in cervical cancer and cervical intraepithelial neoplasia. *Pathol. Int.* **1998**, *48*, 580–585. [CrossRef] [PubMed]
36. Cao, F.; Zhang, W.; Zhang, F.; Han, H.; Xu, J.; Cheng, Y. Prognostic significance of high-risk human papillomavirus and p16(INK4A) in patients with esophageal squamous cell carcinoma. *Int. J. Clin. Exp. Med.* **2014**, *7*, 3430–3438. [PubMed]
37. Nogueira, M.C.; Guedes Neto Ede, P.; Rosa, M.W.; Zettler, E.; Zettler, C.G. Immunohistochemical expression of p16 and p53 in vulvar intraepithelial neoplasia and squamous cell carcinoma of the vulva. *Pathol. Oncol. Res.* **2006**, *12*, 153–157. [CrossRef] [PubMed]
38. Gronhoj Larsen, C.; Gyldenlove, M.; Jensen, D.H.; Therkildsen, M.H.; Kiss, K.; Norrild, B.; Konge, L.; von Buchwald, C. Correlation between human papillomavirus and p16 overexpression in oropharyngeal tumours: A systematic review. *Br. J. Cancer* **2014**, *110*, 1587–1594. [CrossRef] [PubMed]
39. Serup-Hansen, E.; Linnemann, D.; Skovrider-Ruminski, W.; Hogdall, E.; Geertsen, P.F.; Havsteen, H. Human papillomavirus genotyping and p16 expression as prognostic factors for patients with American Joint Committee on Cancer stages I to III carcinoma of the anal canal. *J. Clin. Oncol.* **2014**, *32*, 1812–1817. [CrossRef] [PubMed]
40. Zhao, P.; Hu, Y.C.; Talbot, I.C. Expressing patterns of p16 and CDK4 correlated to prognosis in colorectal carcinoma. *World J. Gastroenterol.* **2003**, *9*, 2202–2206. [CrossRef] [PubMed]

41. Tokugawa, T.; Sugihara, H.; Tani, T.; Hattori, T. Modes of silencing of p16 in development of esophageal squamous cell carcinoma. *Cancer Res.* **2002**, *62*, 4938–4944. [PubMed]
42. Koshiol, J.; Wei, W.Q.; Kreimer, A.R.; Chen, W.; Gravitt, P.; Ren, J.S.; Abnet, C.C.; Wang, J.B.; Kamangar, F.; Lin, D.M.; et al. No role for human papillomavirus in esophageal squamous cell carcinoma in China. *Int. J. Cancer* **2010**, *127*, 93–100. [CrossRef] [PubMed]
43. Teng, H.; Li, X.; Liu, X.; Wu, J.; Zhang, J. The absence of human papillomavirus in esophageal squamous cell carcinoma in East China. *Int. J. Clin. Exp. Pathol.* **2014**, *7*, 4184–4193. [PubMed]
44. Klaes, R.; Friedrich, T.; Spitkovsky, D.; Ridder, R.; Rudy, W.; Petry, U.; Dallenbach-Hellweg, G.; Schmidt, D.; von Knebel, D. Overexpression of p16(INK4A) as a specific marker for dysplastic and neoplastic epithelial cells of the cervix uteri. *Int. J. Cancer* **2001**, *92*, 276–284. [CrossRef] [PubMed]
45. Witkiewicz, A.K.; Knudsen, K.E.; Dicker, A.P.; Knudsen, E.S. The meaning of p16(ink4a) expression in tumors: Functional significance, clinical associations and future developments. *Cell Cycle* **2011**, *10*, 2497–2503. [CrossRef] [PubMed]
46. Bai, P.; Xiao, X.; Zou, J.; Cui, L.; Bui Nguyen, T.M.; Liu, J.; Xiao, J.; Chang, B.; Wu, J.; Wang, H. Expression of p14(ARF), p15(INK4b), p16(INK4a) and skp2 increases during esophageal squamous cell cancer progression. *Exp. Ther. Med.* **2012**, *3*, 1026–1032. [CrossRef] [PubMed]
47. Boffetta, P.; Hashibe, M. Alcohol and cancer. *Lancet Oncol.* **2006**, *7*, 149–156. [CrossRef]
48. Shephard, G.S.; van der Westhuizen, L.; Gatyeni, P.M.; Somdyala, N.I.; Burger, H.M.; Marasas, W.F. Fumonisin mycotoxins in traditional Xhosa maize beer in South Africa. *J. Agric. Food Chem.* **2005**, *53*, 9634–9637. [CrossRef] [PubMed]
49. Hendricks, D.; Parker, M.I. Oesophageal cancer in Africa. *IUBMB Life* **2002**, *53*, 263–268. [CrossRef] [PubMed]
50. Niller, H.H.; Wolf, H.; Minarovits, J. Viral hit and run-oncogenesis: Genetic and epigenetic scenarios. *Cancer Lett.* **2011**, *305*, 200–217. [CrossRef] [PubMed]
51. Campo, M.S.; Moar, M.H.; Sartirana, M.L.; Kennedy, I.M.; Jarrett, W.F. The presence of bovine papillomavirus type 4 DNA is not required for the progression to, or the maintenance of, the malignant state in cancers of the alimentary canal in cattle. *EMBO J.* **1985**, *4*, 1819–1825. [PubMed]
52. Quint, K.D.; Genders, R.E.; de Koning, M.N.; Borgogna, C.; Gariglio, M.; Bouwes Bavinck, J.N.; Doorbar, J.; Feltkamp, M.C. Human β-papillomavirus infection and keratinocyte carcinomas. *J. Pathol.* **2015**, *235*, 342–354. [CrossRef] [PubMed]
53. Viarisio, D.; Gissmann, L.; Tommasino, M. Human papillomaviruses and carcinogenesis: Well-established and novel models. *Curr. Opin. Virol.* **2017**, *26*, 56–62. [CrossRef] [PubMed]
54. Howley, P.M.; Pfister, H.J. β genus papillomaviruses and skin cancer. *Virology* **2015**, *479–480*, 290–296. [CrossRef] [PubMed]
55. Weissenborn, S.J.; Nindl, I.; Purdie, K.; Harwood, C.; Proby, C.; Breuer, J.; Majewski, S.; Pfister, H.; Wieland, U. Human papillomavirus-DNA loads in actinic keratoses exceed those in non-melanoma skin cancers. *J. Investig. Dermatol.* **2005**, *125*, 93–97. [CrossRef] [PubMed]
56. Kusters-Vandevelde, H.V.; de Koning, M.N.; Melchers, W.J.; Quint, W.G.; de Wilde, P.C.; de Jong, E.M.; van de Kerkhof, P.C.; Blokx, W.A. Expression of p14ARF, p16INK4a and p53 in relation to HPV in (pre-)malignant squamous skin tumours. *J. Cell. Mol. Med.* **2009**, *13*, 2148–2157. [CrossRef] [PubMed]
57. Abedi-Ardekani, B.; Hainaut, P. Cancers of the upper gastro-intestinal tract: A review of somatic mutation distributions. *Arch. Iran. Med.* **2014**, *17*, 286–292. [PubMed]
58. Holland, P.M.; Abramson, R.D.; Watson, R.; Gelfand, D.H. Detection of specific polymerase chain reaction product by utilizing the 5′—3′ exonuclease activity of Thermus aquaticus DNA polymerase. *Proc. Natl. Acad. Sci. USA* **1991**, *88*, 7276–7280. [CrossRef] [PubMed]
59. Schmitz, M.; Scheungraber, C.; Herrmann, J.; Teller, K.; Gajda, M.; Runnebaum, I.B.; Durst, M. Quantitative multiplex PCR assay for the detection of the seven clinically most relevant high-risk HPV types. *J. Clin. Virol.* **2009**, *44*, 302–307. [CrossRef] [PubMed]
60. Kleiboeker, S.B. Quantitative assessment of the effect of uracil-DNA glycosylase on amplicon DNA degradation and RNA amplification in reverse transcription-PCR. *Virol. J.* **2005**, *2*, 29. [CrossRef] [PubMed]

11

Protein Expression in Tonsillar and Base of Tongue Cancer and in Relation to Human Papillomavirus (HPV) and Clinical Outcome

Torbjörn Ramqvist [1,*], Anders Näsman [1], Bo Franzén [1], Cinzia Bersani [1], Andrey Alexeyenko [2,3], Susanne Becker [1], Linnea Haeggblom [1], Aeneas Kolev [4], Tina Dalianis [1] and Eva Munck-Wikland [4]

1 Department of Oncology-Pathology, Karolinska Institutet, Karolinska University Hospital, 171 76 Stockholm, Sweden; anders.nasman@ki.se (A.N.); bo.franzen@ki.se (B.F.); cinzia.bersani@ki.se (C.B.); susanne.becker@ki.se (S.B.); linnea.haeggblom@ki.se (L.H.); tina.dalianis@ki.se (T.D.)
2 Department of Microbiology, Tumor and Cell Biology (MTC), Karolinska Institutet, 171 77 Stockholm, Sweden; andrej.alekseenko@scilifelab.se
3 National Bioinformatics Infrastructure Sweden, Science for Life Laboratory, 17121 Solna, Sweden
4 Department of Clinical Science and Technology (CLINTEC), Karolinska Institutet, Karolinska University Hospital, 171 76 Stockholm, Sweden; aeneas.kolev@sll.se (A.K.); eva.munck-afrosenschold-wikland@sll.se (E.M.-W.)
* Correspondence: torbjorn.ramqvist@ki.se

Abstract: Human papillomavirus (HPV) is a major etiological factor for tonsillar and the base of tongue cancer (TSCC/BOTSCC). HPV-positive and HPV-negative TSCC/BOTSCC present major differences in mutations, mRNA expression and clinical outcome. Earlier protein studies on TSCC/BOTSCC have mainly analyzed individual proteins. Here, the aim was to compare a larger set of cancer and immune related proteins in HPV-positive and HPV-negative TSCC/BOTSCC in relation to normal tissue, presence of HPV, and clinical outcome. Fresh frozen tissue from 42 HPV-positive and 17 HPV-negative TSCC/BOTSCC, and corresponding normal samples, were analyzed for expression of 167 proteins using two Olink multiplex immunoassays. Major differences in protein expression between TSCC/BOTSCC and normal tissue were identified, especially in chemo- and cytokines. Moreover, 34 proteins, mainly immunoregulatory proteins and chemokines, were differently expressed in HPV-positive vs HPV-negative TSCC/BOTSCC. Several proteins were potentially related to clinical outcome for HPV-positive or HPV-negative tumors. For HPV-positive tumors, these were mostly related to angiogenesis and hypoxia. Correlation with clinical outcome of one of these, VEGFA, was validated by immunohistochemistry. Differences in immune related proteins between HPV-positive and HPV-negative TSCC/BOTSCC reflect the stronger activity of the immune defense in the former. Angiogenesis related proteins might serve as potential targets for therapy in HPV-positive TSCC/BOTSCC.

Keywords: tonsillar cancer; base of tongue cancer; oropharyngeal cancer; human papillomavirus; proximity extension assay; clinical outcome; protein expression

---

1. Introduction

The incidence of oropharyngeal squamous cell carcinoma (OPSCC) has increased in many Western countries, due to a rise in incidence of human papillomavirus (HPV) positive tonsillar squamous cell carcinoma (TSCC) and base of tongue squamous cell carcinoma (BOTSCC), the two most commonly HPV-positive OPSCC subsites [1–3]. Notably, HPV-positive TSCC/BOTSCC has a much better clinical outcome than the corresponding HPV-negative cancer, with the latter mainly being caused by smoking [4–6]. HPV-positive and HPV-negative TSCC/BOTSCC diverge with regard to chromosomal rearrangements, mutations (e.g., p53), RNA and microRNA expression, and DNA methylation [7–11]. Moreover, some proteins differ in expression between HPV-positive and HPV-negative TSCC/BOTSCC and in their relation to clinical outcome (e.g., p16, CD44) [11–13]. Most studies on expression at the protein level have been performed on specific proteins by immunohistochemistry (IHC), and thus rather few proteins have been analyzed in total. For this reason, there is a lack of data on differences in protein expression between HPV-positive and HPV-negative TSCC/BOTSCC, as well as between these entities and normal tissue and in relation to clinical outcome. Several studies have demonstrated differences with regard to cellular immune response between these tumor types, but they have mainly targeted differences in cell populations, e.g., CD8+, CD4+ and FoxP3+ T-cells, whereas only few immune related proteins, e.g., PD-1 and APM components, have been specifically evaluated [14–19].

Olink multiplex assays are recently developed immunoassays, based on proximity extension assay (PEA) technology, with high sensitivity and reproducibility over a wide range of protein concentrations [20]. Each panel targets 94 cancer and/or immune related proteins, and only requires a small amount of sample. Here, the Olink Oncology II and Immuno-Onc panels were employed to evaluate cancer and immune related proteins in TSCC/BOTSCC in relation to tumor HPV status, normal tissue, and clinical outcome.

2. Results

Fifty-nine TSCC/BOTSCC (Table 1) and their corresponding normal samples were analyzed on Olink Immuno-Oncology and Oncology II panels covering 167 proteins. A heat map combined with a dendrogram demonstrated that 58/59 replicates of the tumor samples were positioned as branches that were directly adjoint to each other, so for future analysis an average between the replicate values was used. The final analysis included 155 proteins in 55 tumor and normal samples. More specifically, the analysis of 12 proteins (IL-2, IL-4, IL-5, IL-10, IL-13, IL-21, IL-35, IFN-β, IFN-γ, PPY, SDF-1, SEZ6L) with values that are below the level of detection (LOD, as defined by Olink) in >70% samples and low values in the others, were not included and four tumor/normal sample pairs, where internal and/or detection controls failed, were excluded.

2.1. Comparison between Tumor and Normal Samples

A heatmap visualization revealed a clear separation between cancer and normal tissue (Figure S1). HPV-positive and HPV-negative cancers were then compared separately with their corresponding normal samples (HPV-positive tumors with normal samples from patients with HPV-positive tumors, etc.). For HPV-positive tumors vs. normal samples, 111/155 proteins showed significant differences and of these 44 proteins, 38 proteins >4-fold upregulated and 6 >2-fold downregulated in the tumors, are presented in Supplementary Table S1 (Bonferroni-adjusted p-values < 0.05, with exception of HK8 for which it was $p = 0.1$). For HPV-negative tumors vs. normal samples 78/155 proteins differed significantly, and 37 proteins (29 >4-fold upregulated and 8 >2-fold downregulated are presented in Supplementary Table S2.

Table 2 presents the 13 proteins with the strongest upregulation in HPV-positive tumors vs. normal samples. The corresponding values for HPV-negative tumors, which are also presented in Table 2, reveal that these proteins were also highly upregulated in those tumors although some (e.g., CCL20 and CXCL11) were apparently less upregulated than in HPV-positive tumors. Most of the proteins in Table 2 were chemokines and cytokines, although e.g., CA9 with >50-fold upregulation is hypoxia related.

Table 1. Patient and tonsillar squamous cell carcinoma (TSCC)/base of tongue squamous cell carcinoma (BOTSCC) characteristics.

Patient and Tumor Characteristics		HPV+TSCC/BOTSCC (n = 42)		HPV-TSCC/BOTSCC (n = 17)		All TSCC/BOTSCC (n = 59)		Validation set (n = 49)	
		n	%	n	%	n	%	n	%
Age	Mean (years)	61.7		60.6		61.4		62.1	
	Median (years)	61		64		61		61	
	Range (years)	46–84		32–83		32–84		42–84	
Diagnose	malignant neoplasm of the base of tongue (C01.9)	9	21%	9	53%	18	31%	13	22%
	malignant neoplasm of the tonsil (C09.0-9)	33	79%	8	47%	41	69%	36	61%
Sex	female	8	19%	3	18%	11	19%	13	22%
	male	34	81%	14	82%	48	81%	36	61%
Tumour differentiation	poorly	21	50%	9	53%	30	51%	29	49%
	moderatley	17	40%	8	47%	25	42%	18	31%
	well	2	5%	0	0%	2	3%	1	2%
	undefined	2	5%	0	0%	2	3%	1	2%
Tumour size	T1	6	14%	3	18%	9	15%	8	14%
	T2	16	38%	2	12%	18	31%	20	34%
	T3	11	26%	5	29%	16	27%	10	17%
	T4	9	21%	7	41%	16	27%	11	19%
Nodal disease	N0	5	12%	6	35%	11	19%	5	8%
	N1	12	29%	1	6%	13	22%	11	19%
	N2a	5	12%	3	18%	8	14%	5	8%
	N2b	15	36%	3	18%	18	31%	21	36%
	N2c	5	12%	2	12%	7	12%	7	12%
	N3	0	0%	1	6%	1	2%	0	0%
	NX	0	0%	1	6%	1	2%	0	0%
Distant metastasis	M0	42	100%	16	94%	58	98%	48	81%
	M1	0	0%	0	0%	0	0%	0	0%
	MX	0	0%	1	6%	1	2%	1	2%
Tumour Stage	I	0	0%	3	18%	3	5%	0	0%
	II	4	10%	1	6%	5	8%	3	5%
	III	11	26%	2	12%	13	22%	10	17%
	IVa	27	64%	9	53%	36	61%	34	58%
	IVb	0	0%	1	6%	1	2%	0	0%
	IVc	0	0%	0	0%	0	0%	1	2%
	Unknown	0	0%	1	6%	1	2%	0	0%
Treatment	Induction chemotherapy and radiation: conventional	4	10%	3	18%	7	12%	8	14%
	Induction chemotherapy and radiation: accelerated	11	26%	9	53%	20	34%	14	24%
	Radiation: conventional	20	48%	4	24%	24	41%	21	36%
	Radiation: accelerated	7	17%	1	6%	8	14%	6	10%
Brachytherapy boost	Not administered	34	81%	12	71%	46	78%	38	64%
	Administered	8	19%	7	41%	15	25%	11	19%
Concomittant Cetuximab	Not administered	35	83%	13	76%	48	81%	42	71%
	Administered	7	17%	4	24%	11	19%	7	12%
Smoking	Never	15	36%	2	12%	17	29%	16	27%
	Former (>15 years ago)	11	26%	1	6%	12	20%	11	19%
	Former (<15 years ago)	7	17%	3	18%	10	17%	11	19%
	Current upon diagnosis	9	21%	11	65%	20	34%	11	19%

Table 2. Proteins with the most prominent differences between TSCC/BOTSCC and normal samples.

Protein	HPV Positive TSCC/BOTSCC vs. Normal		HPV Negative TSCC/BOTSCC vs. Normal	
	Ratio	p-Value	Ratio	p-Value
CA9	55.72	1.10×10^{-21}	50.56	1.80×10^{-8}
CXCL13	37.27	3.70×10^{-14}	21.26	1.40×10^{-5}
CXCL10	36.25	1.40×10^{-19}	15.89	1.50×10^{-6}
CCL4	35.75	9.30×10^{-24}	20.25	1.80×10^{-9}
MMP-12	35.51	4.60×10^{-18}	33.36	3.90×10^{-9}
CCL3	34.78	1.30×10^{-20}	38.85	3.70×10^{-10}
CCL20	25.81	5.40×10^{-13}	11.31	0.00011
CXCL11	22.32	3.90×10^{-18}	11.39	6.30×10^{-6}
IL-6	20.97	3.40×10^{-15}	33.59	3.40×10^{-8}
CXCL9	19.43	2.50×10^{-18}	9.78	7.70×10^{-6}
IL-8	19.16	5.90×10^{-15}	24.42	1.20×10^{-7}
TNFRSF9	18.25	1.50×10^{-15}	11	3.80×10^{-6}
WISP-1	16.34	2.30×10^{-23}	13.64	2.10×10^{-9}

2.2. Predictive Models for HPV-Positive and HPV-Negative Tumors

Predictive models, being potentially useful for distinguishing between cancer and normal tissues were developed by differential protein expression analysis of HPV-positive and HPV-negative tumors against the normal samples (Figure S2). For HPV positive tumors, the following multiple regression model was obtained; $T = -7.3 + 0.62 \times \text{WISP1} + 0.44 \times \text{CXCL10} + 0.39 \times \text{CCL7} - 0.3 \times \text{ICOSLG} + 0.27 \times \text{CA9} - 0.27 \times \text{IL-18} + 0.18 \times \text{VEGFA} - 0.15 \times \text{FR}\alpha - 0.073 \times \text{CD207} + 0.064 \times \text{TCL1A} + 0.04 \times \text{CX3CL1}$, where the protein names denote the respective expression variables, T is the 'tumor/normal' status, and the first term is the intercept of the linear equation. Since the expression values were in the same range after the Olink normalization, the regression coefficients convey proteins' importance of the specific proteins in this model.

For HPV negative tumors, the model was simpler and shared with the former model for all of the proteins, except CCL4, ANG2, and CCL3: $T = -1.6 + 0.47 \times \text{CCL7} - 0.35 \times \text{IL-18} + 0.22 \times \text{CCL4} + 0.17 \times \text{CA9} + 0.13 \times \text{ANG2} + 0.12 \times \text{WISP1} + 0.072 \times \text{CCL3}$.

The algorithm that produced these models aimed at a tradeoff between the models' simplicity and precision. Thus, these protein terms are the least redundant sets sufficient for distinguishing between tumor and normal samples given the absence/presence of the HPV infection. Despite the perfect separation between the tumor and normal samples within this dataset, robustness and reproducibility of the models should be validated using independently collected data in future studies.

2.3. Comparison between HPV-Positive and HPV-Negative Tumors

As noted above, the difference in expression levels between the groups of HPV-positive and HPV-negative tumors was generally less pronounced than between those of tumors vs. normal tissues. Table 3 presents the 34 proteins with significantly different expression in HPV-positive and HPV-negative tumors (unadjusted $p < 0.05$ and FDR < 0.25). Boxplots presenting the expression levels in Mucin-16, FasL, CD8A, and HK14 for HPV-positive and HPV-negative tumors are shown in Figure 1A.

Table 3. Protein with significant differences in expression between HPV positive and negative TSCC and BOTSCC.

Protein	Ratio HPV-Positive/HPV-Negative (Linear)	p-Value	False Discovery Rate (FDR)
MUC-16	4.89	0.0016	0.062
CXCL17	2.73	0.025	0.162
CCL20	2.51	0.0046	0.101
CD8A	2.23	0.024	0.162
FASL	2.23	0.0065	0.101
CCL4	2.20	0.00093	0.057
CXCL10	2.19	0.037	0.186
PD-L1	2.13	0.0024	0.062
IL-12	2.10	0.025	0.162
LY9	1.88	0.026	0.162
CD27	1.83	0.025	0.162
PDCD1	1.62	0.031	0.173
KLRD1	1.61	0.018	0.148
DKN1A	1.53	0.031	0.173
GZMH	1.52	0.046	0.209
NCR1	1.51	0.016	0.148
IL12RB1	1.45	0.015	0.148
CPE	1.45	0.047	0.209
IL-7	1.37	0.016	0.148
VEGFC	0.89	0.046	0.209
ITGAV	0.78	0.017	0.148
HO-1	0.74	0.0096	0.120
GZMB	0.74	0.01	0.120
FURIN	0.73	0.035	0.186

Table 3. *Cont.*

Protein	Ratio HPV-Positive/HPV-Negative (Linear)	*p*-Value	False Discovery Rate (FDR)
TXLNA	0.72	0.027	0.162
GPNMB	0.70	0.0054	0.101
TNFRSF12A	0.66	0.046	0.209
GAL-9	0.59	0.0011	0.057
ADAMTS15	0.59	0.009	0.120
HK8	0.45	0.037	0.186
TFPI2	0.35	0.0061	0.101
AR	0.34	0.0024	0.062
WIF-1	0.24	0.018	0.148
HK14	0.16	0.00032	0.050

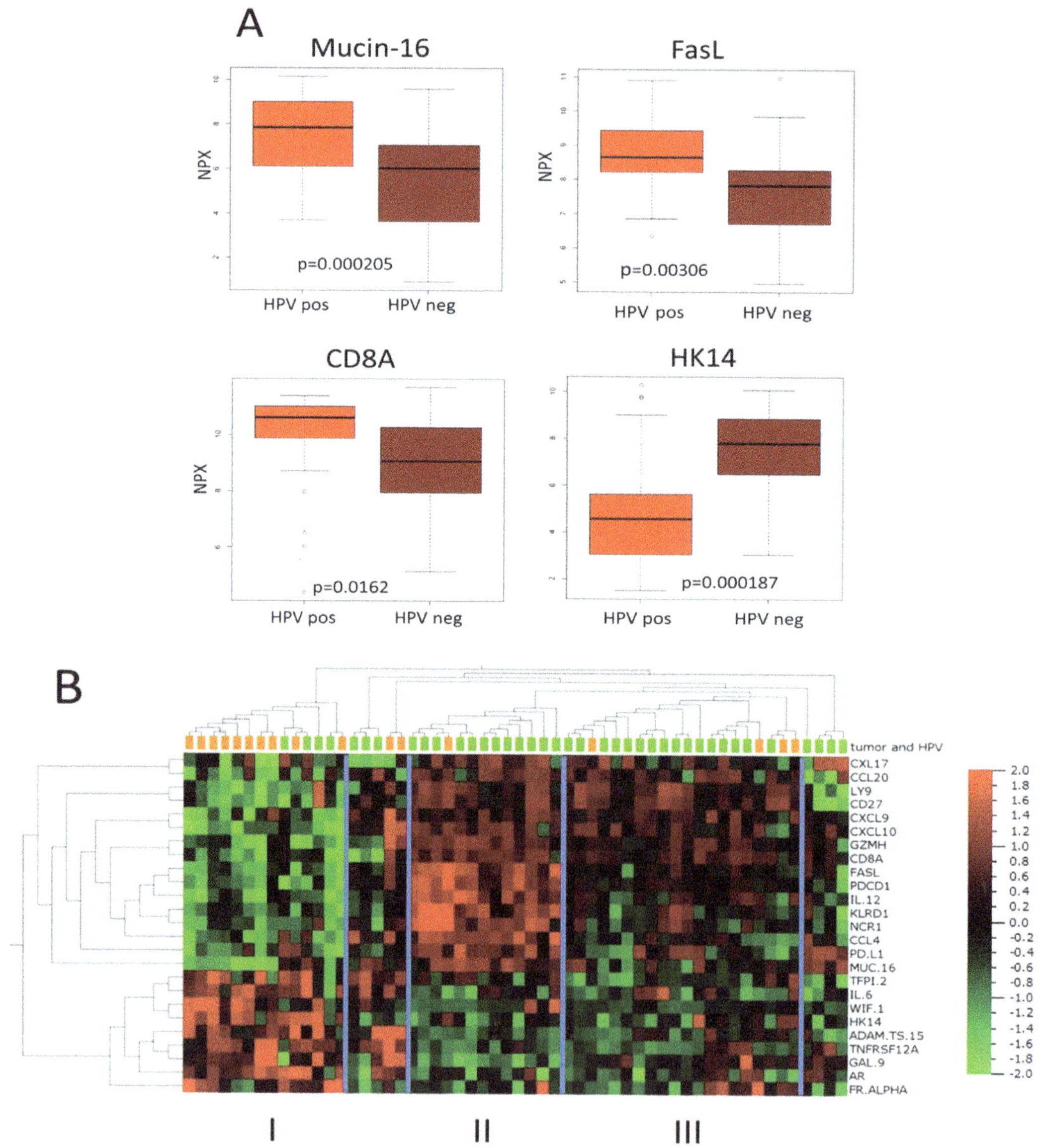

Figure 1. Protein expression in human papillomavirus (HPV) positive and negative TSCC/BOTSCC. (**A**) Boxplots presenting the expression of Mucin-16 (MUC16), Fas ligand (FasL), CD8A, and Human Kallikrein 14 (HK14) in HPV positive and negative TSCC/BOTSCC. NPX-values in 2-log scale. (**B**) Heatmap presenting the expression of 22 proteins with significant differences and >1.5 fold difference between HPV positive and negative TSCC/BOTSCC. Green, HPV positive and orange, HPV negative tumors. Three tentative clusters of samples are marked as I, II, and III.

Out of nine proteins with >2-fold higher in expression in HPV-positive tumors, than in HPV-negative tumors, most were surface immunoregulatory proteins that were present on immune or tumor cells (e.g., CD8A, PD-L1, FasL) or chemokines. Notably, however Mucin-16, with a nearly five-fold difference in expression, is a glycoprotein that is present in mucosal surfaces. In contrast, five proteins with >2-fold lower expression in HPV-positive tumors than in HPV-negative tumors were not immune related proteins, e.g., two kallikreins, HK8 and 14, and Wnt inhibitory factor 1 (WIF-1).

In a heatmap, presenting proteins differentially expressed in the HPV-positive as compared with the HPV-negative tumors, most HPV-positive tumors separated into two out of three main clusters (cluster II and III, Figure 1B). These two clusters differed with regard to expression of several membrane molecules related to the immune defense e.g., PD-L1, PD-1, FasL, NCR1, and KLRD1, whereas e.g., CD8A, CD27, CXCL9, and CXCL10 had similar and high expression levels. The third, cluster I, grouped together with the majority of HPV-negative samples, and was characterized by low values of many immune related proteins such as CD8A and the chemokines CXCL9, 10, 17, and CCL20, and by higher values of e.g., Gal9, HK14, ADAMTS15, and IL6.

2.4. Protein Expression in Relation to Clinical Outcome

Proteins with formally significant expression differences (unadjusted $p < 0.05$), possibly associated to clinical outcome, are depicted for HPV-positive and HPV-negative tumors in Tables 4 and 5. Tumor samples from patients, dead of other causes within 4-years, were excluded in this analysis.

Table 4. Proteins related to recurrence in HPV positive TSCC and BOTSCC.

Protein	Ratio * (Linear)	*p*-Value
DLL1	8.38	0.000422
ESM-1	5.76	0.00286
TNFRSF19	5.41	0.000145
VEGFA	4.01	0.00241
CYR61	3.64	0.0044
CCL7	3.23	0.0402
PLGF	2.89	0.0127
MIC-A/B	2.67	0.0452
ANG-2	2.33	0.00218
TNFRSF21	2.20	0.0411
IGF1R	1.78	0.00847
CSF-1	1.74	0.0361
GPNMB	1.62	0.0347
ICOSLG	1.41	0.0441
IFN-GAMMA-R1	1.40	0.037

* Ratio between tumors with recurrence vs tumors from patients with 4-year disease free survival.

Table 5. Proteins related to recurrence in HPV negative TSCC and BOTSCC.

Protein	Ratio * (Linear)	*p*-Value
WFDC2	4.11	0.000339
EGF	4.03	0.0239
MIA	2.84	0.00962
VEGFR-2	0.76	0.0275
EPHA2	0.68	0.0167
DKN1A	0.58	0.0382
ANG2	0.52	0.00641
TNFRSF4	0.51	0.0449
TRAIL	0.50	0.045
CXCL11	0.26	0.0462
IL-1-ALPHA	0.26	0.00104

* Ratio between tumors with recurrence vs tumors from patients with 4-year disease free survival.

Notably, despite the noticeable fold change ratios, the p-value levels were not always sufficiently low (e.g., only TNFRS19 (Troy) was significant after Bonferroni correction for multiple testing). This could be explained by the relatively small sample sizes and these findings require further validation in an extended cohort. For HPV-positive tumors, 15 proteins were related to recurrence after therapy, all with a higher expression in the primary tumors with later recurrence (Table 4). The 13 proteins with >2-fold higher expression in tumors with recurrence, included proteins that were at least partially related to angiogenesis (DLL1, ESM1, VEGFA, CYR61, and PlGF), and two proteins belonging to the TNF receptors, TNFRSF19 and TNFRSF21. Noteworthy, the IGF receptor IGF1R showed increased expression levels in tumors with recurrence, while no proteins had significantly lower expression. Boxplots for TNFRSF19, VEGFA, ESM-1, and Cyr61 in HPV-positive tumors with/without recurrence are presented in Figure 2A.

In the HPV-negative tumors, 11 proteins were related to recurrence with only three (WFDC2, EGF, and MIA) showing a >2-fold higher expression, while the majority had a lower expression, with IL1α, CXCL11, and TRAIL being the most prominent (Table 5). Expression levels of some proteins that were related to recurrence in HPV negative tumors are presented in boxplots (Figure 2B).

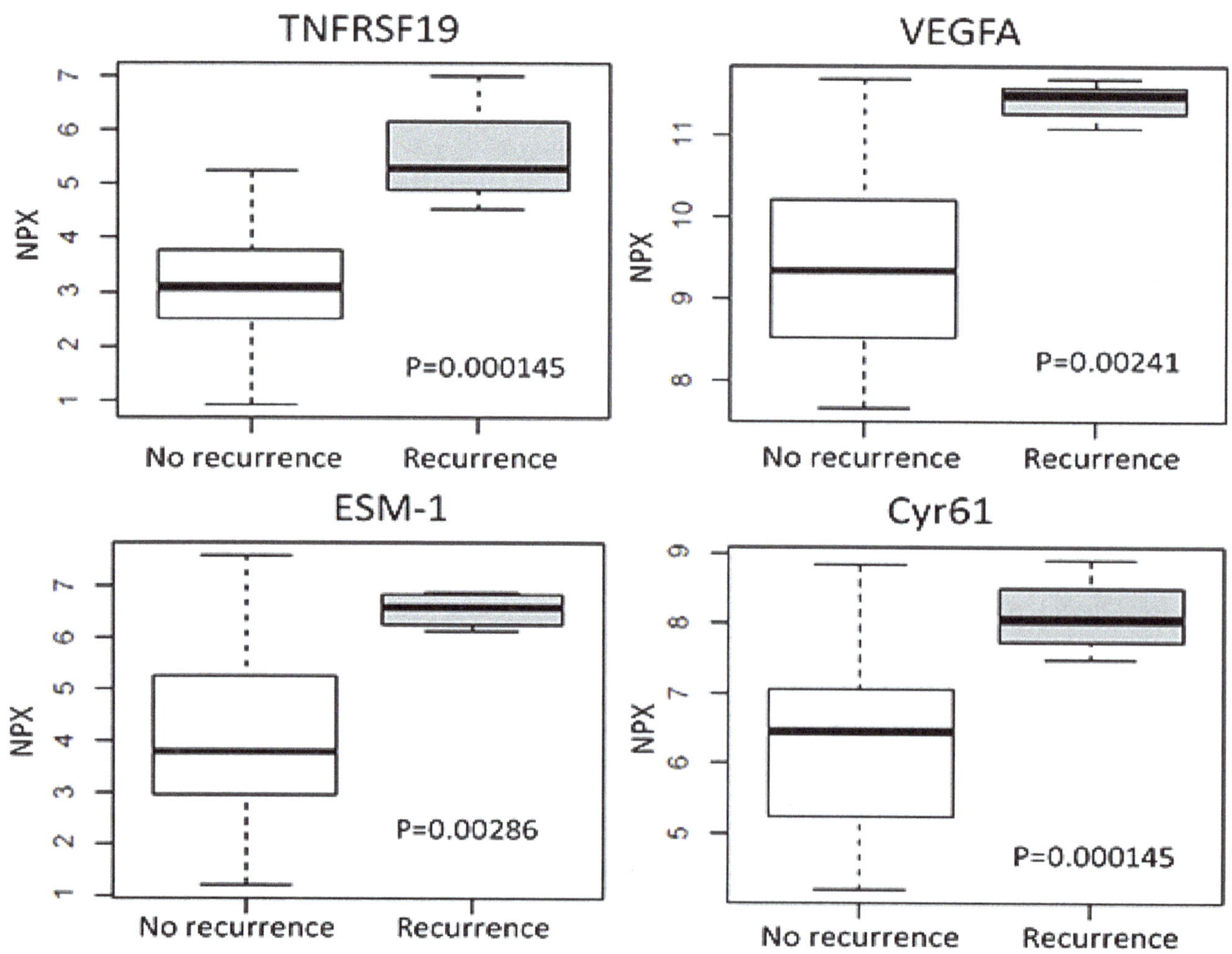

(A)

Figure 2. *Cont.*

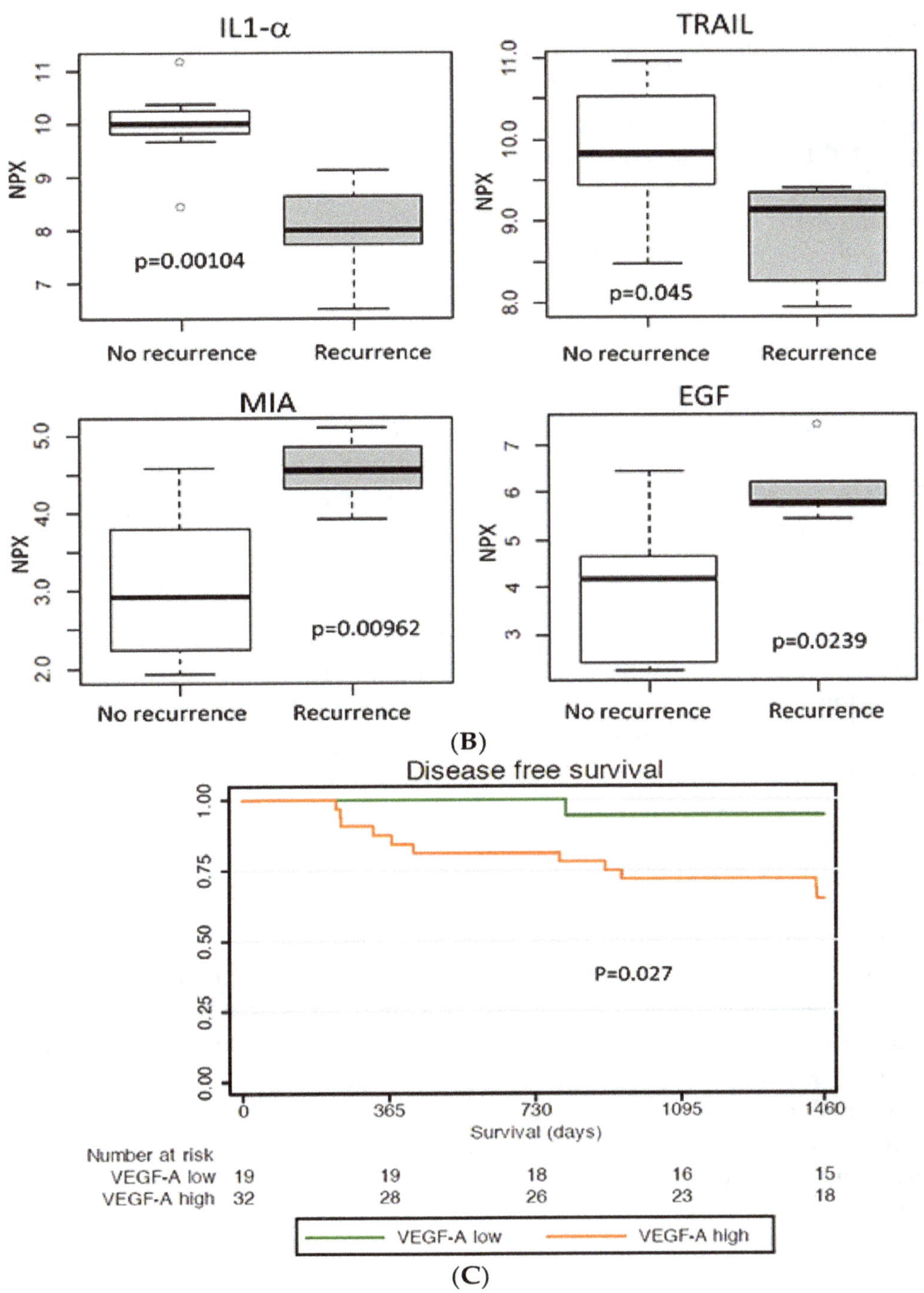

Figure 2. Protein expression in relation to clinical outcome. (**A**) Boxplots presenting the expression of in HPV positive oropharyngeal squamous cell carcinoma (OPSCC) with recurrence vs those with four-year tumor free survival. (**B**) Boxplots presenting the expression of specific proteins in HPV negative OPSCC with recurrence and or death with tumor vs those with four-year tumor free survival. NPX-values in two-log scale. (**C**) Kaplan–Meier curves of disease free survival for 49 patients dichotomized based on tumor VEGFA expression in the validation set. Numbers below the diagram refer to the number of patients at risk at the specific time point.

2.5. Proteins Related to High CD8A Expression

CD8+ positive tumor infiltrating lymphocytes (CD8+ TILs) have earlier been related to improved prognosis for both HPV-positive and HPV-negative TSCC/BOTSCC, although the latter in general have a lower number of CD8+ TILs [21,22]. For this reason, a specific evaluation of proteins that were related to the expression of CD8A, present in the panel, was of special interest.

This evaluation was performed by correcting for % tumor cells in the sample both for all of the tumor samples taken together and then separately for HPV-positive and HPV-negative tumors. Proteins with $p < 0.01$ for all of these analysis are presented in Table S3. When all tumor samples were evaluated together the correlation of GZMH, ABL1, IL-7, LY9, IL-12, SCAMP3, and CD244 were highly significant ($p < 1 \times 10^{-6}$). The correlates usually remained significant even after the fraction of tumor cells was accounted for as a covariate, indicating that the analysis was largely unaffected by this factor.

2.6. Protein Expression in Relation to Tumor T-Stage

Protein expression was also evaluated in relation to tumor T-stage, by comparing the pooled set of T1/T2 with that of T3/T4, for HPV-positive and HPV-negative tumors separately. Significant differences are presented in Table S4. For HPV-positive tumors, the relation between protein expression and T-stage was minor, with <2-fold ratios of the five proteins with significant differences. For HPV-negative tumors more proteins demonstrated profound differences, with MMP7, CXCL17, and CEACAM5 exhibiting >4-fold higher expression at stage T3/T4 vs. stage T1/T2.

2.7. Validation of VEGFA Expression in Relation to Prognosis

To validate the expression of VEGFA in relation to clinical outcome a set of 49 TSCC/BOTSCC biopsies was evaluated for VEGFA expression by IHC. The tumors were then dichotomized based on the VEGFA intensity and analyzed in relation to disease free and progression free survival. Strong VEGFA expression was found to be correlated to poor disease-free survival (DSF), $p = 0.027$, Figure 2C, whereas for progression-free survival (PFS), this tendency failed to reach significance ($p = 0.107$).

3. Discussion

In this study, 55 TSCC/BOTSCC biopsies and corresponding normal samples were analyzed for the expression of 155 cancer and immune related proteins, and differences in protein expression between tumor and normal tissue as well as between HPV-positive and HPV-negative TSCC/BOTSCC, were identified. Furthermore, proteins potentially related to clinical outcome, for HPV-positive and HPV-negative tumors, respectively, were identified. One of these proteins, VEGFA, was validated and was shown to be related to disease-free survival for HPV-positive TSCC/BOTSCC.

The proteins included in the Olink panels utilized in the present study, had been selected as directly or indirectly cancer related and involved in processes such as angiogenesis, cell-cell signaling, cell-cycle control, tumor-immunity, chemotaxis, apoptosis, and cell killing (www.olink.com/products). Therefore, it was not surprising that many differences in protein expression between tumor and normal tissue were found. That fewer (78/155) proteins differed in expression between HPV-negative tumors as compared to HPV-positive tumors (111/155) vs. their corresponding normal samples was likely a combined effect of a smaller cohort (i.e., lower statistical power) and/or a less active immune defense in the former.

The most upregulated proteins in the tumor tissue were mainly chemokines and cytokines, indicating an increased activity of the immune response, as especially shown for chemokines CXCL9, 10, 11, 13, and CCL3, 4 and 20, and cytokines IL-6 and 8. A recent analysis of 20 different cytokines and chemokines (some also included here) in culture supernatants harvested from HNSCC tumor tissue-derived cell suspensions, showed, in accordance with the present study, high tumor related levels of CXCL9, CXCL10, and CCL20 and decreased levels of CCL5 [23]. In parallel to this study, higher expression of CXCL9 and CXCL10 was noted in HPV-positive tumors than in HPV-negative tumors. Furthermore, in a study on melanoma a 12-chemokine signature that was related to immune infiltrates included high expression of CXCL9, 10, 11, 13, and CCL 3 and 4, similarly to TSCC/BOTSCC in the present study [24].

Proteins with a distinct lower expression in HPV-positive TSCC/BOTSCC vs. normal samples included kallikreins KLK13, HK8, and HK11. Kallikreins have mostly been investigated as serum markers in prostate cancer, but also in some other cancer types e.g., lung cancer where HK8 has been found to be related to favorable outcome and shown to suppress tumor invasiveness in lung cancer, while little is known about their expression in HNSCC and TSCC/BOTSCC [25].

Proteins with higher expression in HPV-positive vs. HPV-negative TSCC/BOTSCC were mainly immune related. Several were surface proteins expressed on immune cells, e.g., CD8A, and PD-1, or on cancer cells and affecting immune activity, e.g., PD-L1, FASL, and notably PD-1 and its ligand PD-L1 were similarly upregulated. A higher CD8A expression was expected, since increased numbers of CD8+, FoxP3+, and CD4+ TILs in HPV-positive vs. HPV-negative OPSCC have been noted earlier, where in particular a higher number of CD8+TILs is linked to a better clinical outcome, irrespective of the HPV status of the tumor [19,21,22].

PD-1 and PD-L1 exhibited a higher expression in HPV-positive when compared to HPV-negative tumors in this study. Interaction of PD-1, expressed on the surface of activated T-cells, B-cells, and macrophages, and PD-L1, expressed on immune and cancer cells can suppress the activity of CD8+ T-cell mediated immune response, and have therefore received attention for immunotherapy [26,27]. Recent studies investigating PD-L1 and PD-1 expression in OPSCC gave partly contradictory results, with two studies showing higher PD-L1 mRNA levels in HPV-positive cancer, while the third study reported the opposite result [19,23,28]. Notably, since PD-L1 is expressed on both cancer and tumor infiltrating immune cells, this study, similarly to global mRNA analysis, does not differentiate the expression between these cell types.

In parallel with the higher immune infiltration in HPV-positive than in HPV-negative tumors, some cytokines and chemokines e.g., IL12, CCL4, CCL20, CXCL10, CXCL17 had higher expression in the former, with IL12, notably being an important regulator of T-cell and NK-cells cytotoxicity, and its receptor, IL12R1-β1, was also upregulated in the HPV-positive tumors.

Finally, Mucin-16 (also known as MUC-16), a transmembrane protein that functions as a barrier to bacterial infections, protects cancer cells from being killed by immune cells and contains the CA125 peptide [29], demonstrated a nearly 5-fold difference between HPV-positive and HPV-negative tumors. Mucin-16/CA125 has especially been studied in ovarian cancer where it is used both as a diagnostic and predictive marker and to evaluate the response to therapy in this tumor type [29]. The reason behind the increased expression in HPV-positive vs. HPV-negative TSCC/BOTSCC is yet unclear and needs further investigation.

Few other studies have analyzed a large number of proteins simultaneously in OPSCC in relation to tumor HPV status and/or clinical outcome. Proteomic profiling of OPSCC by mass spectrometry disclosed e.g., enrichment of E2F1 and E2F4 in HPV-positive OPSCC, while reverse-phase protein array profiling revealed differences in e.g., P13K/AKT/mTOR and receptor kinase pathways [30,31].

Here, most HPV-positive tumors separated into clusters, mainly due to differences in the expression of immune related proteins on the cell surface of tumor infiltrating immune cells, e.g., PD-1, FasL, NCR1, and KLRD1 or soluble cytokines/chemokines e.g., IL12 and CCL4 (Figure 1B). High expression of FasL, PD-1, and KLRD1 indicates a high infiltration of several parts of the immune defense with PD-1, mainly expressed on T-, B-cells and macrophages, and NCR1 and KLRD1 mainly expressed on NK-cells [32,33]. Notably, CD8A expression did not differ between cluster II and III indicating no major differences in numbers of CD8+ T-cells between these clusters, in contrast to that observed between HPV-positive and HPV-negative tumors. Similarly, several of the chemokines differing in the expression between HPV-positive and HPV-negative tumors do not differ between these clusters either, e.g., CXCL9 and 10.

Global mRNA expression between HPV-positive and HPV-negative OPSCC has also shown differences in proteins not included in the present study e.g., CDKN2A, NF-KB, and STAT3, making comparisons difficult [34–36]. Notably, in one study HPV-positive OPSCC was split into two subtypes, a Classical subtype (CL)-HPV and an Inflamed/mesenchymal subtype (IMS)-HPV, where the latter was characterized by higher expression of e.g., immune response genes, such as CD8A, ICOS, LAG3, and HLA-DRA, related to CD8 T-cell infiltration [36]. Here, out of these, only CD8A was analyzed and it did not present differences between the two major clusters although HPV-positive tumors in a third cluster (I), dominated by HPV-negative tumors, presented a low amount of CD8A expression.

In this study, some of the proteins with the most pronounced potential correlation to clinical outcome in HPV-positive TSCC/BOTSCC were at least partly related to angiogenesis e.g., DLL1, ESM1 (endocan), VEGFA, CYR61, and PlGF. Higher VEGFA expression in the tumor cells was confirmed by IHC to be associated with poorer DSF. High VEGFA expression, especially in combination with high EGFR expression has earlier been linked to local recurrence in TSCC in one study, and likewise in oral and laryngeal cancer, but not in a third study, including HPV-positive and HPV-negative OPSCC [37–40]. Notably, VEGFA induces the expression of ESM1, which is a mediator of the angiogenic effect of VEGFA [41]. Moreover, ESM1 has been related to shorter survival in e.g., breast, liver, and nasopharyngeal cancer, but has not been investigated in OPSCC [42,43].

Angiogenesis is closely linked to hypoxia, and VEGFA together with CA9, are also markers for hypoxia, both being regulated by HIF1α and are involved in the induction of members of the Notch-pathway where DLL1 functions as a ligand for Notch1 [44]. Also, PlGF, which is a ligand for VEGFA, is induced by hypoxia [45].

Noteworthy, the IGF receptor IGF1R also demonstrated a higher expression in tumors that recurred. IGF1R is potentially targetable for therapy and can inhibit the effect of anti-EGFR therapy by acting on the same downstream pathway [46,47]. Since current TSCC and BOTSCC therapy often includes EGFR inhibitors, there is a potential need to combine such treatment with anti-IGF1R therapy. IGF1R has also previously been linked to poor prognosis in HPV-negative OPSCC [48].

There is a need for new targets for therapy of TSCC and BOTSCC irrespective of HPV-status. Recently, earlier treatment that entailed conventional radiotherapy and/or surgery has been intensified by increased radiotherapy, chemoradiotherapy, and/or EGFR-inhibitors. Intensified treatment leaves the patients with more side effects, but has not improved survival for patients with HPV-positive cancer in Stockholm, Sweden [49]. Treatment for recurrent TSCC and BOTSCC is a real challenge and the results are poor. Proteins related to recurrence in the present study, such as e.g., IGF1R and TNFRSF19/Troy, may possibly be utilized as such targets.

Protein expression was also analyzed in relation to tumor T-stage and major differences related to T-stage were found for some proteins in HPV-negative tumors, but not for HPV-positive tumors. Although T-stage has been shown to be related to clinical outcome [49] the proteins with the strongest relation to T-stage for HPV-negative tumors, MMP7, CSCL17, and CEACAM5, were not found to be related to survival. There are limitations in the present study. Only 59 tumors were included, and when HPV-positive and especially HPV-negative tumors were analyzed separately or compared, random correlations might have been obtained. In addition, both the treatment and T-stage of the tumors included in the cohort were heterogeneous, as presented in Table 1. For proteins related to clinical outcome these issues may have had an effect on the results obtained, especially given the few events. Thus, the result presented here must be interpreted with caution.

In summary, when comparing protein expression in TSCC/BOTSCC with normal tissue, the most prominent differences were found for chemo-and cytokines. Between HPV-positive vs. HPV-negative tumors, most of the differences were detected in the immune related proteins as well as cyto-and chemokines. Some proteins were tentatively related to clinical outcome. For HPV-positive tumors, such proteins were mostly related to hypoxia and angiogenesis, and some may be potentially targetable and need to be further evaluated.

4. Material and Methods

4.1. Patients and Tumor Biopsies

Fifty-nine study specific pre-treatment tumor biopsies of TSCC (ICD-10 code C09.0-9) and BOTSCC (ICD-10 code C01.9), and adjacent normal tissue, from patients treated 2002–2011 at the Karolinska University Hospital were snap frozen and stored at −70 °C until the cutting of the samples. Corresponding diagnostic formalin fixed paraffin embedded (FFPE) pre-treatment biopsies from the tumors had previously been analyzed for HPV DNA and p16INK4A (p16) overexpression, where being positive for both was defined as HPV-positive status [11]. Patient and tumor characteristics are presented in Table 1, where the AJCC/UICC 7th ed is used for staging. Validation of VEGFA expression was performed on FFPE TSCC/BOTSCC pretreatment biopsies from patients treated 2000–2010 at the Karolinska University Hospital (Table 1). The study was performed according to permissions 02-009 (4 February 2002) and 2009/1278-31/4 (2 September 2009) from the Regional Ethics Committee, Karolinska Institutet. Informed consent was obtained from all of the patients.

4.2. Sample Preparation

Biopsies were cut frozen, embedded in optimal cutting temperature compound (OCT), and six cuts/biopsy were made, 1 × 5 μm, 4 × 20 μm and 1 × 5 μm. The first and last were used for haematoxylin-eosin-staining and evaluation of tumor content by an experienced pathologist. All of the tumor samples included had at least 40% tumor cells and 40/59 had >70% tumor cells (taking tumor infiltrating immune cells into account). For tumors, the remaining cuts (20 μm each) were paired 2 and 2, with each pair dissolved and analyzed separately, to obtain two replicates. For normal tissues, the remaining cuts were all pooled and dissolved together. Samples were then dissolved in different amounts of RIPA-buffer (50 mM Tris-HCl pH 7.4, 150 mM NaCl, 1 mM EDTA, 1 % Triton X-100, 0.1% Sodium deoxycholate with protease inhibitors) in relation to the area of the cuts to obtain a standardized protein concentration. Dissolved samples were kept frozen at −70 °C until analysis on the Olink platform.

4.3. Analysis on Olink Panels

Sample aliquots were analyzed with two Olink multiplex immunoassays, Immuno-Oncology and Oncology II, (Olink Bioscience, Uppsala, Sweden) at the Clinical Biomarkers facility, Science for Life Laboratory, Uppsala University, with each panel evaluating the concentration of 92 different proteins. In total, 167 unique proteins were analyzed, since 17 proteins overlapped between the two assays (Supplementary Table S5). The 17 common proteins served as a validation of the results between the assays. Concentrations of each protein were reported as normalized protein concentration (NPX) in a 2-log scale, and limit of detection (LOD) was defined as three standard deviations above background [20]. The assays also included two internal controls and a detection control. Quality control and data pre-processing (including normalization) of PEA data was made according to the manufacturer's recommended procedures. NPX-values for proteins and samples included in the evaluations are presented in Supplementary Table S6.

4.4. Evaluation of Data from Olink Panels

In order to enable as much tools of parametric statistics, such as Pearson linear correlation, *t*-test, regression modeling, the expression profiles were rendered normally distributed by taking log2 of the NPX values. The multiple regression model (Figure S2) was created using R package glmnet (available from http://web.stanford.edu/~hastie/glmnet/glmnet_alpha.html) under alpha = 1 and other parameters set to their defaults. *p*-values from *t*-tests were adjusted to report either family-wise error rates (Bonferroni correction) or the false discovery rate [50].

4.5. Heatmaps

Heatmap in Figure 1B was generated using Qlucore Omics Explorer 3.2 (Qlucore, Lund, Sweden). Heatmap in Figure S1 was created with R package heatmaply (https://cran.r-project.org/package=heatmaply).

4.6. Box Plots of Protein Expression

Using the default settings in R function boxplot, the boxes contain data points within the 25–75th percentile intervals (i.e., in the 2nd and 3rd quartiles). The maximal whisker length, MWL, is defined as 1.5 times the box length. Whiskers can extend to either the MWL or to the maximal available data point when the latter is below MWL. Separate markers thus correspond to data points that extend off the box by more than the MWL value.

4.7. Immunohistochemistry for VEGFA

Evaluation of VEGFA expression was performed only in samples that were both HPV DNA and p16 positive and the validation set is presented in Table 1. More specifically, 4 μm FFPE sections were de-paraffinized in xylene and rehydrated with ethanol. Blocking for endogenous peroxidase was performed using 1% bovine serum albumin in TBS. The sections were then incubated with VEGF antibody C1 (Santa Cruz Biotechnology, Heidelberg, Germany), 1:100 overnight at +4 °C, washed three times in TBS, incubated 30 min with Biotinylated horse-anti-mouse antibody diluted 1:200 in TBS with 0.2% Triton-X, and washed three times in TBS. After incubation with ABC-peroxidase (Vector Laboratories, Burlington, CA, USA) for 30 min at RT and three washes in TBS the staining was developed with DAB-kit (Vector Laboratories) for 3 min. Counterstaining was done with haematoxylin and the sections were dehydrated using increasing concentrations of ethanol and xylene before mounting.

Staining was evaluated by two researchers, including one pathologist, blinded for all other information about the samples. For cases where the evaluation differed, a consensus was reached. Staining intensity on tumor cells was scored as 0, absent, 1, weak, 2, moderate, and 3, strong. The percentage of stained tumor cells was assessed to the nearest 10%. Cases where the staining was not possible to be evaluated adequately were excluded.

4.8. Evaluation of Survival

For analysis of survival in relation to protein expression in the PEA-assay, tumors were analyzed separately for HPV-positive and HPV-negative tumors and tumors from patients being tumor-free and alive after four-years were compared with tumors from patients with recurrence within and/or dead with tumor within four-years. There were five recurrences/death with tumor among patients with HPV-negative tumors and four recurrences among patients with HPV-positive tumors.

For analysis of survival in relation to VEGF expression by IHC, survival was measured in days from diagnosis until an event occurred, or until four years after diagnosis when the patients were censored. Disease recurrence (TSCC or BOTSCC) was used as event for calculation of disease free survival (DFS) and disease recurrence or death with tumor was used when calculating progression-free survival (PFS). Kaplan-Meier estimator was used for DFS and PFS, and the differences in the survival of patients were tested using the logrank test. The Cox proportional hazards model was used for the calculation of the adjusted and unadjusted hazard ratios (HRs). Calculations and analyses were performed using SPSS Statistics, Version 21.0 (IBM Corp., Armonk, NY, USA).

Acknowledgments: The authors would like to acknowledge support of the Clinical biomarker facility at SciLifeLab Sweden for providing assistance in protein analysis. In particular we would like to acknowledge the assistance by Jenny Andersson and Cecilia Kriegholm at this facility. We thank the National Bioinformatics Infrastructure Sweden (NBIS) for providing computational assistance. We thank Inger Bodin for excellent technical assistance. This work was supported by: The Cancer Society in Stockholm (151253), Stockholm County ALF

(20160292), The Swedish Cancer Society (160259), Emil och Wera Cornells Stiftelse, Karolinska Institutet and The Swedish Cancer and Allergy Foundation (73). The funding sources had no involvement in any part of the study.

Author Contributions: Conception and design (Torbjörn Ramqvist, Eva Munck-Wikland, Bo Franzén, Tina Dalianis). Experimental procedures (Torbjörn Ramqvist, Anders Näsman, Bo Franzén, Cinzia Bersani, Susanne Becker, Linnea Haeggblom). Acquisition of clinical samples and clinical data (Eva Munck-Wikland, Susanne Becker, Aeneas Kolev). Analysis of data (Andrey Alexeyenko, Torbjörn Ramqvist, Anders Näsman). Manuscript writing (Torbjörn Ramqvist, Eva Munck-Wikland, Tina Dalianis, Andrey Alexeyenko, Anders Näsman, Cinzia Bersani). All authors approved the final manuscript.

References

1. Haeggblom, L.; Ramqvist, T.; Tommasino, M.; Dalianis, T.; Nasman, A. Time to change perspectives on hpv in oropharyngeal cancer. A systematic review of hpv prevalence per oropharyngeal sub-site the last 3 years. *Papillomavirus Res.* **2017**, *4*, 1–11. [CrossRef] [PubMed]
2. Marklund, L.; Näsman, A.; Ramqvist, T.; Dalianis, T.; Munck-Wikland, E.; Hammarstedt, L. Prevalence of human papillomavirus and survival in oropharyngeal cancer other than tonsil or base of tongue cancer. *Cancer Med.* **2012**, *1*, 82–88. [CrossRef] [PubMed]
3. Ramqvist, T.; Dalianis, T. Oropharyngeal cancer epidemic and human papillomavirus. *Emerg. Infect. Dis.* **2010**, *16*, 1671–1677. [CrossRef] [PubMed]
4. Dahlstrand, H.; Dahlgren, L.; Lindquist, D.; Munck-Wikland, E.; Dalianis, T. Presence of human papillomavirus in tonsillar cancer is a favourable prognostic factor for clinical outcome. *Anticancer Res.* **2004**, *24*, 1829–1835. [PubMed]
5. Lindquist, D.; Romanitan, M.; Hammarstedt, L.; Nasman, A.; Dahlstrand, H.; Lindholm, J.; Onelov, L.; Ramqvist, T.; Ye, W.; Munck-Wikland, E.; et al. Human papillomavirus is a favourable prognostic factor in tonsillar cancer and its oncogenic role is supported by the expression of e6 and e7. *Mol. Oncol.* **2007**, *1*, 350–355. [CrossRef] [PubMed]
6. Gillison, M.L.; D'Souza, G.; Westra, W.; Sugar, E.; Xiao, W.; Begum, S.; Viscidi, R. Distinct risk factor profiles for human papillomavirus type 16-positive and human papillomavirus type 16-negative head and neck cancers. *J. Natl. Cancer Inst.* **2008**, *100*, 407–420. [CrossRef] [PubMed]
7. Slebos, R.J.; Yi, Y.; Ely, K.; Carter, J.; Evjen, A.; Zhang, X.; Shyr, Y.; Murphy, B.M.; Cmelak, A.J.; Burkey, B.B.; et al. Gene expression differences associated with human papillomavirus status in head and neck squamous cell carcinoma. *Clin. Cancer Res.* **2006**, *12*, 701–709. [CrossRef] [PubMed]
8. Pyeon, D.; Newton, M.A.; Lambert, P.F.; den Boon, J.A.; Sengupta, S.; Marsit, C.J.; Woodworth, C.D.; Connor, J.P.; Haugen, T.H.; Smith, E.M.; et al. Fundamental differences in cell cycle deregulation in human papillomavirus-positive and human papillomavirus-negative head/neck and cervical cancers. *Cancer Res.* **2007**, *67*, 4605–4619. [CrossRef] [PubMed]
9. Dahlgren, L.; Mellin, H.; Wangsa, D.; Heselmeyer-Haddad, K.; Bjornestal, L.; Lindholm, J.; Munck-Wikland, E.; Auer, G.; Ried, T.; Dalianis, T. Comparative genomic hybridization analysis of tonsillar cancer reveals a different pattern of genomic imbalances in human papillomavirus-positive and -negative tumors. *Int. J. Cancer* **2003**, *107*, 244–249. [CrossRef] [PubMed]
10. Lajer, C.B.; Garnaes, E.; Friis-Hansen, L.; Norrild, B.; Therkildsen, M.H.; Glud, M.; Rossing, M.; Lajer, H.; Svane, D.; Skotte, L.; et al. The role of mirnas in human papilloma virus (hpv)-associated cancers: Bridging between hpv-related head and neck cancer and cervical cancer. *Br. J. Cancer* **2012**, *106*, 1526–1534. [CrossRef] [PubMed]
11. Smeets, S.J.; Hesselink, A.T.; Speel, E.J.; Haesevoets, A.; Snijders, P.J.; Pawlita, M.; Meijer, C.J.; Braakhuis, B.J.; Leemans, C.R.; Brakenhoff, R.H. A novel algorithm for reliable detection of human papillomavirus in paraffin embedded head and neck cancer specimen. *Int. J. Cancer* **2007**, *121*, 2465–2472. [CrossRef] [PubMed]
12. Nasman, A.; Nordfors, C.; Grun, N.; Munck-Wikland, E.; Ramqvist, T.; Marklund, L.; Lindquist, D.; Dalianis, T. Absent/weak cd44 intensity and positive human papillomavirus (hpv) status in oropharyngeal squamous cell carcinoma indicates a very high survival. *Cancer Med.* **2013**, *2*, 507–518. [CrossRef] [PubMed]
13. Nasman, A.; Bersani, C.; Lindquist, D.; Du, J.; Ramqvist, T.; Dalianis, T. Human papillomavirus and potentially relevant biomarkers in tonsillar and base of tongue squamous cell carcinoma. *Anticancer Res.* **2017**, *37*, 5319–5328. [PubMed]

14. Badoual, C.; Hans, S.; Rodriguez, J.; Peyrard, S.; Klein, C.; Agueznay Nel, H.; Mosseri, V.; Laccourreye, O.; Bruneval, P.; Fridman, W.H.; et al. Prognostic value of tumor-infiltrating cd4+ t-cell subpopulations in head and neck cancers. *Clin. Cancer Res.* **2006**, *12*, 465–472. [CrossRef] [PubMed]
15. Nasman, A.; Andersson, E.; Marklund, L.; Tertipis, N.; Hammarstedt-Nordenvall, L.; Attner, P.; Nyberg, T.; Masucci, G.V.; Munck-Wikland, E.; Ramqvist, T.; et al. Hla class i and ii expression in oropharyngeal squamous cell carcinoma in relation to tumor hpv status and clinical outcome. *PLoS ONE* **2013**, *8*, e77025. [CrossRef] [PubMed]
16. Nasman, A.; Andersson, E.; Nordfors, C.; Grun, N.; Johansson, H.; Munck-Wikland, E.; Massucci, G.; Dalianis, T.; Ramqvist, T. Mhc class i expression in hpv positive and negative tonsillar squamous cell carcinoma in correlation to clinical outcome. *Int. J. Cancer* **2013**, *132*, 72–81. [CrossRef] [PubMed]
17. Tertipis, N.; Haeggblom, L.; Grun, N.; Nordfors, C.; Nasman, A.; Dalianis, T.; Ramqvist, T. Reduced expression of the antigen processing machinery components tap2, lmp2, and lmp7 in tonsillar and base of tongue cancer and implications for clinical outcome. *Transl. Oncol.* **2015**, *8*, 10–17. [CrossRef] [PubMed]
18. Tertipis, N.; Haeggblom, L.; Nordfors, C.; Grun, N.; Nasman, A.; Vlastos, A.; Dalianis, T.; Ramqvist, T. Correlation of lmp10 expression and clinical outcome in human papillomavirus (hpv) positive and hpv-negative tonsillar and base of tongue cancer. *PLoS ONE* **2014**, *9*, e95624. [CrossRef] [PubMed]
19. Oguejiofor, K.; Galletta-Williams, H.; Dovedi, S.J.; Roberts, D.L.; Stern, P.L.; West, C.M. Distinct patterns of infiltrating cd8+ t cells in hpv+ and cd68 macrophages in hpv-oropharyngeal squamous cell carcinomas are associated with better clinical outcome but pd-l1 expression is not prognostic. *Oncotarget* **2017**, *8*, 14416–14427. [CrossRef] [PubMed]
20. Assarsson, E.; Lundberg, M.; Holmquist, G.; Bjorkesten, J.; Thorsen, S.B.; Ekman, D.; Eriksson, A.; Rennel Dickens, E.; Ohlsson, S.; Edfeldt, G.; et al. Homogenous 96-plex pea immunoassay exhibiting high sensitivity, specificity, and excellent scalability. *PLoS ONE* **2014**, *9*, e95192. [CrossRef] [PubMed]
21. Nasman, A.; Romanitan, M.; Nordfors, C.; Grun, N.; Johansson, H.; Hammarstedt, L.; Marklund, L.; Munck-Wikland, E.; Dalianis, T.; Ramqvist, T. Tumor infiltrating cd8$^+$ and foxp3$^+$ lymphocytes correlate to clinical outcome and human papillomavirus (hpv) status in tonsillar cancer. *PLoS ONE* **2012**, *7*, e38711. [CrossRef] [PubMed]
22. Nordfors, C.; Grun, N.; Tertipis, N.; Ahrlund-Richter, A.; Haeggblom, L.; Sivars, L.; Du, J.; Nyberg, T.; Marklund, L.; Munck-Wikland, E.; et al. Cd8$^+$ and cd4$^+$ tumour infiltrating lymphocytes in relation to human papillomavirus status and clinical outcome in tonsillar and base of tongue squamous cell carcinoma. *Eur. J. Cancer* **2013**, *49*, 2522–2530. [CrossRef] [PubMed]
23. Partlova, S.; Boucek, J.; Kloudova, K.; Lukesova, E.; Zabrodsky, M.; Grega, M.; Fucikova, J.; Truxova, I.; Tachezy, R.; Spisek, R.; et al. Distinct patterns of intratumoral immune cell infiltrates in patients with hpv-associated compared to non-virally induced head and neck squamous cell carcinoma. *Oncoimmunology* **2015**, *4*, e965570. [CrossRef] [PubMed]
24. Messina, J.L.; Fenstermacher, D.A.; Eschrich, S.; Qu, X.; Berglund, A.E.; Lloyd, M.C.; Schell, M.J.; Sondak, V.K.; Weber, J.S.; Mule, J.J. 12-chemokine gene signature identifies lymph nod $\times$ 10$-$like structures in melanoma: Potential for patient selection for immunotherapy? *Sci. Rep.* **2012**, *2*, 765. [CrossRef] [PubMed]
25. Sher, Y.P.; Chou, C.C.; Chou, R.H.; Wu, H.M.; Wayne Chang, W.S.; Chen, C.H.; Yang, P.C.; Wu, C.W.; Yu, C.L.; Peck, K. Human kallikrein 8 protease confers a favorable clinical outcome in non-small cell lung cancer by suppressing tumor cell invasiveness. *Cancer Res.* **2006**, *66*, 11763–11770. [CrossRef] [PubMed]
26. Ma, W.; Gilligan, B.M.; Yuan, J.; Li, T. Current status and perspectives in translational biomarker research for pd-1/pd-l1 immune checkpoint blockade therapy. *J. Hematol. Oncol.* **2016**, *9*, 47. [CrossRef] [PubMed]
27. Honeychurch, J.; Cheadle, E.J.; Dovedi, S.J.; Illidge, T.M. Immuno-regulatory antibodies for the treatment of cancer. *Expert Opin. Biol. Ther.* **2015**, *15*, 787–801. [CrossRef] [PubMed]
28. Hong, A.M.; Vilain, R.E.; Romanes, S.; Yang, J.; Smith, E.; Jones, D.; Scolyer, R.A.; Lee, C.S.; Zhang, M.; Rose, B. Pd-l1 expression in tonsillar cancer is associated with human papillomavirus positivity and improved survival: Implications for anti-pd1 clinical trials. *Oncotarget* **2016**, *7*, 77010–77020. [CrossRef] [PubMed]
29. Felder, M.; Kapur, A.; Gonzalez-Bosquet, J.; Horibata, S.; Heintz, J.; Albrecht, R.; Fass, L.; Kaur, J.; Hu, K.; Shojaei, H.; et al. Muc16 (ca125): Tumor biomarker to cancer therapy, a work in progress. *Mol. Cancer* **2014**, *13*, 129. [CrossRef] [PubMed]

30. Slebos, R.J.; Jehmlich, N.; Brown, B.; Yin, Z.; Chung, C.H.; Yarbrough, W.G.; Liebler, D.C. Proteomic analysis of oropharyngeal carcinomas reveals novel hpv-associated biological pathways. *Int. J. Cancer* **2013**, *132*, 568–579. [CrossRef] [PubMed]
31. Sewell, A.; Brown, B.; Biktasova, A.; Mills, G.B.; Lu, Y.; Tyson, D.R.; Issaeva, N.; Yarbrough, W.G. Revers × 10−phase protein array profiling of oropharyngeal cancer and significance of pik3ca mutations in hpv-associated head and neck cancer. *Clin. Cancer Res.* **2014**, *20*, 2300–2311. [CrossRef] [PubMed]
32. Pazina, T.; Shemesh, A.; Brusilovsky, M.; Porgador, A.; Campbell, K.S. Regulation of the functions of natural cytotoxicity receptors by interactions with diverse ligands and alterations in splice variant expression. *Front. Immunol.* **2017**, *8*, 369. [CrossRef] [PubMed]
33. Freud, A.G.; Mundy-Bosse, B.L.; Yu, J.; Caligiuri, M.A. The broad spectrum of human natural killer cell diversity. *Immunity* **2017**, *47*, 820–833. [CrossRef] [PubMed]
34. Mirghani, H.; Ugolin, N.; Ory, C.; Lefevre, M.; Baulande, S.; Hofman, P.; St Guily, J.L.; Chevillard, S.; Lacave, R. A predictive transcriptomic signature of oropharyngeal cancer according to hpv16 status exclusively. *Oral Oncol.* **2014**, *50*, 1025–1034. [CrossRef] [PubMed]
35. Gaykalova, D.A.; Manola, J.B.; Ozawa, H.; Zizkova, V.; Morton, K.; Bishop, J.A.; Sharma, R.; Zhang, C.; Michailidi, C.; Considine, M.; et al. Nf-kappab and stat3 transcription factor signatures differentiate hpv-positive and hpv-negative head and neck squamous cell carcinoma. *Int. J. Cancer* **2015**, *137*, 1879–1889. [CrossRef] [PubMed]
36. Keck, M.K.; Zuo, Z.; Khattri, A.; Stricker, T.P.; Brown, C.D.; Imanguli, M.; Rieke, D.; Endhardt, K.; Fang, P.; Bragelmann, J.; et al. Integrative analysis of head and neck cancer identifies two biologically distinct hpv and three non-hpv subtypes. *Clin. Cancer Res.* **2015**, *21*, 870–881. [CrossRef] [PubMed]
37. Wilkie, M.D.; Emmett, M.S.; Santosh, S.; Lightbody, K.A.; Lane, S.; Goodyear, P.W.; Sheard, J.D.; Boyd, M.T.; Pritchard-Jones, R.O.; Jones, T.M. Relative expression of vascular endothelial growth factor isoforms in squamous cell carcinoma of the head and neck. *Head Neck* **2016**, *38*, 775–781. [CrossRef] [PubMed]
38. Fei, J.; Hong, A.; Dobbins, T.A.; Jones, D.; Lee, C.S.; Loo, C.; Al-Ghamdi, M.; Harnett, G.B.; Clark, J.; O'Brien, C.J.; et al. Prognostic significance of vascular endothelial growth factor in squamous cell carcinomas of the tonsil in relation to human papillomavirus status and epidermal growth factor receptor. *Ann. Surg. Oncol.* **2009**, *16*, 2908–2917. [CrossRef] [PubMed]
39. Zhang, L.P.; Chen, H.L. Increased vascular endothelial growth factor expression predicts a worse prognosis for laryngeal cancer patients: A meta-analysis. *J. Laryngol. Otol.* **2017**, *131*, 44–50. [CrossRef] [PubMed]
40. Zhao, S.F.; Yang, X.D.; Lu, M.X.; Sun, G.W.; Wang, Y.X.; Zhang, Y.K.; Pu, Y.M.; Tang, E.Y. Prognostic significance of vegf immunohistochemical expression in oral cancer: A meta-analysis of the literature. *Tumour Biol.* **2013**, *34*, 3165–3171. [CrossRef] [PubMed]
41. Roudnicky, F.; Poyet, C.; Wild, P.; Krampitz, S.; Negrini, F.; Huggenberger, R.; Rogler, A.; Stohr, R.; Hartmann, A.; Provenzano, M.; et al. Endocan is upregulated on tumor vessels in invasive bladder cancer where it mediates vegf-a-induced angiogenesis. *Cancer Res.* **2013**, *73*, 1097–1106. [CrossRef] [PubMed]
42. Sagara, A.; Igarashi, K.; Otsuka, M.; Kodama, A.; Yamashita, M.; Sugiura, R.; Karasawa, T.; Arakawa, K.; Narita, M.; Kuzumaki, N.; et al. Endocan as a prognostic biomarker of tripl × 10−negative breast cancer. *Breast Cancer Res. Treat.* **2017**, *161*, 269–278. [CrossRef] [PubMed]
43. Huang, X.; Chen, C.; Wang, X.; Zhang, J.Y.; Ren, B.H.; Ma, D.W.; Xia, L.; Xu, X.Y.; Xu, L. Prognostic value of endocan expression in cancers: Evidence from meta-analysis. *Oncol. Targets Ther.* **2016**, *9*, 6297–6304. [CrossRef] [PubMed]
44. Irshad, K.; Mohapatra, S.K.; Srivastava, C.; Garg, H.; Mishra, S.; Dikshit, B.; Sarkar, C.; Gupta, D.; Chandra, P.S.; Chattopadhyay, P.; et al. A combined gene signature of hypoxia and notch pathway in human glioblastoma and its prognostic relevance. *PLoS ONE* **2015**, *10*, e0118201. [CrossRef] [PubMed]
45. Tudisco, L.; Orlandi, A.; Tarallo, V.; De Falco, S. Hypoxia activates placental growth factor expression in lymphatic endothelial cells. *Oncotarget* **2017**, *8*, 32873–32883. [CrossRef] [PubMed]
46. van der Veeken, J.; Oliveira, S.; Schiffelers, R.M.; Storm, G.; van Bergen En Henegouwen, P.M.; Roovers, R.C. Crosstalk between epidermal growth factor receptor-and insulin-like growth factor-1 receptor signaling: Implications for cancer therapy. *Curr. Cancer Drug Targets* **2009**, *9*, 748–760. [CrossRef] [PubMed]
47. Wang, Y.; Yuan, J.L.; Zhang, Y.T.; Ma, J.J.; Xu, P.; Shi, C.H.; Zhang, W.; Li, Y.M.; Fu, Q.; Zhu, G.F.; et al. Inhibition of both egfr and igf1r sensitized prostate cancer cells to radiation by synergistic suppression of DNA homologous recombination repair. *PLoS ONE* **2013**, *8*, e68784. [CrossRef] [PubMed]

48. Matsumoto, F.; Fujimaki, M.; Ohba, S.; Kojima, M.; Yokoyama, J.; Ikeda, K. Relationship between insulin-like growth factor-1 receptor and human papillomavirus in patients with oropharyngeal cancer. *Head Neck* **2015**, *37*, 977–981. [CrossRef] [PubMed]
49. Bersani, C.; Mints, M.; Tertipis, N.; Haeggblom, L.; Sivars, L.; Ahrlund-Richter, A.; Vlastos, A.; Smedberg, C.; Grun, N.; Munck-Wikland, E.; et al. A model using concomitant markers for predicting outcome in human papillomavirus positive oropharyngeal cancer. *Oral Oncol.* **2017**, *68*, 53–59. [CrossRef] [PubMed]
50. Benjamini, Y.; Hochberg, Y. Controlling the false discovery rate -a practical and powerful approach to multiple testing. *J. Roy. Stat. Soc. B Met.* **1995**, *57*, 289–300.

12

The Impact of Human Papilloma Viruses, Matrix Metallo-Proteinases and HIV Protease Inhibitors on the Onset and Progression of Uterine Cervix Epithelial Tumors

Giovanni Barillari [1,*], Paolo Monini [2], Cecilia Sgadari [2] and Barbara Ensoli [2]

1 Department of Clinical Sciences and Translational Medicine, University of Rome Tor Vergata, 1 via Montpellier, 00133 Rome, Italy

2 National HIV/AIDS Research Center, Istituto Superiore di Sanità, 299 viale Regina Elena, 00161 Rome, Italy; paolo.monini@iss.it (P.M.); cecilia.sgadari@iss.it (C.S.); barbara.ensoli@iss.it (B.E.)

* Correspondence: barillar@uniroma2.it

Abstract: Infection of uterine cervix epithelial cells by the Human Papilloma Viruses (HPV) is associated with the development of dysplastic/hyperplastic lesions, termed cervical intraepithelial neoplasia (CIN). CIN lesions may regress, persist or progress to invasive cervical carcinoma (CC), a leading cause of death worldwide. CIN is particularly frequent and aggressive in women infected by both HPV and the Human Immunodeficiency Virus (HIV), as compared to the general female population. In these individuals, however, therapeutic regimens employing HIV protease inhibitors (HIV-PI) have reduced CIN incidence and/or clinical progression, shedding light on the mechanism(s) of its development. This article reviews published work concerning: (i) the role of HPV proteins (including HPV-E5, E6 and E7) and of matrix-metalloproteinases (MMPs) in CIN evolution into invasive CC; and (ii) the effect of HIV-PI on events leading to CIN progression such as basement membrane and extracellular matrix invasion by HPV-positive CIN cells and the formation of new blood vessels. Results from the reviewed literature indicate that CIN clinical progression can be monitored by evaluating the expression of MMPs and HPV proteins and they suggest the use of HIV-PI or their derivatives for the block of CIN evolution into CC in both HIV-infected and uninfected women.

Keywords: HPV; uterine CIN; uterine cervical carcinoma; MMP; HIV-PI

---

1. Role of the HPV-E5, E6 and E7 Proteins in the Development of Uterine Cervical Pre-Cancer and Cancer Lesions

Infection by human papilloma virus (HPV) is frequent in sexually active women and plays a driving role in the development of proliferative/dysplastic or tumor lesions of the uterine cervix [1]. In particular, HPV is present in almost the totality of uterine cervical carcinoma (CC), which is the third most frequent malignancy in women worldwide [1].

The uterine cervix is composed by two different types of epithelium. Specifically, the endocervix is lined by a simple glandular epithelium, while the ectocervix is lined by a squamous epithelium constituted by superficial and deep (basal) layers [2]. At puberty, the endocervical glandular epithelium proximal to the ectocervix is replaced by the stratified squamous epithelium typical of the ectocervix. This trans-differentiated area is known as the "transformation zone" [2].

The proper stratification of uterine ectocervical epithelium and "transformation zone" is maintained through a tight control of epithelial cell growth, differentiation and locomotion. Specifically, epithelial cells originate in the deep layers, maturate and migrate to the superficial layer where they fully differentiate and desquamate in a continuous, self-renewing process [2].

This dynamic equilibrium is compromised by HPV which, in the presence of cervical micro-wounds, can access the deep layers and enter immature, proliferating cells [3,4]. There, the circular, double-stranded HPV DNA is maintained as an extra-chromosomal element (episome) which replicates via the synthetic machinery of host cell genome [3,4]. As such, HPV forces infected immature cervical cells to remain in a proliferative state and impedes their terminal differentiation. In particular, during HPV DNA replication, the HPV early genes, including those coding for the E5, E6 or E7 proteins, are transcribed [4]. Although HPV-E1 and E2 proteins support viral replication in the basal cells at a low copy number [5], HPV-E5, E6 or E7 are produced at levels sufficient to functionally impair host cell growth-regulatory or differentiation factors [4,6].

In particular, the HPV-E5 protein interferes with the signaling pathways of epidermal growth factor (EGF), a powerful inducer of epithelial cell survival, growth and locomotion [6]. Specifically, E5 promotes the sustained activation of EGF receptor, resulting in the triggering of protein kinases including the serine/threonine kinase AKT which, in turn, leads to epithelial cell survival and proliferation [6]. In this context, HPV-E5 down-regulates the expression of the p21 or p27 cell cycle inhibitors and counteracts cell death induced by the Tumor Necrosis Factor-Related Apoptosis-Inducing Ligand (TRAIL) or Fas-Fas ligand signaling [6]. Noteworthy, the amplification of EGF signaling promoted by HPV-E5 also inhibits epithelial cell differentiation via a reduction of keratinocyte growth factor expression [7].

Concerning HPV-E7, it binds and inactivates the retinoblastoma tumor suppressor protein (pRb), hence preventing infected cells from exiting the cell cycle and differentiating (Figure 1) [8].

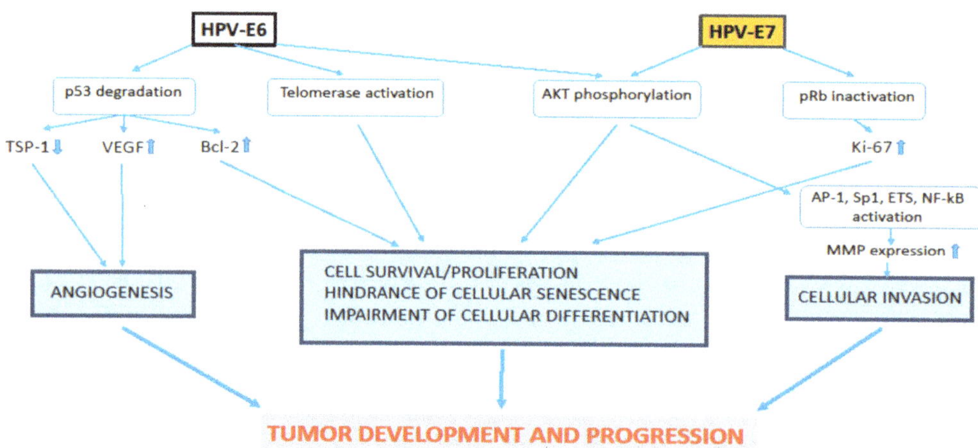

Figure 1. Tumorigenic effects of the E6 or E7 proteins of high-risk human papilloma viruses (HR-HPV). Arrows symbolize directions of connections. The E6 protein promotes p53 degradation, and this is followed by the down-regulation of thrombospondin (TSP)-1, or the up-regulation of vascular endothelial cell growth factor (VEGF) which, in turn, lead to angiogenesis. At the same time, E7 inactivates the retinoblastoma protein (pRb), and this increases Ki-67 protein levels. Noteworthy, both E6 and E7 induce AKT phosphorylation. This, together with the activation of telomerase or the up-regulation of Bcl-2 promoted by E6, and the increase of Ki-67 induced by E7, causes cell survival and proliferation. Moreover, the phosphorylation of AKT triggered by either E6 or E7 activates transcription factors including Activator Protein (AP)-1, Sp (Specificity protein)-1, ETS or Nuclear Factor-kappa B (NF-κB) which, in turn, induce matrix-metalloproteinase (MMP) expression and cellular invasion. Cell survival/proliferation/invasion and angiogenesis promoted by the E6 or E7 protein of HR-PV lead to tumor development and progression.

At the same time, HPV-E6 directs the host cell tumor suppressor protein p53 toward degradation through the cellular proteasome (Figure 1) [9]. In doing so, E6 increases the intracellular levels of the anti-apoptosis Bcl-2 protein which is normally repressed by p53 (Figure 1) [9]. In addition, E6 triggers the activity of telomerase, an enzyme that prevents replicative senescence by stabilizing the length of the chromosomes' end (Figure 1) [9].

Moreover, the HPV-E5, E6 or E7 proteins stimulate the cyclooxygenase (COX)-2 inflammatory pathway which, as for EGF, counteracts the apoptosis of epithelial cells and promotes their proliferation [6].

Altogether, these activities of E5, E6 or E7 render HPV-infected cells insensitive to the regulatory signals of normal growth and block apoptosis otherwise occurring due to the sustained, uncontrolled cellular proliferation [6,8,9].

In about 80% of the cases, HPV-induced cell proliferation remains subclinical, augmenting cervical epithelium thickness or causing benign flat warts [10]. In the other cases, HPV-E5, E6 and E7 accelerate the growth rate of immature cervical basal cells and their migration to the superficial layer. This leads to the development of squamous intraepithelial lesions (SIL) [2,11].

According to the Bethesda system for reporting cervical cytological diagnosis, there are two kinds of SIL: low grade-SIL (L-SIL), which is the histologic correlate of productive HPV infection and high grade-SIL (H-SIL), that represents the early stages of HPV-induced carcinogenesis [2,11].

L-SIL corresponds histologically to grade 1 cervical intraepithelial neoplasia (CIN1), a mild dysplasia characterized by proliferating immature basal cells constituting 1/3 of the cervical epithelium. Because of HPV replication, superficial cells display koilocytosis (cave-like vacuoles around enlarged, hyper-chromatic nuclei) and altered keratinization [2,11].

H-SIL comprises grade 2 and 3 CIN (CIN2 and CIN3). The former is a moderate dysplasia in which basal cell types represent the 2/3 of cervical epithelium and koilocytosis and dyskeratosis are more evident than in CIN1; whereas, CIN3 is a severe dysplasia, characterized by proliferating basal-like cells having abnormal nuclei and atypical mitoses, which occupy the 2/3 of the entire epithelium, including the superficial layers [2,11].

Generally, an effective immune response to HPV develops within months after infection, resulting in viral clearance [3,10]. Thereafter, p53 and pRb function in the basal layers is restored and epithelial cell growth and differentiation return to normality. This generally occurs with low-risk HPV types, such as HPV6 or HPV11, whose DNA remains episomal [10].

In contrast, persistent infection with high risk (HR)-HPV, including HPV16 and HPV18, can be followed by the integration of viral DNA into the host cell genome [5]. The probability of having persistent HR-HPV infection augments progressively with higher viral load [12] and it is favored by E5, E6 or E7 activities permitting HR-HPV escape from host immune surveillance. In particular, both the E5 and E7 proteins of HR-HPV can impair major histocompatibility complex (MHC)-restricted presentation of viral peptides to cytotoxic or helper T lymphocytes [13,14]. Specifically, E7 down-regulates the expression of MHC class I molecules, while E5 arrests them in the Golgi apparatus, reduces their transport to the cell surface and/or diminishes MHC class II expression [13,14]. Moreover, either the E6 or E7 protein of HR-HPV inhibits the production of the immune-stimulatory, anti-viral interferons and reduces the synthesis of proteins attracting antigen-presenting macrophages, dendritic or Langerhans cells to the infected area [13]. In addition, HR-HPV-E5, E6 or E7 can trigger molecular pathways leading to the shift from cellular to humoral immune response, the silencing of inflammation, the impairment of natural killer cell or natural killer T cell activity and the recruitment of immune suppressive regulatory T cells [13,14].

HPV integration into cellular DNA causes the deletion of the *HPV-E2* gene and the consequent overexpression of *HPV-E6 or E7* [5]. As for *HPV-E2*, also the *E5* gene of HPV is often deleted upon the integration of HPV DNA in the host cell genome [6]. Thus, at variance with *E5*, the *E6* and *E7* genes are permanently expressed during HPV infection, being indispensable for the maintenance of the transformed cell phenotype. For this reason, E6 and E7 are considered as the main transforming proteins of HPV [7–9].

In fact, as a result of *HPV-E6* or *E7* overexpression caused by *E2* gene deletion, the disturbance of cervical epithelial cell maturation and stratification is exacerbated [5]. In this context, cellular key mitotic checkpoints are impaired, leading to genomic instability, accumulation of secondary mutations and aneuploidy in infected cells [15–18]. Subsequently, the entire cervical epithelium is replaced by poorly differentiated cells displaying abnormal nuclei and atypical mitoses [2,11]. Later on, some of these cells acquire a "spindle" morphology and degrade the epithelial basement membrane, giving rise to the onset of an invasive cancer, whose predominant histological type is squamous cell carcinoma [2,11].

Interestingly, CC develops mainly in uterine cervical "transformation zone", which is rich in immature, highly proliferating and HPV-sensitive basal cells [2]. Noteworthy, as for other tumor settings [19], CIN evolution into a true malignancy is accompanied by the formation of new blood vessels (angiogenesis) at the stromal/epithelial junction of CIN lesions [20,21]. Specifically, endothelial cells lining the lumen of the pre-existing vessels invade the vascular basement membrane, sprout, proliferate and migrate in the extra-vascular space, where they organize into hollow cords permitting blood influx [20,21]. These newly formed vessels nourish the growing tumor and provide additional routes for CC metastasis [20,21]. Accordingly, higher intra-tumor vessel density is associated with CC aggressiveness or recurrence and poorer patient survival [22,23].

It is of note that HPV infection has an important role also in CC-associated neovascularization. In particular, following p53 degradation promoted by HPV-E6, p53-induced genes encoding for angiogenesis inhibitors, such as thrombospondin (TSP)-1, are no longer transcribed; whereas, the p53-repressed genes of angiogenic factors, including vascular endothelial growth factor (VEGF), are up-regulated (Figure 1) [21]. Of interest, also HPV-E5 can promote VEGF expression and this is due to E5 capability of triggering both EGF and COX-2 signaling [6].

However, it should be highlighted that HPV infection progresses to cancer only in a small percentage of cases and that CIN lesions can also stabilize or regress [24]. In particular, the natural history of CIN1 includes regression (60% of cases), persistence (30%) and progression to CIN3 (10%) [24]. The like-hood of CIN2 regression is 45%, persisting 30% and progressing to CIN3 or invasive CC are 20% and 5%, respectively [24]. Concerning CIN3, about 35% of cases regress, while 10–15% evolve into invasive CC [24].

The risk of CIN progression to invasive CC is increased by the use of oral contraceptives, smoking, early age at first sexual intercourse, multiple sexual partners, repeated parity and co-infections [25–27]. To this regard, women infected by both HR-HPV and the human immunodeficiency virus (HIV)-1 have a higher incidence of uterine CIN and CC, as compared to their HIV-negative counterparts [28–35]. In addition, HR-HPV/HIV-doubly infected women have lower regression rates from high-grade to low-grade CIN, or from low-grade CIN to normal epithelium [31] and faster progression from low-grade to high-grade CIN [28,35]. Consistently, the median age of HIV-positive CC patients is much lower than in HIV-negative CC patients [36]. Furthermore, CIN recurrence after treatment is particularly frequent in HR-HPV/HIV-doubly infected women [31]. Because of these findings, uterine CC is considered an Acquired Immune Deficiency Syndrome (AIDS)-defining disease [37].

Indeed, both the incidence and the progression rates of cervical lesions increase with the impairment of immune functions promoted by HIV, as indicated by the decrease in $CD4^+$ T cell counts [28,35,38]. Certainly, the lack of an effective immune response to HR-HPV may favor its persistence, which is the main risk factor for CC development [1,38]. Nevertheless, HIV-1 is likely to have also a direct role in CIN progression to CC. In particular, results from in vitro studies indicate that the HIV-1 trans-activator (Tat) can up-regulate HR-HPV E6 or E7 expression, thereby decreasing the protein levels of cellular onco-suppressors and accelerating epithelial cell growth [39–42]. In addition, HIV-1 Tat may also favor the angiogenic switch of high-grade CIN, either because of its direct angiogenic effects [43] or, again, via the up-regulation of HR-HPV E6 or E7 expression [21].

Due to the dynamism of cervical lesions, cytological testing (PAP test) or histopathology is not always sufficient to assess the risk of CIN progression or its regression. Thus, morphological examination of cervical samples must be accompanied by the detection of risk biomarkers for high-grade CIN or CC, including the presence of HR-HPV DNA [44] and HR-HPV E6 and E7 overexpression [45,46]. Additional analyses are employed to monitor cellular (host) proteins whose synthesis in cervical lesions is de-regulated upon p53 and pRb functional inactivation. These include the cyclin-dependent kinase inhibitor p16INK4a and the proliferation antigen Ki-67 (Figure 1) [47–49]. In particular, the progression of CIN lesions can be predicted by the combination of low pRb/p53 and high Ki-67/p16INK4a expression in the basal layers of the cervical epithelium [47–49]. Recent data suggest evaluating the profile of specific cellular micro-RNAs that have an important role in cervical carcinogenesis and are modulated by the E5, E6 or E7 proteins of HR-HPV [50]. Further prognostic information is provided by $CD4^+$ T cell counts in HPV/HIV doubly-infected women [35,36,38] and by an increase in the expression and/or activity of the cellular matrix-metalloproteinases (MMPs), which occurs in both HIV-positive and HIV-negative women.

2. The MMPs and Their Role in the Progression of Uterine Cervical Pre-Cancer and Cancer Lesions

MMPs are a family of peptidases defined by the presence of a zinc ion at the catalytic site and by the capability of degrading the protein components of the basement membrane and extracellular matrix (ECM) [51–53].

Human MMPs consist of distinct proteases which are classified in soluble and membrane-anchored [51–53].

Soluble MMPs contain an NH_2-terminal secretion leader sequence and they include collagenases, stromelysins, matrilysins, the metalloelastase and gelatinases. Among the latter are MMP-2 and MMP-9, two powerful enzymes which can degrade a wide variety of basement membrane and ECM molecules [51–53].

The membrane-anchored MMPs (also called membrane type MMPs, MT-MMPs) display a COOH-terminal trans-membrane domain linked to a cytoplasmic tail and comprise 6 members [54]. The best known is MT1-MMP which, in addition to breaking down the ECM, is the main activator of MMP-2 [54].

Thus, ECM molecules are not the only target of MMPs, which can modify or digest many other extracellular or intracellular proteins, including enzymes, cytokines, chemokines and adhesion or growth factor receptors [52,55].

Because of these properties, MMPs are deeply involved in tissue remodeling or repair, in the regulation of cell growth, survival or differentiation and in reactive processes such as immunity and inflammation [56].

In physiological conditions, MMP expression is absent or very low in most tissues, being induced or up-regulated only during reactive/reparative processes [51–53,57]. In these events, growth factors, hormones or cytokines trigger signaling molecules that, in turn, activate transcription factors such as Activator Protein (AP)-1, ETS, Specificity protein (Sp)-1, and/or Nuclear Factor-kappa B (NF-κB) [58–68]. These activation pathways depend on the different cell types synthesizing MMPs and are influenced by cell-to-cell contacts and cell-ECM interactions [51–53].

In order to keep under control their potent activity, soluble MMPs are secreted as latent zymogens (proMMPs), which are converted into the active form around their producing cells, this providing a spatial control of MMP function [51,52].

Active soluble MMPs, in turn, can be inactivated/degraded by other MMPs or members of other classes of proteases, according to a multi-faced regulatory circuit [51,52].

Moreover, the function of active MMPs is controlled by endogenous inhibitors, which include serum globulins and the Tissue Inhibitors of MMPs (TIMPs) [51,52,69]. The latter are present in most human tissues, where they inhibit MMPs with a 1:1 molar stoichiometry, mainly via the binding of their amino-terminal to the MMP catalytic site [69]. To date, four TIMPs have been identified: they share about 40% sequence homology and hinder all the human MMPs, albeit with variable affinity.

In particular, TIMP-3 has the broadest inhibitory spectrum [69], while TIMP-1 or TIMP-2 antagonize mostly MMP-9 or MMP-2 activity, respectively [70].

As for soluble MMPs, also the MT-MMPs are synthesized as latent zymogens (proMT-MMPs): inside the cell, the pro-hormone convertase furin converts proMT-MMPs into active MT-MMPs, which are then expressed on the cell surface, where they can complex TIMPs [54]. TIMP capability of inhibiting MT-MMP activity and the fact that in the absence of TIMPs the MT-MMPs quickly undergo autolysis, reduces the number of fully active MT-MMP molecules on the cell membrane [54].

In summary, consistent with its potency and pleiotropic effects, the activity of MMPs is tightly regulated at both the transcriptional and post-transcriptional level. This occurs in physiological conditions; whereas, inflammatory/degenerative diseases or tumors are characterized by an abnormal expression of MMPs and/or their deregulated function [58,71].

In particular, cancer cells synthesize MMPs following the activation of oncogenes, the inactivation of onco-suppressor proteins, the stimulation by growth factors or inflammatory mediators, the generation of reactive oxygen species, and/or the presence of hypoxia [72–76]. Since MMPs are produced in tumor tissues also by fibroblasts and infiltrating inflammatory cells [58,71,77,78], MMP expression and/or activity are increased in tumor lesions as compared to neighbor normal tissues [79–88].

Most notably, MMPs play an important role in tumor progression. Specifically, they digest cell surface molecules mediating cell-to-cell or cell-to-ECM adhesion (e.g., cadherins or integrins), thus allowing tumor cells to detach from adjacent cells and the ECM [58,71]. Then, MMPs degrade basement membranes and the ECM, preparing a route for cancer cell locomotion and tissue invasion [58,71]. In doing so, MMPs allow cancer cells to reach local blood or lymphatic vessels and penetrate their wall, leading to metastases [58,71]. Following their arrest in the lumen of small vessels, cancer cells penetrate again the vessel wall through MMP action and, in case they escape immune surveillance, give rise to secondary tumors [58,71]. Noteworthy, during all these steps, MMPs sustain cancer cells by releasing sequestered, ECM-bound growth factors, activating growth factor receptors, triggering cell survival-promoting signaling pathways and degrading mediators of apoptosis (e.g., FAS ligand) or anti-tumor immunity (e.g., MHC) [58].

MMPs play an important role also in tumor-associated angiogenesis, by breaking down the blood vessel basement membrane and perivascular ECM, mobilizing endothelial cell precursors from bone marrow and recruiting them in the newly forming vessels [89–95].

It has to be highlighted, however, that in addition to their tumorigenic effects, MMPs can also exert anti-tumor activities. For example, MMP-13 protects against melanoma metastases [96], although its presence is prognostic of poor survival in gastric cancer patients [87]. Moreover, MMP-8 expression is associated with improved survival of tongue cancer patients [97], even though MMP-8 serum levels correlate with colorectal cancer stage and distant metastases [88].

These contrasting activities of MMPs appear to depend on the cellular source of MMP expression, the extracellular environment, the type of tumor and the stage of the disease [98]. In particular, the protective role of some MMPs in specific types and clinical stages of tumor is likely to result from MMP occasional/paradoxical capability of inhibiting tumor angiogenesis via the degradation of angiogenic growth factor receptors, and/or the generation of anti-angiogenesis molecules deriving from the cleavage of ECM components [94,99].

In conclusion, the balance between MMP pro- and anti-tumorigenic effects is critical for the prognosis of the patients. Therefore, evaluating the different MMPs which are expressed in a particular type and/or stage of a specific tumor and patient is of extreme importance.

Among the various members of the MMP family, MMP-9, MMP-2 or MT1-MMP are especially involved in the development or progression of uterine cervical tumors, as indicated by the finding that their RNA or protein expression and their activity are extremely low or absent in normal uterine cervix and low-grade CIN, are well detectable in high-grade CIN and they are very high in invasive CC [100–112]. In addition, expression of these MMPs parallels the angiogenic switch occurring during the evolution of high-grade CIN into CC and positively correlates with vessel density in tumor lesions [100,108,109]. Finally, MMP-9, MMP-2 or MT1-MMP levels in patients' plasma or cervical smears correlate with poor prognosis and higher tumor stage [103,113].

Noteworthy, in normal or transformed epithelial cells the E5, E6 or E7 protein of tumorigenic, HR-HPV down-regulate TIMPs and/or up-regulate MMP-2, MMP-9 or MT1-MMP expression or activity, thereby increasing epithelial cell invasiveness (Figure 1) [114–120]. These phenomena are likely to depend on E5, E6 or E7 capability of triggering the serine/threonine kinase AKT [6,121,122]. In fact, phosphorylated AKT activates transcription factors promoting *mmp* gene expression (Figure 1) [65–68].

3. Effects of HIV-Protease Inhibitors on Pre-Cancer and Cancer Lesions of the Uterine Cervix

Given the major role of MMPs in tumor-associated cellular invasion and angiogenesis, many MMP inhibitors (MMPIs) have been developed in the past years and exploited as anti-cancer interventions.

Among them are natural MMPIs, such as shark cartilage extract or soy isoflavonoid [123,124] and synthetic MMPIs comprising tetracycline derivatives, bisphosphonates and peptides chelating the zinc ion of MMP catalytic site [125–127].

Preclinical studies have shown that both synthetic and natural MMPIs are effective against a variety of tumors, included uterine CC [125,128–134]. However, clinical trials performed with some of the abovementioned MMPIs have failed due to dose-limiting side effects, inefficacy and/or lack of specificity [123,124,135–138]. In fact, because of the high similarity in the sequence of MMP catalytic domains, first-generation MMPIs have acted against multiple members of the MMP family, hence inhibiting either the therapeutic targets or anti-targets of a specific tumor [136,138].

In spite of these discouraging results and considering MMP role in tumor onset or progression, efforts are currently being made in order to design new MMPIs for oncological therapy. In particular, to specifically counteract the MMPs that are produced by a given tumor at a specific stage of development [98], novel strategies for MMP inhibition are exploiting the blockage of secondary substrate-binding sites (exosites), which show remarkable differences among the various MMPs [139,140].

However, inhibition of MMP activity may not be sufficient, as this strategy does not compromise the binding of MMPs to their cellular targets and the consequent signaling [58,71,91,141,142]. Thus, antibody-based inhibitors are being explored [143]. Alternatively, novel MMPIs could antagonize the molecular pathways leading to MMP expression. In this regard, experimental and clinical evidence points to the HIV-protease inhibitors (HIV-PI) as effective MMPIs.

The HIV-PI are a class of drugs which block the active site of HIV aspartyl protease, hence impeding the cleavage of HIV Gag-Pol poly-protein precursor and, consequently, the production of mature, infectious HIV particles (Figure 2) [144].

HIV-PI are administered to HIV-infected people together with the HIV reverse transcriptase inhibitors (HIV-RTI) and/or drugs counteracting the HIV integrase [145]. This combination therapy has been defined as highly active antiretroviral therapy (HAART) [145]. In fact, by potently suppressing viral replication and promoting immune reconstitution, HAART has strongly reduced AIDS-related morbidity and mortality, rendering HIV infection a manageable, chronic disease [146].

Most disappointingly, however, long-term treatment with HIV-PI may cause undesirable side effects. This is because, in addition to the HIV protease, HIV-PI can also target host molecules. Among the off-targets of HIV-PI is the cellular proteasome, a multi-subunit protease complex which controls the turnover of intracellular proteins (Figure 2) [147–152]. One of the many consequences resulting from the functional impairment of the cellular proteasome promoted by the HIV-PI is the intracellular accumulation of the sterol regulatory element-binding protein (SREBP) 1 transcription factor [153,154]. As a consequence, SREBP1-targeted lipogenic enzymes are increased, this augmenting lipid plasma levels and altering body fat distribution in treated patients [154–157]. Another unexpected target of HIV-PI is the glucose transporter (GLUT)-4 (Figure 2), whose inhibition by HIV-PI halts glucose uptake by adipocytes, eventually causing insulin-resistance and diabetes [158].

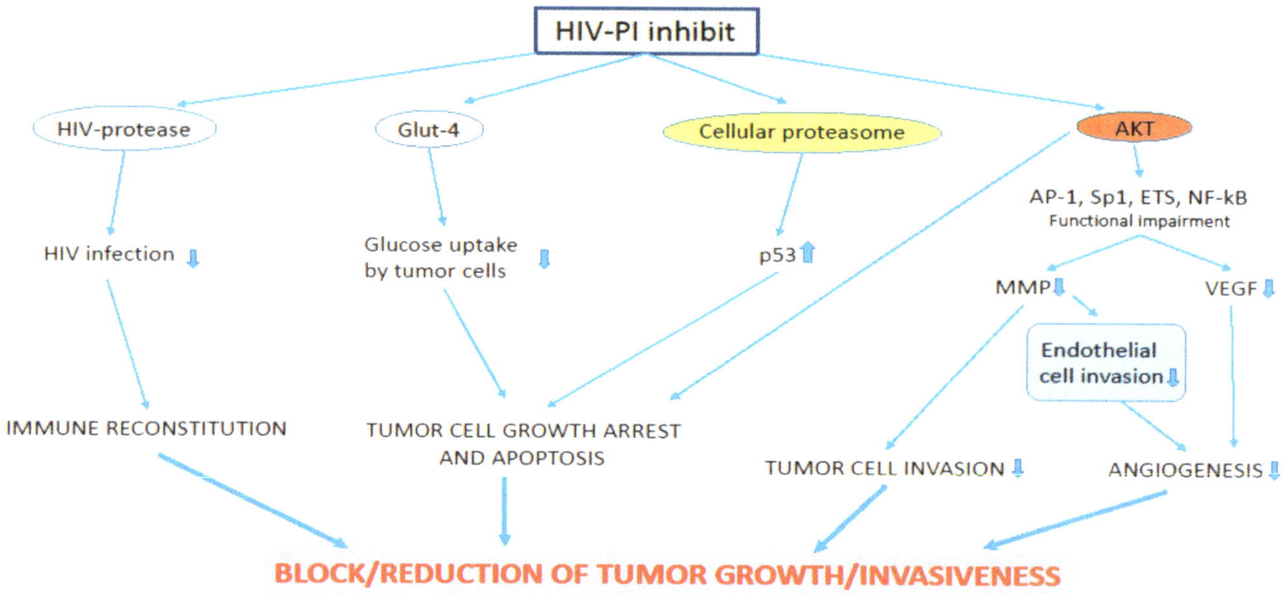

Figure 2. Anti-tumor activities of Human Immunodeficiency Virus (HIV)-protease inhibitors (PI). Arrows symbolize directions of connections. HIV-PI hinder the activity of the: (i) HIV aspartyl protease, hence impeding the production of infectious viral particles and promoting immune reconstitution; (ii) glucose transporter (GLUT)-4, thus impairing glucose uptake by tumor cells; (iii) cellular proteasome, therefore causing p53 protein intracellular accumulation; iv) AKT, this leading to the functional impairment of the Activator Protein (AP)-1, Sp (Specificity protein)-1, ETS or Nuclear Factor-kappa B (NF-κB) transcription factors, the down-regulation of matrix-metalloproteinase (MMP) or vascular endothelial growth factor (VEGF) expression, and the inhibition of angiogenesis or tumor cell invasion. Altogether, these activities of HIV-PI can block or reduce tumor growth and invasiveness.

Consequently and considering the need of life-long HAART administration, HIV-PI are often substituted by combined nucleoside and non-nucleoside HIV-RTI, respectively targeting the catalytic and allosteric sites of HIV reverse transcriptase [145].

Either HIV-PI-based or HIV-RTI-based HAART is effective at suppressing HIV infection, rapidly reducing plasma viremia and increasing the number of $CD4^+$ T cells in HIV-infected individuals [146]. However, the absence of HIV-PI in HAART deprives patients of the beneficial activities of these drugs which, in addition to the improvement of T cell function [159], comprise the inhibition of AIDS-associated opportunistic infections [160].

Moreover, HIV-infected individuals undergoing HAART have experienced a reduced incidence and/or an increased regression of AIDS-associated tumors including Kaposi's sarcoma (KS) and non-Hodgkin lymphoma (NHL) [161–165]. In contrast, the incidence of progressed uterine CC in HIV-positive women has not decreased since the introduction of HAART [166]. Nevertheless, similarly to KS or NHL, use of HAART in HIV-infected women has decreased the incidence of CIN, caused its regression, and/or lowered its recurrence rates after excision [29–32,167–171].

The restoration of the immune responses promoted by HAART is likely to have a major role in inhibiting the development of AIDS-associated KS, NHL or CIN (Figure 2) [172]. However, in some HAART-treated, HIV-positive patients the decrease in tumor incidence, relapse or progression rates is not accompanied by the recovery of $CD4^+$ T cell number and/or the reduction of viral load [172]. This has suggested that drugs present in HAART could exert anti-tumor effects independently of their capability of suppressing viral replication and reconstituting the immune system.

With reference to uterine CIN or CC, epidemiological studies analyzed all together data collected from patients undergoing different HAART regimens [29–32,166–171]. In contrast, studies concerning AIDS-associated KS have provided better clues of HAART anti-tumor effects. In particular, these studies have reported that complete remission of KS is more frequent in HIV-PI than in HIV-RTI-treated individuals, although both therapeutic regimens can promote the regression of KS lesions [164]. Moreover, a relapse of KS after its resolution has been described in patients switching from HIV-PI-based to HIV-RTI-based HAART [173].

In agreement with these findings, results from preclinical (in vitro and in vivo) and clinical studies indicate that HIV-PI can directly inhibit tumor cell survival, growth or invasion and angiogenesis in experimental models devoid of HIV and immune cells.

In particular, due to their inhibitory effect on the function of the cellular proteasome, HIV-PI cause an increase in the intracellular amounts of cell cycle inhibitors or onco-suppressors, hence leading to the growth arrest and/or death of a variety of human tumor cells (Figure 2) [148,150,174–182]. Additional anti-cancer cytostatic and/or cytotoxic effects result from HIV-PI capability of inhibiting GLUT-4, hence impairing glucose uptake by tumor cells (Figure 2) [158,183].

As to the uterine cervix, the HIV-PI indinavir (IDV), ritonavir (RTV) or lopinavir have been shown to reduce the viability of HR-HPV-transformed epithelial cells [151]. This is accompanied by a stable increase in the intracellular levels of the proteasome-targeted, growth-arresting and pro-apoptosis p53 protein [151]. However, this activity requires high HIV-PI concentrations, that are not (or only transiently) present in the plasma of treated patients [184].

In the matter of HIV-PI therapeutic (low) amounts, the HIV-PI saquinavir (SQV) or RTV halts the proliferation of cells derived from CIN but not CC lesions, with no significant effect on cell survival or proteasome capability of degrading p53 [185]. In view of these in vitro findings, one may conclude that at low HIV-PI tissue concentrations a CIN lesion persists, although it does not grow. Therefore, in order to efficiently counteract the malignant evolution of high-grade CIN, an HIV-PI should be combined with a classical proteasome inhibitor, as previously done to induce apoptosis of CC-derived cells [186]. Nevertheless, a recent clinical study conducted in HIV-negative women has shown that HIV-PI can by themselves promote the regression or the complete remission of high-grade CIN lesions [187].

Preclinical work has suggested that this may depend on HIV-PI capability of blocking epithelial basement membrane invasion by the CIN cells, as this event starts CIN progression to CC [2,11].

In particular, SQV or RTV at levels as detected in the plasma of treated patients can efficiently inhibit EGF-promoted invasion of primary keratinocytes bearing episomal HR-HPV DNA and derived from low-grade CIN lesions of HIV-negative women [185,188]. For both SQV and RTV, this effect is accompanied by a significant reduction of MMP-2 or MMP-9 activity released by the CIN cells in response to EGF [185,188].

These in vitro results are consistent with those from animal studies which have indicated that HIV-PI therapeutic concentrations inhibit the growth of human tumors in animal models via a proteasome-independent reduction of cancer cell invasion and MMP activity (Figure 2) [189].

In agreement with the fact that the HIV protease does not belong to MMP functional class, HIV-PI cannot directly antagonize MMP activity [190]. Rather, SQV or RTV therapeutic levels down-regulate the expression of MMP-9 induced in CIN cells by EGF and, to a lesser extent, CIN cell constitutive expression of MMP-2 [185,188]. This finding is consistent with HIV-PI capability of inhibiting MMP expression by both tumor and normal cells [150,152,189,191,192] and either in HIV-positive or in HIV-negative individuals [193,194].

EGF induces MMP-9 expression by epithelial cells via the phosphorylation of the serine/threonine kinase AKT [195] and the consequent activation of Fos-related antigen (Fra)-1, a member of the AP-1 transcriptional complex [67]. Strikingly, exposure of CIN cells to the same SQV or RTV doses reducing EGF-promoted cellular invasion and MMP-9 expression, also inhibits AKT phosphorylation and Fra-1 nuclear localization triggered in these cells by EGF [188]. These findings are consistent with results obtained in other experimental models, which indicate that inhibition of EGF-induced AKT phosphorylation, or Fra-1 silencing, reduces cellular invasion via the down-regulation of MMP-9 expression [67,195].

Indeed, SQV or RTV capability of inhibiting EGF-promoted AKT/Fra-1 activation and MMP-9 expression in low-grade CIN cells may have relevant clinical implications. In fact, the levels of EGF, MMP-9 and phosphorylated AKT are extremely low or undetectable in the normal uterine cervix, becoming appreciable in low-grade CIN and further increasing during its evolution to high-grade CIN and CC [103,107,113,196–198].

4. The Anti-Angiogenesis Effect of the HIV-PI and its Possible Impact on Pre-Cancer and Cancer Lesions of the Uterine Cervix

As for the CIN cells, HIV-PI can impair AKT phosphorylation in several tumor cell types (Figure 2) [150,182,199–204]. In this regard, it has to highlighted that the activation of AKT is pivotal in the acquisition and maintenance of cancer hallmarks, including cellular immortalization, deregulated growth and invasiveness, as well as in angiogenesis [65,67,68,205–207].

Because of its role in the development and clinical progression of uterine cervical lesions [22,23,100], angiogenesis is a rational target for the treatment of high-grade CIN and/or CC. Accordingly, antagonists of angiogenic growth factors have been found active against recurrent CC [20]. However, these antagonists are very expensive [20] and their therapeutic use may lead to drug resistance [208]. This has prompted the exploitation of drugs counteracting other players of the angiogenic process.

In this context, the HIV-PI nelfinavir reduces the production of the angiogenic VEGF (Figure 2) [179,204,209]. Moreover, SQV or IDV blocks MMP activity in endothelial cells, hence halting the first step of angiogenesis that is endothelial cell invasion of vessel basement membrane (Figure 2) [191]. As a consequence, nelfinavir, SQV or IDV efficiently counteract the formation of new blood vessels in HIV-free in vivo models [189,191,209,210]. Consistently, IDV can promote the regression or stabilization of proliferative vascular lesions also in HIV-negative KS patients [194].

In vitro work has shown that IDV concentrations as found in plasma of treated HIV-positive or negative patients inhibit the conversion of latent MMP-2 zymogen into the active enzyme, while having little or no impact on the synthesis of latent MMP-2 and no direct effects against MMP-2 catalytic activity [190,191].

The main mediator of latent MMP-2 conversion into active MMP-2 is the MT1-MMP [54]. Noteworthy, in vitro work has shown that IDV therapeutic amounts impair the binding of the Sp1 transcription factor to the promoter region of the *mt1-mmp* gene, thus down-regulating MT1-MMP RNA and protein expression by endothelial cells [210].

It has to be noted that MT1-MMP can directly degrade the blood vessel basement membrane even in the absence of MMP-2 [211]. Moreover, by modulating the expression of endothelial cell adhesion molecules, MT1-MMP is important to endothelial cell organization in a capillary network [211]. Thus, the reduction of MT1-MMP expression provides a molecular mechanism for IDV anti-angiogenesis effect, rendering this anti-HIV drug a promising antagonist of the formation of new blood vessels which accompanies CIN evolution into CC. Notwithstanding, the finding that long term-treatment with HIV-PI can damage blood vessels [212] deserves careful consideration.

5. HIV-PI and HPV

As discussed above, the E6 and E7 proteins of HR-HPV exert tumorigenic and pro-angiogenesis activities, including the inactivation of the growth-suppressive pRb or p53 proteins, the phosphorylation of AKT and the promotion of MMP expression (Figure 1) [8,9,114–116,118–122]. These activities of HR-HPV-E6 or E7, in turn, lead to cellular growth and invasiveness (Figure 1). Nevertheless, the anti-proliferative and anti-invasive effects exerted in CIN cells by SQV or RTV therapeutic amounts are not accompanied by the down-regulation of HPV-E6 or E7 expression [188]. This is consistent with the clinical finding that HIV-PI promote CIN regression without clearing HPV infection [167], albeit other studies have obtained opposite results [32,187].

In this regard, in vitro work indicates that therapeutic concentrations of HIV-PI are active against CIN cells containing episomal HPV DNA, while having little or no effect on CC cells bearing integrated HPV-DNA [185,188].

Therefore, one may speculate that the activity of HIV-PI against HPV-associated cervical lesions could depend on the integration status of HPV-DNA. This hypothesis is corroborated by molecular studies indicating that HIV-PI selectively accumulate in the nuclei of HPV-E6 expressing cells and trigger the synthesis of anti-viral proteins [213,214].

Thus, HIV-PI could counteract the early stage of HPV infection, when cervical epithelial cells are not heavily transformed, HPV-DNA is episomic and the expression of E6, E7 and other HPV proteins is low. In this specific setting, HIV-PI could also favor the clearance of episomal HPV DNA. This suggests that HIV-PI should be used early in the clinical management of CIN and, possibly, prior to transition to CC.

6. Concluding Remarks and Future Directions

HPV-associated uterine CC is a serious health problem which can be prevented firstly by anti-HPV vaccination [215] and, secondly, through screening programs for the detection and surveillance of early lesions and the removal of high-grade CIN [216].

These prevention programs are feasible and largely applied in the developed but not in the developing world [217]. Consequently, low-income countries are now coping with a burden of uterine CC, especially in association with HIV infection [217].

With reference to CIN therapeutic approaches, however, it has to be noted that surgical removal of high-grade lesions may lead to uterine ulcers, hemorrhages, cicatrices and pre-term labor [218]. In this context, results from preclinical and clinical studies point to HIV-PI as a promising intervention in the management of high-grade CIN lesions of both HIV-infected and uninfected women. In fact, HIV-PI directly inhibit events leading to CIN onset and progression including the growth or invasion of HPV-positive epithelial cells [151,185,188] and angiogenesis [189,191,210]. These effects are independent of the anti-HIV and immune-reconstituting activities of HIV-PI. Consistently, as for HIV-positive patients, HIV-PI causes the regression or the complete remission of high-grade CIN lesions also in HIV-negative women [187].

Indeed, HIV-PI counteract key players of carcinogenesis, such as the cellular proteasome, the serine/threonine kinase AKT and the MMPs (Figure 2) [147–152,182,185,188–192,199–204,210]. As the cellular proteasome has a role in regulating AKT phosphorylation and MMP expression [152,219,220], a connection among the pathways targeted by the HIV-PI may exist.

At the present time, novel and tailor-made inhibitors of the cellular proteasome, AKT or MMPs are under evaluation [139,140,143,221–223], since the first generation of these drugs has given unsatisfactory results [123,124,135–138,224,225].

In this regard, however, it should be considered that differently from novel proteasome, AKT or MMP inhibitors, HIV-PI are in use since many years and information on their pharmacokinetic, tissue distribution and toxicity profile is large [184,226].

Regarding the use of HIV-PI in oncology, this should entail tumor-customized therapies. In fact, each member of the HIV-PI class of drugs seems to have distinct effects on MMPs, while each tumor produces different MMPs, depending on its type and stage of progression.

As to uterine CIN, MMP-9, MT1-MMP and MMP-2 have a key role in its evolution into CC [100–113]. Noteworthy, either SQV or RTV down-regulates MMP-9 expression, thereby blocking CIN cell invasiveness [185,188]; whereas, IDV reduces MT1-MMP synthesis, thus impairing MMP-2 functional activation and angiogenesis [191,210].

Therefore, in order to promote the regression of high-grade CIN, administrating a combination of different HIV-PI may be advisable and effective, as indicated by a recent clinical study [187].

Although the systemic use of HIV-PI has been reported to cause the regression or the complete remission of KS or CIN in the absence of adverse side effects [187,194], topical application of these drugs may be safer and effective, against high-grade CIN. This therapeutic strategy can be applicable worldwide.

Future work will identify HIV-PI chemical groups responsible for their anti-tumor effects, hence leading to the synthesis of more active, less toxic derivatives.

In this context, it has to be highlighted that HIV-PI inhibitory effect on MMPs impairs events which are key for high-grade CIN evolution into CC, namely CIN and endothelial cell invasion of the basement membranes. Therefore, labeled probes utilizing MMP substrate specificity for imaging purposes [227–229] could allow the monitoring of the therapeutic response to HIV-PI or their future derivatives.

Author Contributions: G.B. wrote the paper; B.E., P.M. and C.S. substantively revised it.

Acknowledgments: Supported by the Italian Ministry of Health (Rome, Italy; grant no. OR/0906) and the Italian Ministry of Education, University and Research (Rome, Italy; grant no. RSA/0906). The authors thank M. Falchi (National HIV/AIDS Research Center, Rome, Italy) for assistance in preparing the figures.

References

1. Forman, D.; de Martel, C.; Lacey, C.J.; Soerjomataram, I.; Lortet-Tieulent, J.; Bruni, L.; Vignat, J.; Ferlay, J.; Bray, F.; Plummer, M.; et al. Global burden of human papillomavirus and related diseases. *Vaccine* **2012**, *30*, F12–F23. [CrossRef] [PubMed]
2. Jenkins, D. Histopathology and cytopathology of cervical cancer. *Dis. Markers* **2007**, *23*, 199–212. [CrossRef] [PubMed]
3. zur Hausen, H. Papillomaviruses and cancer: From basic studies to clinical application. *Nat. Rev. Cancer* **2002**, *2*, 342–350. [CrossRef] [PubMed]
4. Hebner, C.M.; Laimins, L.A. Human papillomaviruses: Basic mechanisms of pathogenesis and oncogenicity. *Rev. Med. Virol.* **2006**, *16*, 83–97. [CrossRef] [PubMed]
5. McBride, A.A.; Warburton, A. The role of integration in oncogenic progression of HPV-associated cancers. *PLoS Pathog.* **2017**, *13*, e1006211. [CrossRef] [PubMed]
6. Kim, M.; Kim, H.S.; Kim, S.; Oh, J.; Han, J.Y.; Lim, J.M.; Juhnn, Y.; Song, Y. Human papillomavirus type 16 E5 oncoprotein as a new target for cervical cancer treatment. *Biochem. Pharmacol.* **2010**, *80*, 1930–1935. [CrossRef] [PubMed]
7. Wasson, C.W.; Morgan, E.L.; Müller, M.; Ross, R.L.; Hartley, M.; Roberts, S.; Macdonald, A. Human papillomavirus type 18 E5 oncogene supports cell cycle progression and impairs epithelial differentiation by modulating growth factor receptor signalling during the virus life cycle. *Oncotarget* **2017**, *8*, 103581–103600. [CrossRef] [PubMed]
8. Roman, A.; Munger, K. The papillomavirus E7 proteins. *Virology* **2013**, *445*, 138–168. [CrossRef] [PubMed]
9. Vande Pol, S.B.; Klingelhutz, A.J. Papillomavirus E6 oncoproteins. *Virology* **2013**, *445*, 115–137. [CrossRef] [PubMed]
10. Doorbar, J. Molecular biology of human papillomavirus infection and cervical cancer. *Clin. Sci.* **2006**, *110*, 525–541. [CrossRef] [PubMed]
11. Petignat, P.; Roy, M. Diagnosis and management of cervical Cancer. *BMJ* **2007**, *335*, 765–768. [CrossRef] [PubMed]
12. Mittal, S.; Basu, P.; Muwonge, R.; Banerjee, D.; Ghosh, I.; Sengupta, MM.; Das, P.; Dey, P.; Mandal, R.; Panda, C.; et al. Risk of high-grade precancerous lesions and invasive cancers in high-risk HPV-positive women with normal cervix or CIN 1 at baseline-A population-based cohort study. *Int. J. Cancer* **2017**, *140*, 1850–1859. [CrossRef] [PubMed]
13. Grabowska, A.K.; Riemer, A.B. The invisible enemy—How human papillomaviruses avoid recognition and clearance by the host immune system. *Open Virol. J.* **2012**, *6*, 249–256. [CrossRef] [PubMed]
14. De Freitas, A.C.; de Oliveira, T.H.A.; Barros, M.R., Jr.; Venuti, A. hrHPV E5 oncoprotein: Immune evasion and related immunotherapies. *J. Exp. Clin. Cancer Res.* **2017**, *36*, 71–86. [CrossRef] [PubMed]
15. White, A.E.; Livanos, E.M.; Tlsty, T.D. Differential disruption of genomic integrity and cell cycle regulation in normal human fibroblasts by the HPV oncoproteins. *Genes Dev.* **1994**, *8*, 666–677. [CrossRef] [PubMed]

16. Thomas, J.T.; Laimins, L.A. Human papillomavirus oncoproteins E6 and E7 independently abrogate the mitotic spindle checkpoint. *J. Virol.* **1998**, *72*, 1131–1137. [PubMed]
17. Southern, S.A.; Noya, F.; Meyers, C.; Broker, T.R.; Chow, L.T.; Herrington, C.S. Tetrasomy is induced by human papillomavirus type 18 E7 gene expression in keratinocyte raft cultures. *Cancer Res.* **2001**, *61*, 4858–4863. [PubMed]
18. Duensing, S.; Munger, K. Centrosome abnormalities and genomic instability induced by human papillomavirus oncoproteins. *Prog. Cell Cycle Res.* **2003**, *5*, 383–391. [PubMed]
19. Takahashi, Y.; Ellis, L.M.; Mai, M. The angiogenic switch of human colon cancer occurs simultaneous to initiation of invasion. *Oncol. Rep.* **2003**, *10*, 9–13. [CrossRef] [PubMed]
20. Monk, B.J.; Willmott, L.J.; Sumner, D.A. Antiangiogenesis agents in metastatic or recurrent cervical cancer. *Gynecol. Oncol.* **2010**, *116*, 181–186. [CrossRef] [PubMed]
21. Krill, L.S.; Krishnansu, S.T. Exploring the therapeutic rationale for angiogenesis blockade in cervical cancer. *Clin. Ther.* **2015**, *37*, 9–19. [CrossRef] [PubMed]
22. Saijo, Y.; Furumoto, H.; Yoshida, K.; Nishimura, M.; Irahara, M. Clinical Significance of Vascular Endothelial Growth Factor Expression and Microvessel Density in Invasive Cervical Cancer. *J. Med. Investig.* **2015**, *62*, 154–160. [CrossRef] [PubMed]
23. Zheng, W.; Xiong, Y.H.; Han, J.; Guo, Z.X.; Li, Y.H.; Li, A.H.; Pei, X.Q. Contrast-enhanced ultrasonography of cervical carcinoma: Perfusion pattern and relationship with tumour angiogenesis. *Br. J. Radiol.* **2016**, *89*, 20150887. [CrossRef] [PubMed]
24. Ostor, A.G. Natural history of cervical intraepithelial neoplasia: A critical review. *Int. J. Gynecol. Pathol.* **1993**, *12*, 186–192. [CrossRef] [PubMed]
25. Bosch, F.X.; Lorincz, A.; Munoz, N.; Meijer, C.J.; Shah, K.V. The causal relation between human papillomavirus and cervical cancer. *J. Clin. Pathol.* **2002**, *55*, 244–265. [CrossRef] [PubMed]
26. Moreno, V.; Bosch, F.X.; Muñoz, N.; Meijer, C.J.; Shah, K.V.; Walboomers, J.M.; Herrero, R.; Franceschi, S.; International Agency for Research on Cancer; Multicentric Cervical Cancer Study Group. Effect of oral contraceptives on risk of cervical cancer in women with human papillomavirus infection: The IARC multicentric case-control study. *Lancet* **2002**, *359*, 1085–1092. [CrossRef]
27. Girianelli, V.R.; Azevedo, E.; Silva, G.; Thuler, L.C. Factors associated with the risk of progression to precursor lesions or cervical cancer in women with negative cytologic findings. *Int. J. Gynaecol. Obstet.* **2009**, *107*, 228–231. [CrossRef] [PubMed]
28. Six, C.; Heard, I.; Bergeron, C.; Orth, G.; Poveda, J.D.; Zagury, P.; Cesbron, P.; Crenn-Hébert, C.; Pradinaud, R.; Sobesky, M.; et al. Comparative prevalence, incidence and short-term prognosis of cervical squamous intraepithelial lesions amongst HIV-positive and HIV-negative women. *AIDS* **1998**, *12*, 1047–1056. [CrossRef] [PubMed]
29. Robinson, W., 3rd. Invasive and preinvasive cervical neoplasia in human immunodeficiency virus-infected women. *Semin. Oncol.* **2000**, *27*, 463–470. [PubMed]
30. Sellors, J.; Lewis, K.; Kidula, N.; Muhombe, K.; Tsu, V.; Herdman, C. Screening and management of precancerous lesions to prevent cervical cancer in low-resource settings. *Asian Pac. J. Cancer Prev.* **2003**, *4*, 277–280. [PubMed]
31. Omar, T.; Schwartz, S.; Hanrahan, C.; Modisenyane, T.; Tshabangu, N.; Golub, J.E.; McIntyre, J.A.; Gray, G.E.; Mohapi, L.; Martinson, N.A. Progression and regression of premalignant cervical lesions in HIV-infected women from Soweto: A prospective cohort. *AIDS* **2011**, *25*, 87–94. [CrossRef] [PubMed]
32. Blitz, S.; Baxter, J.; Raboud, J.; Walmsley, S.; Rachlis, A.; Smaill, F.; Ferenczy, A.; Coutlée, F.; Hankins, C.; Money, D. Canadian Women's HIV Study Group: Evaluation of HIV and highly active antiretroviral therapy on the natural history of human papillomavirus infection and cervical cytopathologic findings in HIV-positive and high-risk HIV-negative women. *J. Infect. Dis.* **2013**, *208*, 454–462. [CrossRef] [PubMed]
33. Denslow, S.A.; Rositch, A.F.; Firnhaber, C.; Ting, J.; Smith, J.S. Incidence and progression of cervical lesions in women with HIV: A systematic global review. *Int. J. STD AIDS* **2014**, *25*, 163–177. [CrossRef] [PubMed]
34. Obiri-Yeboah, D.; Akakpo, P.K.; Mutocheluh, M.; Adjei-Danso, E.; Allornuvor, G.; Amoako-Sakyi, D.; Adu-Sarkodie, Y.; Mayaud, P. Epidemiology of cervical human papillomavirus (HPV) infection and squamous intraepithelial lesions (SIL) among a cohort of HIV-infected and uninfected Ghanaian women. *BMC Cancer* **2017**, *17*, 688. [CrossRef] [PubMed]

35. Nappi, L.; Carriero, C.; Bettocchi, S.; Herrero, J.; Vimercati, A.; Putignano, G. Cervical squamous intraepithelial lesions of low-grade in HIV-infected women: Recurrence, persistence and progression, in treated and untreated women. *Eur. J. Obstet. Gynecol. Reprod. Biol.* **2005**, *121*, 226–232. [CrossRef] [PubMed]
36. Ntekim, A.; Campbell, O.; Rothenbacher, D. Optimal management of cervical cancer in HIV-positive patients: A systematic review. *Cancer Med.* **2015**, *4*, 1381–1393. [CrossRef] [PubMed]
37. Levine, A.M. AIDS-related malignancies: The emerging epidemic. *J. Natl. Cancer Inst.* **1993**, *85*, 1382–1397. [CrossRef] [PubMed]
38. Kang, M.; Cu-Uvin, S. Association of HIV viral load and CD4 cell count with human papillomavirus detection and clearance in HIV-infected women initiating highly active antiretroviral therapy. *HIV Med.* **2012**, *13*, 372–378. [CrossRef] [PubMed]
39. Tornesello, M.L.; Buonaguro, F.M.; Beth-Giraldo, E.; Giraldo, G. Human immunodeficiency virus type 1 tat gene enhances human papillomavirus early gene expression. *Intervirology* **1993**, *36*, 57–64. [CrossRef] [PubMed]
40. Nyagol, J.; Leucci, E.; Onnis, A.; De Falco, G.; Tigli, C.; Sanseverino, F.; Torriccelli, M.; Palummo, N.; Pacenti, L.; Santopietro, R.; et al. The effects of HIV-1 Tat protein on cell cycle during cervical carcinogenesis. *Cancer Biol. Ther.* **2006**, *5*, 684–690. [CrossRef] [PubMed]
41. Kim, R.H.; Yochim, J.M.; Kang, M.K.; Shin, K.H.; Christensen, R.; Park, N.H. HIV-1 Tat enhances replicative potential of human oral keratinocytes harboring HPV-16 genome. *Int. J. Oncol.* **2008**, *33*, 777–782. [PubMed]
42. Barillari, G.; Palladino, C.; Bacigalupo, I.; Leone, P.; Falchi, M.; Ensoli, B. Entrance of the Tat protein of HIV-1 into human uterine cervical carcinoma cells causes up-regulation of HPV-E6 expression and a decrease in p53 protein levels. *Oncol. Lett.* **2016**, *12*, 2389–2394. [CrossRef] [PubMed]
43. Barillari, G.; Ensoli, B. Angiogenic effects of extracellular HIV-1 Tat protein and its role in the pathogenesis of AIDS-associated Kaposi's sarcoma. *Clin. Microbiol. Rev.* **2002**, *15*, 310–326. [CrossRef] [PubMed]
44. Tornesello, M.L.; Buonaguro, L.; Giorgi-Rossi, P.; Buonaguro, F.M. Viral and cellular Biomarkers in the diagnosis of cervical intraepithelial neoplasia and cancer. *Biomed. Res. Int.* **2013**, *2013*, 519619. [CrossRef] [PubMed]
45. Ren, C.; Zhu, Y.; Yang, L.; Zhang, X.; Liu, L.; Ren, C. Diagnostic performance of HPV E6/E7 mRNA assay for detection of cervical high-grade intraepithelial neoplasia and cancer among women with ASCUS Papanicolaou smears. *Arch. Gynecol. Obstet.* **2017**, *297*, 425–432. [CrossRef] [PubMed]
46. Shi, W.J.; Liu, H.; Wu, D.; Tang, Z.H.; Shen, Y.C.; Guo, L. E6/E7 proteins are potential markers for the screening and diagnosis of cervical pre-cancerous lesions and cervical cancer in a Chinese population. *Oncol. Lett.* **2017**, *14*, 6251–6258. [CrossRef] [PubMed]
47. Liang, C.W.; Lin, M.C.; Hsiao, C.H.; Lin, Y.T.; Kuo, K.T. Papillary squamous intraepithelial lesions of the uterine cervix: Human papillomavirus-dependent changes in cell cycle expression and cytologic features. *Hum. Pathol.* **2010**, *41*, 326–335. [CrossRef] [PubMed]
48. Bucchi, L.; Cortecchia, S.; Galanti, G.; Sgadari, C.; Costa, S.; De Lillo, M.; Caparra, L.; Barillari, G.; Monini, P.; Nannini, R.; et al. Follow-up study of patients with cervical intraepithelial neoplasia grade 1 overexpressing p16ink4a. *Int. J. Gynecol. Cancer* **2013**, *23*, 1663–1669. [CrossRef]
49. Calil, L.N.; Edelweiss, M.I.; Meurer, L.; Igansi, C.N.; Bozzetti, M.C. p16 INK4a and Ki67 expression in normal, dysplastic and neoplastic uterine cervical epithelium and human papillomavirus (HPV) infection. *Pathol. Res. Pract.* **2014**, *210*, 482–487. [CrossRef] [PubMed]
50. Pedroza-Torres, A.; López-Urrutia, E.; García-Castillo, V.; Jacobo-Herrera, N.; Herrera, L.A.; Peralta-Zaragoza, O.; López-Camarillo, C.; Cantú De Leon, D.; Fernández-Retana, J.; Cerna-Cortés, J.F.; et al. MicroRNAs in cervical cancer: Evidences for a miRNA profile deregulated by HPV and its impact on radio-resistance. *Molecules* **2014**, *19*, 6263–6281. [CrossRef] [PubMed]
51. Klein, T.; Bischoff, R. Physiology and pathophysiology of matrix metalloproteases. *Amino Acids* **2011**, *41*, 271–290. [CrossRef] [PubMed]
52. Löffek, S.; Schilling, O.; Franzke, C.W. Biological role of matrix metalloproteinases: A critical balance. *Eur. Respir. J.* **2011**, *38*, 191–208. [CrossRef] [PubMed]
53. Rohani, M.G.; Parks, W.C. Matrix remodeling by MMPs during wound repair. *Matrix Biol.* **2015**, *44–46*, 113–121. [CrossRef] [PubMed]

54. Itoh, Y. Membrane-type matrix metalloproteinases: Their functions and regulations. *Matrix Biol.* **2015**, *46*, 207–223. [CrossRef] [PubMed]
55. Hua, Y.; Nair, S. Proteases in cardiometabolic diseases: Pathophysiology, molecular mechanisms and clinical applications. *Biochem. Biophys. Acta* **2015**, *1852*, 195–208. [CrossRef] [PubMed]
56. Khokha, R.; Murthy, A.; Weiss, A. Metalloproteinases and their natural inhibitors in inflammation and immunity. *Nat. Rev. Immunol.* **2013**, *13*, 649–665. [CrossRef] [PubMed]
57. Garcia-Irigoyen, O.; Carotti, S.; Latasa, M.U.; Uriarte, I.; Fernández-Barrena, M.G.; Elizalde, M.; Urtasun, R.; Vespasiani-Gentilucci, U.; Morini, S.; Banales, J.M.; et al. Matrix metalloproteinase-10 expression is induced during hepatic injury and plays a fundamental role in liver tissue repair. *Liver Int.* **2014**, *34*, 257–270. [CrossRef] [PubMed]
58. Yadav, L.; Puri, N.; Rastogi, V.; Satpute, P.; Ahmad, R.; Kaur, G. Matrix metalloproteinases and cancer—Roles in threat and therapy. *Asian Pac. J. Cancer Prev.* **2014**, *15*, 1085–1091. [CrossRef] [PubMed]
59. Ågren, M.S.; Schnabel, R.; Christensen, L.H.; Mirastschijski, U. Tumor necrosis factor-α-accelerated degradation of type I collagen in human skin is associated with elevated matrix metalloproteinase (MMP)-1 and MMP-3 ex vivo. *Eur. J. Cell Biol.* **2015**, *94*, 12–21. [CrossRef] [PubMed]
60. Yan, H.; Xin, S.; Wang, H.; Ma, J.; Zhang, H.; Wei, H. Baicalein inhibits MMP-2 expression in human ovarian cancer cells by suppressing the p38MAPK-dependent NF-κB signaling pathway. *Anticancer Drugs* **2015**, *26*, 649–656. [CrossRef] [PubMed]
61. Pan, H.C.; Jiang, Q.; Yu, Y.; Mei, J.P.; Cui, Y.K.; Zhao, W.J. Quercetin promotes cell apoptosis and inhibits the expression of MMP-9 and fibronectin via the AKT and ERK signalling pathways in human glioma cells. *Neurochem. Int.* **2015**, *80*, 60–71. [CrossRef] [PubMed]
62. Nemoto, E.; Gotoh, K.; Tsuchiya, M.; Sakisaka, Y.; Shimauchi, H. Extracellular ATP inhibits IL-1-induced MMP-1 expression through the action of CD39/nucleotidase triphosphate dephosphorylase-1 on human gingival fibroblasts. *Int. Immunopharmacol.* **2013**, *17*, 513–518. [CrossRef] [PubMed]
63. Gilet, A.; Zou, F.; Boumenir, M.; Frippiat, J.P.; Thornton, S.N.; Lacolley, P.; Ropars, A. Aldosterone up-regulates MMP-9 and MMP-9/NGAL expression in human neutrophils through p38, ERK1/2 and PI3K pathways. *Exp. Cell Res.* **2015**, *331*, 152–163. [CrossRef] [PubMed]
64. Chambers, M.; Kirkpatrick, G.; Evans, M.; Gorski, G.; Foster, S.; Borghaei, R.C. IL-4 inhibition of IL-1 induced Matrix metalloproteinase-3 (MMP-3) expression in human fibroblasts involves decreased AP-1 activation via negative crosstalk involving of Jun N-terminal kinase (JNK). *Exp. Cell Res.* **2013**, *319*, 1398–1408. [CrossRef] [PubMed]
65. Peña, E.; de la Torre, R.; Arderiu, G.; Slevin, M.; Badimon, L. mCRP triggers angiogenesis by inducing F3 transcription and TF signalling in microvascular endothelial cells. *Tromb Haemost* **2017**, *117*, 357–370. [CrossRef] [PubMed]
66. Lin, H.Y.; Chen, Y.S.; Wang, K.; Chien, H.W.; Hsieh, Y.H.; Yang, S.F. Fisetin inhibits epidermal growth factor-induced migration of ARPE-19 cells by suppression of AKT activation and Sp1-dependent MMP-9 expression. *Mol. Vis.* **2017**, *23*, 900–910. [PubMed]
67. Adiseshaiah, P.; Vaz, M.; Machireddy, N.; Kalvakolanu, D.V.; Reddy, S.P. A. Fra-1-dependent, matrix metalloproteinase driven EGFR activation promotes human lung epithelial cell motility and invasion. *J. Cell. Physiol.* **2008**, *216*, 405–412. [CrossRef] [PubMed]
68. Zhang, G.J.; Zhao, J.; Jiang, M.L.; Zhang, L.C. ING5 inhibits cell proliferation and invasion in esophageal squamous cell carcinoma through regulation of the Akt/NF-κB/MMP-9 signaling pathway. *Biochem. Biophys. Res. Commun.* **2018**, *496*, 387–393. [CrossRef] [PubMed]
69. Brew, K.; Nagase, H. The tissue inhibitors of metalloproteinases (TIMPs): An ancient family with structural and functional diversity. *Biochim. Biophys. Acta* **2010**, *1803*, 55–71. [CrossRef] [PubMed]
70. Yabluchanskiy, A.; Ma, Y.; Iyer, R.P.; Hall, M.E.; Lindsey, M.L. Matrix metalloproteinase-9: Many shades of function in cardiovascular disease. *Physiology* **2013**, *28*, 391–403. [CrossRef] [PubMed]
71. Folgueras, A.R.; Pendas, A.M.; Sanchez, L.M.; Lopez-Otin, C. Matrix metalloproteinases in cancer: From new functions to improved inhibition strategies. *Int. J. Dev. Biol.* **2004**, *48*, 411–424. [CrossRef] [PubMed]
72. Hanzawa, M.; Shindoh, M.; Higashino, F.; Yasuda, M.; Inoue, N.; Hida, K.; Ono, M.; Kohgo, T.; Nakamura, M.; Notani, K.; et al. Hepatocyte growth factor up-regulates E1AF that induces oral squamous cell carcinoma cell invasion by activating matrix metalloproteinase genes. *Carcinogenesis* **2000**, *21*, 1079–1085. [CrossRef] [PubMed]

73. Jacob-Ferreira, A.L.; Schulz, R. Activation of intracellular matrix metalloproteinase-2 by reactive oxygen-nitrogen species: Consequences and therapeutic strategies in the heart. *Arch. Biochem. Biophys.* **2013**, *540*, 82–93. [CrossRef] [PubMed]
74. Tang, Y.L.; Liu, X.; Gao, S.Y.; Feng, H.; Jiang, Y.P.; Wang, S.S.; Yang, J.; Jiang, J.; Ma, X.R.; Tang, Y.J.; et al. WIP1 stimulates migration and invasion of salivary adenoid cystic carcinoma by inducing MMP-9 and VEGF-C. *Oncotarget* **2015**, *6*, 9031–9044. [CrossRef] [PubMed]
75. Shin, D.H.; Dier, U.; Melendez, J.A.; Hempel, N. Regulation of MMP-1 expression in response to hypoxia is dependent on the intracellular redox status of metastatic bladder cancer cells. *Biochim. Biophys. Acta* **2015**, *1852*, 2593–2602. [CrossRef] [PubMed]
76. Zong, L.; Li, J.; Chen, X.; Chen, K.; Li, W.; Li, X.; Zhang, L.; Duan, W.; Lei, J.; Xu, Q.; et al. Lipoxin A4 Attenuates Cell Invasion by Inhibiting ROS/ERK/MMP Pathway in Pancreatic Cancer. *Oxid. Med. Cell. Longev.* **2016**, 6815727. [CrossRef]
77. Fullár, A.; Kovalszky, I.; Bitsche, M.; Romani, A.; Schartinger, V.H.; Sprinzl, G.M.; Riechelmann, H.; Dudás, J. Tumor cell and carcinoma-associated fibroblast interaction regulates matrix metalloproteinases ant their inhibitors in oral squamous cell carcinoma. *Exp. Cell Res.* **2012**, *318*, 1517–1527. [CrossRef] [PubMed]
78. Zhao, L.; Wu, Y.; Xie, X.D.; Chu, Y.F.; Li, J.Q.; Zheng, L. c-Met identifies a population of matrix metalloproteinase 9-producing monocytes in peritumoural stroma of hepatocellular carcinoma. *J. Pathol.* **2015**, *237*, 319–329. [CrossRef] [PubMed]
79. Mashhadiabbas, F.; Mahjour, F.; Mahjour, S.B.; Fereidooni, F.; Hosseini, F.S. The immunohistochemical characterization of MMP-2, MMP-10, TIMP-1, TIMP-2 and podoplanin in oral squamous cell carcinoma. *Oral Surg. Oral Med. Pathol Oral. Radiol.* **2012**, *114*, 240–250. [CrossRef] [PubMed]
80. Hui, P.; Xu, X.; Xu, L.; Hui, G.; Wu, S.; Lan, Q. Expression of MMP14 in invasive pituitary adenomas: Relationship to invasion and angiogenesis. *Int. J. Clin. Exp. Pathol.* **2015**, *8*, 3556–3567. [PubMed]
81. Herszényi, L.; Barabás, L.; Hritz, I.; István, G.; Tulassay, Z. Impact of proteolytic enzymes in colorectal cancer development and progression. *World J. Gastroenterol.* **2014**, *20*, 13246–13257. [CrossRef] [PubMed]
82. Figueira, R.C.; Gomes, L.R.; Neto, J.S.; Silva, F.C.; Silva, I.D.; Sogayar, M.C. Correlation between MMPs and their inhibitors in breast cancer tumor tissue specimens and in cell lines with different metastatic potential. *BMC Cancer* **2009**, *9*. [CrossRef] [PubMed]
83. Hu, X.; Li, D.; Zhang, W.; Zhou, J.; Tang, B.; Li, L. Matrix metalloproteinase-9 expression correlates with prognosis and involved in ovarian cancer cell invasion. *Arch. Gynecol. Obstet.* **2012**, *286*, 1537–1543. [CrossRef] [PubMed]
84. Gao, Z.H.; Tretiakova, M.S.; Liu, W.H.; Gong, C.; Farris, P.D.; Hart, J. Association of E-cadherin, matrix metalloproteinases with the progression and metastasis of hepatocellular carcinoma. *Mod. Pathol.* **2006**, *19*, 533–540. [CrossRef] [PubMed]
85. Giannopoulos, G.; Pavlakis, K.; Parasi, A.; Kavatzas, N.; Tiniakos, D.; Karakosta, A.; Tzanakis, N.; Peros, G. The expression of matrix metalloproteinase-2 and -9 and their tissue inhibitor 2 in pancreatic ductal and ampullary carcinoma and their relation to angiogenesis and clinicopathological parameters. *Anticancer Res.* **2008**, *28*, 1875–1881. [PubMed]
86. Groblewska, M.; Siewko, M.; Mroczko, B.; Szmitkowski, M. The role of matrix metalloproteinases (MMPs) and their inhibitors (TIMPs) in the development of esophageal cancer. *Folia Histochem. Cytobiol.* **2012**, *50*, 12–19. [CrossRef] [PubMed]
87. Sheibani, S.; Mahmoudian, R.A.; Abbaszadegan, M.R.; Chamani, J.; Memar, B.; Gholamin, M. Expression analysis of matrix metalloproteinase-13 in human gastric cancer in the presence of Helicobacter Pylori infection. *Cancer Biomark.* **2017**, *18*, 349–356. [CrossRef] [PubMed]
88. Väyrynen, J.P.; Vornanen, J.; Tervahartiala, T.; Sorsa, T.; Bloigu, R.; Salo, T.; Tuomisto, A.; Mäkinen, M.J. Serum MMP-8 levels increase in colorectal cancer and correlate with disease course and inflammatory properties of primary tumors. *Int. J. Cancer* **2012**, *131*, 463–474. [CrossRef] [PubMed]
89. Rundhaug, J.E. Matrix metalloproteinases and angiogenesis. *J. Cell. Mol. Med.* **2005**, *9*, 267–285. [CrossRef] [PubMed]
90. Zhang, J.L.; Chen, G.W.; Liu, Y.C.; Wang, P.Y.; Wang, X.; Wan, Y.L.; Zhu, J.; Gao, H.Q.; Yin, J.; Wang, W.; et al. Secreted protein acidic and rich in cysteine (SPARC) suppresses angiogenesis by down-regulating the expression of VEGF and MMP-7 in gastric cancer. *PLoS ONE* **2012**, *7*, e44618. [CrossRef] [PubMed]

91. Chen, Q.; Jin, M.; Yang, F.; Zhu, J.; Xiao, Q.; Zhang, L. Matrix metalloproteinases: Inflammatory regulators of cell behavior in vascular formation and remodeling. *Mediat. Inflamm.* **2013**, e928315. [CrossRef] [PubMed]
92. Fang, C.; Wen, G.; Zhang, L.; Lin, L.; Moore, A.; Wu, S.; Ye, S.; Xiao, Q. An important role of matrix metalloproteinase-8 in angiogenesis in vitro and in vivo. *Cardiovasc. Res.* **2013**, *99*, 146–155. [CrossRef] [PubMed]
93. Gu, Y.; Ke, G.; Wang, L.; Gu, Q.; Zhou, E.; He, Q.; Wang, S. Silencing Matrix Metalloproteinases 9 and 2 Inhibits Human Retinal Microvascular Endothelial Cell Invasion and Migration. *Ophthalmic Res.* **2015**, *55*, 70–75. [CrossRef] [PubMed]
94. Yadav, L.; Puri, N.; Rastogi, V.; Satpute, P.; Sharma, V. Tumour Angiogenesis and Angiogenic Inhibitors: A Review. *Clin. Diagn. Res.* **2015**, *9*, XE01–XE05. [CrossRef] [PubMed]
95. Hu, J.; Ni, S.; Cao, Y.; Zhang, T.; Wu, T.; Yin, X.; Lang, Y.; Lu, H. The Angiogenic Effect of microRNA-21 Targeting TIMP3 through the Regulation of MMP2 and MMP9. *PLoS ONE* **2016**, *11*, e0149537. [CrossRef]
96. Fukuda, H.; Mochizuki, S.; Abe, H.; Okano, H.J.; Hara-Miyauchi, C.; Okano, H.; Yamaguchi, N.; Nakayama, M.; D'Armiento, J.; Okada, Y. Host-derived MMP-13 exhibits a protective role in lung metastasis of melanoma cells by local endostatin production. *Br. J. Cancer* **2011**, *105*, 1615–1624. [CrossRef] [PubMed]
97. Korpi, J.T.; Kervinen, V.; Mäklin, H.; Väänänen, A.; Lahtinen, M.; Läärä, E.; Ristimäki, A.; Thomas, G.; Ylipalosaari, M.; Aström, P.; et al. Collagenase-2 (matrix metalloproteinase-8) plays a protective role in tongue cancer. *Br. J. Cancer* **2008**, *98*, 766–775. [CrossRef] [PubMed]
98. Overall, C.M.; Kleifeld, O. Tumour microenvironment—Opinion: Validating matrix metalloproteinases as drug targets and anti-targets for cancer therapy. *Nat. Rev. Cancer* **2006**, *6*, 227–239. [CrossRef] [PubMed]
99. Abdul Muneer, P.M.; Alikunju, S.; Szlachetka, A.M.; Haorah, J. The mechanisms of cerebral vascular dysfunction and neuroinflammation by MMP-mediated degradation of VEGFR-2 in alcohol ingestion. *Arterioscler. Thromb. Vasc. Biol.* **2012**, *32*, 1167–1177. [CrossRef] [PubMed]
100. Wang, H.; Zhang, X.; Huang, L.; Li, J.; Qu, S.; Pan, F. Matrix metalloproteinase-14 expression and its prognostic value in cervical carcinoma. *Cell Biochem. Biophys.* **2014**, *70*, 729–734. [CrossRef] [PubMed]
101. Zhai, Y.; Hotary, K.B.; Nan, B.; Bosch, F.X.; Muñoz, N.; Weiss, S.J.; Cho, K.R. Expression of membrane type 1 matrix metalloproteinase is associated with cervical carcinoma progression and invasion. *Cancer Res.* **2005**, *65*, 6543–6550. [CrossRef] [PubMed]
102. Sheu, B.C.; Lien, H.C.; Ho, H.N.; Lin, H.H.; Chow, S.N.; Huang, S.C.; Hsu, S.M. Increased expression and activation of gelatinolytic matrix metalloproteinases is associated with the progression and recurrence of human cervical cancer. *Cancer Res.* **2003**, *63*, 6537–6542. [PubMed]
103. Matheus, E.R.; Zonta, M.A.; Discacciati, M.G.; Paruci, P.; Velame, F.; Cardeal, L.B.; Barros, S.B.; Pignatari, A.C.; Maria-Engler, S.S. MMP-9 expression increases according to the grade of squamous intraepithelial lesion in cervical smears. *Diagn. Cytopathol.* **2014**, *42*, 827–833. [CrossRef] [PubMed]
104. Nasr, M.; Ayyad, S.B.; El-Lamie, I.K.; Mikhail, M.Y. Expression of matrix metalloproteinase-2 in preinvasive and invasive carcinoma of the uterine cervix. *Gynaecol. Oncol.* **2005**, *26*, 199–202.
105. Davidson, B.; Goldberg, I.; Kopolovic, J.; Lerner-Geva, L.; Gotlieb, W.H.; Weis, B.; Ben-Baruch, G.; Reich, R. Expression of matrix metalloproteinase-9 in squamous cell carcinoma of the uterine cervix-clinicopathologic study using immunohistochemistry and mRNA in situ hybridization. *Gynecol. Oncol.* **1999**, *72*, 380–386. [CrossRef] [PubMed]
106. Branca, M.; Ciotti, M.; Giorgi, C.; Santini, D.; Di Bonito, L.; Costa, S.; Benedetto, A.; Bonifacio, D.; Di Bonito, P.; Paba, P.; et al. HPV-Pathogen ISS Study Group. Matrix metalloproteinase-2 (MMP-2) and its tissue inhibitor (TIMP-2) are prognostic factors in cervical cancer, related to invasive disease but not to high-risk human papillomavirus (HPV) or virus persistence after treatment of CIN. *Anticancer Res.* **2006**, *26*, 1543–1556. [PubMed]
107. Talvensaari-Mattila, A.; Turpeenniemi-Hujanen, T. Matrix metalloproteinase 9 in the uterine cervix during tumor progression. *Int. J. Gynaecol. Obstet.* **2006**, *92*, 83–84. [CrossRef] [PubMed]
108. Van Trappen, P.O.; Ryan, A.; Carroll, M.; Lecoeur, C.; Goff, L.; Gyselman, V.G.; Young, B.D.; Lowe, D.G.; Pepper, M.S.; Shepherd, J.H.; et al. A model for co-expression pattern analysis of genes implicated in angiogenesis and tumor cell invasion in cervical cancer. *Br. J. Cancer* **2002**, *87*, 537–544. [CrossRef] [PubMed]
109. Asha Nair, S.; Karunagaran, D.; Nair, M.B.; Sudhakaran, P.R. Changes in matrix metalloproteinases and their endogenous inhibitors during tumor progression in the uterine cervix. *J. Cancer Res. Clin. Oncol.* **2003**, *129*, 123–131. [CrossRef] [PubMed]

110. Gaiotto, M.A.; Focchi, J.; Ribalta, J.L.; Stávale, J.N.; Baracat, E.C.; Lima, G.R.; Guerreiro da Silva, I.D. Comparative study of MMP-2 (matrix metalloproteinase 2) immune expression in normal uterine cervix, intraepithelial neoplasias and squamous cells cervical carcinoma. *Am. J. Obstet. Gynecol.* **2004**, *190*, 1278–1282. [CrossRef] [PubMed]
111. Lu, Z.; Chen, H.; Zheng, X.M.; Chen, M.L. Expression and clinical significance of high risk human papillomavirus and invasive gene in cervical carcinoma. *Asian Pac. J. Trop. Med.* **2017**, *10*, 195–200. [CrossRef] [PubMed]
112. Fernandes, T.; de Angelo-Andrade, L.A.; Morais, S.S.; Pinto, G.A.; Chagas, C.A.; Maria-Engler, S.S.; Zeferino, L.C. Stromal cells play a role in cervical cancer progression mediated by MMP-2 protein. *Eur. J. Gynaecol. Oncol.* **2008**, *29*, 341–344. [PubMed]
113. Yang, S.F.; Wang, P.H.; Lin, L.Y.; Ko, J.L.; Chen, G.D.; Yang, J.S.; Lee, H.S.; Hsieh, Y.S. A significant elevation of plasma level of matrix metalloproteinase-9 in patients with high-grade intraepithelial neoplasia and early squamous cell carcinoma of the uterine cervix. *Reprod. Sci.* **2007**, *14*, 710–718. [CrossRef] [PubMed]
114. Smola-Hess, S.; Pahne, J.; Mauch, C.; Zigrino, P.; Smola, H.; Pfister, H.J. Expression of membrane type 1 matrix metalloproteinase in papillomavirus-positive cells: Role of the human papillomavirus (HPV) 16 and HPV8 E7 gene products. *J. Gen. Virol.* **2005**, *86*, 1291–1296. [CrossRef] [PubMed]
115. Da Silva Cardeal, L.B.; Brohem, C.A.; Correa, T.C.; Winnischofer, S.M.; Nakano, F.; Boccardo, E.; Villa, L.L.; Sogayar, M.C.; Maria-Engler, S.S. Higher expression and activity of metalloproteinases in human cervical carcinoma cell lines is associated with HPV presence. *Biochem. Cell Biol.* **2006**, *84*, 713–719. [CrossRef] [PubMed]
116. Yoshida, S.; Kajitani, N.; Satsuka, A.; Nakamura, H.; Sakai, H. Ras modifies proliferation and invasiveness of cells expressing human papillomavirus oncoproteins. *J. Virol.* **2008**, *82*, 8820–8827. [CrossRef] [PubMed]
117. Barbaresi, S.; Cortese, M.S.; Quinn, J.; Ashrafi, G.H.; Graham, S.V.; Campo, M.S. Effects of human papillomavirus type 16 E5 deletion mutants on epithelial morphology: Functional characterization of each trans-membrane domain. *J. Gen. Virol.* **2010**, *91*, 521–530. [CrossRef] [PubMed]
118. Cardeal, L.B.; Boccardo, E.; Termini, L.; Rabachini, T.; Andreoli, M.A.; di Loreto, C.; Longatto Filho, A.; Villa, L.L.; Maria-Engler, S.S. HPV16 oncoproteins induce MMPs/RECK-TIMP-2 imbalance in primary keratinocytes: Possible implications in cervical carcinogenesis. *PLoS ONE* **2012**, *7*, e33585. [CrossRef] [PubMed]
119. Kaewprag, J.; Umnajvijit, W.; Ngamkham, J.; Ponglikitmongkol, M. HPV16 oncoproteins promote cervical cancer invasiveness by upregulating specific matrix metalloproteinases. *PLoS ONE* **2013**, *8*, e71611. [CrossRef] [PubMed]
120. Zhu, D.; Ye, M.; Zhang, W. E6/E7 oncoproteins of high risk HPV-16 upregulate MT1-MMP, MMP-2 and MMP-9 and promote the migration of cervical cancer cells. *Int. J. Clin. Exp. Pathol.* **2015**, *8*, 4981–4989. [PubMed]
121. Menges, C.W.; Baglia, L.A.; Lapoint, R.; McCance, D.J. Human papillomavirus type 16 E7 up-regulates AKT activity through the retinoblastoma protein. *Cancer Res.* **2006**, *66*, 5555–5559. [CrossRef] [PubMed]
122. Spangle, J.M.; Münger, K. The human papillomavirus type 16 E6 oncoprotein activates mTORC1 signalling and increases protein synthesis. *J. Virol.* **2010**, *84*, 9398–9407. [CrossRef] [PubMed]
123. Lu, C.; Lee, J.J.; Komaki, R.; Herbst, R.S.; Feng, L.; Evans, W.K.; Choy, H.; Desjardins, P.; Esparaz, B.T.; Truong, M.T.; et al. Chemoradiotherapy with or without AE-941 in stage III non-small cell lung cancer: A randomized phase III trial. *J. Natl. Cancer Inst.* **2010**, *102*, 859–865. [CrossRef] [PubMed]
124. El-Rayes, B.F.; Philip, P.A.; Sarkar, F.H.; Shields, A.F.; Ferris, A.M.; Hess, K.; Kaseb, A.O.; Javle, M.M.; Varadhachary, G.R.; Wolff, R.A.; et al. A phase II study of isoflavones, erlotinib and gemcitabine in advanced pancreatic cancer. *Investig. New Drugs* **2011**, *29*, 694–699. [CrossRef] [PubMed]
125. Gu, Y.; Lee, H.M.; Roemer, E.J.; Musacchia, L.; Golub, L.M.; Simon, S.R. Inhibition of tumor cell invasiveness by chemically modified tetracyclines. *Curr. Med. Chem.* **2001**, *8*, 261–270. [CrossRef] [PubMed]
126. Lai, T.J.; Hsu, S.F.; Li, T.M.; Hsu, H.C.; Lin, J.G.; Hsu, C.J.; Chou, M.C.; Lee, M.C.; Yang, S.F.; Fong, Y.C. Alendronate inhibits cell invasion and MMP-2 secretion in human chondrosarcoma cell line. *Acta Pharmacol. Sin.* **2007**, *28*, 1231–1235. [CrossRef] [PubMed]
127. Scatena, R. Prinomastat, a hydroxamate-based matrix metalloproteinase inhibitor. A novel pharmacological approach for tissue remodeling-related diseases. *Expert Opin. Investig. Drugs* **2000**, *9*, 2159–2165. [CrossRef] [PubMed]

128. Eccles, S.A.; Box, G.M.; Court, W.J.; Bone, E.A.; Thomas, W.; Brown, P.D. Control of lymphatic and hematogenous metastasis of a rat mammary carcinoma by the matrix metalloproteinase inhibitor batimastat (BB-94). *Cancer Res.* **1996**, *56*, 2815–2822. [PubMed]

129. Price, A.; Shi, Q.; Morris, D.; Wilcox, M.E.; Brasher, P.M.; Rewcastle, N.B.; Shalinsky, D.; Zou, H.; Appelt, K.; Johnston, R.N.; et al. Marked inhibition of tumor growth in a malignant glioma tumor model by a novel synthetic matrix metalloproteinase inhibitor AG3340. *Clin. Cancer Res.* **1999**, *5*, 845–854. [PubMed]

130. Shevrin, D.H.; Gorny, K.I.; Rosol, T.J.; Kukreja, S.C. Effect of etidronate disodium on the development of bone lesions in an animal model of bone metastasis using the human prostate cancer cell line PC-3. *Prostate* **1991**, *19*, 149–154. [CrossRef] [PubMed]

131. Weber, M.H.; Lee, J.; Orr, F.W. The effect of Neovastat (AE-941) on an experimental metastatic bone tumor model. *Int. J. Oncol.* **2002**, *20*, 299–303. [CrossRef] [PubMed]

132. Giraudo, E.; Inoue, M.; Hanahan, D. An amino-bisphosphonate targets MMP-9-expressing macrophages and angiogenesis to impair cervical carcinogenesis. *J. Clin. Investig.* **2004**, *114*, 623–633. [CrossRef] [PubMed]

133. Wang, S.Y.; Yang, K.W.; Hsu, Y.T.; Chang, C.L.; Yang, Y.C. The differential inhibitory effects of genistein on the growth of cervical cancer cells in vitro. *Neoplasma* **2001**, *48*, 227–233. [PubMed]

134. Iwasaki, M.; Nishikawa, A.; Fujimoto, T.; Akutagawa, N.; Manase, K.; Endo, T.; Yoshida, K.; Maekawa, R.; Yoshioka, T.; Kudo, R. Anti-invasive Effect of MMI-166, a New Selective Matrix Metalloproteinase Inhibitor, in Cervical Carcinoma Cell Lines. *Gynecol. Oncol.* **2002**, *85*, 103–107. [CrossRef] [PubMed]

135. Bramhall, S.R.; Schulz, J.; Nemunaitis, J.; Brown, P.D.; Baillet, M.; Buckels, J.A. A double-blind placebo-controlled, randomised study comparing gemcitabine and marimastat with gemcitabine and placebo as first line therapy in patients with advanced pancreatic cancer. *Br. J. Cancer* **2002**, *87*, 161–167. [CrossRef] [PubMed]

136. Pavlaki, M.; Zucker, S. Matrix metalloproteinase inhibitors (MMPIs): The beginning of phase I or the termination of phase III clinical trials. *Cancer Metastasis Rev.* **2003**, *22*, 177–203. [CrossRef] [PubMed]

137. Bissett, D.; O'Byrne, K.J.; von Pawel, J.; Gatzemeier, U.; Price, A.; Nicolson, M.; Mercier, R.; Mazabel, E.; Penning, C.; Zhang, M.H.; et al. Phase III study of matrix metalloproteinase inhibitor prinomastat in non-small-cell lung cancer. *J. Clin. Oncol.* **2005**, *23*, 842–849. [CrossRef] [PubMed]

138. Dormán, G.; Cseh, S.; Hajdú, I.; Barna, L.; Kónya, D.; Kupai, K.; Kovács, L.; Ferdinandy, P. Matrix metalloproteinase inhibitors: A critical appraisal of design principles and proposed therapeutic utility. *Drugs* **2010**, *70*, 949–964. [CrossRef] [PubMed]

139. Amar, S.; Fields, G.B. Potential clinical implications of recent matrix metalloproteinase inhibitor design strategies. *Expert Rev. Proteom.* **2015**, *12*, 445–447. [CrossRef] [PubMed]

140. Fields, G.B. New strategies for targeting matrix metalloproteinases. *Matrix Biol.* **2015**, *44–46*, 239–246. [CrossRef] [PubMed]

141. Rims, C.R.; McGuire, J.K. Matrilysin (MMP-7) catalytic activity regulates β-catenin localization and signaling activation in lung epithelial cells. *Exp. Lung Res.* **2014**, *40*, 126–136. [CrossRef] [PubMed]

142. Valacca, C.; Tassone, E.; Mignatti, P. TIMP-2 Interaction with MT1-MMP Activates the AKT Pathway and Protects Tumor Cells from Apoptosis. *PLoS ONE* **2015**, *10*, e0136797. [CrossRef] [PubMed]

143. Shay, G.; Lynch, C.C.; Fingleton, B. Moving targets: Emerging roles for MMPs in cancer progression and metastasis. *Matrix Biol.* **2015**, *44–46*, 200–206. [CrossRef] [PubMed]

144. Flexner, C. HIV-protease inhibitors. *N. Engl. J. Med.* **1998**, *338*, 1281–1292. [CrossRef] [PubMed]

145. Riddler, S.A.; Haubrich, R.; Di Rienzo, A.G.; Peeples, L.; Powderly, W.G.; Klingman, K.L.; Garren, K.W.; George, T.; Rooney, J.F.; Brizz, B.; et al. AIDS Clinical Trials Group Study A5142 Team. Class-sparing regimens for initial treatment of HIV-1 infection. *N. Engl. J. Med.* **2008**, *358*, 2095–2106. [CrossRef] [PubMed]

146. UNAIDS. *Global Report: UNAIDS Report on the Global AIDS Epidemic 2010*; UNAIDS: Geneva, Switzerland, 2010.

147. André, P.; Groettrup, M.; Klenerman, P.; de Giuli, R.; Booth, B.L., Jr; Cerundolo, V.; Bonneville, M.; Jotereau, F.; Zinkernagel, R.M.; Lotteau, V. An inhibitor of HIV-1 protease modulates proteasome activity, antigen presentation and T cell responses. *Proc. Natl. Acad. Sci. USA* **1998**, *95*, 13120–13124. [CrossRef] [PubMed]

148. Gaedicke, S.; Firat-Geier, E.; Constantiniu, O.; Lucchiari-Hartz, M.; Freudenberg, M.; Galanos, C.; Niedermann, G. Antitumor effect of the human immunodeficiency virus protease inhibitor ritonavir: Induction of tumor-cell apoptosis associated with perturbation of proteasomal proteolysis. *Cancer Res.* **2002**, *62*, 6901–6908. [PubMed]
149. Pajonk, F.; Himmelsbach, J.; Riess, K.; Sommer, A.; McBride, W.H. The human immunodeficiency virus (HIV)-1 protease inhibitor saquinavir inhibits proteasome function and causes apoptosis and radio-sensitization in non-HIV-associated human cancer cells. *Cancer Res.* **2002**, *62*, 5230–5235. [PubMed]
150. Yang, Y.; Ikezoe, T.; Nishioka, C.; Bandobashi, K.; Takeuchi, T.; Adachi, Y.; Kobayashi, M.; Takeuchi, S.; Koeffler, H.P.; Taguchi, H. NFV, an HIV-1 protease inhibitor, induces growth arrest, reduced Akt signalling, apoptosis and docetaxel sensitisation in NSCLC cell lines. *Br. J Cancer* **2006**, *95*, 1653–1662. [CrossRef] [PubMed]
151. Hampson, L.; Kitchener, H.C.; Hampson, I.N. Specific HIV protease inhibitors inhibit the ability of HPV16 E6 to degrade p53 and selectively kill E6-dependent cervical carcinoma cells in vitro. *Antivir. Ther.* **2006**, *11*, 813–825. [PubMed]
152. De Barros, S.; Zakaroff-Girard, A.; Lafontan, M.; Galitzky, J.; Bourlier, V. Inhibition of human preadipocyte proteasomal activity by HIV protease inhibitors or specific inhibitor lactacystin leads to a defect in adipogenesis, which involves matrix metalloproteinase-9. *J. Pharmacol. Exp. Ther.* **2007**, *320*, 291–299. [CrossRef] [PubMed]
153. Bastard, J.P.; Caron, M.; Vidal, H.; Jan, V.; Auclair, M.; Vigouroux, C.; Luboinski, J.; Laville, M.; Maachi, M.; Girard, P.M.; et al. Association between altered expression of adipogenic factor SREBP1 in lipoatrophic adipose tissue from HIV-1-infected patients and abnormal adipocyte differentiation and insulin resistance. *Lancet* **2002**, *359*, 1026–1031. [CrossRef]
154. Riddle, T.M.; Kuhel, D.G.; Woollett, L.A.; Fichtenbaum, C.J.; Hui, D.Y. HIV protease inhibitor induces fatty acid and sterol biosynthesis in liver and adipose tissues due to the accumulation of activated sterol regulatory element-binding proteins in the nucleus. *J. Biol. Chem.* **2001**, *276*, 37514–37519. [CrossRef] [PubMed]
155. Lenhard, J.M.; Croom, D.K.; Weiel, J.E.; Winegar, D.A. HIV protease inhibitors stimulate hepatic triglyceride synthesis. *Arterioscler. Thromb. Vasc. Biol.* **2000**, *20*, 2625–2629. [CrossRef] [PubMed]
156. Hui, D.Y. Effects of HIV protease inhibitor therapy on lipid metabolism. *Prog. Lipid Res.* **2003**, *42*, 81–92. [CrossRef]
157. Tran, H.; Robinson, S.; Mikhailenko, I.; Strickland, D.K. Modulation of the LDL receptor and LRP levels by HIV protease inhibitors. *J. Lipid. Res.* **2003**, *44*, 1859–1869. [CrossRef] [PubMed]
158. Murata, H.; Hruz, P.W.; Mueckler, M. The mechanism of insulin resistance caused by HIV protease inhibitor therapy. *J. Biol. Chem.* **2000**, *275*, 20251–20254. [CrossRef] [PubMed]
159. Lehmann, C.; Jung, N.; Hofmann, A.; Cornely, O.A.; Wyen, C.; Fätkenheuer, G.; Hartmann, P. Nucleoside-free boosted double PI regimen: Significant CD4+ T-cell recovery in patients with poor immunologic response despite virologic suppression. *Curr. HIV Res.* **2008**, *6*, 555–559. [CrossRef] [PubMed]
160. Cassone, A.; De Bernardis, F.; Torosantucci, A.; Tacconelli, E.; Tumbarello, M.; Cauda, R. In vitro and in vivo anticandidal activity of human immunodeficiency virus protease inhibitors. *J. Infect. Dis.* **1999**, *180*, 448–453. [CrossRef] [PubMed]
161. Krischer, J.; Rutschmann, O.; Hirschel, B.; Vollenweider-Roten, S.; Saurat, J.H.; Pechère, M. Regression of Kaposi's sarcoma during therapy with HIV-1 protease inhibitors: A prospective pilot study. *J. Am. Acad. Dermatol.* **1998**, *38*, 594–598. [CrossRef]
162. Lebbé, C.; Blum, L.; Pellet, C.; Blanchard, G.; Vérola, O.; Morel, P.; Danne, O.; Calvo, F. Clinical and biological impact of antiretroviral therapy with protease inhibitors on HIV-related Kaposi's sarcoma. *AIDS* **1998**, *12*, F45–F49. [CrossRef] [PubMed]
163. Niehues, T.; Horneff, G.; Megahed, M.; Schroten, H.; Wahn, V. Complete regression of AIDS-related Kaposi's sarcoma in a child treated with highly active antiretroviral therapy. *AIDS* **1999**, *13*, 1148–1149. [CrossRef] [PubMed]
164. Gill, J.; Bourboulia, D.; Wilkinson, J.; Hayes, P.; Cope, A.; Marcelin, A.G.; Calvez, V.; Gotch, F.; Boshoff, C.; Gazzard, B. Prospective study of the effects of antiretroviral therapy on Kaposi sarcoma—Associated herpesvirus infection in patients with and without Kaposi sarcoma. *J. Acquir. Immune Defic. Syndr.* **2002**, *31*, 384–390. [CrossRef] [PubMed]

165. Hallit, R.R.; Afridi, M.; Sison, R.; Szabela, M.E.; Bajaj, N.; Alchaa, R.; Hallit, S.; Awkar, N.; Boghossian, J.; Slim, J. AIDS-related lymphoma: Resolution with antiretroviral therapy alone. *J. Int. Assoc. Provid. AIDS Care* **2014**, *13*, 313–315. [CrossRef] [PubMed]
166. Cobucci, R.N.; Lima, P.H.; de Souza, P.C.; Costa, V.V.; Cornetta Mda, C.; Fernandes, J.V.; Gonçalves, A.K. Assessing the impact of HAART on the incidence of defining and non-defining AIDS cancers among patients with HIV/AIDS: A systematic review. *J. Infect. Public Health* **2015**, *8*, 1–10. [CrossRef] [PubMed]
167. Heard, I.; Schmitz, V.; Costagliola, D.; Orth, G.; Kazatchkine, M.D. Early regression of cervical lesions in HIV-seropositive women receiving highly active antiretroviral therapy. *AIDS* **1998**, *12*, 1459–1464. [CrossRef] [PubMed]
168. Firnhaber, C.; Westreich, D.; Schulze, D.; Williams, S.; Siminya, M.; Michelow, P.; Levin, S.; Faesen, M.; Smith, J.S. Highly active antiretroviral therapy and cervical dysplasia in HIV-positive women in South Africa. *J. Int. AIDS Soc.* **2012**, *15*, e17382. [CrossRef] [PubMed]
169. Adler, D.H.; Kakinami, L.; Modisenyane, T.; Tshabangu, N.; Mohapi, L.; De Bruyn, G.; Martinson, N.A.; Omar, T. Increased regression and decreased incidence of human papillomavirus-related cervical lesions among HIV-infected women on HAART. *AIDS* **2012**, *26*, 1645–1652. [CrossRef] [PubMed]
170. Soncini, E.; Zoncada, A.; Condemi, V.; Antoni, A.D.; Bocchialini, E.; Soregotti, P. Reduction of the risk of cervical intraepithelial neoplasia in HIV-infected women treated with highly active antiretroviral therapy. *Acta Biomed.* **2007**, *78*, 36–40. [PubMed]
171. Minkoff, H.; Ahdieh, L.; Massad, L.S.; Anastos, K.; Watts, D.H.; Melnick, S.; Muderspach, L.; Burk, R.; Palefsky, J. The effect of highly active antiretroviral therapy on cervical cytologic changes associated with oncogenic HPV among HIV-infected women. *AIDS* **2001**, *15*, 2157–2164. [CrossRef] [PubMed]
172. Monini, P.; Sgadari, C.; Toschi, E.; Barillari, G.; Ensoli, B. Antitumour effects of antiretroviral therapy. *Nat. Rev. Cancer* **2004**, *4*, 861–875. [CrossRef] [PubMed]
173. Bani-Sadr, F.; Fournier, S.; Molina, J.M. Relapse of Kaposi's sarcoma in HIV-infected patients switching from a protease inhibitor to a non-nucleoside reverse transcriptase inhibitor-based highly active antiretroviral therapy regimen. *AIDS* **2003**, *17*, 1580–1581. [CrossRef] [PubMed]
174. Bono, C.; Karlin, L.; Harel, S.; Mouly, E.; Labaume, S.; Galicier, L.; Apcher, S.; Sauvageon, H.; Fermand, J.P.; Bories, J.C.; et al. The human immunodeficiency virus-1 protease inhibitor nelfinavir impairs proteasome activity and inhibits the proliferation of multiple myeloma cells in vitro and in vivo. *Haematologica* **2012**, *97*, 1101–1109. [CrossRef] [PubMed]
175. Brüning, A.; Burger, P.; Vogel, M.; Rahmeh, M.; Gingelmaiers, A.; Friese, K.; Lenhard, M.; Burges, A. Nelfinavir induces the unfolded protein response in ovarian cancer cells, resulting in ER vacuolization, cell cycle retardation and apoptosis. *Cancer Biol. Ther.* **2009**, *8*, 226–232. [CrossRef] [PubMed]
176. Chow, W.A.; Guo, S.; Valdes-Albini, F. Nelfinavir induces liposarcoma apoptosis and cell cycle arrest by up-regulating sterol regulatory element binding protein-1. *Anticancer Drugs* **2006**, *17*, 891–903. [CrossRef] [PubMed]
177. Dewan, M.Z.; Tomita, M.; Katano, H.; Yamamoto, N.; Ahmed, S.; Yamamoto, M.; Sata, T.; Mori, N.; Yamamoto, N. An HIV protease inhibitor, ritonavir targets the nuclear factor-kappaB and inhibits the tumor growth and infiltration of EBV-positive lymphoblastoid B cells. *Int. J. Cancer* **2009**, *124*, 622–629. [CrossRef] [PubMed]
178. Gills, J.J.; Lopiccolo, J.; Tsurutani, J. Nelfinavir, A lead HIV protease inhibitor, is a broad-spectrum, anticancer agent that induces endoplasmic reticulum stress, autophagy and apoptosis in vitro and in vivo. *Clin. Cancer Res.* **2007**, *13*, 5183–5194. [CrossRef] [PubMed]
179. Ikezoe, T.; Saito, T.; Bandobashi, K.; Yang, Y.; Koeffler, H.P.; Taguchi, H. HIV-1 protease inhibitor induces growth arrest and apoptosis of human multiple myeloma cells via inactivation of signal transducer and activator of transcription 3 and extracellular signal-regulated kinase 1/2. *Mol. Cancer Ther.* **2004**, *3*, 473–479. [PubMed]
180. Jiang, W.; Mikochik, P.J.; Ra, J.H.; Lei, H.; Flaherty, K.T.; Winkler, J.D.; Spitz, F.R. HIV protease inhibitor nelfinavir inhibits growth of human melanoma cells by induction of cell cycle arrest. *Cancer Res.* **2007**, *67*, 1221–1227. [CrossRef] [PubMed]
181. Timeus, F.; Crescenzio, N.; Doria, A.; Foglia, L.; Pagliano, S.; Ricotti, E.; Fagioli, F.; Tovo, P.A.; Cordero di Montezemolo, L. In vitro anti-neuroblastoma activity of saquinavir and its association with imatinib. *Oncol Rep.* **2012**, *27*, 734–740. [CrossRef] [PubMed]

182. Yang, Y.; Ikezoe, T.; Takeuchi, T.; Adachi, Y.; Ohtsuki, Y.; Takeuchi, S.; Koeffler, H.P.; Taguchi, H. HIV-1 protease inhibitor induces growth arrest and apoptosis of human prostate cancer LNCaP cells in vitro and in vivo in conjunction with blockade of androgen receptor STAT3 and AKT signaling. *Cancer Sci.* **2005**, *96*, 425–433. [CrossRef] [PubMed]
183. Wei, C.; Bajpai, R.; Sharma, H.; Heitmeier, M.; Jain, A.D.; Matulis, S.M.; Nooka, A.K.; Mishra, R.K.; Hruz, P.W.; Schiltz, G.E.; et al. Development of GLUT4-selective antagonists for multiple myeloma therapy. *Eur. J. Med. Chem.* **2017**, *139*, 573–586. [CrossRef] [PubMed]
184. Justesen, U.S. Protease inhibitor plasma concentrations in HIV antiretroviral therapy. *Dan. Med. Bull.* **2008**, *55*, 165–185. [PubMed]
185. Barillari, G.; Iovane, A.; Bacigalupo, I.; Palladino, C.; Bellino, S.; Leone, P.; Monini, P.; Ensoli, B. Ritonavir or saquinavir impairs the invasion of cervical intraepithelial neoplasia cells via a reduction of MMP expression and activity. *AIDS* **2012**, *26*, 909–919. [CrossRef] [PubMed]
186. Brüning, A.; Vogel, M.; Mylonas, I.; Friese, K.; Burges, A. Bortezomib targets the caspase-like proteasome activity in cervical cancer cells, triggering apoptosis that can be enhanced by nelfinavir. *Curr. Cancer Drug Targets* **2011**, *11*, 799–809. [CrossRef] [PubMed]
187. Hampson, L.; Maranga, I.O.; Masinde, M.S.; Oliver, A.W.; Batman, G.; He, X.; Desai, M.; Okemwa, P.M.; Stringfellow, H.; Martin-Hirsch, P.; et al. A Single-Arm, Proof-Of-Concept Trial of Lopimune (Lopinavir/Ritonavir) as a Treatment for HPV-Related Pre-Invasive Cervical Disease. *PLoS ONE* **2016**, *11*, e0147917. [CrossRef] [PubMed]
188. Bacigalupo, I.; Palladino, C.; Leone, P.; Toschi, E.; Sgadari, C.; Ensoli, B.; Barillari, G. Inhibition of MMP-9 expression by ritonavir or saquinavir is associated with inactivation of the AKT/Fra-1 pathway in cervical intraepithelial neoplasia cells. *Oncol. Lett.* **2017**, *13*, 2903–2908. [CrossRef] [PubMed]
189. Toschi, E.; Sgadari, C.; Malavasi, L.; Bacigalupo, I.; Chiozzini, C.; Carlei, D.; Compagnoni, D.; Bellino, S.; Bugarini, R.; Falchi, M.; et al. Human immunodeficiency virus protease inhibitors reduce the growth of human tumors via a proteasome-independent block of angiogenesis and matrix metalloproteinases. *Int. J. Cancer* **2011**, *128*, 82–93. [CrossRef] [PubMed]
190. Sgadari, C.; Monini, P.; Barillari, G.; Ensoli, B. Use of HIV protease inhibitors to block Kaposi's sarcoma and tumor growth. *Lancet Oncol.* **2003**, *4*, 537–547. [CrossRef]
191. Sgadari, C.; Barillari, G.; Toschi, E.; Carlei, D.; Bacigalupo, I.; Baccarini, S.; Palladino, C.; Leone, P.; Bugarini, R.; Malavasi, L.; et al. HIV protease inhibitors are potent anti-angiogenic molecules and promote regression of Kaposi sarcoma. *Nat. Med.* **2002**, *8*, 225–232. [CrossRef] [PubMed]
192. Liuzzi, G.M.; Mastroianni, C.M.; Latronico, T.; Mengoni, F.; Fasano, A.; Lichtner, M.; Vullo, V.; Riccio, P. Anti-HIV drugs decrease the expression of matrix metalloproteinases in astrocytes and microglia. *Brain* **2004**, *127*, 398–407. [CrossRef] [PubMed]
193. Latronico, T.; Liuzzi, G.M.; Riccio, P.; Lichtner, M.; Mengoni, F.; D'Agostino, C.; Vullo, V.; Mastroianni, C.M. Antiretroviral therapy inhibits matrix metalloproteinase-9 from blood mononuclear cells of HIV-infected patients. *AIDS* **2007**, *21*, 677–684. [CrossRef] [PubMed]
194. Monini, P.; Sgadari, C.; Grosso, M.G.; Bellino, S.; Di Biagio, A.; Toschi, E.; Bacigalupo, I.; Sabbatucci, M.; Cencioni, G.; Salvi, E.; et al. For the Concerted Action on Kaposi's Sarcoma. Clinical course of classic Kaposi's sarcoma in HIV-negative patients treated with the HIV protease inhibitor indinavir. *AIDS* **2009**, *23*, 534–538. [CrossRef] [PubMed]
195. Hsieh, C.Y.; Tsai, P.C.; Tseng, C.H.; Chen, Y.L.; Chang, L.S.; Lin, S.R. Inhibition of EGF/EGFR activation with naphtho[1,2-b] furan-4,5-dione blocks migration and invasion of MDA-MB-231 cells. *Toxicol. In Vitro* **2013**, *27*, 1–10. [CrossRef] [PubMed]
196. Mathur, S.P.; Mathur, R.S.; Rust, P.F.; Young, R.C. Human papilloma virus (HPV)-E6/E7 and epidermal growth factor receptor (EGF-R) protein levels in cervical cancer and cervical intraepithelial neoplasia (CIN). *Am. J. Reprod. Immunol.* **2001**, *46*, 280–287. [CrossRef] [PubMed]
197. Bertelsen, B.I.; Steine, S.J.; Sandvei, R.; Molven, A.; Laerum, O.D. Molecular analysis of the PI3K-AKT pathway in uterine cervical neoplasia: Frequent PIK3CA amplification and AKT phosphorylation. *Int. J. Cancer* **2006**, *118*, 1877–1883. [CrossRef] [PubMed]
198. Du, C.X.; Wang, Y. Expression of P-Akt, NFkappa B and their correlation with human papillomavirus infection in cervical carcinoma. *Eur. J. Gynaecol. Oncol.* **2012**, *33*, 274–277. [PubMed]

199. Batchu, R.B.; Gruzdyn, O.V.; Bryant, C.S.; Qazi, A.M.; Kumar, S.; Chamala, S.; Kung, S.T.; Sanka, R.S.; Puttagunta, U.S.; Weaver, D.W.; et al. Ritonavir-mediated induction of apoptosis in pancreatic cancer occurs via the RB/E2F-1 and AKT pathways. *Pharmaceuticals* **2014**, *7*, 46–57. [CrossRef] [PubMed]
200. Gills, J.J.; Lopiccolo, J.; Dennis, P.A. Nelfinavir, a new anti-cancer drug with pleiotropic effects and many paths to autophagy. *Autophagy* **2008**, *4*, 107–109. [CrossRef] [PubMed]
201. Gupta, A.K.; Cerniglia, G.J.; Mick, R.; McKenna, W.G.; Muschel, R.J. HIV protease inhibitors block Akt signaling and radiosensitize tumor cells both in vitro and in vivo. *Cancer Res.* **2005**, *65*, 8256–8265. [CrossRef] [PubMed]
202. Kraus, M.; Müller-Ide, H.; Rückrich, T.; Bader, J.; Overkleeft, H.; Driessen, C. Ritonavir, nelfinavir, saquinavir and lopinavir induce proteotoxic stress in acute myeloid leukemia cells and sensitize them for proteasome inhibitor treatment at low micromolar drug concentrations. *Leuk. Res.* **2014**, *38*, 383–392. [CrossRef] [PubMed]
203. Plastaras, J.P.; Vapiwala, N.; Ahmed, M.S.; Gudonis, D.; Cerniglia, G.J.; Feldman, M.D.; Frank, I.; Gupta, A.K. Validation and toxicity of PI3K/Akt pathway inhibition by HIV protease inhibitors in humans. *Cancer Biol. Ther.* **2008**, *7*, 628–635. [CrossRef] [PubMed]
204. Pore, N.; Gupta, A.K.; Cerniglia, G.J.; Maity, A. HIV protease inhibitors decrease VEGF/HIF-1alpha expression and angiogenesis in glioblastoma cells. *Neoplasia* **2006**, *8*, 889–895. [CrossRef] [PubMed]
205. Gills, J.J.; Dennis, P.A. Perifosine: Update on a novel Akt inhibitor. *Curr. Oncol. Rep.* **2009**, *11*, 102–110. [CrossRef] [PubMed]
206. Chuang, C.W.; Pan, M.R.; Hou, M.F.; Hung, W.C. Cyclooxygenase-2 up-regulates CCR7 expression via AKT-mediated phosphorylation and activation of Sp1 in breast cancer cells. *J. Cell. Physiol.* **2013**, *228*, 341–348. [CrossRef] [PubMed]
207. Liu, J.X.; Luo, M.Q.; Xia, M.; Wu, Q.; Long, S.M.; Hu, Y.; Gao, G.C.; Yao, X.L.; He, M.; Su, H.; et al. Marine compound catunaregin inhibits angiogenesis through the modulation of phosphorylation of akt and eNOS in vivo and in vitro. *Mar. Drugs* **2014**, *12*, 2790–2801. [CrossRef] [PubMed]
208. Bottsford-Miller, J.N.; Coleman, R.L.; Sook, A.K. Resistance and escape from anti-angiogenesis therapy: Clinical implications and future strategies. *J. Clin. Oncol.* **2012**, *30*, 4026–4034. [CrossRef] [PubMed]
209. Pore, N.; Gupta, A.K.; Cerniglia, G.J.; Jiang, Z.; Bernhard, E.J.; Evans, S.M.; Koch, C.J.; Hahn, S.M.; Maity, A. Nelfinavir down-regulates hypoxia-inducible factor 1alpha and VEGF expression and increases tumor oxygenation: Implications for radiotherapy. *Cancer Res.* **2006**, *66*, 9252–9259. [CrossRef] [PubMed]
210. Barillari, G.; Iovane, A.; Bacigalupo, I.; Labbaye, C.; Chiozzini, C.; Sernicola, L.; Quaranta, M.T.; Falchi, M.; Sgadari, C.; Ensoli, B. The HIV protease inhibitor indinavir down-regulates the expression of the pro-angiogenic MT1-MMP by human endothelial cells. *Angiogenesis* **2014**, *7*, 831–838. [CrossRef] [PubMed]
211. Van Hinsbergh, V.W.; Engelse, M.A.; Quax, P.H. Pericellular proteases in angiogenesis and vasculogenesis. *Arterioscler. Tromb. Vasc. Biol.* **2006**, *26*, 716–728. [CrossRef] [PubMed]
212. Hakeem, A.; Bhatti, S.; Cilingiroglu, M. The spectrum of atherosclerotic coronary artery disease in HIV patients. *Curr. Atheroscler. Rep.* **2010**, *12*, 119–124. [CrossRef] [PubMed]
213. Kim, D.H.; Jarvis, R.M.; Allwood, J.W.; Batman, G.; Moore, R.E.; Marsden-Edwards, E.; Hampson, L.; Hampson, I.N.; Goodacre, R. Raman chemical mapping reveals site of action of HIV protease inhibitors in HPV16 E6 expressing cervical carcinoma cells. *Anal. Bioanal. Chem.* **2010**, *398*, 3051–3061. [CrossRef] [PubMed]
214. Batman, G.; Oliver, A.W.; Zehbe, I.; Richard, C.; Hampson, L.; Hampson, I.N. Lopinavir up-regulates expression of the antiviral protein ribonuclease L in human papillomavirus-positive cervical carcinoma cells. *Antivir. Ther.* **2011**, *16*, 515–525. [CrossRef] [PubMed]
215. Pils, S.; Joura, E.A. From the monovalent to the ninevalent HPV vaccine. *Clin. Microbiol. Infect.* **2015**, *21*, 827–833. [CrossRef] [PubMed]
216. Apgar, B.S.; Kaufman, A.J.; Bettcher, C.; Parker-Featherstone, E. Gynecologic procedures: Colposcopy, treatment of cervical intraepithelial neoplasia and endometrial assessment. *Am. Fam. Phys.* **2013**, *87*, 836–843.
217. Sigfrid, L.; Murphy, G.; Haldane, V.; Chuah, F.L.H.; Ong, S.E.; Cervero-Liceras, F.; Watt, N.; Alvaro, A.; Otero-Garcia, L.; Balabanova, D.; et al. Integrating cervical cancer with HIV healthcare services: A systematic review. *PLoS ONE* **2017**, *12*, e0181156. [CrossRef] [PubMed]
218. Armarnik, S.; Sheiner, E.; Piura, B.; Meirovitz, M.; Zlotnik, A.; Levy, A. Obstetric outcome following cervical conization. *Arch. Gynecol. Obstet.* **2011**, *283*, 765–769. [CrossRef] [PubMed]

219. Ikebe, T.; Takeuchi, H.; Jimi, E.; Beppu, M.; Shinohara, M.; Shirasuna, K. Involvement of proteasomes in migration and matrix metalloproteinase-9 production of oral squamous cell carcinoma. *Int. J. Cancer* **1998**, *77*, 578–585. [CrossRef]
220. Li, J.Y.; Huang, W.Q.; Tu, R.H.; Zhong, G.Q.; Luo, B.B.; He, Y. Resveratrol rescues hyperglycemia-induced endothelial dysfunction via activation of Akt. *Acta Pharmacol. Sin.* **2017**, *38*, 182–191. [CrossRef] [PubMed]
221. Makinoshima, H.; Umemura, S.; Suzuki, A.; Nakanishi, H.; Maruyama, A.; Udagawa, H.; Mimaki, S.; Matsumoto, S.; Niho, S.; Ishii, G.; et al. Metabolic determinants of sensitivity to phosphatidylinositol 3-kinase pathway inhibitor in small-cell lung carcinoma. *Cancer Res.* **2018**. [CrossRef] [PubMed]
222. Pulido, M.; Roubaud, G.; Cazeau, A.L.; Mahammedi, H.; Vedrine, L.; Joly, F.; Mourey, L.; Pfister, C.; Goberna, A.; Lortal, B.; et al. Safety and efficacy of temsirolimus second line treatment for patients with recurrent bladder cancer. *BMC Cancer* **2018**, *8*, 194. [CrossRef] [PubMed]
223. Guo, K.Y.; Han, L.; Li, X.; Yang, A.V.; Lu, J.; Guan, S.; Li, H.; Yu, Y.; Zhao, Y.; Yang, J.; et al. Novel proteasome inhibitor delanzomib sensitizes cervical cancer cells to doxorubicin-induced apoptosis via stabilizing tumor suppressor proteins in the p53 pathway. *Oncotarget* **2017**, *8*, 114123–114135. [CrossRef] [PubMed]
224. Cao, B.; Li, J.; Mao, X. Dissecting bortezomib: Development, application, adverse effects and future direction. *Curr. Pharm. Des.* **2013**, *19*, 3190–3200. [CrossRef] [PubMed]
225. Herschbein, L.; Liesveld, J.L. Dueling for dual inhibition: Means to enhance effectiveness of PI3K/Akt/mTOR inhibitors in AML. *Blood Rev.* **2017**. [CrossRef] [PubMed]
226. Lv, Z.; Chu, Y.; Wang, Y. HIV protease inhibitors: A review of molecular selectivity and toxicity. *HIV AIDS* **2015**, *7*, 95–104. [CrossRef]
227. Akers, W.J.; Xu, B.; Lee, H.; Sudlow, G.P.; Fields, G.B.; Achilefu, S.; Edwards, W.B. Detection of MMP-2 and MMP-9 activity in vivo with a triple-helical peptide optical probe. *Bioconjugate Chem.* **2012**, *23*, 656–663. [CrossRef] [PubMed]
228. Matusiak, N.; van Waarde, A.; Bischoff, R.; Oltenfreiter, R.; van de Wiele, C.; Dierckx, R.A.; Elsinga, P.H. Probes for non-invasive matrix metalloproteinase-targeted imaging with PET and SPECT. *Curr. Pharm. Des.* **2013**, *19*, 4647–4672. [CrossRef] [PubMed]
229. Shimizu, Y.; Temma, T.; Hara, I.; Makino, A.; Kondo, N.; Ozeki, E.; Ono, M.; Saji, H. In vivo imaging of membrane type-1 matrix metalloproteinase with a novel activatable near-infrared fluorescence probe. *Cancer Sci.* **2014**, *105*, 1056–1062. [CrossRef] [PubMed]

13

Agnoprotein is an Essential Egress Factor during BK Polyomavirus Infection

Margarita-Maria Panou [1,†], Emma L. Prescott [1,†], Daniel L. Hurdiss [1], Gemma Swinscoe [1], Michael Hollinshead [2], Laura G. Caller [2], Ethan L. Morgan [1], Louisa Carlisle [2], Marietta Müller [1], Michelle Antoni [1], David Kealy [1], Neil A. Ranson [1], Colin M. Crump [2] and Andrew Macdonald [1,*]

1 Faculty of Biological Sciences and Astbury Centre for Structural and Molecular Biology, University of Leeds, Leeds LS2 9JT, UK; bs13mmp@leeds.ac.uk (M.-M.P.); e.l.prescott@leeds.ac.uk (E.L.P.); bs10d2h@leeds.ac.uk (D.L.H.); ed10ges@leeds.ac.uk (G.S.); bs10elm@leeds.ac.uk (E.L.M.); m.muller@leeds.ac.uk (M.M.); bsman@leeds.ac.uk (M.A.); bsdjk@leeds.ac.uk (D.K.); n.a.ranson@leeds.ac.uk (N.A.R.)

2 Department of Pathology, University of Cambridge, Tennis Court Road, Cambridge CB2 1QP, UK; msh57@cam.ac.uk (M.H.); lgw27@cam.ac.uk (L.G.C.); c.w.wasson@leeds.ac.uk (L.C.); cmc56@cam.ac.uk (C.M.C.)

* Correspondence: a.macdonald@leeds.ac.uk

† These authors contributed equally to the manuscript.

Abstract: BK polyomavirus (BKPyV; hereafter referred to as BK) causes a lifelong chronic infection and is associated with debilitating disease in kidney transplant recipients. Despite its importance, aspects of the virus life cycle remain poorly understood. In addition to the structural proteins, the late region of the BK genome encodes for an auxiliary protein called agnoprotein. Studies on other polyomavirus agnoproteins have suggested that the protein may contribute to virion infectivity. Here, we demonstrate an essential role for agnoprotein in BK virus release. Viruses lacking agnoprotein fail to release from host cells and do not propagate to wild-type levels. Despite this, agnoprotein is not essential for virion infectivity or morphogenesis. Instead, agnoprotein expression correlates with nuclear egress of BK virions. We demonstrate that the agnoprotein binding partner α-soluble N-ethylmaleimide sensitive fusion (NSF) attachment protein (α-SNAP) is necessary for BK virion release, and siRNA knockdown of α-SNAP prevents nuclear release of wild-type BK virions. These data highlight a novel role for agnoprotein and begin to reveal the mechanism by which polyomaviruses leave an infected cell.

Keywords: polyomavirus; agnoprotein; virus exit

---

1. Introduction

Polyomaviruses are small, non-enveloped viruses that use mammals, fish and birds as their hosts [1–4]. Currently, thirteen human polyomaviruses have been discovered and a number are linked to disease [5–7]. The first two human polyomaviruses discovered, BK polyomavirus and JC polyomavirus (JCPyV; hereafter referred to as JC), were named after the index case patients upon their discovery more than 40 years ago [8,9] and cause disease in immunosuppressed patients. JC is the causative agent of the lethal brain disease progressive multifocal leukoencephalopathy.

BK is an opportunistic pathogen, and is associated with several diseases in the immunosuppressed [10]. Primary infection with BK typically occurs in childhood, after which the virus establishes a chronic infection in the kidneys in approximately 80% of adults [11]. Whilst reactivation of BK does occur in healthy individuals, this is usually associated with asymptomatic low-level urinary shedding [12]. However, in the immunosuppressed, reactivation of BK is far more serious, resulting in increased urinary shedding because of increased replication in the absence of a competent immune response [13,14]. Such uncontrolled replication is ultimately linked with severe health problems, including polyomavirus-associated nephropathy (PVAN) and hemorrhagic cystitis in kidney and bone marrow transplant patients, respectively [15,16]. Up to 10% of kidney transplant patients experience PVAN, and of these, up to 90% may go on to lose their graft [17]. The incidence of BK-associated disease is rising due to the increase in transplants, and the use of more powerful immunosuppressive drugs to support such patients [11]. Despite the clinical impact of BK-associated disease, no anti-viral drugs that specifically target BK, or indeed any human polyomavirus, are currently available. Rather, generic anti-viral agents such as Cidofovir can be used, however, these have low efficacy and are themselves associated with nephrotoxicity [18]. Treatment is typically limited to a reduction in immunosuppression, which runs the risk of transplant rejection [19]. A better understanding of the BK life cycle is therefore needed in order to identify new targets for anti-viral therapy.

BK and JC polyomaviruses are closely related to the prototypic primate polyomavirus simian vacuolating agent 40 (SV40) [1]. Their ~5000 bp double-stranded DNA (dsDNA) genome is divided into three functional units consisting of early and late coding regions, separated by a non-coding control region (NCCR) [6]. The NCCR contains the origin of virus replication as well as enhancer and regulatory regions that control virus transcription. In kidney transplant recipients, circulating BK strains undergo re-arrangement of the NCCR region and this is thought to play an important role in disease [20]. The early region encodes the small (sT) and large (LT) tumor antigens, essential for virus transcription and replication. The late region encodes for the major (VP1) and minor capsid (VP2/VP3) proteins, which form the structural components of the BK virion [4,21], as well as the non-structural auxiliary agnoprotein.

Agnoprotein is a small, highly basic protein encoded by only a minority of polyomaviruses [22]. Whilst none of the recently discovered human polyomaviruses encode an agnoprotein, their remains a striking diversity of agnoprotein sequence and size within the mammalian polyomaviruses containing an agnoprotein open reading frame [1,22]. Amongst this diversity, the agnoproteins of BK, JC and SV40 share a high degree of sequence identity, particularly within the amino-terminal half of the protein (up to 83% identity between BK and JC), suggesting a conservation of function. Agnoprotein is predominantly expressed within the cytoplasm and perinuclear regions of infected cells during the later stages of the polyomavirus life cycle [23]. More recently, agnoprotein has also been shown to co-localise with lipid droplets in BK infected primary renal tubular epithelial cells [24], however, the physiological relevance of this is currently unclear. The agnoproteins of BK, JC and SV40 are phosphorylated when expressed in cells, and studies have shown that this phosphorylation plays a critical role in the respective virus life cycle [22,23,25,26]. Despite these observations, mechanistic insight into the role of agnoprotein phosphorylation is lacking.

Whilst the precise function of BK agnoprotein is currently not known, studies in JC and SV40 have produced contradictory findings [22,27]. A number of studies have shown that changes in agnoprotein expression, either from deletion of the open reading frame (ORF) or mutation of its start codon, impact on expression of other virus proteins [28–32]. Given the abundant expression of agnoprotein at the later stages of the polyomavirus life cycle, a role in virion assembly, morphogenesis and release has also been suggested. In SV40, agnoprotein expression might be required for correct localization of the VP1 major capsid protein [33], and cells infected with SV40 virus lacking agnoprotein release progeny virions deficient in DNA content [34,35]. Similar findings have been reported for JC virus [35], however, loss of agnoprotein has also been correlated with a defect in virus release [28]. Studies using clinical isolates of BK virus containing deletions within the agnogene indicate that agnoprotein expression correlates with virion infectivity [36]. The reasons for such wide-ranging phenotypes associated with agnoprotein deficiency remain unclear.

In this study we aimed to increase our understanding of the role of this enigmatic protein in the BK life cycle by generating a mutation in the start codon of the agnogene in the disease-associated Dunlop strain of BK virus. Using a primary renal proximal tubular epithelial cell culture model system, we found that loss of agnoprotein led to a profound reduction in virion release and impaired virus propagation in culture. In contrast with previous findings we show that these virions are infectious but remain trapped within the nucleus of an infected cell. We implicate an agnoprotein binding partner, α-SNAP, as an essential BK egress factor. Together, these data demonstrate that agnoprotein is required for the release of infectious BK virions.

2. Results

2.1. Loss of Agnoprotein Increases BK Transcription and Protein Expression

Agnoprotein is thought to be essential at several stages in the polyomavirus life cycle. To investigate this, we generated an agnoprotein knockout mutant in the clinically relevant Dunlop strain of BK. In this ΔAgno mutant, site directed mutagenesis was employed to replace the start codon (ATG) with a stop codon (TAG) (Figure 1A). Sequencing of the entire Dunlop genome confirmed the introduction of the mutation and established that no secondary mutations had been introduced. Equal amounts of wild type (WT) and ΔAgno genomes were transfected into primary renal proximal tubular epithelial (RPTE) cells, a physiologically relevant cell model for BK infection, and levels of BK protein expression determined at 72 h post transfection. Western blot analysis demonstrated production of early (LT) and late (VP1, VP2/VP3) proteins from both BK WT and ΔAgno genomes, and as expected only BK WT produced agnoprotein (Figure 1B). Interestingly, ΔAgno exhibited a consistent increase in virus protein expression compared to WT. Quantitative reverse transcriptase PCR (qRT-PCR) was used to determine if the increased BK protein expression was due to changes in virus gene transcription. Primer sets were used to amplify LT to detect early transcripts and VP1 to detect late transcripts. Levels of both transcripts were higher in ΔAgno transfected RPTE cells compared to WT BK control, suggesting that loss of agnoprotein correlates with an increase in early and late BK transcription. Given the role of LT in virus genome replication, we reasoned that increased expression of LT might result in increased virus replication. Indeed, in the absence of agnoprotein, levels of virus genome were higher than WT BK. Together, these data suggest that agnoprotein might play a role in the negative regulation of virus transcription and genome replication.

2.2. Agnoprotein Is Required for BK Virus Release

To further investigate the role of agnoprotein, we performed a virus growth assay. RPTE cells were transfected with WT BK or ΔAgno genomes and the number of VP1 capsid protein positive cells determined using Incucyte Zoom software (Essen BioScience, Ann Arbor, MI, USA) [37]. Whilst numbers of VP1 positive cells were similar at three days post transfection, the number of VP1 positive cells in ΔAgno transfected cells was significantly lower at six days post transfection, suggesting that virus dissemination was impaired in the absence of agnoprotein (Figure 2A). Levels of VP1 were then measured from harvested cells and culture media supernatant at 48 and 72 h time points (Figure 2B). In agreement with our previous observations, VP1 levels were higher in the cell lysate of ΔAgno transfected RPTE cells compared to WT BK (Figure 2B). Low levels of VP1 protein were also detectable by Western blot in the media supernatant of WT BK transfected RPTE cells 48 h after transfection, and levels increased at the 72 h time point. In contrast, VP1 was undetected at 48 h in the supernatants of cells transfected with ΔAgno, and remained lower than the WT at the 72 h time point (Figure 2B). To rule out potential non-specific effects of transfection, RPTE cells were infected with 1 IU/cell WT and ΔAgno viruses and incubated for 72 h, and the cell lysate and culture media harvested separately. The infectious virus titer from each fraction was then determined by fluorescent focus assay (Figure 2C). Whilst there was a small decrease in cell-associated infectious virus from ΔAgno infected cells, the proportion of virus released was approximately 10 fold reduced (Figure 2C). Recently, the broad-spectrum anion channel inhibitor 4,4′-Diisothiocyanatostilbene-2,2′-disulphonate (DIDS) has been shown to impair the release of BK virus particles from RPTE cells [38]. Whilst the molecular basis by which DIDS prevents BK release is currently not known, DIDS has been shown to prevent enterovirus 71 (EV71) release by targeting the virus encoded 2B protein [39]. EV71 2B is a small hydrophobic protein belonging to the viroporin family of membrane permeabilizing proteins [40,41]. Given that JC agnoprotein has been described as a viroporin, we sought to determine whether BK agnoprotein might be the target for the inhibitory activity of DIDS. To investigate this, RPTE cells were infected with WT BK or ΔAgno, and DIDS added to cells 48 h post infection. At 72 h post infection cell-associated and culture media supernatant samples was harvested separately and used to infect fresh RPTE cells, from which the infectious titer of cell-associated and released BK virus was determined by a fluorescence focus assay [38]. Incubation with 50 μM DIDS resulted in an approximately ten-fold decrease in the proportion of released WT BK virus (Figure 2D). Increasing the dose of DIDS to 100 μM further reduced the proportion of released virus. The proportion of released virus from ΔAgno infected cells was 10-fold lower than WT BK control, and this was reduced further after DIDS treatment, in a concentration dependent manner. Together, these data show that agnoprotein is important for BK virus release, although via a pathway that is independent of the target of the DIDS compound.

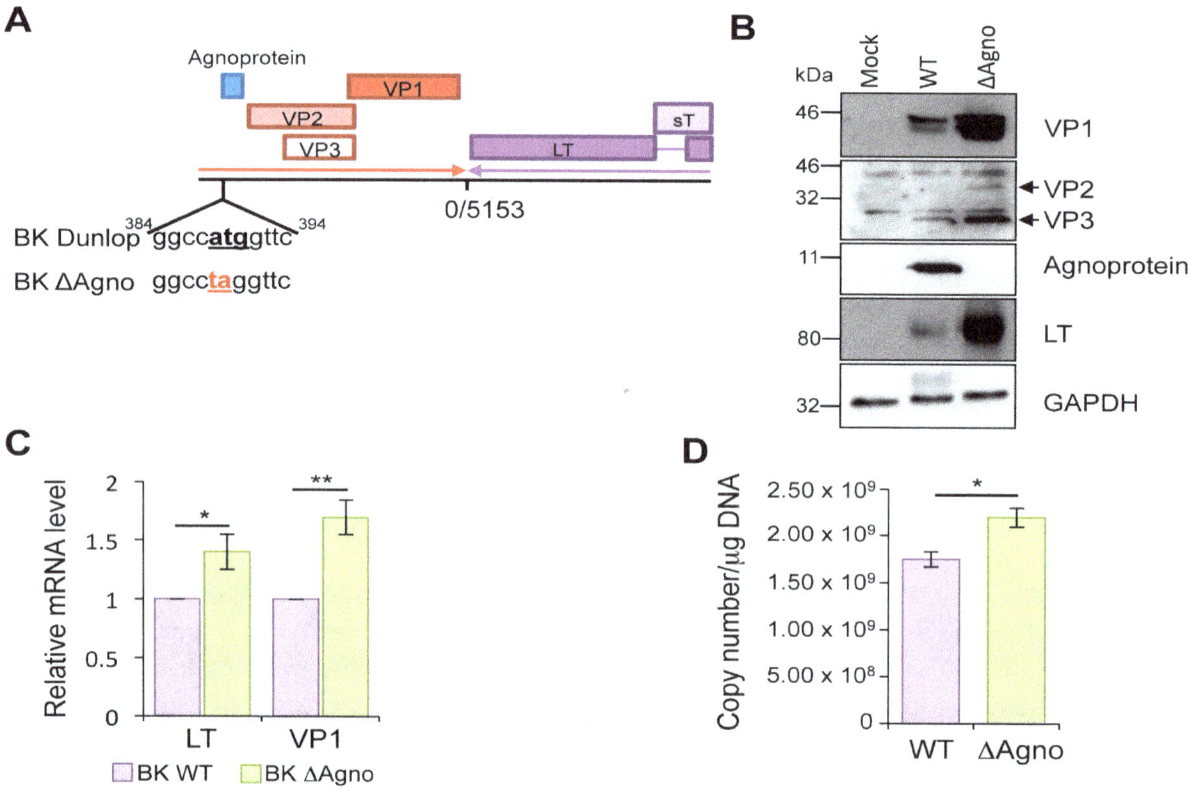

Figure 1. Loss of agnoprotein increases BK gene expression. (**A**) Schematic illustration of the BK Dunlop genome including the agnoprotein sequence mutated to generate the ΔAgno virus. Agnoprotein start codon in bold and base changes underlined in red; (**B**) Lysates from RPTE cells transfected with BK WT and ΔAgno genomes were probed with antibodies against early (LT) and late (VP1-3 and agnoprotein) proteins. Glyceraldehyde 3-phosphate dehydrogenase (GAPDH) was included as a protein loading control. Loss of agnoprotein correlated with increased expression of other virus protein products; (**C**) Levels of early (LT) and late (VP1) mRNA transcripts were measured from RPTE cells containing BK WT or ΔAgno genomes. Levels of virus transcript were increased in the absence of agnoprotein; (**D**) Virus genome replication was measured by qPCR in RPTE cells containing BK WT and ΔAgno virus. Genome replication was increased in the absence of agnoprotein. All experiments are representative of at least three independent experimental repeats. Significance of changes were analyzed by Student's *t*-test and indicated by * $p < 0.05$, ** $p < 0.01$.

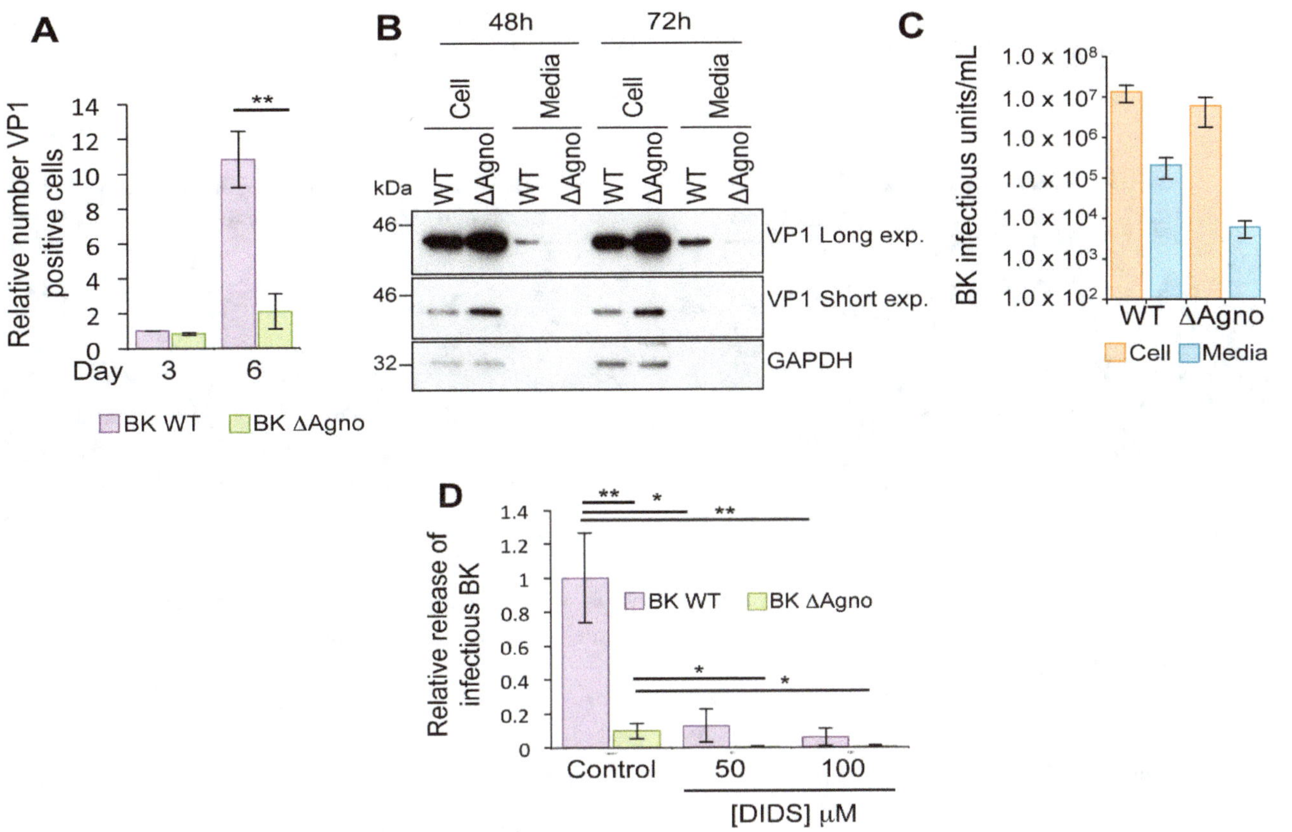

Figure 2. Agnoprotein facilitates virion release and enhances virus propagation. (**A**) RPTE cells transfected with BK WT and ΔAgno genomes were incubated over a 6-day time course, and levels of VP1 protein expression determined by indirect immunofluorescence using Incucyte Zoom software (Essen BioScience, Ann Arbor, MI, USA). Levels of VP1 expression are shown relative to the Day 3 BK WT sample. Significance of the changes were analyzed by Student's *t*-test and indicated by ** $p < 0.01$; (**B**) BK virus lacking agnoprotein fails to release virus into the cell culture media. Whole cell lysates and media samples from RPTE cells transfected with BK WT or ΔAgno genomes were analyzed at 48 and 72 h post-transfection for the VP1 capsid protein. GAPDH served as a protein loading control for the whole cell lysates; (**C**) RPTE cells were infected with BK WT and ΔAgno and cell-associated and media fractions harvested separately. Fluorescence focus assay was then performed to determine the IU/mL^{-1} of virus in the cells and supernatant; (**D**) Effect of the anion channel blocker DIDS is independent of agnoprotein. RPTE cells were infected with BK WT or ΔAgno and treated with dimethyl sulphoxide (DMSO) only (control) or 50–100 μM DIDS at 48 h post infection. Media and cell-associated fractions were harvested separately at 72 h post infection. Infectious virus titers were quantified by fluorescence focus assay on naïve RPTE cells and the proportion of total infectivity released into the media for each condition was calculated. Levels of released infectivity are represented as relative to the untreated BK WT samples. The graph corresponds to an average of three experimental repeats. Significance was analyzed by Student's *t*-test and is indicated by an asterix * $p < 0.05$, ** $p < 0.01$.

2.3. *Agnoprotein Is Not Required for the Production of BK Virions*

Previous negatively-stained electron microscopy (nsEM) analysis of a JC virus ΔAgno mutant revealed virions which were of a similar size to WT particles but appeared less regular or less ordered [42]. To investigate whether BK agnoprotein might also influence virion morphology, virions were purified from the media and cell lysates of WT and ΔAgno transfected cells using a modification to previously described protocols [4,43], by centrifugation in an isopycnic cesium chloride gradient. nsEM analysis of purified virions revealed polyhedral particles with a diameter of 45–50 nm (Figure 3), indistinguishable from WT BK Dunlop virions purified using the same protocol. These findings show that the inability of the ΔAgno to propagate an infection is unlikely to be due to defects in virion assembly or infectivity but more likely due to a defect(s) in virion release.

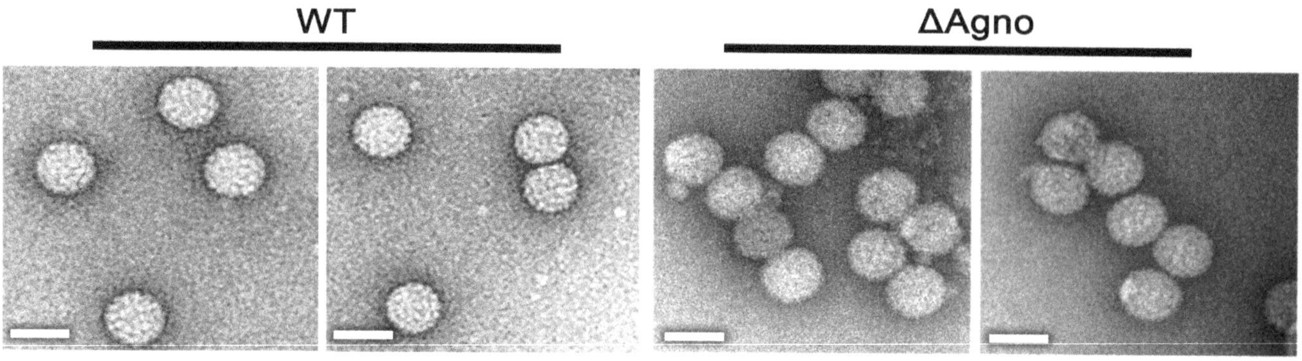

Figure 3. Loss of agnoprotein does not impair BK virion assembly. Negative stain electron micrograph of BK WT and ΔAgno virions following centrifugation through an isopycnic caesium chloride gradient. Scale bars 50 nm.

2.4. Agnoprotein Is Required for the Nuclear Egress of BK Particles

The data accumulated suggested that the ΔAgno mutant was defective with regard to virion release. This prompted us to monitor the different steps of virus release at the single cell level by electron microscopy. Virions with the distinctive morphology of a polyomavirus were readily detected in nuclear, cytoplasmic and plasma membrane compartments of RPTE cells infected with WT BK virus (Figure 4). In contrast, virions were exclusively detected in the nuclei of ΔAgno infected cells. No cytoplasmic or plasma membrane localized virions were detected after the examination of numerous ΔAgno infected RPTE cells (n = 40), whereas at least 98% of WT BK virus infected cells had clear cytoplasmic and/or plasma membrane localized virions. We extended this analysis by performing a biochemical fractionation of RPTE cells. GAPDH and Histone H3 served as markers for the cytoplasmic and nuclear fractions, respectively. Whilst similar levels of VP1 protein were observed in both fractions in WT BK expressing cells (Figure 4B,C), in agreement with the electron microscopy data, we observed a significant enrichment for VP1 protein in the nuclear fraction of ΔAgno expressing cells. Together, these data indicate that agnoprotein is required for the release of BK virions from the nucleus of RPTE cells.

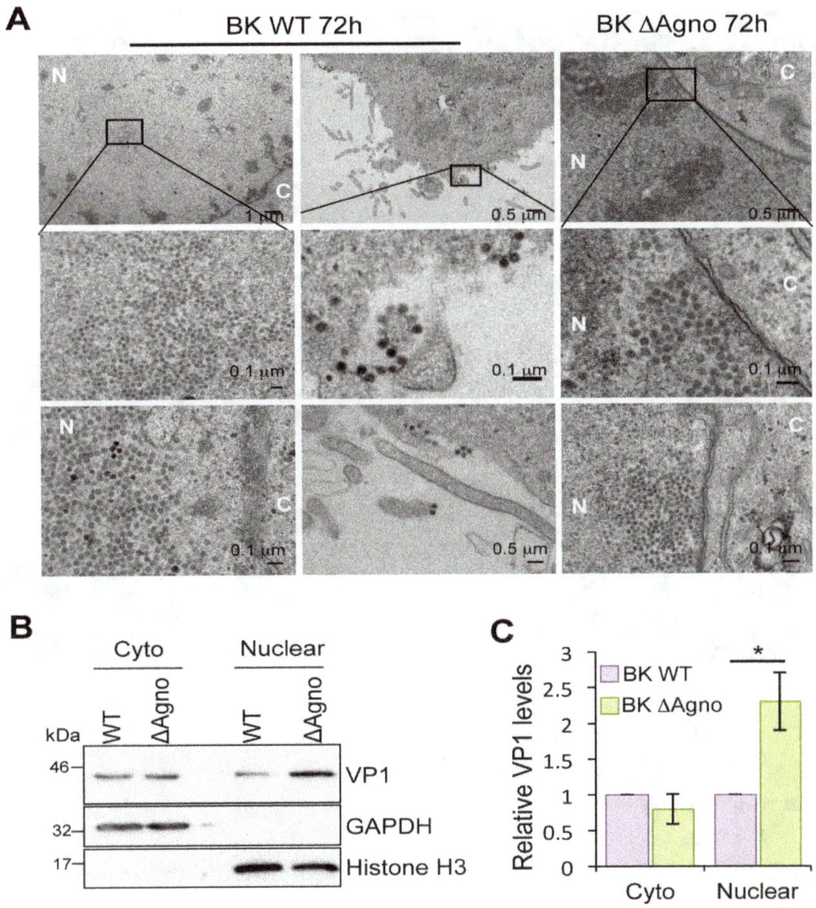

Figure 4. Agnoprotein facilitates nuclear release of BK virions. (**A**) Electron microscopy analysis of BK WT and ΔAgno infected RPTE cells (n = 40 cells). Boxed areas in the upper panel are shown at higher magnification in the middle panels. Viral particles of about 40 nm in diameter were found in the nuclei of BK WT and ΔAgno transfected cells. Nuclei (N) and cytoplasm (C) are labeled. Scale bars are shown in the panels; (**B**) Cell fractionation of RPTE cells transfected with BK WT or ΔAgno genomes. Fractions were probed with for VP1 expression. Antibodies detecting GAPDH and Histone H3 served as markers for the cytoplasm and nuclear fractions; (**C**) Quantification of the Western blot data was performed using ImageJ software (1.8.0_101, NIH, USA) on the VP1 positive bands and is represented relative to BK WT VP1. The graph corresponds to an average from three independent experimental repeats. Significance was analyzed by Student's t-test and is indicated by an asterix * p <0.05.

2.5. *Agnoprotein Does Not Cause Gross Destabilization of the Nuclear Membrane*

Exogenous expression of JC agnoprotein has been shown to uncouple interactions between proteins within the nuclear lamina, which might facilitate nuclear release of virions [44]. To investigate whether BK agnoprotein expression is also associated with a disruption of the nuclear membrane architecture, immunofluorescence microscopy was performed on markers of the nuclear membrane. Overall, staining with an antibody against Lamin B, a structural component of the inner nuclear membrane, revealed an absence of the nuclear envelope invaginations previously associated with JC agnoprotein expression [44]. Lamin B localization was unaffected in both WT and ΔAgno containing cells (Figure 5). We also noted subtle differences in the localization of VP1 expressed from WT and ΔAgno genomes. VP1 expressed in BK WT containing cells had a pronounced perinuclear localization, whereas VP1 expressed by ΔAgno appeared diffuse. Similar observations have been observed in cells infected with an SV40 agnoprotein mutant [45]. These subtle differences were not consistent between experiments so it is unclear whether they reflect a true effect of agnoprotein on VP1 localization.

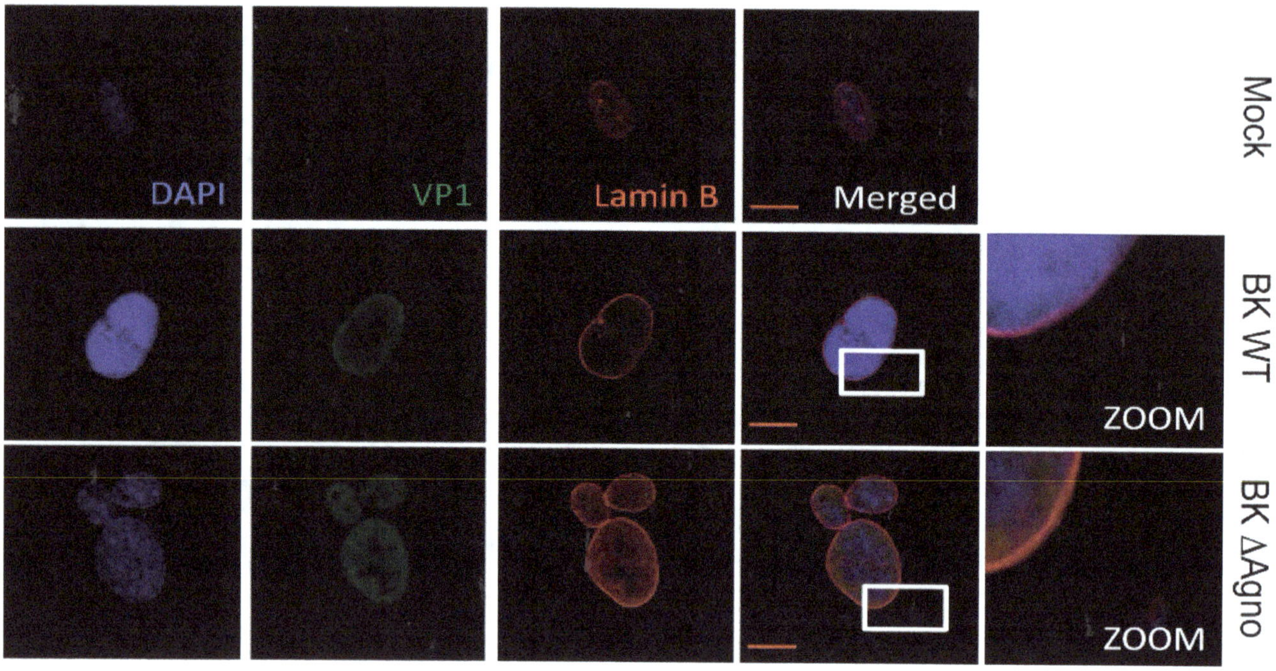

Figure 5. Lamin B localization is not altered by agnoprotein. Immunofluorescence staining of RPTE cells 72 h post transfection with BK WT or ΔAgno genomes. Cells were incubated with antibodies against VP1 and Lamin B and a secondary antibodies. Alexa Fluor 488 chicken anti-mouse and Alexa Fluor 594 chicken anti-rabbit. 4′,6-diamidino-2-phenylindole (DAPI) was used to indicate cell nuclei. Representative images are shown from at least three independent experimental repeats and white frames indicate area shown in the zoomed image. Scale bar 10 μm.

2.6. Host α-SNAP Is Necessary for BK Egress

In the absence of gross perturbation of the nuclear envelope, we investigated the role of cellular proteins in the nuclear egress activity of agnoprotein. Few BK agnoprotein interacting proteins have been identified [22]. Amongst these, α-soluble N-ethylmaleimide sensitive fusion (NSF) attachment protein (α-SNAP) was of interest given its role in vesicular trafficking [46]. The interaction between agnoprotein and α-SNAP was first confirmed using recombinant glutathione S-transferase (GST)-agnoprotein proteins produced in bacteria. No interaction was seen when GST alone was incubated with mammalian cell lysates containing α-SNAP (Figure 6A). In contrast, an interaction was observed with GST-BK agnoprotein, confirming previous findings [46]. A similar interaction with α-SNAP was observed with JC virus agnoprotein, indicating that α-SNAP might be a common agnoprotein binding partner. The consequences of an interaction between BK agnoprotein and α-SNAP for the BK life cycle have not been studied. To elucidate a potential role for α-SNAP in BK virion egress, we depleted α-SNAP from RPTE cells transfected with BK genomes using a pool of validated siRNA. A scrambled siRNA served as a control for potential off target effects (Figure 6B). To determine whether α-SNAP might function within the BK egress pathway we analyzed the sub-cellular localization of BK virions in α-SNAP depleted cells. RPTE cells were infected with BK WT virus prior to transfection with α-SNAP or scramble control siRNA to avoid any potential impact of α-SNAP loss on virus entry. Cells were subsequently analyzed by electron microscopy and the presence of virions in nuclear, cytoplasmic and extracellular compartments was scored for 50 cells of each condition. In scramble siRNA transfected samples, 70% of the cells counted were positive for the presence of cytoplasmic and/or plasma membrane localized virions. In contrast, in α-SNAP depleted cells only 6% of the cells counted had detectable particles in the cytoplasm (Figure 6B).

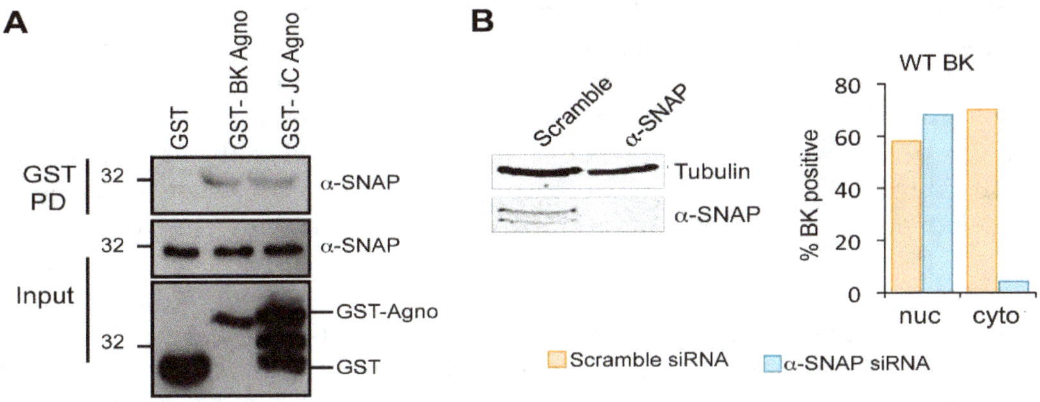

Figure 6. The agnoprotein binding partner α-SNAP is required for BK virion release. (**A**) Recombinant GST-agnoprotein interacts with α-SNAP. Bacterial expressed GST-agnoproteins from BK and JC virus bound to glutathione-agarose beads were incubated with RPTE cell lysates. GST alone served as a negative control. Bound samples were probed with an anti-α-SNAP antibody; (**B**) Quantification of transmission electron microscopy data. RPTE cells infected with BK WT were treated with siRNA targeting α-SNAP or a scrambled control and electron microscopy used to quantify the numbers cells demonstrating BK virions in nuclear (nuc) and cytoplasmic (cyto) compartments from 50 cells. Associated Western blots for α-SNAP to confirm effective knockdown. Tubulin serves as a loading control.

3. Discussion

Despite intensive research, the mechanisms by which polyomavirus particles are released during infection remain poorly understood. It is broadly believed that as non-enveloped viruses, polyomaviruses exit from an infected cell by a process of cell lysis. However, non-specific disintegration of a cell, and release of its potentially inflammatory milieu, might be considered detrimental to the establishment of a chronic virus infection. Rather, a process of controlled virus release would be preferable to avoid immune detection. Evidence for the existence of non-lytic release of polyomaviruses exists for SV40 [47] and has recently been shown for BK [38]. Despite these observations, the role of virus proteins in the release of BK virus remains poorly described.

Here we describe the agnoprotein as a critical factor for the shuttling of progeny BK virions from the nucleus, the site of polyomavirus replication and assembly, to the cytoplasm for release. Identified nearly two decades ago, the agnoprotein is expressed by a limited number of human polyomaviruses. Studies have produced contradictory findings, confounding our understanding of the contribution of this small auxiliary protein to the life cycles of polyomaviruses. Using a BK genome containing a mutation, which converted the agnogene start codon into a stop codon, our data generated from transfected genomes or from virus infection studies demonstrates that loss of agnoprotein correlates with a reduction of virus secretion into the extracellular environment. This deficit in release resulted in reduced virus propagation and an accumulation of virions within infected cells. Transfection studies also highlighted a concomitant increase in BK transcript levels and genome replication in cells lacking agnoprotein. Whether this was due to potential negative regulation of LT function by agnoprotein, as has been reported in JC virus [32], or the consequence of an interaction with the host proliferating cell nuclear antigen (PCNA) protein [48] was not tested further.

Release of BK virus has recently been shown to be sensitive to the actions of the broad spectrum anion channel blocker DIDS [38]. In addition to cellular targets, DIDS can block the channel activity of the enterovirus 2B protein [39]. Many viruses encode small hydrophobic proteins, termed viroporins, that form pore-like structures similar to 2B [40]. JC agnoprotein is a viroporin, essential for JC virion release [28,49]. It is plausible that BK agnoprotein also performs a viroporin function to aid in virion release. Despite this possibility, addition of DIDS further reduced virion release in ΔAgno infected cells, suggesting an agnoprotein independent target for this compound. These data also imply that agnoprotein-independent egress pathways exist that must contain the cellular target of DIDS.

Whilst some studies have suggested that loss of agnoprotein impairs polyomavirus maturation and infectivity, our data clearly demonstrates that virions produced in the absence of agnoprotein are infectious and retain WT morphology. Instead, our results are consistent with the notion that loss of agnoprotein blocks the physical release of BK virions from infected cells, rather than affecting virion infectivity. These observations raised the question of where virions are localized within an infected cell in the absence of agnoprotein. We observed virions throughout the cell in BK WT infected cells, with high concentrations of virions in the nucleus but clear localization of virions in cytoplasmic compartments and at the plasma membrane. In contrast, while we could also observe high concentrations of virions in the nucleus of ΔAgno infected cells, virtually no virions were identified in the cytoplasm. Given the lack of gross impact on nuclear membrane morphology at the time-points analyzed, we reasoned that agnoprotein might recruit host factors to promote virion nuclear egress. Whilst a number of host interacting partners have been identified for JC virus agnoprotein, the BK agnoprotein interactome is less understood [22]. We focused on α-SNAP because of its critical role in vesicular trafficking [46,50,51]. α-SNAP is a known BK agnoprotein binding protein, however, its role in the virus life cycle has not been studied. Knockdown of α-SNAP conferred an agnoprotein knockout phenotype on WT BK by preventing nuclear virion egress. Importantly, loss of α-SNAP had no cumulative impact on the ΔAgno phenotype, suggesting that both proteins may function within the same egress pathway (data not shown). Whilst our data implicates BK agnoprotein and α-SNAP in virion egress, how virus is transported from the nucleus to the cytoplasm remains to be understood. A crucial area of future work will be to determine the route and mode of virion transport and to define the precise function of α-SNAP within this process. In infected polarized epithelial cells, SV40 virions have been observed within cytoplasmic membrane reticular structures, contiguous with the nuclear membrane and ER [47]. Moreover, whilst studying the effects of DIDS on the virus lifecycle, BK virions were noted in cytoplasmic vacuoles and in LAMP-1 positive vesicles, implicating the secretory system in virus release [38]. Given that α-SNAP is an integral regulator of endoplasmic reticulum to Golgi trafficking, it is tempting to speculate that BK may usurp this host secretory pathway to traffic virions from the nucleus to exterior of the cell for release. In support of this idea, treatment with the ionophore monensin impaired the release of SV40 from polarized epithelial cells and resulted in an accumulation of virions in the cytoplasmic reticular structures [47]. As part of our ongoing studies, it will also be of interest to determine whether SV40 and JC virus utilize similar processes for virion release. Loss of agnoprotein imparts an egress defect in both viruses, and JC agnoprotein is known to interact with components of the trafficking apparatus [49]. In this study we demonstrated that α-SNAP is an interacting partner for JC agnoprotein, and as such it may also be required during virus release.

In summary, our data show clearly that agnoprotein is a key virus-encoded regulator of BK virus release, and through an interaction with α-SNAP aids in an active egress pathway. Our findings provide further evidence for a virus regulated release mechanism.

4. Methods and Materials

4.1. Cell Culture

BK virus stocks were generated in Vero cells, which were maintained in Dulbeco's Modified Eagle's medium (DMEM) supplemented with 10% fetal calf serum (FCS) and 50 IU/mL penicillin/streptomycin. Primary renal proximal tubular epithelial (RPTE) cells (Lonza, Basel, Switzerland) were cultured in renal epithelial growth media with the REGM Bulletkit supplements (Lonza) at 37 °C with 5% CO_2 in a humidified incubator as described [4].

4.2. Generation of an Agnoprotein Knockout BK Dunlop Genome

A BK knockout genome was created by site directed mutagenesis of the pGEM7-Dunlop plasmid (a gift from Michael Imperiale, University of Michigan) using the QuikChange site directed mutagenesis kit (Thermo Fisher, Waltham, MA, USA) and the primer pair $^{5'}$CCA GTT AAA CTG GAC AAA GGC CTA GGT TCT GCG CCA GCT GTC ACG$^{'3}$ and $^{5'}$CGT GAC AGC TGG CGC AGA ACC TAG GCC TTT GTC CAG TTT AAC TGG$^{'3}$ (Agilent Technologies, Santa Clara, CA, USA). The entire genome was subsequently sequenced to confirm the introduction of the mutation and ensure that secondary mutations had not arisen.

4.3. Transfection of Virus Genomes

Cells were transfected with WT BK Dunlop or ΔAgno genomes using the NanoJuice transfection kit (Merck Millipore, Darmstadt, Germany) according to the manufacturer's instructions. The transfection mixture was removed and replaced with fresh media 8 h post-transfection.

4.4. Virus Culture and Purification

BK Dunlop was cultured and purified on a cesium chloride linear gradient as previously described [4]. RPTE cells were infected at approximately 50% confluency with purified virus in Opti-MEM (Thermo Fisher, Waltham, MA, USA) and incubated at 4 °C for 1 h with shaking every 15 min. Cells were subsequently transferred to 37 °C after the incubation.

4.5. Cell Infections and Harvesting Virus

For virus release assays, RPTE cells were infected with BK virus at 1 IU cell^{-1}. After 1 h, the medium was removed, the cells gently washed in phosphate buffered saline (PBS) and then fresh medium added. For inhibitor studies, at 48 h post-infection 50–100 μM DIDS or DMSO only was added. At 72 h post infection the culture media was collected and centrifuged for 5 min at 2000× *g* to pellet any cell debris in the media, and then the supernatant transferred to new tubes. This was repeated to ensure no cell debris was present, before centrifuging the supernatant at 100,000× *g* for 2 h to pellet the virus. The media was aspirated and the pellets were resuspended in 1/20th of the original volume. The RPTE cell monolayer was harvested separately in 1 mL of REGM and freeze thawed 3-times to release cell-associated virus. Infectious virus titers in the release and cell-associated fractions were determined by fluorescence focus unit (FFU) assay [38].

4.6. Fluorescent Focus Unit Assay Using IncuCyte ZOOM Analysis

RPTE cells were seeded out into 96-well-plate (2×10^3 cells per well, in a total volume of 100 μL) and incubated for 16 h. Purified BKPyV was serially diluted two-fold into serum-free media (in a total volume of 100 μL per well) and allowed to infect RPTE cells for 2 h at 37 °C. Infected cells were washed once with phosphate-buffered saline (PBS) and fresh media was added. RPTE cells were incubated for 48 h at 37 °C. Cells were fixed with 4% paraformaldehyde for 10 min at room temperature and washed with PBS. Fixed RPTE cells were permeabilised with 0.1% Triton-X100 (Sigma-Aldrich, St-Louis, MI, USA) in PBS, washed and incubated overnight at 4 °C in primary antibody against VP1 protein. Anti-VP1 primary antibody was used at 1:250 dilutions (in PBS with 1% BSA). Cells were further washed and incubated with a fluorophore-488-conjugated chicken anti-mouse secondary antibody (1:250 in PBS with 1% BSA) for 1 h at 37 °C. Finally, RPTE cells stored in PBS and the plate was imaged with the IncuCyte ZOOM instrument (Essen BioScience, Ann Arbor, MI, USA). The software parameters with a 10× objective were used for imaging [37]. The number of positive infected cells per well was calculated. BKPyV titer was measured by multiplying the number of positive-infected cells/well by the corresponding dilution factor and the IU mL^{-1} was determined by calculating the number of infected cells in the entire well from the mean number of infected cells in the 10 fields of view, and then the number of infectious units calculated. [37].

4.7. *Immunofluorescence*

RPTE cells (1×10^5) grown on glass coverslips were fixed with 4% paraformaldehyde for 10 min. RPTE cells were then washed twice in PBS and permeabilized with 0.1% Triton X-100 for 10 min. Non-specific targets were blocked by incubation in blocking buffer (5% BSA in PBS) for 30 min. Cells were incubated with primary antibodies against VP1 (Pab597—a gift from Chris Buck, National Cancer Institute; used 1:250)) and Lamin B1 (Abcam, Cambridge, UK; ab16048) overnight at 4 °C. Cells were washed three times in PBS prior to incubation in secondary antibodies Alexa Fluor 488 chicken anti-mouse and Alexa Fluor 594 chicken anti-rabbit (Invitrogen, Carlsbad, CA, USA) for 1 h at room temperature. Cells were washed three times in PBS prior to mounting onto microscope slides using Prolong Gold Antifade Reagent with DAPI (Thermo Fisher, Waltham, MA, USA). Samples were observed under a Zeiss LSM 700 laser (Carl Zeiss Ltd, Jena, Germany) scanning confocal microscope under an oil-immersion objective lens.

4.8. *Western Blotting*

Triton lysis buffer (10 mM Tris (pH 7.6), 10 mM sodium phosphate, 130 mM NaCl, 1% Triton X-100, 20 mM N-ethylmaleimide, complete protease inhibitor cocktail; Roche, Basel, Switzerland) was used to harvest total cellular protein from the infected cells. Protein concentration was quantified with the Bradford assay (Bio-Rad, Hercules, CA, USA). Lysates were separated by SDS PAGE and following transfer to nitrocellulose membrane were probed with the following antibodies diluted in 5% non-fat dried milk in TBS with 0.1% Tween-20; mouse anti-VP1 pAb-597 (1:5000), rabbit anti-VP2/VP3 (Abcam; ab53983; used 1:1000), mouse anti-Large T antigen (Abcam; ab16879; used at 1:200) and α-SNAP (Santa Cruz 4E4); used at 1:1000) and mouse anti-GAPDH (Santa Cruz, Dallas, TX, USA; used 1:5000).

4.9. *Quantitative PCR*

Total DNA was extracted from infected cells using the E.Z.N.A. Tissue DNA kit (Omega Bio-Tek, Norcross, GE, USA) and 10 ng of DNA was analysed by qPCR using the Quantifast SYBR Green PCR kit (Qiagen, Venlo, The Netherlands) with the following primers against BK Dunlop; BK Forward $^{5'}$TGT GAT TGG GAT TCA GTG CT$^{'3}$ and Reverse $^{5'}$AAG GAA AGG CTG GAT TCT GA$^{'3}$. To extract DNA from released virus, the culture media was collected and centrifuged for 5 min at 2000$\times$ g to pellet any cell debris in the media, and then the supernatant transferred to new tubes. This was repeated to ensure no cell debris was present, before centrifuging the supernatant at 100,000$\times$ g for 2 h to pellet the virus. Virus was treated with RQ1 RNase-free DNase (Promega, Madison, WI, USA) for 30 min at 37 °C to remove any unprotected DNA, and the reaction terminated by the addition of DNase Stop Solution and incubation for 10 min at 65 °C. A serial dilution of the pGEM7-Dunlop plasmid was used to calculate the copy number per microgram of DNA.

4.10. *Quantitative Reverse Transcriptase PCR*

Total RNA was extracted from RPTE cells using the E.Z.N.A Total RNA Kit I (Omega Bio-Tek) following the manufacture's protocol. One µg of the total extracted RNA was reverse transcribed using the iScript^{TM} cDNA Synthesis Kit (Bio-Rad) based on the protocol instructions. Quantitative Real-time PCR was performed using the QuantiFast SYBR Green PCR kit (Qiagen) and specific primers against VP1. The PCR reaction was carried out on a Corbett Rotor-Gene 6000 (Qiagen) following three different steps. The initial activation step for 10 min at 95 °C and a three-step cycle of denaturation of 10 s at 95 °C; the second step of annealing for 15 s at 60 °C and the step of extension for 20 s at 72 °C. All the three different steps were repeated 40 times and concluded by melting curve analysis. U6 was used as normaliser control.

4.11. Electron Microscopy

Negative staining of BK virus particles was carried out as follows, 3.5 μL aliquots of purified wild-type BK or ΔAgno in buffer A were applied to continuous carbon grids that had been glow-discharged for ~30 s in air using a PELCO easiGlow™ (Ted Pella Inc, Redding, CA, USA). The samples were then stained with 1% uranyl acetate solution before being allowed to dry in air for 5 min. Samples were imaged on a Tecnai G^2-Spirit transmission EM (FEI, Eindhoven, The Netherlands) at 120 keV, equipped with a Gatan US1000XP CCD camera (Gatan, Pleasanton, CA, USA). Images of virions were recorded at 30,000× magnification.

4.12. Transmission Electron Microscopy in Cells

Cells were fixed in 0.5% glutaraldehyde in 200 mM sodium cacodylate buffer for 30 min, washed in buffer and secondarily fixed in reduced 1% osmium tetroxide, 1.5% potassium ferricyanide for 60 min. The samples were washed in distilled water and stained overnight at 4 °C in 0.5% magnesium uranyl acetate, washed in distilled water and dehydrated in graded ethanol. The samples were then embedded flat in the dish in Epon resin. Resin filled stubs were placed on embedded cell monolayers and polymerized. Ultrathin sections (typically 50–70 nm) were cut parallel to the dish and examined in a FEI Tecnai20 electron microscope (FEI Co., Eindhoven, The Netherlands) with CCD camera image acquisition.

4.13. Cell Fractionation

RPTE (2×10^5) cells seeded in 6 well plates were scraped into PBS. Cell pellets were resuspended in 100 μL cytoplasmic lysis buffer (100 mM Tris pH 7.5, 100 mM NaCl, 5 mM $MgCl_2$, 0.5% NP-40, complete protease inhibitor cocktail (Roche)) and lysed on ice for 20 min. Then samples were centrifuged at 9600× *g* for 20 min at 4 °C. The supernatant, containing the cytoplasmic fraction, was collected and the pellet washed twice in cytoplasmic lysis buffer and then resuspended in 100 μL RIPA buffer (50 mM Tris pH 7.5, 150 mM NaCl, 1% NP-40, 0.5% sodium deoxycholate, 0.1% SDS, protease inhibitors). Samples were centrifuged at 9600× *g* for 10 min at 4 °C and the nuclear fraction supernatant collected.

Acknowledgments: We thank Michael Imperiale (University of Michigan), Dennis Galloway (Fred Hutchinson Cancer Research Center), Ugo Moens (The Arctic University of Norway) and Chris Buck (National Cancer Institute) for providing essential reagents and advice. We are grateful to Kidney Research UK (RP25/2013, ST4/2014 and RP_022_20170302), Yorkshire Kidney Research Fund, the Medical Research Council (MR/K012665/1 and Ph.D. studentship to Laura G. Caller) and Wellcome Trust (102572/B/13/Z, 1052221/Z/14/Z and 109157/Z/15/Z) for funding this work.

Author Contributions: Conceived the study: Andrew Macdonald; Designed experiments: Neil A. Ranson, Colin M. Crump and Andrew Macdonald; Carried out the study: Margarita-Maria Panou, Emma L. Prescott, Daniel L. Hurdiss, Gemma Swinscoe, Michael Hollinshead, Laura G. Caller, Louisa Carlisle, Ethan L. Morgan, Michelle Antoni, David Kealy and Marietta Müller; Critical analysis and interpretation of data: Margarita-Maria Panou, Emma L. Prescott, Colin M. Crump, Neil A. Ranson and Andrew Macdonald; Drafted the output: Andrew Macdonald; Corrected the output: Neil A. Ranson, Colin M. Crump, Marietta Müller and Andrew Macdonald.

References

1. Buck, C.B.; van Doorslaer, K.; Peretti, A.; Geoghegan, E.M.; Tisza, M.J.; An, P.; Katz, J.P.; Pipas, J.M.; McBride, A.A.; Camus, A.C.; et al. The Ancient Evolutionary History of Polyomaviruses. *PLoS Pathog.* **2016**, *12*, e1005574. [CrossRef] [PubMed]
2. Peretti, A.; FitzGerald, P.C.; Bliskovsky, V.; Pastrana, D.V.; Buck, C.B. Genome Sequence of a Fish-Associated Polyomavirus, Black Sea Bass (*Centropristis striata*) Polyomavirus 1. *Genome Announc.* **2015**, *3*, e01476-14. [CrossRef] [PubMed]
3. Peretti, A.; FitzGerald, P.C.; Bliskovsky, V.; Buck, C.B.; Pastrana, D.V. Hamburger polyomaviruses. *J. Gen. Virol.* **2015**, *96*, 833–839. [CrossRef] [PubMed]

4. Hurdiss, D.L.; Morgan, E.L.; Thompson, R.F.; Prescott, E.L.; Panou, M.M.; Macdonald, A.; Ranson, N.A. New Structural Insights into the Genome and Minor Capsid Proteins of BK Polyomavirus using Cryo-Electron Microscopy. *Structure* **2016**, *24*, 528–536. [CrossRef] [PubMed]
5. Feng, H.; Shuda, M.; Chang, Y.; Moore, P.S. Clonal integration of a polyomavirus in human Merkel cell carcinoma. *Science* **2008**, *319*, 1096–1100. [CrossRef] [PubMed]
6. DeCaprio, J.A.; Garcea, R.L. A cornucopia of human polyomaviruses. *Nat. Publ. Group* **2013**, *11*, 264–276. [CrossRef] [PubMed]
7. Van der Meijden, E.; Janssens, R.W.A.; Lauber, C.; Bouwes Bavinck, J.N.; Gorbalenya, A.E.; Feltkamp, M.C.W. Discovery of a new human polyomavirus associated with *trichodysplasia spinulosa* in an immunocompromized patient. *PLoS Pathog.* **2010**, *6*, e1001024. [CrossRef] [PubMed]
8. Gardner, S.D.; Field, A.M.; Coleman, D.V.; Hulme, B. New human papovavirus (B.K.) isolated from urine after renal transplantation. *Lancet* **1971**, *1*, 1253–1257. [CrossRef]
9. Padgett, B.L.; Walker, D.L.; ZuRhein, G.M.; Eckroade, R.J.; Dessel, B.H. Cultivation of papova-like virus from human brain with progressive multifocal leucoencephalopathy. *Lancet* **1971**, *1*, 1257–1260. [CrossRef]
10. Knowles, W.A. Discovery and epidemiology of the human polyomaviruses BK virus (BKV) and JC virus (JCV). *Adv. Exp. Med. Biol.* **2006**, *577*, 19–45. [PubMed]
11. Bennett, S.M.; Broekema, N.M.; Imperiale, M.J. BK polyomavirus: Emerging pathogen. *Microbes Infect.* **2012**, *14*, 672–683. [CrossRef] [PubMed]
12. Egli, A.; Infanti, L.; Dumoulin, A.; Buser, A.; Samaridis, J.; Stebler, C.; Gosert, R.; Hirsch, H.H. Prevalence of polyomavirus BK and JC infection and replication in 400 healthy blood donors. *J. Infect. Dis.* **2009**, *199*, 837–846. [CrossRef] [PubMed]
13. Egli, A.; Köhli, S.; Dickenmann, M.; Hirsch, H.H. Inhibition of polyomavirus BK-specific T-Cell responses by immunosuppressive drugs. *Transplantation* **2009**, *88*, 1161–1168. [CrossRef] [PubMed]
14. Ahsan, N.; Shah, K.V. Polyomaviruses and human diseases. *Adv. Exp. Med. Biol.* **2006**, *577*, 1–18. [PubMed]
15. Dropulic, L.K.; Jones, R.J. Polyomavirus BK infection in blood and marrow transplant recipients. *Bone Marrow Transpl.* **2008**, *41*, 11–18. [CrossRef] [PubMed]
16. Balba, G.P.; Javaid, B.; Timpone, J.G. BK polyomavirus infection in the renal transplant recipient. *Infect. Dis. Clin. N. Am.* **2013**, *27*, 271–283. [CrossRef] [PubMed]
17. Ramos, E.; Drachenberg, C.B.; Wali, R.; Hirsch, H.H. The decade of polyomavirus BK-associated nephropathy: State of affairs. *Transplantation* **2009**, *87*, 621–630. [CrossRef] [PubMed]
18. Safrin, S.; Cherrington, J.; Jaffe, H. Clinical uses of cidofovir. *Rev. Med. Virol.* **1997**, *7*, 145–156. [CrossRef]
19. Kuypers, D.R.J. Management of polyomavirus-associated nephropathy in renal transplant recipients. *Nat. Rev. Nephrol.* **2012**, *8*, 390–402. [CrossRef] [PubMed]
20. Sharma, P.M.; Gupta, G.; Vats, A.; Shapiro, R.; Randhawa, P.S. Polyomavirus BK non-coding control region rearrangements in health and disease. *J. Med. Virol.* **2007**, *79*, 1199–1207. [CrossRef] [PubMed]
21. Buck, C.B. Exposing the Molecular Machinery of BK Polyomavirus. *Structure* **2016**, *24*, 495. [CrossRef] [PubMed]
22. Gerits, N.; Moens, U. Agnoprotein of mammalian polyomaviruses. *Virology* **2012**, *432*, 316–326. [CrossRef] [PubMed]
23. Rinaldo, C.H.; Traavik, T.; Hey, A. The agnogene of the human polyomavirus BK is expressed. *J. Virol.* **1998**, *72*, 6233–6236. [PubMed]
24. Unterstab, G.; Gosert, R.; Leuenberger, D.; Lorentz, P.; Rinaldo, C.H.; Hirsch, H.H. The polyomavirus BK agnoprotein co-localizes with lipid droplets. *Virology* **2010**, *399*, 322–331. [CrossRef] [PubMed]
25. Johannessen, M.; Myhre, M.R.; Dragset, M.; Tümmler, C.; Moens, U. Phosphorylation of human polyomavirus BK agnoprotein at Ser-11 is mediated by PKC and has an important regulative function. *Virology* **2008**, *379*, 97–109. [CrossRef] [PubMed]
26. Sariyer, I.K.; Khalili, K.; Safak, M. Dephosphorylation of JC virus agnoprotein by protein phosphatase 2A: Inhibition by small t antigen. *Virology* **2008**, *375*, 464–479. [CrossRef] [PubMed]
27. Khalili, K.; White, M.K.; Sawa, H.; Nagashima, K.; Safak, M. The agnoprotein of polyomaviruses: A multifunctional auxiliary protein. *J. Cell Physiol.* **2005**, *204*, 1–7. [CrossRef] [PubMed]
28. Suzuki, T.; Orba, Y.; Okada, Y.; Sunden, Y.; Kimura, T.; Tanaka, S.; Nagashima, K.; Hall, W.W.; Sawa, H. The Human Polyoma JC Virus Agnoprotein Acts as a Viroporin. *PLoS Pathog.* **2010**, *6*, e1000801. [CrossRef] [PubMed]

29. Akan, I.; Sariyer, I.K.; Biffi, R.; Palermo, V.; Woolridge, S.; White, M.K.; Amini, S.; Khalili, K.; Safak, M. Human polyomavirus JCV late leader peptide region contains important regulatory elements. *Virology* **2006**, *349*, 66–78. [CrossRef] [PubMed]
30. Barkan, A.; Welch, R.C.; Mertz, J.E. Missense mutations in the VP1 gene of simian virus 40 that compensate for defects caused by deletions in the viral agnogene. *J. Virol.* **1987**, *61*, 3190–3198. [PubMed]
31. Sedman, S.A.; Good, P.J.; Mertz, J.E. Leader-encoded open reading frames modulate both the absolute and relative rates of synthesis of the virion proteins of simian virus 40. *J. Virol.* **1989**, *63*, 3884–3893. [PubMed]
32. Safak, M.; Khalili, K. Physical and functional interaction between viral and cellular proteins modulate JCV gene transcription. *J. Neurovirol.* **2001**, *7*, 288–292. [PubMed]
33. Carswell, S.; Alwine, J.C. Simian virus 40 agnoprotein facilitates perinuclear-nuclear localization of VP1, the major capsid protein. *J. Virol.* **1986**, *60*, 1055–1061. [PubMed]
34. Carswell, S.; Resnick, J.; Alwine, J.C. Construction and characterization of CV-1P cell lines which constitutively express the simian virus 40 agnoprotein: Alteration of plaquing phenotype of viral agnogene mutants. *J. Virol.* **1986**, *60*, 415–422. [PubMed]
35. Sariyer, I.K.; Saribas, A.S.; White, M.K.; Safak, M. Infection by agnoprotein-negative mutants of polyomavirus JC and SV40 results in the release of virions that are mostly deficient in DNA content. *Virol. J.* **2011**, *8*, 255. [CrossRef] [PubMed]
36. Myhre, M.R.; Olsen, G.-H.; Gosert, R.; Hirsch, H.H.; Rinaldo, C.H. Clinical polyomavirus BK variants with agnogene deletion are non-functional but rescued by trans-complementation. *Virology* **2010**, *398*, 12–20. [CrossRef] [PubMed]
37. Stewart, H.; Bartlett, C.; Ross-Thriepland, D.; Shaw, J.; Griffin, S.; Harris, M. A novel method for the measurement of hepatitis C virus infectious titres using the IncuCyte ZOOM and its application to antiviral screening. *J. Virol. Methods* **2015**, *218*, 59–65. [CrossRef] [PubMed]
38. Evans, G.L.; Caller, L.G.; Foster, V.; Crump, C.M. Anion homeostasis is important for non-lytic release of BK polyomavirus from infected cells. *Open Biol.* **2015**, *5*, 150041. [CrossRef] [PubMed]
39. Xie, S.; Wang, K.; Yu, W.; Lu, W.; Xu, K.; Wang, J.; Ye, B.; Schwarz, W.; Jin, Q.; Sun, B. DIDS blocks a chloride-dependent current that is mediated by the 2B protein of enterovirus 71. *Cell Res.* **2011**, *21*, 1271–1275. [CrossRef] [PubMed]
40. Royle, J.; Dobson, S.J.; Müller, M.; Macdonald, A. Emerging Roles of Viroporins Encoded by DNA Viruses: Novel Targets for Antivirals? *Viruses* **2015**, *7*, 5375–5387. [CrossRef] [PubMed]
41. Scott, C.; Griffin, S.D.C. Viroporins: Structure, function and potential as antiviral targets. *J. Gen. Virol.* **2015**, *96*, 2000–2027. [CrossRef] [PubMed]
42. Suzuki, T.; Semba, S.; Sunden, Y.; Orba, Y.; Kobayashi, S.; Nagashima, K.; Kimura, T.; Hasegawa, H.; Sawa, H. Role of JC virus agnoprotein in virion formation. *Microbiol. Immunol.* **2012**, *56*, 639–646. [CrossRef] [PubMed]
43. Shen, P.S.; Enderlein, D.; Nelson, C.D.S.; Carter, W.S.; Kawano, M.; Xing, L.; Swenson, R.D.; Olson, N.H.; Baker, T.S.; Cheng, R.H.; et al. The structure of avian polyomavirus reveals variably sized capsids, non-conserved inter-capsomere interactions, and a possible location of the minor capsid protein VP4. *Virology* **2011**, *411*, 142–152. [CrossRef] [PubMed]
44. Okada, Y.; Suzuki, T.; Sunden, Y.; Orba, Y.; Kose, S.; Imamoto, N.; Takahashi, H.; Tanaka, S.; Hall, W.W.; Nagashima, K.; et al. Dissociation of heterochromatin protein 1 from lamin B receptor induced by human polyomavirus agnoprotein: Role in nuclear egress of viral particles. *EMBO Rep.* **2005**, *6*, 452–457. [CrossRef] [PubMed]
45. Resnick, J.; Shenk, T. Simian virus 40 agnoprotein facilitates normal nuclear location of the major capsid polypeptide and cell-to-cell spread of virus. *J. Virol.* **1986**, *60*, 1098–1106. [PubMed]
46. Johannessen, M.; Walquist, M.; Gerits, N.; Dragset, M.; Spang, A.; Moens, U. BKV agnoprotein interacts with α-soluble N-ethylmaleimide-sensitive fusion attachment protein, and negatively influences transport of VSVG-EGFP. *PLoS ONE* **2011**, *6*, e24489. [CrossRef] [PubMed]
47. Clayson, E.T.; Brando, L.V.; Compans, R.W. Release of simian virus 40 virions from epithelial cells is polarized and occurs without cell lysis. *J. Virol.* **1989**, *63*, 2278–2288. [PubMed]
48. Gerits, N.; Johannessen, M.; Tümmler, C.; Walquist, M.; Kostenko, S.; Snapkov, I.; van Loon, B.; Ferrari, E.; Hübscher, U.; Moens, U. Agnoprotein of polyomavirus BK interacts with proliferating cell nuclear antigen and inhibits DNA replication. *Virol. J.* **2015**, *12*, 7. [CrossRef] [PubMed]

49. Suzuki, T.; Orba, Y.; Makino, Y.; Okada, Y.; Sunden, Y.; Hasegawa, H.; Hall, W.W.; Sawa, H. Viroporin activity of the JC polyomavirus is regulated by interactions with the adaptor protein complex 3. *Proc. Natl. Acad. Sci. USA* **2013**, *110*, 18668–18673. [CrossRef] [PubMed]
50. Peter, F.; Wong, S.H.; Subramaniam, V.N.; Tang, B.L.; Hong, W. α-SNAP but not gamma-SNAP is required for ER-Golgi transport after vesicle budding and the Rab1-requiring step but before the EGTA-sensitive step. *J. Cell Sci.* **1998**, *111 Pt 17*, 2625–2633. [PubMed]
51. Barnard, R.J.; Morgan, A.; Burgoyne, R.D. Stimulation of NSF ATPase activity by α-SNAP is required for SNARE complex disassembly and exocytosis. *J. Cell Biol.* **1997**, *139*, 875–883. [CrossRef] [PubMed]

14

High-Risk Human Papillomaviral Oncogenes E6 and E7 Target Key Cellular Pathways to Achieve Oncogenesis

Nicole S. L. Yeo-Teh [1,2], Yoshiaki Ito [1,2] and Sudhakar Jha [1,3,*]

1 Cancer Science Institute of Singapore, National University of Singapore, Singapore 117599, Singapore; nicole.yeo@u.nus.edu (N.S.L.Y.-T.); yoshi_ito@nus.edu.sg (Y.I.)
2 NUS Graduate School for Integrative Sciences and Engineering, National University of Singapore, Singapore 117456, Singapore
3 Department of Biochemistry, Yong Loo Lin School of Medicine, National University of Singapore, Singapore 117596, Singapore
* Correspondence: csisjha@nus.edu.sg

Abstract: Infection with high-risk human papillomavirus (HPV) has been linked to several human cancers, the most prominent of which is cervical cancer. The integration of the viral genome into the host genome is one of the manners in which the viral oncogenes E6 and E7 achieve persistent expression. The most well-studied cellular targets of the viral oncogenes E6 and E7 are p53 and pRb, respectively. However, recent research has demonstrated the ability of these two viral factors to target many more cellular factors, including proteins which regulate epigenetic marks and splicing changes in the cell. These have the ability to exert a global change, which eventually culminates to uncontrolled proliferation and carcinogenesis.

Keywords: HPV; human papillomavirus; cervical cancer; viral oncogenes; E6; E7; viral-induced cancers

---

1. Introduction

In 1966, Dr. Peyton Rous was awarded the Nobel Prize in Medicine and Physiology for his discovery of the Rous sarcoma virus, which has the ability to cause sarcoma in chickens, although his discovery was made five decades earlier in 1911 [1]. This was mainly due to the great amount of skepticism that his work was met with, and the rejection of using an avian model to study infectious cancers in human. Since then, there has been a growing interest in identifying the role of viruses in cancer. Currently, it is estimated that of all cancers worldwide, 20–25% have a viral etiology [2]. There are seven known viruses which can cause cancer in humans—DNA tumor viruses comprising of human papillomavirus (HPV), Epstein–Barr virus (EBV), Kaposi's sarcoma-associated herpesvirus (KSHV), Merkel cell polyomavirus (MCV), hepatitis B virus (HBV), and RNA viruses which include hepatitis C virus (HCV) and human T-lymphotropic virus-1 (HTLV-1).

HPVs can also be classified as cutaneous or mucosal type, depending on its target tissue. Cutaneous HPV types infect the basal epithelial cells of the hands and feet (and typically cause plantar warts, common warts, and flat warts (reviewed in [3]), while mucosal types infect the inner lining of tissues, such as the respiratory tract [4], oropharyngeal region [5], or anogenital epithelium. The mucosal types can be further divided to high-risk or low-risk HPV, and it is the high-risk mucosal HPV types which have been found to have close links with cervical cancer.

HPV is a double-stranded circular DNA virus, between 6.8 kb and 8 kb in length, and is the most common sexually-transmitted infection. There are now over 200 subtypes of the virus, which can be categorized as either low-risk or high-risk, depending on their activity on the host cell. The difference between high and low-risk HPV lies mainly in their cellular targets, wherein the targets of high-risk HPV have the ability to cause the formation of tumors while the cellular targets of low-risk HPV are not able to. This review, however, will focus on the cellular pathways targeted by high-risk HPV.

In 1983, the first high-risk HPV type (HPV16) was discovered by a German virologist and a Nobel Prize awardee (in 2008), Professor Harald zur Hausen, when he attempted to find similarities in DNA sequences between patients with cervical cancer [6]. In 61.1% of cervical cancer samples, the extracted DNA hybridized to a HPV16 probe under highly stringent conditions, while lower hybridization rates were obtained from samples from other diseased tissue, alluding to a possible causative function of HPV in cervical cancer. Since then, 13 high-risk types (types 16, 18, 31, 33, 35, 39, 45, 51, 52, 56, 58, 59, and 68) have been discovered, with HPV16 and 18 being the most common in patients with invasive cervical carcinomas.

1.1. HPV in Cancers

Although the presence of the HPV genome has been found in cancers of multiple tissue origins (see references above), the strongest link of HPV has been in cancer of the cervix. Cervical cancer is the fourth most common cancer in females worldwide, behind breast, colon, and lung cancer, with a similar standing in mortality rate [7]. There is variation in the incidence and mortality rate depending on geographical location, and this is primarily due to the difference in education of prevention and availability to screening. An estimated 530,000 cases of invasive cervical cancer are diagnosed globally every year [8]. Of these, 99.7% of samples have tested positive for HPV [9,10]. The absence of HPV in the remaining 0.3% of cervical carcinomas has been attributed to the following factors: inadequate detection methods, presence of unidentified types of HPV, and disruption of the HPV genome during integration events [11].

HPV has also been implicated heavily in head and neck squamous cell carcinoma (HNSCC), with the virus found in up to 25% of patients [12,13]. Since the etiological link between HPV and HNSCC was made in 1983, there has been a significant increase in HPV-positive HNSCC samples detected. Similar to cervical cancer, the most common type of HPV known to infect HNSCC patients is HPV16, accounting for over 80% of HPV-positive HNSCC patients. Other cancers which have been linked to HPV include breast [14–16], anal, vaginal, penile, and vulvar [17]. However, in these cancers, the incidence of HPV is variable, with relatively lower detectable viral load. Specific techniques have been used to detect the low level of HPV in these samples. These techniques include in situ PCR [18,19], histology [14], bead-based Luminex technology [20], hybrid capture 2 (HC2) assay [21], and HPV capture paired with massive paralleled sequencing [22]. The low levels of HPV detected in these patients does not discount HPV from having a causative role. Instead, it has been hypothesized that there are other factors which could drive the cancer in HPV-negative patients. This review, however, will focus on HPV in cervical cancer.

1.2. HPV Genome

There are three distinct regions in the HPV genome: the early gene-coding region (E), late gene-coding region (L), and the long control region (LCR). The early genes (E1, E2, E4–E7) encode for viral replication and regulatory proteins, three of which, E5, E6, and E7, are oncogenic. Depending on their specific function, the early genes are expressed at different stages of the viral life cycle [23,24]. The late genes encode for two structural proteins involved in the formation of the viral capsid, while the LCR contains regulatory elements involved in the control of viral DNA replication and transcription [23,24].

Upon HPV infection of cells in the basal layer, E1 and E2 facilitate viral genome replication at a low-copy number rate. However, it is not until the basal cells differentiate into the suprabasal layer of the epithelia prior to sloughing off that the virus switches to high-copy replication rate. Although the exact mechanism of virion release is not known, it occurs at around the same time as epithelia desquamation. These virions are then capable of infecting other neighboring cells. Despite the fact that integration of the HPV genome can occur in pre-malignant lesions, the proportion of cells which undergo HPV integration increases as the cells progress in malignancy [25]. One of the first proposed mechanisms of integration demonstrated that the breakpoint of circular HPV episome was at the E2 gene [26]. This was of great interest as E2 transcriptionally represses E6 and E7, and therefore, the disruption of E2 upon integration leads to the de-repression of E6 and E7. In cells where the HPV genome was integrated, most viral genes are found to be silenced, with the exception of E6 and E7.

Since the integration of the viral genome into the host is a "dead-end" of the viral life cycle, it seems counterintuitive that the virus would integrate, and in doing so, limiting its infectious potential. However, scientists have rationalized this, and since a virus lacks machinery to replicate its own genome, it needs to hijack its host's replication machinery [27]. The driving oncogenes, E6 and E7, hijack the cellular ubiquitin–proteasome system (UPS) to degrade retinoblastoma protein (pRb), forcing the host cell into S phase, where replication machinery and resources, such as nucleotide pool, are abundant. This allows viral replication to occur [28]. Since unscheduled cell cycle progression would trigger apoptosis in the cell, HPV E6 concurrently degrades p53, which would result in the evasion of apoptosis [29]. Through the constitutive expression of E6 and E7, coupled with aneuploidy that is brought on by uncontrolled cell division, genomic instability ensues. This provides the cells a growth advantage and promotes oncogenesis [30]. Therefore, the ultimate aim of the virus is not to cause carcinogenesis in the host organism. Instead, the formation of tumors is an unwanted, accidental side effect of viral infection.

Despite the growth advantage that follows an integration event, not all HPV-positive cells undergo integration. A study analyzing HPV-positive samples obtained from The Cancer Genome Atlas revealed integration events in all HPV18-positive samples, but only occurred in 76% of HPV16-positive samples [31]. However, it is interesting to note that in samples where the HPV genome is not integrated, there are often epigenetic or genetic changes found on the regulatory regions of E6 and E7 that result in the dysregulation of these two oncogenes [32–34]. This indicates that although cancer progression and the integration of HPV genome are separate events, the dysregulation of E6 and E7 is crucial for tumorigenesis.

1.3. HPV Oncogenes

Although all the genes encoded by the HPV genome are crucial during different stages of replication and the viral life cycle, the two most important genes in high-risk HPV are E6 and E7, also known as the key viral oncogenes. They were termed viral oncogenes due to their targets in the cell and the implications of this dysregulation. As discussed earlier, although the main target of E6 and E7 are p53 [29,35] and pRb respectively, much research has shown that there are many other targets of E6 and E7 which allow high-risk HPV types to be tumorigenic. Studies have shown that HPV16 E6 alone is able to immortalize human mammary epithelial cell, allowing it to overcome mortality stage mechanism M1, that is, evade senescence and exhibit an extended lifespan [36]. However, it was found that the overexpression of only E6 and E7 were insufficient for complete cellular transformation in primary human keratinocytes [37].

E5 is the least-studied of all three oncogenes, and was recently shown to contribute to cell cycle progression and unscheduled host DNA synthesis in differentiating keratinocytes [38]. This was shown to be dependent on the EGFR pathway in a ligand-dependent manner, where it functioned to enhance the activation of the EGFR pathway [39,40]. Overexpression of E5 in transgenic mice resulted in hyperplasia of the epithelia, which was rescued by the expression of a dominant negative form of EGFR [41]. High-risk E5 was also shown to enhance immortalization potential of E6 and E7 in keratinocytes [42]. However, it was found that E5 expression is often lost upon HPV integration into the host genome, and therefore, it is postulated that E5 has an important role in the initial stages of cervical carcinogenesis, but less so in its persistence [38].

In following sections, we will discuss the mechanism by which HPV can regulate cellular pathways, primarily through E6 and E7.

2. Mechanism of E6 and E7 Regulating Cellular Pathways

2.1. Protein Targets Dependent on E6AP

HPV regulates cellular pathways through various mechanisms to exert an effect in the infected cell. A summary of E6 regulating selected cellular factors and the consequences of this disruption is illustrated in Figure 1. E6-associated protein (E6AP) is encoded by the *UBE3A* gene and is the founding member of the HECT (homologous to E6AP carboxyl-terminus) E3 family of ubiquitin ligases. E6, in cooperation with E6AP, is one of the most well-studied mechanisms in which HPV degrades its targets. E6AP is a cellular protein, the dysregulation of which has been implicated in Angelman syndrome (AS), a severe neurodevelopmental disorder [43]. However, in the context of HPV-infected cells, E6AP is hijacked by E6 to ubiquitylate and target cellular proteins for degradation by the UPS.

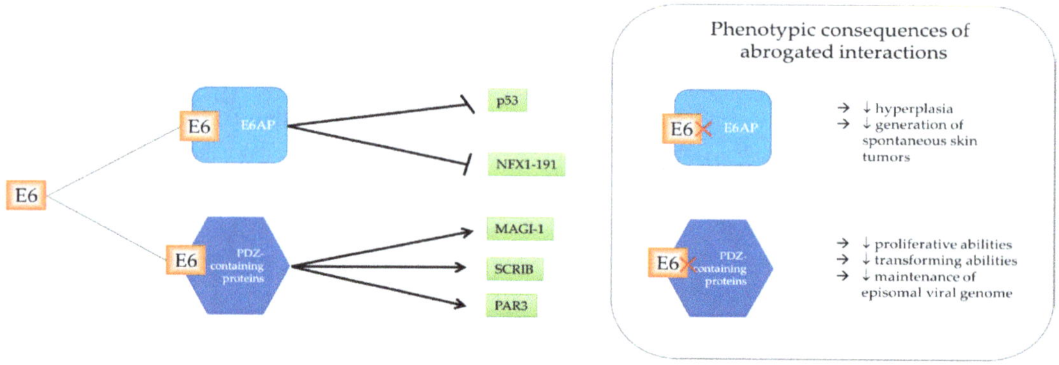

Figure 1. E6 regulates many cellular targets through E6AP and E6′s PDZ-binding motif. Some of these targets are depicted above. Upon the abrogation of the sites on E6 which are involved in the interaction of E6AP and PDZ-containing proteins respectively, there are drastic phenotypic consequences.

The most prominent target of E6AP is the crucial cellular regulator, p53, more candidly known as the *guardian of the genome* [29,44,45]. E6 binds to the LxxLL consensus sequence in the conserved domain of E6AP, and as a heterodimer, is able to degrade p53 [46]. The structure of the interaction between E6, E6AP, and p53 was solved in 2016 [46]. The interaction of E6 and E6AP was shown to be crucial in vivo, with a transgenic mouse harboring an E6 mutant incapable of binding to E6AP demonstrating reduced levels of skin hyperplasia, and lower generation of spontaneous skin tumors when compared to wild-type E6 [47]. Similar studies demonstrating the importance of E6AP we also carried out, with the transient knockdown of E6AP producing an almost identical phenotype to that of E6 depletion [48]. However, there are an increasing number of studies exploring E6AP-independent manners in which HPV can regulate cellular mechanisms.

2.2. PDZ-Domain Family of E6 Interactors

One of the key differences between high-risk and low-risk HPV E6 oncoprotein is the existence of a PDZ (PSD-95/DLG/ZO-1) binding motif (PBM) in high-risk but not low-risk HPV types. The PBM enables the interaction and subsequent degradation of proteins containing a PDZ domain, which is frequently (but not always) located on the C-terminus of proteins [49]. The importance of a PBM on high-risk E6 was demonstrated through a compromised proliferative phenotype in human foreskin keratinocytes (HFKs) containing HPV18 E6ΔPBM when compared to wild-type E6 [50]. Further, cells expressing wild-type E6 demonstrated a selective growth advantage, where extensive passaging resulted in the loss of cells containing E6ΔPBM genome [50]. Interestingly, it was found that both wild-type E6 and E6ΔPBM were able to target and degrade p53, evidenced by a similar response to radiation, due to the lack of p53 and subsequent p21 induction [51].

To date, high-risk E6 is known to interact with at least 14 cellular PDZ-containing proteins, which results in the eventual degradation of the cellular substrate. Interestingly, MAGI-1 is the only cellular target which is degraded by both HPV16 and 18, while the degradation of other cellular substrates is dependent on HPV type [52,53]. MAGI-1 is a member of the MAGI family, which has canonical roles in the regulation of tight junction assembly [54]. Thus, the degradation of MAGI-1 results in the loss of tight junction formation.

Additionally, the interaction of E6 with its PDZ-containing substrates is key in different stages of its viral life cycle, and also in the final stage of oncogenesis. The presence of an intact and functional PBM is seen to be crucial early in the viral life cycle, essential for maintenance of viral copy number and proliferation [55]. Further, E6 PBM is required for the maintenance of episomal viral genome, indicative of its role in viral DNA replication [50]. During oncogenesis, E6's ability to bind to PDZ domain-containing targets was shown to be important for the host cell's transformation abilities, both in vitro [56,57] and in vivo [51].

One of the phenotypic changes of immortalized keratinocytes with HPV18 E6 is the marked changes in cell morphology, organization of microfilament network, and the formation of intercellular adhesion junctions [56]. This was shown to be highly, but not completely dependent on a functional PDZ domain. Part of this phenotype can be attributed to the downregulation of Dlg, a cellular target of E6 PDZ-binding motif.

Another study compared the tumorigenic ability of a E6Δ151 mutant (incapable of binding to PDZ domains) to wild-type E6, and showed that the mutant lost almost all tumorigenic ability [58]. This was surprising, since the mutant defective for p53 degradation showed a phenotype close to wild-type E6. The involvement of E6 PDZ substrates was demonstrated when cells containing E6Δ151 mutant were depleted of SCRIB, MAGI-1, and PAR3. While the single knockdown of SCRIB, MAGI-1, or PAR3 only partially but significantly restored tumorigenic ability, co-depletion of SCRIB and MAGI-1 completely restored tumorigenic ability [58].

The members of the 14-3-3 protein family function as adapter proteins, and interact with many cellular proteins with roles in cell cycle, apoptosis, metabolic control, cytoskeletal maintenance, transcription, and tumor suppression [59]. 14-3-3 zeta (an isoform of 14-3-3) is known to interact with E6 in a phospho-specific manner, only interacting with high-risk E6 when its PBM is phosphorylated. This interaction is of particular interest as it is important in maintaining E6 stability in HeLa cells [60]. However, it is still not known whether the phosphorylation of E6's PBM is required for its interaction with 14-3-3, or whether it is 14-3-3 that modulates the phosphorylation on the PBM of E6.

Although many groups have studied the cellular ubiquitin ligases which are involved in the degradation of cellular PDZ domain-containing proteins, there is little consensus on the mechanism of degradation. Multiple groups have published results showing conflicting data on whether E6AP is involved in the process of degradation, as with Scrib and Bak (targeted by E6AP [61]), while Dlg, MAGI-1, and MUPP1 are not targeted by E6AP [62,63]).

From the evidence presented above, the interaction of E6 with its PDZ domain-containing substrates are very important. Equally as important, perhaps, is the regulation of the PBM on E6. A threonine residue at position 156 within the PBM of E6 is susceptible to phosphorylation by a range of different kinases, which abrogates its ability to fit into the PDZ domain of interactors. Interestingly, it was shown that the PBM of different high-risk HPV types are phosphorylated by different kinases—18E6 by protein kinase A (PKA) and 16E6 by PKA or AKT [60]. However, the phosphorylation of PBM does not always result in the inhibition of interactions.

2.3. Other Protein Targets

Since much research has shown E6's preference to binding to ubiquitin ligases, Poirson et al. screened for novel ubiquitin ligases bound by E6 and E7 [64]. They utilized the *Gaussia princeps* luciferase protein complementation assay (GPCA) to co-express E6 and E7 independently, with a library containing 50% (575 unique protein entries) of the human ubiquitination system. In addition to discovering novel binding partners of E6 and E7, they were also able to study the domains required for interaction. Aside from E6AP, E6 interacting with other ligase components did not require the LxxLL motif, but instead, its PBM on its C-terminus. E7 was also shown to be capable of binding to proteins containing the BTB (BR-C, ttk and bab) domain, RING domain, and deubiquitinating enzymes (DUBs). Furthermore, since the group expressed the HPV oncoproteins from six different HPV types (three high-risk types—HPV16, HPV18, HPV33, one low-risk type—HPV6, and two cutaneous HPV types—HPV8, HPV38), they were able to identify preferential binding of cellular proteins to HPV proteins of different types. In short, they concluded that UPS factors were more differentially bound to E6 than E7, and this was observed across multiple types.

2.4. Epigenetic Targets

E6/E7 Regulating Methylation Status

Several studies have also shown a close relationship between HPV infection of cells and promoter hypermethylation of cellular genes, repressing their transcription [65,66]. The phenomenon of loci-specific methylation has long been observed in cells infected with HPV, with 87% of CpG sites of methylation variable positions (MVP) showing increased methylation in virus-positive compared to virus-negative cells. Through the comparison of a heterogeneous population HNSCC, it was observed that HPV could modulate the epigenome [66].

In support of the increase in global methylation upon infection with E6/E7, human keratinocytes transduced with E6/E7 were subjected to methylation-specific digital karyotyping (MSDK) and there were 34 genes showing increased methylation [67]. From publicly available gene expression data, the authors showed that these 34 genes had a concomitant decrease in expression. Subsequent validations were successfully performed on a subset of these 34 genes. Although not explored in the paper, it would have been interesting to study the genes that underwent hypomethylation. Part of the mechanism was elucidated when it was shown that transient depletion of E6 resulted in a decrease in DNMT1 protein levels in HPV-positive cell lines in a p53-dependent manner [68].

E-cadherin, in particular, is a functionally significant methylation target of HPV due to its importance in the detection of virus in the epithelia. E-cadherin is expressed on the membrane of keratinocytes, and its presence, together with other stimulatory signals, orchestrates the movement of Langerhans cells through the epithelium, which serves as an endogenous surveillance mechanism [69]. Langerhans cells detect, process, and present viral foreign antigens to other immunocytes, which encourages the stimulation of the host immune system to rid the virus from the organism [69]. In HPV-infected cells, the E-cadherin expression was lower [70], resulting in decreased migration of Langerhans cells [71], and therefore, lower virus clearance rate from the epithelia.

Interestingly, both HPV16 E6 and E7 were found to downregulate *E-cadherin* through DNMT1 activity. The overexpression of HPV16 E6, resulted in decreased surface and total protein level of E-cadherin, as well as an increase in DNMT activity [72]. Although *E-cadherin* levels were rescued after the treatment with DNMT inhibitor 5-aza-2′-deoxycytidine, there was no change in methylation status of E-cadherin promoter in the presence and absence of HPV16 E6. Instead, the authors postulated that HPV16 E6 regulating *E-cadherin* levels was due to the increased presence of repressive elements on E-cadherin promoter, instead. The regulation of DNMT1 by E7 was shown to be via a direct interaction, utilizing the CR3 (conserved region 3) zinc-finger domain of E7, one of the three regions conserved with adenovirus E1A [73]. Although the expression of *E-cadherin* was rescued in normal immortalized human keratinocytes (NIKs) positive for HPV upon the treatment with 5-aza-2′-deoxycytidine, there was no significant change in the methylation status of *E-cadherin* promoter [74]. This led the authors to believe that E7 stabilizes DNMT1 via binding to it, and this regulates another intermediary factor (still unknown), which then leads to the silencing of *E-cadherin*.

The direct effect of E7 and DNMT1, however, was seen in on the promoter of *CCNA1*. This interaction was validated in vitro through ChIP experiments where both DNMT1 and HPV16

E7 were found to be localized on the *CCNA1* promoter, resulting in a repression of *CCNA1* [75]. Following this discovery, it was suggested to utilize the methylation status of *CCNA1* as a potential biomarker to distinguish between normal cervix, low-grade, and high-grade squamous intraepithelial lesions [76,77]. *CXCL14* was also identified as a methylation target of E7, with the HPV oncogene resulting in promoter hypermethylation, and thus, silenced expression [78]. Since CXCL14 has a role in angiogenesis suppression, E7 functions to de-repress angiogenesis to support tumor growth.

Despite numerous studies which have demonstrated HPV's ability to increase methylation, there have also been a handful of studies which have demonstrated the decrease in methylation at specific loci. For an example, Yin et al. demonstrated that HPV16 E7 led to an increase in the expression of *STK31* [79], a gene which has been implicated in the maintenance of the undifferentiated state of colon cancer cells [80]. Through the use of bisulfite genome sequencing (BGS) PCR and TA cloning, the promoter of *STK31* was shown to be hypomethylated in HPV-infected cells compared to HPV-negative cells. However, the exact mechanism of E7 causing a hypomethylation of the promoter remains to be elucidated. A comparison of tissues from different grades of cervical lesions showed a global hypomethylation upon cervical cancer progression [81]. Through staining with an antibody against methylated CpG, there were no observable differences between normal tissue, benign lesions, low-grade lesions, and high-grade lesions. However, there was a significant decrease in methylation when tissues from invasive SCC were compared to pre-neoplastic lesions.

2.5. E6/E7 Targeting Epiegnetic Modifiers

In addition to regulating the methylation status of cellular targets, E6/E7 is also able to interact with other epigenetic modulators which dictate the post-translational modifications on histones, which is illustrated in Figure 2.

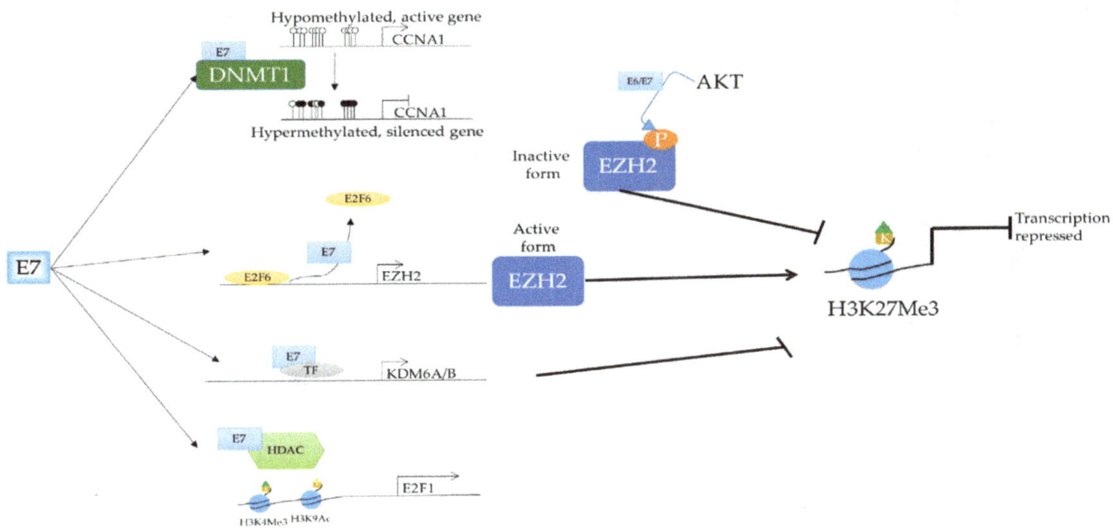

Figure 2. The HPV oncogene E7 is able to regulate multiple epigenetic factors such as DNMT1, HDAC1, and EZH2. E7 associates with DNMT1 on the promoter region of *CCNA1*, resulting in its hypermethylation, and therefore repression. E7 is also able to regulate the repressive H3K27Me3 mark through several of its upstream factors. By binding to the repressive E2F6, E7 is able to de-repress the transcription of the methyltransferase *EZH2*. Conversely, E7 was shown to transcriptionally upregulate *KDM6A/B*, demethylases of H3K27Me3. Further, E6 and E7 has also been shown to recruit AKT to phosphorylate and therefore inactivate EZH2. Lastly, E7 is able to cooperate with HDAC, resulting in the tri-methylation of H3K4 and the acetylation of H3K9 on the promoter of *E2F1* to upregulate the transcription of *E2F1*.

There exists an intricate relationship between HPV and polycomb repressive complex (PRC). The two most prominent PRCs are PRC1 and PRC2, which have functions in epigenetic silencing. It has been shown that the HPV oncoproteins interact with different regulators and components of the PRC2 complex, resulting in an altered epigenetic program and the concomitant dysregulation of downstream expression. The EZH2 methyltransferase is the enzymatic component of PRC2, responsible for the tri-methylation of H3K27, a repressive promoter mark. HPV16 E7 ablates the repressive activity of E2F6, a non-canonical member of the E2F family, and transcriptional repressor of EZH2 [82], resulting in the increase in EZH2 expression [83]. However, the increase in EZH2 expression was not shown to result in an increase in global H3K27Me3; instead, it led to a marginal decrease in global H3K27Me3 [84,85]. To date, there are two explanations for this. Firstly, it is known that KDM6A and KDM6B, two demethylases of H3K27Me3, are transcriptionally upregulated by HPV E7 [85]. Secondly, the activity of EZH2 is negatively regulated by phosphorylation by AKT on serine residue 21 [86]. Since both HPV16 E6 and E7 have been shown to activate AKT, the transcription of EZH2 might be upregulated but the activity suppressed [87,88].

Histone acetylation is one of the key post-translational modifications, as it not only signals for its downstream genes to be activated or repressed, but also determines the 3D chromatin structure, defining its euchromatic or heterochromatic state. This is made possible as the acetylation group is able to neutralize the positive charge of histones, therefore loosening the conformation of chromatin, enabling transcription machinery to bind. Although acetylated chromatin is generally a mark of actively transcribed genes, there are instances where an acetylated histone recruits repressive machinery. An example of this is the acetylated histone being read by Brd4, and thus resulting in a repression of downstream gene expression [89–91].

HPV has been shown to target both groups of enzymes which are responsible for the deposition and the removal of the acetylated marks—histone acetyltransferases (HATs) and histone deacetylases (HDAC). CBP/p300 is a HAT which acetylates all four canonical histones as well as non-histone targets, and is possibly the most well-studied HAT involved in HPV-mediated changes in epigenetic landscape. The interaction of CBP/p300 and E6 was first shown in 1999 through experiments with purified CBP/p300 and E6 on an affinity column [92]. By sequentially abolishing individual domains of the respective proteins, the E6-binding domain on CBP and p300 was found to be a 19 amino-acid sequence which interacted with the C-terminal zinc finger on E6 [92]. The authors termed it a transcriptional adapter motif (TRAM), which was the same motif targeted by the adenovirus E1A protein [93]. Since CBP/p300 has many cellular targets in the cell, it is most advantageous of E6 to target CBP/p300. One key function of CBP/p300 is as a transcriptional coactivator in p53-dependent transcription, and through its inhibition of CBP/p300, HPV16 E6 is able to attenuate p53 transcription in a E6AP-independent manner [92].

Another target of E6 is the *bona fide* tumor suppressor, TIP60 [89]. This HAT was found to be destabilized by both high-risk and low-risk HPV through the E3 ubiquitin ligase, EDD1, in a proteasome-dependent manner [94]. E6 targeting TIP60 is one element of a feedback loop, since TIP60 is responsible for acetylating histone H4 on the regulatory promoter of E6, thereby recruiting Brd4 to suppress E6 transcription [89].

E6 of high-risk HPV was also shown to regulate the activity of histone methyltransferases (HMTs) co-activator associated arginine methyltransferase 1 (CARM1), protein arginine methyltransferase 1 (PRMT1), and SET domain containing lysine methyltransferase 7 (SET7), thereby affecting p53 levels independent of E6AP [95]. Canonically, CARM1 and PRMT1 are recruited to p53-responsive promoters to activate downstream transcription. However, in the presence of E6, the recruitment and subsequent activity of the two HMTs was abrogated independently of E6 degrading p53. SET7 is a lysine-specific histone methyltransferase that has a canonical function of stabilizing p53 by methylating it at K372, therefore preventing its ubiquitylation and subsequent degradation. However, in the presence of E6, SET7's enzymatic activity is dampened, subjecting p53 to degradation independent of E6AP [95].

Interestingly, the activity of E6 in regulating TIP60 and the three abovementioned HMTs was not restricted to high-risk HPV, as a similar observation was made with low-risk HPV.

The E7 oncoprotein of HPV has been shown to regulate multiple cellular HDACs through multiple pathways in manners independent and dependent on pRb. Through its interaction with Mi2β, the ATPase subunit of the NuRD complex, E7, is able to bind to HDAC1 through the oncoprotein's zinc finger domain and attenuate its activity [96,97]. However, when this interaction was abrogated, E7 was no longer able to extend the lifespan of transfected keratinocytes [97].

The family of E2F transcription factors regulate transcription in both positive and negative manners, and have important functions in the cell [98]. E2F2, an activating member of the E2F family, has roles in positively regulating viral genome replication during the process of keratinocyte differentiation, and was also shown to be regulated by HPV 31 E7 [99]. By inhibiting the binding of class I HDACs to the regulatory region of *E2F2*, E7 is able to increase *E2F2* transcription, and therefore promote virus replication. Indeed, the transient depletion of *E2F2* via siRNA resulted in decreased virus replication, although there was no significant change in cell proliferation. The authors concluded that E7 regulating *E2F2* was independent of cell cycle changes mediated by pRb, since the expression of other cell cycle-dependent genes in E2F family did not change.

In a similar manner, E7 was also shown to upregulate the acetylation of H3K9 in a manner dependent on pRb [100]. This was coupled with the tri-methylation of H3K4, a mark of an active promoter, which usually results in the displacement of repressive HDACs from the promoter region [101]. Through mutational analyses, the association with pRb and HDAC was shown to be essential for the post-translational modification of histones, which appeared to be specific to E2F-associated genes.

2.6. RNA Targets of HPV

There are many RNA species in the cell, and among other biologically significant molecules, RNA has been proven to have a range of different but important functions in the cell, such as coding for proteins and in modulating coding genes, as in the case of lncRNAs. As such, it comes as no surprise that HPV, too, targets RNA in cells. Two HPV16 proteins, E2 and E6, were shown to have RNA-binding capabilities, both of which negatively affect the splicing of genes with suboptimal intronic splice sites [102]. Interestingly, on the HPV16 E6 protein, the nuclear localization signal 3 (NLS3) at the C-terminus is involved in RNA binding, while a separate N-terminus suppresses RNA splicing. The HPV16 E2 protein, on the other hand, interacts with RNA through its C-terminal DNA binding domain, but utilizes its N-terminal half and hinge region for splicing suppression. Using recombinant HPV proteins and various RNA substrates, the authors demonstrated that HPV16 E6E7 pre-mRNA, BPV-1 late pre-mRNA, and *doublesex* gene from *Drosophila* (all of which contain suboptimal splice sites) had less efficient splicing in the presence of HPV16 E6, while β-globin was spliced efficiently, since it contained optimal splice sites. The mechanism of splicing regulation was partially elucidated when the same authors showed that E2 and E6 bound and interacted with a small subset of splicing regulatory (SR) proteins, although no functional studies were carried out to ascertain this.

Several years later, another group demonstrated the splicing capabilities of HPV16 E6 overexpression in HPV-negative cells [103]. The changes to the transcriptome and splicing profile were compared upon the transient overexpression of HPV16 E6, and validated in clinical tissue samples. It was noted that there was a modest 56 annotated and 22 novel genes which were differentially expressed. Of these, there were 153 skipped exons, 23 alternative 5′ splice sites, 32 alternative 3′ splice sites, and 20 retained introns, although no functional conclusions were drawn.

MicroRNAs (miRNAs) are an abundant class of short RNAs which have the ability to recruit machinery that degrade mature RNA strands, therefore affecting their stability. Depending on its sequence and seed site recognition, different miRNAs have different cellular targets in the cell, with a myriad of consequences. Many studies have been published on miRNAs regulated by HPV, but only a handful of groups have studied miRNAs encoded by HPV itself. To our knowledge,

Qian et al. was the first group to experimentally detect miRNAs encoded by HPV [104]. They utilized SOLiD 4 sequencing technology to identify short RNAs in HPV16, HPV38, and HPV68 infected cells, and mirSeqNovel to identify miRNAs encoded specifically by HPV after mapping it to the viral genome. Nine putative miRNAs were detected, of which four were successfully validated via TaqMan assays in cell lines and cervical tissue samples. The authors focused on two miRNAs (HPV16-miR-H1 and HPV16-miR-H2) encoded by HPV16, since it is the most abundant in patients. These miRNAs mapped back to the E1 gene and LCR, respectively. The cellular targets of the two abovementioned miRNAs were predicted based on seed sequence, and were implicated in important host cell processes, such as cell cycle, immune functions, neoplastic development, focal adhesion, cell migration, epithelium development, and cancer. One very interesting cellular gene targeted by both HPV16-miR-H1 and HPV16-miR-H2 is *CYP26B1*, which encodes for a protein essential in retinoic acid (RA) metabolism [105]. It has been previously shown that RA is capable of regulating the differentiation of epithelial cells, and more importantly, inhibit the growth of HeLa cells in vitro [106]. Although not validated in this study, it would be an exciting avenue to pursue. Other studies have focused on cellular miRNAs whose expression is regulated by E6 and E7, with predicted roles in cell proliferation, apoptosis, and differentiation [107,108].

3. Cellular Consequences Affected by E6 and E7

The infection of a cell with HPV can cause many cellular changes through mechanisms detailed above. Here, we will highlight the main cellular consequences which are driven by E6 and E7 and depicted in Figure 3.

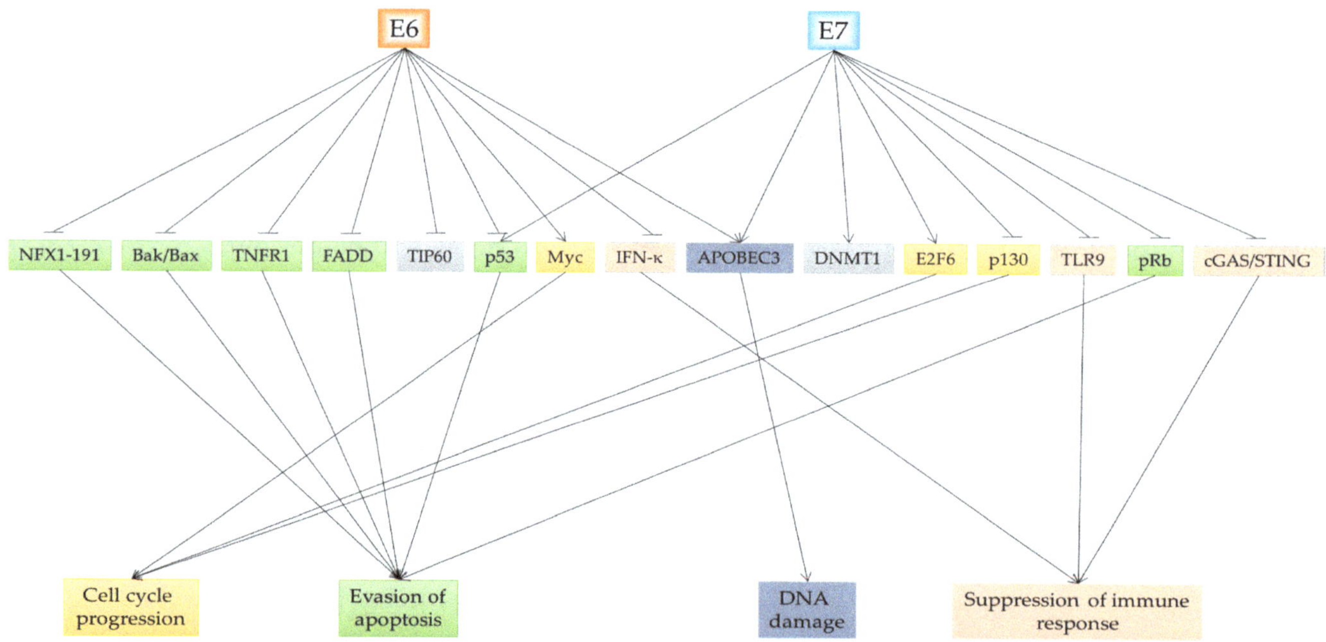

Figure 3. HPV oncogenes E6 and E7 target a plethora of cellular factors. This eventuates in cell cycle progression, evasion of apoptosis, DNA damage, and the suppression of the host cell immune response.

3.1. Cell Cycle Changes and Growth Promotion

Evading checkpoints in the cell cycle is paramount for the process of tumorigenesis to occur. Thus, HPV targets key cellular factors which are involved in checkpoints in this crucial process. The most well-characterized targets of high-risk HPV—p53 and pRb are both key players in cell cycle dysregulation. Through the degradation of these two tumor suppressors, HPV-infected cells are able to bypass cell cycle checkpoints to maintain a proliferative, transformed phenotype. Researchers have utilized different molecular techniques to downregulate the expression of E6 and E7, such as

CRISPR [109], transcription activator-like effector nuclease (TALENs) [110], artificial miRNAs [111], and siRNA [112], and have come to a similar conclusion that growth is markedly inhibited, and apoptosis/senescence is induced [113]. The HPV oncoproteins bring about these growth stimulatory changes through targeting and downregulating several key pro-apoptotic factors.

As previously mentioned, p53 is targeted by only high-risk types in a proteasome-dependent manner. Similarly, high-risk E7 binds to cullin 2 (CUL2) ubiquitin ligase complex to ubiquitylate and subsequently degrade pRb, initiating a cascade of downstream effects. The binding of E7 to Rb results in the de-repression the E2F family of transcription factors, and signals the cell to enter the S phase, and thus follow through with replication, despite insufficient resources in the cell.

Through the dysregulation of the proteins involved in these checkpoints, E6 and E7 can lead to uncontrolled cell growth and proliferation. The aberrant regulation of factors involved in cell cycle regulation in HPV-infected cells is so consistent that several groups have suggested using this as a diagnostic marker. Its potential was shown in a study of tissue microarrays by Conesa-Zamora et al. [114]. Using 144 fixed cervical tissue specimens, pathologists successfully detected p16, Ki-67, ProEx C (novel marker of MCM2 and TOP2A proteins), and p53 in high grade squamous intraepithelial lesions, of which p16, ProEx C, and Cyclin D1 correlated well with the severity of the lesion.

Through microarray analysis comparing normal human keratinocytes and differentiating cells harboring E6/E7, genes involved in G_2–M phase transition were found to be upregulated, such as Plk1, Aurora-A, Nek2, and Cdk1, amongst others [115]. Due to Plk1′s important function in not only the G_2–M transition, but also activation of cyclin B/Cdk1 and centrosome maturation, and regulation of the anaphase-promoting complex [116–118], the authors explored the mechanism of regulation of Plk1. Through mutational analyses, it was shown that Plk1′s regulation was dependent on E6′s ability to degrade p53, as well as E7′s repressive effect on pRb.

Myc is inarguably the most potent oncogene in the human cell, regulating the expression of 10–15% of cellular genes [119], and is involved in cell proliferation, apoptosis, and cellular transformation. *Myc* has been shown to be activated in more than half of cancer cases, and therefore, its overexpression is now accepted as one of the hallmarks of cancer (reviewed in [120]). *Myc* overexpression in cervical cancer is no exception, and the overexpression is brought about mainly by amplified gene expression, and detectable by PCR of cervical tissue scrapes of normal cervical tissue and HPV-infected tissues [121]. It was also observed that *c-myc* copy number increased along with histological grade, indicating that *c-myc* has a role in cellular transformation [121]. It was shown through in situ hybridization experiments that in some cases, the integration site of the HPV genome was in the same chromosomal band as *c-myc* or *n-myc*, 8q24, and 2p24, respectively, and the integration resulted in the cellular oncogene being structurally altered and/or amplified [122]. However, other studies have shown that the HPV integration site is not the only factor involved in *Myc* overexpression and thus, the mechanism of *Myc* amplification still remains to be elucidated.

In addition to HPV being a factor which led to the amplification of the locus, E6/E7 have also been found to regulate both the expression and the activity of Myc through several distinct mechanisms. Although both high and low-risk HPV can bind to Myc, only high-risk HPV can activate Myc, providing evidence for differences between high and low-risk phenotype [123]. Firstly, it has been shown that E6 is involved in the phosphorylation of Myc, increasing its stability in the cell [124]. However, this only occurs to a subpopulation of Myc, with different subpopulations of Myc having different stability and functions [125]. The exact mechanism of the phosphorylation, however, still stands to be confirmed.

Secondly, it was shown via co-immunoprecipitation assays that E6 and Myc are part of a complex, and is able to have functional consequences, such as the activation of *TERT* [126]. Thirdly, it was shown that E7 can bind to and activate c-Myc-mediated transcription activity [123]. Although both high and low-risk HPV E7 was shown to bind to c-Myc, only high-risk E7 can enhance c-Myc transcription activation. Using the *TERT* promoter as a model, the authors utilized immunoprecipitation assays to demonstrate that c-Myc bound to the *TERT* promoter in a manner dependent on E7 [123].

Treatment with Myc antagonists significantly abrogated E6-mediated transcription of *TERT* [124]. Fourthly, research has shown alternative methods in which E6 can utilize Myc to upregulate *TERT* expression. Zhang et al. showed that in addition to binding to Myc, E6 can also bind directly to Max, although only in the presence of Myc, suggesting that E6 can bind to Max when it is part of the Myc/Max heterodimer [124]. The final method in which HPV can regulate Myc is in a manner reverse to the abovementioned mechanisms. It has been shown that the family of Myc proteins also have a growth inhibitory role by inducing apoptosis when cell proliferation is inhibited [127]. It is hypothesized that through its degradation of Myc, E6 is able to evade Myc-driven apoptosis to maintain viral infection and proliferation [128]. However, this does not occur very often since tumor cells are mostly in their proliferative stage.

TERT is one of the central regulators in cancer, its expression is key for continued proliferation and to evade replicative senescence. As such, it is targeted and activated by HPV oncoproteins by several different methods. The promoter of *TERT* contains binding sites for the transcription factors Myc and Sp1, and it was shown that when either Myc or Sp1 binding sites were mutated, there was marginal decrease in *TERT* expression [129]. However, when all Myc and Sp1 binding sites were mutated, there was complete ablation of *TERT* expression. As described above, E6 and Myc have a very intricate relationship in the cell, and cooperatively, E6 and Myc are able to activate the *TERT* promoter [130]. It was also recently demonstrated that TIP60, destabilized by E6, acetylates Sp1 and ablates its binding to the *TERT* promoter [131]. In the presence of E6, TIP60 is targeted for proteasomal degradation by EDD1 [94], allowing Sp1 to bind to the *TERT* promoter, and therefore, leading to its activation. Secondly, NFX1-91, a repressor of *TERT*, has been shown to be a substrate of E6/E6AP [132,133]. Through the degradation of NFX1-91 in a proteasome-dependent manner, *TERT* expression is de-repressed in the presence of E6 and E6AP. Thirdly, it was demonstrated that E6 was able to post-translationally modify the histones at the promoter of *TERT*, the overexpression of which led to an increase in the activating H3K4Me3 mark with a concomitant decrease in the repressive H3K9Me2 mark [124]. Further ChIP assays also demonstrated that in cells overexpressing E6, there was increased serine 2 phosphorylation of Pol II at the *TERT* promoter, indicating increased transcription of *TERT*. A separate study by McMurray et al. demonstrated that in the absence of E6, a repressive complex containing USF1 and/or USF2 was localized on the *TERT* promoter, which was demonstrated by ChIP assays [134]. However, when E6 was transiently introduced into the system, this repressive complex was displaced from the promoter region, and Myc subsequently bound to the E-boxes on the *TERT* promoter. This led to a concomitant increase in TERT activity. However, the exact mechanism in which E6 affects the binding of these regulatory factors are not known.

3.2. Evading Apoptosis

Apoptosis is the scheduled death of the cell, and often occurs when the cell accumulates DNA damage. There are several methods in which E6 and E7 can block apoptosis. Although the HPV oncogene E7 sensitizes cells to p53-mediated apoptosis through the abrogation of pRb, this is overcome by E6-mediated degradation of p53.

Bak and Bax are members of the Bcl-2 protein family which have the canonical function of inducing apoptosis via the mitochondrial pathway. Upon sensing a variety of stress signals, Bak and Bax change conformation and assemble into oligomeric complexes in the mitochondrial outer membrane to form pores in the outer membrane, which subsequently lead to cellular apoptosis (reviewed in [135]). As key pro-apoptotic factors in the cell, it is only natural that Bak and Bax are targeted by both high and low-risk HPV E6, albeit via different mechanisms. Bak is targeted by both high and low-risk HPV types independent of p53 [136]. Interestingly, Bak was found to interact with both E6 and E6AP independent of each other, suggesting that Bak is a cellular target of E6AP, which is augmented by the presence of E6 [136]. The interaction of E6 with Bak is functional, where HPV18 E6 was shown to inhibit Bak-induced apoptosis in an HPV-negative cell line in the presence of exogenous E6. Although Bax has also been shown to be a target of E6, this interaction is mediated by p53 [137]. Upon the silencing of high-risk E6 via RNA interference (RNAi), p53 levels are rescued, reactivating the *PUMA* promoter, thus allowing Bax to be activated and translocated to the mitochondrial membrane.

Interestingly, studies have shown that E7 alone can induce apoptosis in a manner dependent on tumor necrosis factor [138]. This is hypothesized to be due to E7′s ability to stabilize the p53 protein which is abrogated in the presence of HPV E6 [139,140]. It was shown that E7 stabilizes p53 through a mechanism independent of p19 (ARF) [141]. The repressive complex DREAM (dimerization partner, RB-like, E2F4, and MuvB) is known to be downstream of p53, and has also been identified as a substrate of E7 [142–144]. The main function of the DREAM complex is to repress genes during quiescence. Upon re-entry into the cell cycle, DREAM complex, together with p130 and the repressive E2F4, dissociates from the promoter regions of cell cycle genes, and allows activating machinery to bind. DREAM is regulated by p53, and by binding to a myriad of different regions, such as cycle-dependent elements (CDEs), cell cycle genes homology regions (CHRs), CHR-like elements (CLE) and E2F sites, the DREAM complex is directed to promoters of cell cycle genes, leading to its repression [145,146]. It was shown that HPV16 E7 can bind to p130 of the DREAM complex via its LxCxE motif, resulting in proteasomal degradation, and therefore, abrogation of the DREAM complex [147]. Transcriptional repression mediated by DREAM opens up another avenue of p53 transcriptional activation, as p53 is canonically known to have only an activating function. One example of a gene of whose expression was de-repressed in a manner dependent on E7 and DREAM is *PLK4*, repressed by the p53–p21–DREAM complex. However, in the presence of E7, its expression is de-repressed [148].

This mechanism of E7 regulating cell cycle genes was verified when Fischer et al. analyzed publicly available gene expression datasets and found that the DREAM genes were the main group of genes deregulated by E7 [149]. This disruption was found to be indispensable for cell cycle progression in cervical cancer cells [142]. Interestingly, E7 can also target cyclin-dependent kinase (CDK) inhibitor p21, a crucial target of p53 required for its activity [150,151].

Inflammatory cytokines, such as TNFα, are secreted upon sensing of a viral infection, and this activates death receptors on the surface of the cell, such as TNF receptor 1 (TNFR1), TNF-related apoptosis-inducing ligand (TRAIL) receptors, and FAS. Through binding to TNFR1, E6 can inhibit the formation of the death-inducing signaling complex, blocking apoptosis from occurring [152]. Further, E6 also binds to protein FAS-associated protein with death domain (FADD) and caspase 8 to block downstream cell death [153,154].

3.3. Immune Response

Since HPV are infectious agents, it makes perfect evolutionary sense that one of the abilities of HPV is to disrupt innate immunity, since it is the first line of host defense against pathogenic infections. Viral nucleic acids (such as HPV) are sensed by pathogen recognition receptors (PRRs), such as TLR9 [155]. Canonically, TLR9 is able to recognize unmethylated CpG motifs on double-stranded DNA from viruses such as HPV [156], EBV [157], and HSV [158]. This initiates a cascade that eventually leads to the activation of host immune defense via the production of type I IFN and pro-inflammatory cytokines [159]. Research has recently shown that HPV16 is able to repress TLR9 transcription through the HPV oncoprotein E7, forming an inhibitory transcriptional complex comprising of NF-κBp50–p65 and ERα [160]. This complex recruits the histone demethylase JARID1B and histone deacetylase HDAC1 to the TLR9 promoter, resulting in a repression of its transcription. This is biologically significant, as the downstream interferon induction was negatively affected, muting the cellular effect of the viral infection.

A recent development in this field was the discovery that HPV oncogene E7 binds to STING, an adapter protein downstream of the intracellular DNA sensor, cGAS [161]. The cGAS–STING pathway is crucial in the detection of intracellular DNA and the initiation of the immune response in response to viral infection. The authors found that via the LXCXE motif on the E7 protein, the same motif involved in the degradation of pRb, E7 is able to antagonize DNA sensing, and therefore suppress an immune response post-infection.

Genes further downstream in the immunity signaling cascade were also observed to be regulated by HPV, such as the network of genes correlated with IL-1β [162]. These chemotactic and pro-inflammatory genes were discovered when a genome-wide screen was performed on undifferentiated keratinocytes containing episomal copies of high-risk HPV16 and 18. Along the same lines, HPV16, 18 and 31 were also found to repress the expression of genes involved in IFN signaling (*STAT1*), antiviral genes (*IFIT1* and *MX1*), pathogen recognition receptors (*TLR3*, *RIG-I*, and *MDA5*), and pro-apoptotic genes (*TRAIL* and *XAF1*) [163]. This was mediated by IFN-κ, found to be downregulated by E6 with the promoter of IFN-κ found to be methylated by DNMT1, recruited by E6 [164].

3.4. DNA Damage

Genomic instability is a hallmark of cancer cells, and malignant cells driven by HPV is no exception. It has been shown that through multiple mechanisms, HPV oncogenes E6 and E7 are able to independently cause DNA damage and chromosomal aberrations. This was shown through DNA breakage detection-fluorescence in situ hybridization (DNA-FISH) which detects DNA damage on a global, unbiased level [165]. Upon silencing of E6 and E7 in HeLa cells with lentiviral shRNA, there was a significant decrease in DNA damage, confirmed by alkaline comet assay. One mechanism that brings about chromosomal instability is stress during DNA replication, also known as replication stress. In the presence of HPV infection, E7 targets pRb for degradation, and this forces the cell to proceed with replication, despite an insufficient pool of nucleotides in the cell [166]. This eventuates in replication stress, and thus, genome instability. This phenotype was rescued when nucleosides were exogenously provided to HPV-infected cells, or by activating *c-myc*, which signals the increased transcription of nucleotide biosynthesis genes.

Another mechanism of genomic instability in malignant tumors is centrosome abnormalities, which potentially leads to defective/multiple mitotic spindle pole formations, resulting in chromosomal mis-segregation and genomic instability [167]. HPV16 E7 has been shown to induce abnormal centrosome synthesis, while E6 from the same type brings about nuclear atypia (multiple irregular nuclei) and accumulation of centrosomes [168]. Interestingly, the mechanism that E6 and E7 utilizes are disparate, from the simple observation that the overexpression of E7 results in an almost immediate increase in centrosomes, while centrosome abnormalities driven by E6 were only observed after several weeks in culture [169]. The mechanisms were later elucidated, and demonstrated that it is through the degradation of pRb by E7 that results in aberrant centriole synthesis [168]. Additionally, expression of both HPV16 E6 and E7 overwhelmed the spindle checkpoint control, allowing cells to enter into anaphase despite having multiple spindle poles [170]. The authors also found that HPV16 E7 could induce anaphase bridge formation, which typically occur after extensive chromosomal breaks, as well as inducing PARP formation, independent of E6.

One other target of high-risk E6/E7 is apolipoprotein B mRNA editing enzyme catalytic-polypeptide-like proteins 3 (APOBEC3), a family of cytidine deaminase proteins. The APOBEC3 family consists of seven different proteins, of which two are of interest in the field of HPV-induced cancers. APOBEC3A and APOBEC3B have been found to be highly expressed in HPV-positive samples, regulated by both E6 and E7, which will be detailed later [171,172]. The primary function of APOBEC3A/B enzymes is in the deamination of cytosine residues—that is, removing a crucial amino group from cytosine, converting it to an uracil. This occurs in both cellular genome and the viral genome, which eventually results in mutations (reviewed in [173]). APOBEC3-dependent mutagenesis on the viral genomes often result in the inhibition of replication, such as in the case of HIV-1 [174–176] and HPV [171]. Despite this, expression of APOBEC3 was found to be upregulated in HPV-positive samples [171,172]. This occurs through several different methods. Firstly, through the degradation of p53, HPV16 E6 upregulates TEAD1/4, a family of transcription factors involved in the regulation of multiple cellular targets, of which APOBEC3B is one [177]. Through ChIP experiments, the authors demonstrated that in normal immortalized human keratinocytes (NIKs) exogenously expressing E6, there was higher localization of TEAD4 on the promoter of APOBEC3B, resulting in increased transcription of APOBEC3B [177]. The second mechanism in which HPV is shown to result in increased levels of APOBEC3 is through the binding of HPV16 E7 to APOBEC3 via the CUL2 binding motif on APOBEC3A [178]. The formation of this complex prevented the degradation of APOBEC3A, and interestingly, despite this binding, APOBEC3A enzyme was found to retain its catalytic activity.

The dual function of APOBEC3A and APOBEC3B in restricting HPV, yet causing somatic mutations is probably one of the captivating reasons why researchers have focused on studying APOBEC. Interestingly, it has been found that small DNA viruses, HPV included, have an underrepresentation of TC dinucleotides [179], which are the substrates of the target-specific APOBEC3A enzymes. As such, HPV is partially resistant to the viral restriction acted upon by APOBEC3A and APOBEC3B. By upregulating these two enzymes, E6 and E7 are, instead, able to cause cancer mutagenesis. As an example, mutations in the oncogenic driver *PIK3CA* gene were found to be more prevalent in HPV-positive head and neck cancers (HNCs) compared to HPV-negative patients [180,181]. Further analyses revealed that all of the *PIK3CA* mutations in HPV-positive HNCs were caused by GA-to-AA mutations, while only half were caused by the same mutations in HPV-negative HNCs.

4. Therapeutics against HPV

4.1. Vaccinations

In June 2006, the US Food and Drug Administration (FDA) approved the first ever cervical cancer vaccine, Gardasil [182]. This vaccine contains virus-like particles (VLP) of HPV types 6, 11, 16, and 18, and was intended for females 9–26 years of age to prevent cervical cancer, and also HPV types that cause non-malignant lesions. This has been proven to be highly effective, reducing the incidence of genital warts in Australia from 12% to 5% in women, and 12% to 9% in heterosexual men, the latter attributed to lower exposure to the virus, since their sexual partners were no longer carriers of the virus [183]. As such, in 2013, Australia was the first country to begin vaccinating their male population in an effort to curb the transmission of HPV by males [184]. A second vaccination against HPV was approved by the FDA, Cervarix, a bivalent vaccine containing VLPs of high-risk types 16 and 18 (reviewed in [185]). The third vaccine to be approved by the FDA is 9-valent/Gardasil-9, which contains VLP covered by Gardasil, as well as five other oncogenic types-HPV type 31, 33, 42, 52, and 58. Clinical trials concluded in 2015 showed that when compared to Gardasil, there was lower incidence of high grade cervical, vulvar, and vaginal disease related to HPV type 31, 33, 45, 52, and 58 in the group with Gardasil-9. However, there was no difference in antibody responses to HPV types 6, 11, 16, and 18, of which both groups are vaccinated against [186].

In addition, there is an ongoing Phase III Clinical Trial funded by Shanghai Zerun Biotechnology Co. Ltd. (Clinical Trials Identifier: NCT02733068). This vaccine contains VLPs from L1 proteins from HPV types 16 and 18, and the trial is due to be completed in November of 2020.

4.2. Treatment

Although there have been numerous studies showing positive growth inhibitory effects upon the downregulation of the HPV driving oncogenes, most of the groups utilized RNAi, which had a transient effect on the cells. This temporary modulation is sufficient for mechanistic studies to be carried out, but insufficient to completely ablate E6 and E7 in vivo, to trigger apoptosis. Thus, mechanisms which utilize permanent genome editing capabilities are more promising for cervical cancer patients in the clinic.

There is an ongoing Phase I clinical trial by Hu and colleagues, in which they are utilizing TALEN and CRISPR/Cas9 targeting HPV16 E6/E7 or HPV18 E6/E7 to treat cervical cancer patients [109,110]. This study is expected to conclude in January of 2019 (Clinical Trial identifier: NCT03057912).

Additionally, other groups have developed molecules which block the activity of either E6, E7, or E6AP. This was made possible by the use of peptides, organic compounds, RNA molecules, nucleotide analog, small molecules, compound, zinc-ejecting inhibitor, heparin-like molecules or naturally-derived biopolymers (reviewed in [187]). There has also been a recent interest in the development of natural compounds derived from plants which have shown the ability to reduce the viral infection in patients with HPV-positive cervical cancer. In particular, a polyherbal formulation, administration of "Praneem" to patients positive for high-risk HPV16 for a period of 30 days led to non-detection of the virus via PCR-based methods [188].

Despite the great deal of research carried out, many of these therapies are still years away from making it into the clinic, and most late stage cervical cancer patients resort to hysterectomy and chemoradiation as treatment options with the most promise. Recently, a new targeted therapy was developed, which aims to curb the process of angiogenesis. Bevacizumab (Avastin®) is a monoclonal antibody against VEGF, the treatment of which has shown great promise in several types of cancer [189,190]. Clinicians have begun to include Bevacizumab into their treatment regime in combination with existing approved drugs.

5. Conclusions

HPV has evolved over the years to target many aspects of the host machinery, affecting ubiquitin ligases, epigenetic writers, miRNA, splicing regulators, amongst others. Through these interactions, the potent HPV oncogenes E6 and E7 are able to exert their effects in the cell to affect multiple cellular pathways. They are able to do so aside from the well-known targets p53 and pRb. We have shown that HPV targets major cellular pathways in multiple ways, creating "back-up plans" for itself to ensure that the cellular pathways are, indeed, dysregulated. The resultant effect is the continued proliferation of the host cell and replication of the virus to infect neighboring cells, thus contributing to carcinogenesis. Although there have been limited developments in targeted therapies, there are several vaccines in the market which are able to prevent the development of high-risk HPV infection.

Abbreviations

AS	Angelman syndrome
APOBEC3	apolipoprotein B mRNA editing enzyme catalytic-polypeptide-like proteins 3
BGS	bisulfite genome sequencing
BTB	BR-C, ttk and bab
CARM1	co-activator associated arginine methyltransferase 1
CDEs	cycle-dependent elements
CDK	cyclin-dependent kinase
CHRs	cell cycle genes homology regions
CLE	CHR-like elements
CUL2	cullin 2
DNA-FISH	DNA breakage detection-fluorescence in situ hybridization
DREAM	DP, RB-like, E2F4, and MuvB
DUBs	deubiquitinating enzymes
E6AP	E6-associated protein
EBV	Epstein–Barr virus
FADD	FAS-associated protein with death domain
FDA	Food and Drug Administration
GPCA	*Gaussia princeps* luciferase protein complementation assay

HAT	histone acetyltransferase
HBV	hepatitis B virus
HC2	hybrid capture 2
HCV	hepatitis C virus
HDAC	histone deacetylases
HECT	homologous to E6AP carboxyl-terminus
HFKs	human foreskin keratinocytes
HMTs	histone methyltransferases
HNCs	head and neck cancers
HNSCC	head and neck squamous cell carcinoma
HPV	human papillomavirus
HSV	herpes simplex virus
HTLV-1	human T-lymphotropic virus-1
KSHV	Kaposi's sarcoma-associated herpesvirus
LCR	long control region
MCV	Merkel cell polyomavirus
miRNAs	microRNAs
MSDK	methylation-specific digital karyotyping
MVP	methylation variable positions
NIKs	normal immortalized human keratinocytes
NLS3	nuclear localization signal 3
P/CAF	p300/CBP-associated factor
PBM	PDZ-binding motif
PKA	protein kinase A
pRb	retinoblastoma protein
PRC	polycomb repressive complex
PRMT1	protein arginine methyltransferase 1
PRRs	pathogen recognition receptors
RNAi	RNA interference
SCC	squamous cell carcinoma
SET7	SET domain containing lysine methyltransferase 7
SR	splicing regulatory
SSA	single-strand annealing
TALEN	transcription activator-like effector nuclease
TNFR1	TNF receptor 1
TRAIL	TNF-related apoptosis-inducing ligand
TRAM	transcriptional adapter motif
UPS	ubiquitin–proteasome system
VLP	virus-like particles

References

1. Rous, P. A Sarcoma of the Fowl Transmissible by an Agent Separable from the Tumor Cells. *J. Exp. Med.* **1911**, *13*, 397–411. [CrossRef] [PubMed]
2. Pagano, J.S.; Blaser, M.; Buendia, M.A.; Damania, B.; Khalili, K.; Raab-Traub, N.; Roizman, B. Infectious agents and cancer: Criteria for a causal relation. *Semin. Cancer Biol.* **2004**, *14*, 453–471. [CrossRef] [PubMed]
3. Egawa, N.; Doorbar, J. The low-risk papillomaviruses. *Virus Res.* **2017**, *231*, 119–127. [CrossRef] [PubMed]
4. Carifi, M.; Napolitano, D.; Morandi, M.; Dall'Olio, D. Recurrent respiratory papillomatosis: Current and future perspectives. *Ther. Clin. Risk Manag.* **2015**, *11*, 731–738. [CrossRef] [PubMed]
5. Taberna, M.; Mena, M.; Pavon, M.A.; Alemany, L.; Gillison, M.L.; Mesia, R. Human papillomavirus-related oropharyngeal cancer. *Ann. Oncol.* **2017**, *28*, 2386–2398. [CrossRef] [PubMed]

6. Durst, M.; Gissmann, L.; Ikenberg, H.; zur Hausen, H. A papillomavirus DNA from a cervical carcinoma and its prevalence in cancer biopsy samples from different geographic regions. *Proc. Natl. Acad. Sci. USA* **1983**, *80*, 3812–3815. [CrossRef] [PubMed]
7. Society, A.A.C. *Global Cancer Facts & Figures*, 3rd ed.; American Cancer Society: Atlanta, GA, USA, 2015.
8. Arbyn, M.; Castellsague, X.; de Sanjose, S.; Bruni, L.; Saraiya, M.; Bray, F.; Ferlay, J. Worldwide burden of cervical cancer in 2008. *Ann. Oncol.* **2011**, *22*, 2675–2686. [CrossRef] [PubMed]
9. Bosch, F.X.; Manos, M.M.; Munoz, N.; Sherman, M.; Jansen, A.M.; Peto, J.; Schiffman, M.H.; Moreno, V.; Kurman, R.; Shah, K.V.; et al. Prevalence of human papillomavirus in cervical cancer: A worldwide perspective. *J. Natl. Cancer Inst.* **1995**, *87*, 796–802. [CrossRef] [PubMed]
10. Walboomers, J.M.M.; Jacobs, M.V.; Manos, M.M.; Bosch, F.X.; Kummer, J.A.; Shah, K.V.; Snijders, P.J.F.; Peto, J.; Meijer, C.J.L.M.; Munoz, N. Human papillomavirus is a necessary cause of invasive cervical cancer worldwide. *J. Pathol.* **1999**, *189*, 12–19. [CrossRef]
11. Walboomers, J.; Meijer, C. Do HPV-negative cervical carcinomas exist? *J. Pathol.* **1997**, *181*, 253–254. [CrossRef]
12. Tanaka, T.I.; Alawi, F. Human Papillomavirus and Oropharyngeal Cancer. *Dent. Clin. N. Am.* **2018**, *62*, 111–120. [CrossRef] [PubMed]
13. Spence, T.; Bruce, J.; Yip, K.W.; Liu, F.F. HPV Associated Head and Neck Cancer. *Cancers* **2016**, *8*. [CrossRef] [PubMed]
14. Heng, B.; Glenn, W.K.; Ye, Y.; Tran, B.; Delprado, W.; Lutze-Mann, L.; Whitaker, N.J.; Lawson, J.S. Human papilloma virus is associated with breast cancer. *Br. J. Cancer* **2009**, *101*, 1345–1350. [CrossRef] [PubMed]
15. Gannon, O.M.; Antonsson, A.; Bennett, I.C.; Saunders, N.A. Viral infections and breast cancer—A current perspective. *Cancer Lett.* **2018**, *420*, 182–189. [CrossRef] [PubMed]
16. Lawson, J.S.; Salmons, B.; Glenn, W.K. Oncogenic Viruses and Breast Cancer: Mouse Mammary Tumor Virus (MMTV), Bovine Leukemia Virus (BLV), Human Papilloma Virus (HPV), and Epstein-Barr Virus (EBV). *Front. Oncol.* **2018**, *8*, 1. [CrossRef] [PubMed]
17. Gillison, M.L.; Chaturvedi, A.K.; Lowy, D.R. HPV prophylactic vaccines and the potential prevention of noncervical cancers in both men and women. *Cancer* **2008**, *113*, 3036–3046. [CrossRef] [PubMed]
18. Glenn, W.K.; Heng, B.; Delprado, W.; Iacopetta, B.; Whitaker, N.J.; Lawson, J.S. Epstein-Barr virus, human papillomavirus and mouse mammary tumour virus as multiple viruses in breast cancer. *PLoS ONE* **2012**, *7*, e48788. [CrossRef] [PubMed]
19. Salman, N.A.; Davies, G.; Majidy, F.; Shakir, F.; Akinrinade, H.; Perumal, D.; Ashrafi, G.H. Association of High Risk Human Papillomavirus and Breast cancer: A UK based Study. *Sci. Rep.* **2017**, *7*, 43591. [CrossRef] [PubMed]
20. ElAmrani, A.; Gheit, T.; Benhessou, M.; McKay-Chopin, S.; Attaleb, M.; Sahraoui, S.; El Mzibri, M.; Corbex, M.; Tommasino, M.; Khyatti, M. Prevalence of mucosal and cutaneous human papillomavirus in Moroccan breast cancer. *Papillomavirus Res.* **2018**, *5*, 150–155. [CrossRef] [PubMed]
21. Wang, Y.W.; Zhang, K.; Zhao, S.; Lv, Y.; Zhu, J.; Liu, H.; Feng, J.; Liang, W.; Ma, R.; Wang, J. HPV Status and Its Correlation with BCL2, p21, p53, Rb, and Survivin Expression in Breast Cancer in a Chinese Population. *BioMed Res. Int.* **2017**, *2017*, 6315392. [CrossRef] [PubMed]
22. Wang, T.; Zeng, X.; Li, W.; Zhu, H.; Wang, G.; Liu, X.; Lv, Y.; Wu, J.; Zhuang, X.; Zhang, J.; et al. Detection and analysis of human papillomavirus (HPV) DNA in breast cancer patients by an effective method of HPV capture. *PLoS ONE* **2014**, *9*, e90343. [CrossRef] [PubMed]
23. Ghittoni, R.; Accardi, R.; Hasan, U.; Gheit, T.; Sylla, B.; Tommasino, M. The biological properties of E6 and E7 oncoproteins from human papillomaviruses. *Virus Genes* **2010**, *40*, 1–13. [CrossRef] [PubMed]
24. Zur Hausen, H. Papillomavirus infections, a major cause of human cancers. *Biochim. Biophys. Acta Rev. Cancer* **1996**, *1288*, F55–F78. [CrossRef]
25. Shukla, S.; Mahata, S.; Shishodia, G.; Pande, S.; Verma, G.; Hedau, S.; Bhambhani, S.; Kumari, A.; Batra, S.; Basir, S.F.; et al. Physical state & copy number of high risk human papillomavirus type 16 DNA in progression of cervical cancer. *Indian J. Med. Res.* **2014**, *139*, 531–543. [PubMed]
26. Thierry, F.; Yaniv, M. The BPV1-E2 trans-acting protein can be either an activator or a repressor of the HPV18 regulatory region. *EMBO J.* **1987**, *6*, 3391–3397. [PubMed]

27. McBride, A.A.; Warburton, A. The role of integration in oncogenic progression of HPV-associated cancers. *PLoS Pathog.* **2017**, *13*, e1006211. [CrossRef] [PubMed]
28. Munger, K.; Werness, B.A.; Dyson, N.; Phelps, W.C.; Harlow, E.; Howley, P.M. Complex formation of human papillomavirus E7 proteins with the retinoblastoma tumor suppressor gene product. *EMBO J.* **1989**, *8*, 4099–4105. [PubMed]
29. Scheffner, M.; Werness, B.A.; Huibregtse, J.M.; Levine, A.J.; Howley, P.M. The E6 oncoprotein encoded by human papillomavirus types 16 and 18 promotes the degradation of p53. *Cell* **1990**, *63*, 1129–1136. [CrossRef]
30. Jeon, S.; Allen-Hoffmann, B.L.; Lambert, P.F. Integration of human papillomavirus type 16 into the human genome correlates with a selective growth advantage of cells. *J. Virol.* **1995**, *69*, 2989–2997. [PubMed]
31. Cancer Genome Atlas Research Network; Albert Einstein College of Medicine; Analytical Biological Services; Barretos Cancer Hospital; Baylor College of Medicine; Beckman Research Institute of City of Hope; Buck Institute for Research on Aging; Canada's Michael Smith Genome Sciences Centre; Harvard Medical School; Helen, F.; et al. Integrated genomic and molecular characterization of cervical cancer. *Nature* **2017**, *543*, 378–384. [CrossRef]
32. Chaiwongkot, A.; Vinokurova, S.; Pientong, C.; Ekalaksananan, T.; Kongyingyoes, B.; Kleebkaow, P.; Chumworathayi, B.; Patarapadungkit, N.; Reuschenbach, M.; von Knebel Doeberitz, M. Differential methylation of E2 binding sites in episomal and integrated HPV 16 genomes in preinvasive and invasive cervical lesions. *Int. J. Cancer* **2013**, *132*, 2087–2094. [CrossRef] [PubMed]
33. Dong, X.P.; Stubenrauch, F.; Beyer-Finkler, E.; Pfister, H. Prevalence of deletions of YY1-binding sites in episomal HPV 16 DNA from cervical cancers. *Int. J. Cancer* **1994**, *58*, 803–808. [CrossRef] [PubMed]
34. Bhattacharjee, B.; Sengupta, S. CpG methylation of HPV 16 LCR at E2 binding site proximal to P97 is associated with cervical cancer in presence of intact E2. *Virology* **2006**, *354*, 280–285. [CrossRef] [PubMed]
35. Werness, B.A.; Levine, A.J.; Howley, P.M. Association of human papillomavirus types 16 and 18 E6 proteins with p53. *Science* **1990**, *248*, 76–79. [CrossRef] [PubMed]
36. Shay, J.W.; Wright, W.E.; Brasiskyte, D.; Van der Haegen, B.A. E6 of human papillomavirus type 16 can overcome the M1 stage of immortalization in human mammary epithelial cells but not in human fibroblasts. *Oncogene* **1993**, *8*, 1407–1413. [PubMed]
37. Richard, C.; Lanner, C.; Naryzhny, S.N.; Sherman, L.; Lee, H.; Lambert, P.F.; Zehbe, I. The immortalizing and transforming ability of two common human papillomavirus 16 E6 variants with different prevalences in cervical cancer. *Oncogene* **2010**, *29*, 3435–3445. [CrossRef] [PubMed]
38. Wasson, C.W.; Morgan, E.L.; Muller, M.; Ross, R.L.; Hartley, M.; Roberts, S.; Macdonald, A. Human papillomavirus type 18 E5 oncogene supports cell cycle progression and impairs epithelial differentiation by modulating growth factor receptor signalling during the virus life cycle. *Oncotarget* **2017**, *8*, 103581–103600. [CrossRef] [PubMed]
39. Straight, S.W.; Hinkle, P.M.; Jewers, R.J.; McCance, D.J. The E5 oncoprotein of human papillomavirus type 16 transforms fibroblasts and effects the downregulation of the epidermal growth factor receptor in keratinocytes. *J. Virol.* **1993**, *67*, 4521–4532. [PubMed]
40. Pim, D.; Collins, M.; Banks, L. Human papillomavirus type 16 E5 gene stimulates the transforming activity of the epidermal growth factor receptor. *Oncogene* **1992**, *7*, 27–32. [PubMed]
41. Genther Williams, S.M.; Disbrow, G.L.; Schlegel, R.; Lee, D.; Threadgill, D.W.; Lambert, P.F. Requirement of epidermal growth factor receptor for hyperplasia induced by E5, a high-risk human papillomavirus oncogene. *Cancer Res.* **2005**, *65*, 6534–6542. [CrossRef] [PubMed]
42. Stoppler, M.C.; Straight, S.W.; Tsao, G.; Schlegel, R.; McCance, D.J. The E5 gene of HPV-16 enhances keratinocyte immortalization by full-length DNA. *Virology* **1996**, *223*, 251–254. [CrossRef] [PubMed]
43. Kishino, T.; Lalande, M.; Wagstaff, J. UBE3A/E6-AP mutations cause Angelman syndrome. *Nat. Genet.* **1997**, *15*, 70–73. [CrossRef] [PubMed]
44. Huibregtse, J.M.; Scheffner, M.; Howley, P.M. A cellular protein mediates association of p53 with the E6 oncoprotein of human papillomavirus types 16 or 18. *EMBO J.* **1991**, *10*, 4129–4135. [PubMed]
45. Scheffner, M.; Huibregtse, J.M.; Vierstra, R.D.; Howley, P.M. The HPV-16 E6 and E6-AP complex functions as a ubiquitin-protein ligase in the ubiquitination of p53. *Cell* **1993**, *75*, 495–505. [CrossRef]
46. Martinez-Zapien, D.; Ruiz, F.X.; Poirson, J.; Mitschler, A.; Ramirez, J.; Forster, A.; Cousido-Siah, A.; Masson, M.; Vande Pol, S.; Podjarny, A.; et al. Structure of the E6/E6AP/p53 complex required for HPV-mediated degradation of p53. *Nature* **2016**, *529*, 541–545. [CrossRef] [PubMed]

47. Nguyen, M.; Song, S.; Liem, A.; Androphy, E.; Liu, Y.; Lambert, P.F. A mutant of human papillomavirus type 16 E6 deficient in binding alpha-helix partners displays reduced oncogenic potential in vivo. *J. Virol.* **2002**, *76*, 13039–13048. [CrossRef] [PubMed]
48. Kelley, M.L.; Keiger, K.E.; Lee, C.J.; Huibregtse, J.M. The global transcriptional effects of the human papillomavirus E6 protein in cervical carcinoma cell lines are mediated by the E6AP ubiquitin ligase. *J. Virol.* **2005**, *79*, 3737–3747. [CrossRef] [PubMed]
49. Songyang, Z.; Fanning, A.S.; Fu, C.; Xu, J.; Marfatia, S.M.; Chishti, A.H.; Crompton, A.; Chan, A.C.; Anderson, J.M.; Cantley, L.C. Recognition of unique carboxyl-terminal motifs by distinct PDZ domains. *Science* **1997**, *275*, 73–77. [CrossRef] [PubMed]
50. Delury, C.P.; Marsh, E.K.; James, C.D.; Boon, S.S.; Banks, L.; Knight, G.L.; Roberts, S. The role of protein kinase A regulation of the E6 PDZ-binding domain during the differentiation-dependent life cycle of human papillomavirus type 18. *J. Virol.* **2013**, *87*, 9463–9472. [CrossRef] [PubMed]
51. Nguyen, M.L.; Nguyen, M.M.; Lee, D.; Griep, A.E.; Lambert, P.F. The PDZ ligand domain of the human papillomavirus type 16 E6 protein is required for E6's induction of epithelial hyperplasia in vivo. *J. Virol.* **2003**, *77*, 6957–6964. [CrossRef] [PubMed]
52. Kranjec, C.; Banks, L. A systematic analysis of human papillomavirus (HPV) E6 PDZ substrates identifies MAGI-1 as a major target of HPV type 16 (HPV-16) and HPV-18 whose loss accompanies disruption of tight junctions. *J. Virol.* **2011**, *85*, 1757–1764. [CrossRef] [PubMed]
53. Glaunsinger, B.A.; Lee, S.S.; Thomas, M.; Banks, L.; Javier, R. Interactions of the PDZ-protein MAGI-1 with adenovirus E4-ORF1 and high-risk papillomavirus E6 oncoproteins. *Oncogene* **2000**, *19*, 5270–5280. [CrossRef] [PubMed]
54. Murata, M.; Kojima, T.; Yamamoto, T.; Go, M.; Takano, K.; Chiba, H.; Tokino, T.; Sawada, N. Tight junction protein MAGI-1 is up-regulated by transfection with connexin 32 in an immortalized mouse hepatic cell line: cDNA microarray analysis. *Cell Tissue Res.* **2005**, *319*, 341–347. [CrossRef] [PubMed]
55. Lee, C.; Laimins, L.A. Role of the PDZ domain-binding motif of the oncoprotein E6 in the pathogenesis of human papillomavirus type 31. *J. Virol.* **2004**, *78*, 12366–12377. [CrossRef] [PubMed]
56. Watson, R.A.; Thomas, M.; Banks, L.; Roberts, S. Activity of the human papillomavirus E6 PDZ-binding motif correlates with an enhanced morphological transformation of immortalized human keratinocytes. *J. Cell Sci.* **2003**, *116*, 4925–4934. [CrossRef] [PubMed]
57. Kiyono, T.; Hiraiwa, A.; Fujita, M.; Hayashi, Y.; Akiyama, T.; Ishibashi, M. Binding of high-risk human papillomavirus E6 oncoproteins to the human homologue of the Drosophila discs large tumor suppressor protein. *Proc. Natl. Acad. Sci. USA* **1997**, *94*, 11612–11616. [CrossRef] [PubMed]
58. Yoshimatsu, Y.; Nakahara, T.; Tanaka, K.; Inagawa, Y.; Narisawa-Saito, M.; Yugawa, T.; Ohno, S.I.; Fujita, M.; Nakagama, H.; Kiyono, T. Roles of the PDZ-binding motif of HPV 16 E6 protein in oncogenic transformation of human cervical keratinocytes. *Cancer Sci.* **2017**, *108*, 1303–1309. [CrossRef] [PubMed]
59. Dougherty, M.K.; Morrison, D.K. Unlocking the code of 14-3-3. *J. Cell Sci.* **2004**, *117*, 1875–1884. [CrossRef] [PubMed]
60. Boon, S.S.; Banks, L. High-risk human papillomavirus E6 oncoproteins interact with 14-3-3zeta in a PDZ binding motif-dependent manner. *J. Virol.* **2013**, *87*, 1586–1595. [CrossRef] [PubMed]
61. Nakagawa, S.; Huibregtse, J.M. Human scribble (Vartul) is targeted for ubiquitin-mediated degradation by the high-risk papillomavirus E6 proteins and the E6AP ubiquitin-protein ligase. *Mol. Cell. Biol.* **2000**, *20*, 8244–8253. [CrossRef] [PubMed]
62. Pim, D.; Thomas, M.; Javier, R.; Gardiol, D.; Banks, L. HPV E6 targeted degradation of the discs large protein: Evidence for the involvement of a novel ubiquitin ligase. *Oncogene* **2000**, *19*, 719–725. [CrossRef] [PubMed]
63. Grm, H.S.; Banks, L. Degradation of hDlg and MAGIs by human papillomavirus E6 is E6-AP-independent. *J. Gen. Virol.* **2004**, *85*, 2815–2819. [CrossRef] [PubMed]
64. Poirson, J.; Biquand, E.; Straub, M.L.; Cassonnet, P.; Nomine, Y.; Jones, L.; van der Werf, S.; Trave, G.; Zanier, K.; Jacob, Y.; et al. Mapping the interactome of HPV E6 and E7 oncoproteins with the Ubiquitin-Proteasome System. *FEBS J.* **2017**. [CrossRef] [PubMed]
65. Sartor, M.A.; Dolinoy, D.C.; Jones, T.R.; Colacino, J.A.; Prince, M.E.; Carey, T.E.; Rozek, L.S. Genome-wide methylation and expression differences in HPV(+) and HPV(−) squamous cell carcinoma cell lines are consistent with divergent mechanisms of carcinogenesis. *Epigenetics* **2011**, *6*, 777–787. [CrossRef] [PubMed]

66. Lechner, M.; Fenton, T.; West, J.; Wilson, G.; Feber, A.; Henderson, S.; Thirlwell, C.; Dibra, H.K.; Jay, A.; Butcher, L.; et al. Identification and functional validation of HPV-mediated hypermethylation in head and neck squamous cell carcinoma. *Genome Med* **2013**, *5*, 15. [CrossRef] [PubMed]
67. Steenbergen, R.D.; Ongenaert, M.; Snellenberg, S.; Trooskens, G.; van der Meide, W.F.; Pandey, D.; Bloushtain-Qimron, N.; Polyak, K.; Meijer, C.J.; Snijders, P.J.; et al. Methylation-specific digital karyotyping of HPV16E6E7-expressing human keratinocytes identifies novel methylation events in cervical carcinogenesis. *J. Pathol.* **2013**, *231*, 53–62. [CrossRef] [PubMed]
68. Au Yeung, C.L.; Tsang, W.P.; Tsang, T.Y.; Co, N.N.; Yau, P.L.; Kwok, T.T. HPV-16 E6 upregulation of DNMT1 through repression of tumor suppressor p53. *Oncol. Rep.* **2010**, *24*, 1599–1604. [PubMed]
69. Giannini, S.L.; Hubert, P.; Doyen, J.; Boniver, J.; Delvenne, P. Influence of the mucosal epithelium microenvironment on Langerhans cells: Implications for the development of squamous intraepithelial lesions of the cervix. *Int. J. Cancer* **2002**, *97*, 654–659. [CrossRef] [PubMed]
70. Denk, C.; Hulsken, J.; Schwarz, E. Reduced gene expression of E-cadherin and associated catenins in human cervical carcinoma cell lines. *Cancer Lett.* **1997**, *120*, 185–193. [CrossRef]
71. Caberg, J.H.; Hubert, P.; Herman, L.; Herfs, M.; Roncarati, P.; Boniver, J.; Delvenne, P. Increased migration of Langerhans cells in response to HPV16 E6 and E7 oncogene silencing: Role of CCL20. *Cancer Immunol. Immunother.* **2009**, *58*, 39–47. [CrossRef] [PubMed]
72. D'Costa, Z.J.; Jolly, C.; Androphy, E.J.; Mercer, A.; Matthews, C.M.; Hibma, M.H. Transcriptional repression of E-cadherin by human papillomavirus type 16 E6. *PLoS ONE* **2012**, *7*, e48954. [CrossRef] [PubMed]
73. Burgers, W.A.; Blanchon, L.; Pradhan, S.; de Launoit, Y.; Kouzarides, T.; Fuks, F. Viral oncoproteins target the DNA methyltransferases. *Oncogene* **2007**, *26*, 1650–1655. [CrossRef] [PubMed]
74. Laurson, J.; Khan, S.; Chung, R.; Cross, K.; Raj, K. Epigenetic repression of E-cadherin by human papillomavirus 16 E7 protein. *Carcinogenesis* **2010**, *31*, 918–926. [CrossRef] [PubMed]
75. Chalertpet, K.; Pakdeechaidan, W.; Patel, V.; Mutirangura, A.; Yanatatsaneejit, P. Human papillomavirus type 16 E7 oncoprotein mediates CCNA1 promoter methylation. *Cancer Sci.* **2015**, *106*, 1333–1340. [CrossRef] [PubMed]
76. Chujan, S.; Kitkumthorn, N.; Siriangkul, S.; Mutirangura, A. CCNA1 promoter methylation: A potential marker for grading Papanicolaou smear cervical squamous intraepithelial lesions. *Asian Pac. J. Cancer Prev.* **2014**, *15*, 7971–7975. [CrossRef] [PubMed]
77. Yang, N.; Nijhuis, E.R.; Volders, H.H.; Eijsink, J.J.; Lendvai, A.; Zhang, B.; Hollema, H.; Schuuring, E.; Wisman, G.B.; van der Zee, A.G. Gene promoter methylation patterns throughout the process of cervical carcinogenesis. *Cell. Oncol.* **2010**, *32*, 131–143. [CrossRef] [PubMed]
78. Cicchini, L.; Westrich, J.A.; Xu, T.; Vermeer, D.W.; Berger, J.N.; Clambey, E.T.; Lee, D.; Song, J.I.; Lambert, P.F.; Greer, R.O.; et al. Suppression of Antitumor Immune Responses by Human Papillomavirus through Epigenetic Downregulation of CXCL14. *MBio* **2016**, *7*. [CrossRef] [PubMed]
79. Yin, F.F.; Wang, N.; Bi, X.N.; Yu, X.; Xu, X.H.; Wang, Y.L.; Zhao, C.Q.; Luo, B.; Wang, Y.K. Serine/threonine kinases 31(STK31) may be a novel cellular target gene for the HPV16 oncogene E7 with potential as a DNA hypomethylation biomarker in cervical cancer. *Virol. J.* **2016**, *13*, 60. [CrossRef] [PubMed]
80. Fok, K.L.; Chung, C.M.; Yi, S.Q.; Jiang, X.; Sun, X.; Chen, H.; Chen, Y.C.; Kung, H.F.; Tao, Q.; Diao, R.; et al. STK31 maintains the undifferentiated state of colon cancer cells. *Carcinogenesis* **2012**, *33*, 2044–2053. [CrossRef] [PubMed]
81. Missaoui, N.; Hmissa, S.; Dante, R.; Frappart, L. Global DNA methylation in precancerous and cancerous lesions of the uterine cervix. *Asian Pac. J. Cancer Prev.* **2010**, *11*, 1741–1744. [PubMed]
82. McLaughlin-Drubin, M.E.; Huh, K.W.; Munger, K. Human papillomavirus type 16 E7 oncoprotein associates with E2F6. *J. Virol.* **2008**, *82*, 8695–8705. [CrossRef] [PubMed]
83. Holland, D.; Hoppe-Seyler, K.; Schuller, B.; Lohrey, C.; Maroldt, J.; Durst, M.; Hoppe-Seyler, F. Activation of the enhancer of zeste homologue 2 gene by the human papillomavirus E7 oncoprotein. *Cancer Res.* **2008**, *68*, 9964–9972. [CrossRef] [PubMed]
84. Hyland, P.L.; McDade, S.S.; McCloskey, R.; Dickson, G.J.; Arthur, K.; McCance, D.J.; Patel, D. Evidence for alteration of EZH2, BMI1, and KDM6A and epigenetic reprogramming in human papillomavirus type 16 E6/E7-expressing keratinocytes. *J. Virol.* **2011**, *85*, 10999–11006. [CrossRef] [PubMed]

85. McLaughlin-Drubin, M.E.; Crum, C.P.; Munger, K. Human papillomavirus E7 oncoprotein induces KDM6A and KDM6B histone demethylase expression and causes epigenetic reprogramming. *Proc. Natl. Acad. Sci. USA* **2011**, *108*, 2130–2135. [CrossRef] [PubMed]
86. Cha, T.L.; Zhou, B.P.; Xia, W.; Wu, Y.; Yang, C.C.; Chen, C.T.; Ping, B.; Otte, A.P.; Hung, M.C. Akt-mediated phosphorylation of EZH2 suppresses methylation of lysine 27 in histone H3. *Science* **2005**, *310*, 306–310. [CrossRef] [PubMed]
87. Menges, C.W.; Baglia, L.A.; Lapoint, R.; McCance, D.J. Human papillomavirus type 16 E7 up-regulates AKT activity through the retinoblastoma protein. *Cancer Res.* **2006**, *66*, 5555–5559. [CrossRef] [PubMed]
88. Spangle, J.M.; Munger, K. The human papillomavirus type 16 E6 oncoprotein activates mTORC1 signaling and increases protein synthesis. *J. Virol.* **2010**, *84*, 9398–9407. [CrossRef] [PubMed]
89. Jha, S.; Vande Pol, S.; Banerjee, N.S.; Dutta, A.B.; Chow, L.T.; Dutta, A. Destabilization of TIP60 by human papillomavirus E6 results in attenuation of TIP60-dependent transcriptional regulation and apoptotic pathway. *Mol. Cell* **2010**, *38*, 700–711. [CrossRef] [PubMed]
90. Wu, S.Y.; Lee, A.Y.; Hou, S.Y.; Kemper, J.K.; Erdjument-Bromage, H.; Tempst, P.; Chiang, C.M. Brd4 links chromatin targeting to HPV transcriptional silencing. *Genes Dev.* **2006**, *20*, 2383–2396. [CrossRef] [PubMed]
91. Sakamaki, J.I.; Wilkinson, S.; Hahn, M.; Tasdemir, N.; O'Prey, J.; Clark, W.; Hedley, A.; Nixon, C.; Long, J.S.; New, M.; et al. Bromodomain Protein BRD4 Is a Transcriptional Repressor of Autophagy and Lysosomal Function. *Mol. Cell* **2017**, *66*, 517–532.e9. [CrossRef] [PubMed]
92. Zimmermann, H.; Degenkolbe, R.; Bernard, H.U.; O'Connor, M.J. The human papillomavirus type 16 E6 oncoprotein can down-regulate p53 activity by targeting the transcriptional coactivator CBP/p300. *J. Virol.* **1999**, *73*, 6209–6219. [PubMed]
93. O'Connor, M.J.; Zimmermann, H.; Nielsen, S.; Bernard, H.U.; Kouzarides, T. Characterization of an E1A-CBP interaction defines a novel transcriptional adapter motif (TRAM) in CBP/p300. *J. Virol.* **1999**, *73*, 3574–3581. [PubMed]
94. Subbaiah, V.K.; Zhang, Y.; Rajagopalan, D.; Abdullah, L.N.; Yeo-Teh, N.S.; Tomaic, V.; Banks, L.; Myers, M.P.; Chow, E.K.; Jha, S. E3 ligase EDD1/UBR5 is utilized by the HPV E6 oncogene to destabilize tumor suppressor TIP60. *Oncogene* **2015**, *35*, 2062–2074. [CrossRef] [PubMed]
95. Hsu, C.H.; Peng, K.L.; Jhang, H.C.; Lin, C.H.; Wu, S.Y.; Chiang, C.M.; Lee, S.C.; Yu, W.C.; Juan, L.J. The HPV E6 oncoprotein targets histone methyltransferases for modulating specific gene transcription. *Oncogene* **2012**, *31*, 2335–2349. [CrossRef] [PubMed]
96. Brehm, A.; Nielsen, S.J.; Miska, E.A.; McCance, D.J.; Reid, J.L.; Bannister, A.J.; Kouzarides, T. The E7 oncoprotein associates with Mi2 and histone deacetylase activity to promote cell growth. *EMBO J.* **1999**, *18*, 2449–2458. [CrossRef] [PubMed]
97. Longworth, M.S.; Laimins, L.A. The binding of histone deacetylases and the integrity of zinc finger-like motifs of the E7 protein are essential for the life cycle of human papillomavirus type 31. *J. Virol.* **2004**, *78*, 3533–3541. [CrossRef] [PubMed]
98. Ren, B.; Cam, H.; Takahashi, Y.; Volkert, T.; Terragni, J.; Young, R.A.; Dynlacht, B.D. E2F integrates cell cycle progression with DNA repair, replication, and G(2)/M checkpoints. *Genes Dev.* **2002**, *16*, 245–256. [CrossRef] [PubMed]
99. Longworth, M.S.; Wilson, R.; Laimins, L.A. HPV31 E7 facilitates replication by activating E2F2 transcription through its interaction with HDACs. *EMBO J.* **2005**, *24*, 1821–1830. [CrossRef] [PubMed]
100. Zhang, B.; Laribee, R.N.; Klemsz, M.J.; Roman, A. Human papillomavirus type 16 E7 protein increases acetylation of histone H3 in human foreskin keratinocytes. *Virology* **2004**, *329*, 189–198. [CrossRef] [PubMed]
101. Nishioka, K.; Chuikov, S.; Sarma, K.; Erdjument-Bromage, H.; Allis, C.D.; Tempst, P.; Reinberg, D. Set9, a novel histone H3 methyltransferase that facilitates transcription by precluding histone tail modifications required for heterochromatin formation. *Genes Dev.* **2002**, *16*, 479–489. [CrossRef] [PubMed]
102. Bodaghi, S.; Jia, R.; Zheng, Z.M. Human papillomavirus type 16 E2 and E6 are RNA-binding proteins and inhibit in vitro splicing of pre-mRNAs with suboptimal splice sites. *Virology* **2009**, *386*, 32–43. [CrossRef] [PubMed]
103. Xu, J.; Fang, Y.; Qin, J.; Chen, X.; Liang, X.; Xie, X.; Lu, W. A transcriptomic landscape of human papillomavirus 16 E6-regulated gene expression and splicing events. *FEBS Lett.* **2016**, *590*, 4594–4605. [CrossRef] [PubMed]

104. Qian, K.; Pietila, T.; Ronty, M.; Michon, F.; Frilander, M.J.; Ritari, J.; Tarkkanen, J.; Paulin, L.; Auvinen, P.; Auvinen, E. Identification and validation of human papillomavirus encoded microRNAs. *PLoS ONE* **2013**, *8*, e70202. [CrossRef] [PubMed]
105. MacLean, G.; Abu-Abed, S.; Dolle, P.; Tahayato, A.; Chambon, P.; Petkovich, M. Cloning of a novel retinoic-acid metabolizing cytochrome P450, Cyp26B1, and comparative expression analysis with Cyp26A1 during early murine development. *Mech. Dev.* **2001**, *107*, 195–201. [CrossRef]
106. Borutinskaite, V.V.; Navakauskiene, R.; Magnusson, K.E. Retinoic acid and histone deacetylase inhibitor BML-210 inhibit proliferation of human cervical cancer HeLa cells. *Ann. N. Y. Acad. Sci.* **2006**, *1091*, 346–355. [CrossRef] [PubMed]
107. Gu, W.; An, J.; Ye, P.; Zhao, K.N.; Antonsson, A. Prediction of conserved microRNAs from skin and mucosal human papillomaviruses. *Arch. Virol.* **2011**, *156*, 1161–1171. [CrossRef] [PubMed]
108. Honegger, A.; Schilling, D.; Bastian, S.; Sponagel, J.; Kuryshev, V.; Sultmann, H.; Scheffner, M.; Hoppe-Seyler, K.; Hoppe-Seyler, F. Dependence of intracellular and exosomal microRNAs on viral E6/E7 oncogene expression in HPV-positive tumor cells. *PLoS Pathog.* **2015**, *11*, e1004712. [CrossRef] [PubMed]
109. Hu, Z.; Yu, L.; Zhu, D.; Ding, W.; Wang, X.; Zhang, C.; Wang, L.; Jiang, X.; Shen, H.; He, D.; et al. Disruption of HPV16-E7 by CRISPR/Cas system induces apoptosis and growth inhibition in HPV16 positive human cervical cancer cells. *BioMed Res. Int.* **2014**, *2014*, 612823. [CrossRef] [PubMed]
110. Hu, Z.; Ding, W.; Zhu, D.; Yu, L.; Jiang, X.; Wang, X.; Zhang, C.; Wang, L.; Ji, T.; Liu, D.; et al. TALEN-mediated targeting of HPV oncogenes ameliorates HPV-related cervical malignancy. *J. Clin. Investig.* **2015**, *125*, 425–436. [CrossRef] [PubMed]
111. Bonetta, A.C.; Mailly, L.; Robinet, E.; Trave, G.; Masson, M.; Deryckere, F. Artificial microRNAs against the viral E6 protein provoke apoptosis in HPV positive cancer cells. *Biochem. Biophys. Res. Commun.* **2015**, *465*, 658–664. [CrossRef] [PubMed]
112. Lea, J.S.; Sunaga, N.; Sato, M.; Kalahasti, G.; Miller, D.S.; Minna, J.D.; Muller, C.Y. Silencing of HPV 18 oncoproteins With RNA interference causes growth inhibition of cervical cancer cells. *Reprod. Sci.* **2007**, *14*, 20–28. [CrossRef] [PubMed]
113. Hall, A.H.; Alexander, K.A. RNA interference of human papillomavirus type 18 E6 and E7 induces senescence in HeLa cells. *J. Virol.* **2003**, *77*, 6066–6069. [CrossRef] [PubMed]
114. Conesa-Zamora, P.; Domenech-Peris, A.; Orantes-Casado, F.J.; Ortiz-Reina, S.; Sahuquillo-Frias, L.; Acosta-Ortega, J.; Garcia-Solano, J.; Perez-Guillermo, M. Effect of human papillomavirus on cell cycle-related proteins p16, Ki-67, Cyclin D1, p53, and ProEx C in precursor lesions of cervical carcinoma: A tissue microarray study. *Am. J. Clin. Pathol.* **2009**, *132*, 378–390. [CrossRef] [PubMed]
115. Patel, D.; Incassati, A.; Wang, N.; McCance, D.J. Human papillomavirus type 16 E6 and E7 cause polyploidy in human keratinocytes and up-regulation of G2-M-phase proteins. *Cancer Res.* **2004**, *64*, 1299–1306. [CrossRef] [PubMed]
116. Glover, D.M.; Hagan, I.M.; Tavares, A.A. Polo-like kinases: A team that plays throughout mitosis. *Genes Dev.* **1998**, *12*, 3777–3787. [CrossRef] [PubMed]
117. Nigg, E.A. Polo-like kinases: Positive regulators of cell division from start to finish. *Curr. Opin. Cell Biol.* **1998**, *10*, 776–783. [CrossRef]
118. Peters, J.M. The anaphase-promoting complex: Proteolysis in mitosis and beyond. *Mol. Cell* **2002**, *9*, 931–943. [CrossRef]
119. Knoepfler, P.S. Myc goes global: New tricks for an old oncogene. *Cancer Res.* **2007**, *67*, 5061–5063. [CrossRef] [PubMed]
120. Gabay, M.; Li, Y.; Felsher, D.W. MYC activation is a hallmark of cancer initiation and maintenance. *Cold Spring Harb. Perspect. Med.* **2014**, *4*. [CrossRef] [PubMed]
121. Abba, M.C.; Laguens, R.M.; Dulout, F.N.; Golijow, C.D. The c-myc activation in cervical carcinomas and HPV 16 infections. *Mutat. Res.* **2004**, *557*, 151–158. [CrossRef] [PubMed]
122. Couturier, J.; Sastre-Garau, X.; Schneider-Maunoury, S.; Labib, A.; Orth, G. Integration of papillomavirus DNA near myc genes in genital carcinomas and its consequences for proto-oncogene expression. *J. Virol.* **1991**, *65*, 4534–4538. [PubMed]
123. Wang, Y.W.; Chang, H.S.; Lin, C.H.; Yu, W.C. HPV-18 E7 conjugates to c-Myc and mediates its transcriptional activity. *Int. J. Biochem. Cell Biol.* **2007**, *39*, 402–412. [CrossRef] [PubMed]

124. Zhang, Y.; Dakic, A.; Chen, R.; Dai, Y.; Schlegel, R.; Liu, X. Direct HPV E6/Myc interactions induce histone modifications, Pol II phosphorylation, and hTERT promoter activation. *Oncotarget* **2017**, *8*, 96323–96339. [CrossRef] [PubMed]
125. Tworkowski, K.A.; Salghetti, S.E.; Tansey, W.P. Stable and unstable pools of Myc protein exist in human cells. *Oncogene* **2002**, *21*, 8515–8520. [CrossRef] [PubMed]
126. Veldman, T.; Liu, X.; Yuan, H.; Schlegel, R. Human papillomavirus E6 and Myc proteins associate in vivo and bind to and cooperatively activate the telomerase reverse transcriptase promoter. *Proc. Natl. Acad. Sci. USA* **2003**, *100*, 8211–8216. [CrossRef] [PubMed]
127. McMahon, S.B. MYC and the control of apoptosis. *Cold Spring Harb. Perspect. Med.* **2014**, *4*, a014407. [CrossRef] [PubMed]
128. Gross-Mesilaty, S.; Reinstein, E.; Bercovich, B.; Tobias, K.E.; Schwartz, A.L.; Kahana, C.; Ciechanover, A. Basal and human papillomavirus E6 oncoprotein-induced degradation of Myc proteins by the ubiquitin pathway. *Proc. Natl. Acad. Sci. USA* **1998**, *95*, 8058–8063. [CrossRef] [PubMed]
129. Oh, S.T.; Kyo, S.; Laimins, L.A. Telomerase Activation by Human Papillomavirus Type 16 E6 Protein: Induction of Human Telomerase Reverse Transcriptase Expression through Myc and GC-Rich Sp1 Binding Sites. *J. Virol.* **2001**, *75*, 5559–5566. [CrossRef] [PubMed]
130. Veldman, T.; Horikawa, I.; Barrett, J.C.; Schlegel, R. Transcriptional activation of the telomerase hTERT gene by human papillomavirus type 16 E6 oncoprotein. *J. Virol.* **2001**, *75*, 4467–4472. [CrossRef] [PubMed]
131. Rajagopalan, D.; Pandey, A.K.; Xiuzhen, M.C.; Lee, K.K.; Hora, S.; Zhang, Y.; Chua, B.H.; Kwok, H.S.; Bhatia, S.S.; Deng, L.W.; et al. TIP60 represses telomerase expression by inhibiting Sp1 binding to the TERT promoter. *PLoS Pathog.* **2017**, *13*, e1006681. [CrossRef] [PubMed]
132. Liu, X.; Dakic, A.; Zhang, Y.; Dai, Y.; Chen, R.; Schlegel, R. HPV E6 protein interacts physically and functionally with the cellular telomerase complex. *Proc. Natl. Acad. Sci. USA* **2009**, *106*, 18780–18785. [CrossRef] [PubMed]
133. Gewin, L.; Myers, H.; Kiyono, T.; Galloway, D.A. Identification of a novel telomerase repressor that interacts with the human papillomavirus type-16 E6/E6-AP complex. *Genes Dev.* **2004**, *18*, 2269–2282. [CrossRef] [PubMed]
134. McMurray, H.R.; McCance, D.J. Human papillomavirus type 16 E6 activates TERT gene transcription through induction of c-Myc and release of USF-mediated repression. *J. Virol.* **2003**, *77*, 9852–9861. [CrossRef] [PubMed]
135. Westphal, D.; Dewson, G.; Czabotar, P.E.; Kluck, R.M. Molecular biology of Bax and Bak activation and action. *Biochim. Biophys. Acta* **2011**, *1813*, 521–531. [CrossRef] [PubMed]
136. Thomas, M.; Banks, L. Human papillomavirus (HPV) E6 interactions with Bak are conserved amongst E6 proteins from high and low risk HPV types. *J. Gen. Virol.* **1999**, *80*, 1513–1517. [CrossRef] [PubMed]
137. Vogt, M.; Butz, K.; Dymalla, S.; Semzow, J.; Hoppe-Seyler, F. Inhibition of Bax activity is crucial for the antiapoptotic function of the human papillomavirus E6 oncoprotein. *Oncogene* **2006**, *25*, 4009–4015. [CrossRef] [PubMed]
138. Basile, J.R.; Zacny, V.; Munger, K. The cytokines tumor necrosis factor-alpha (TNF-alpha) and TNF-related apoptosis-inducing ligand differentially modulate proliferation and apoptotic pathways in human keratinocytes expressing the human papillomavirus-16 E7 oncoprotein. *J. Biol. Chem.* **2001**, *276*, 22522–22528. [CrossRef] [PubMed]
139. Demers, G.W.; Halbert, C.L.; Galloway, D.A. Elevated wild-type p53 protein levels in human epithelial cell lines immortalized by the human papillomavirus type 16 E7 gene. *Virology* **1994**, *198*, 169–174. [CrossRef] [PubMed]
140. Jones, D.L.; Thompson, D.A.; Munger, K. Destabilization of the RB tumor suppressor protein and stabilization of p53 contribute to HPV type 16 E7-induced apoptosis. *Virology* **1997**, *239*, 97–107. [CrossRef] [PubMed]
141. Seavey, S.E.; Holubar, M.; Saucedo, L.J.; Perry, M.E. The E7 oncoprotein of human papillomavirus type 16 stabilizes p53 through a mechanism independent of p19(ARF). *J. Virol.* **1999**, *73*, 7590–7598. [PubMed]
142. Nor Rashid, N.; Yusof, R.; Watson, R.J. Disruption of repressive p130-DREAM complexes by human papillomavirus 16 E6/E7 oncoproteins is required for cell-cycle progression in cervical cancer cells. *J. Gen. Virol.* **2011**, *92*, 2620–2627. [CrossRef] [PubMed]
143. Nor Rashid, N.; Yusof, R.; Watson, R.J. Disruption of pocket protein dream complexes by E7 proteins of different types of human papillomaviruses. *Acta Virol.* **2013**, *57*, 447–451. [CrossRef] [PubMed]

144. DeCaprio, J.A. Human papillomavirus type 16 E7 perturbs DREAM to promote cellular proliferation and mitotic gene expression. *Oncogene* **2014**, *33*, 4036–4038. [CrossRef] [PubMed]
145. Mannefeld, M.; Klassen, E.; Gaubatz, S. B-MYB is required for recovery from the DNA damage-induced G2 checkpoint in p53 mutant cells. *Cancer Res.* **2009**, *69*, 4073–4080. [CrossRef] [PubMed]
146. Quaas, M.; Muller, G.A.; Engeland, K. p53 can repress transcription of cell cycle genes through a p21(WAF1/CIP1)-dependent switch from MMB to DREAM protein complex binding at CHR promoter elements. *Cell Cycle* **2012**, *11*, 4661–4672. [CrossRef] [PubMed]
147. Zhang, B.; Chen, W.; Roman, A. The E7 proteins of low- and high-risk human papillomaviruses share the ability to target the pRB family member p130 for degradation. *Proc. Natl. Acad. Sci. USA* **2006**, *103*, 437–442. [CrossRef] [PubMed]
148. Fischer, M.; Quaas, M.; Wintsche, A.; Muller, G.A.; Engeland, K. Polo-like kinase 4 transcription is activated via CRE and NRF1 elements, repressed by DREAM through CDE/CHR sites and deregulated by HPV E7 protein. *Nucl. Acids Res.* **2014**, *42*, 163–180. [CrossRef] [PubMed]
149. Fischer, M.; Uxa, S.; Stanko, C.; Magin, T.M.; Engeland, K. Human papilloma virus E7 oncoprotein abrogates the p53-p21-DREAM pathway. *Sci. Rep.* **2017**, *7*, 2603. [CrossRef] [PubMed]
150. Fischer, M.; Grossmann, P.; Padi, M.; DeCaprio, J.A. Integration of TP53, DREAM, MMB-FOXM1 and RB-E2F target gene analyses identifies cell cycle gene regulatory networks. *Nucl. Acids Res.* **2016**, *44*, 6070–6086. [CrossRef] [PubMed]
151. Fischer, M. p21 governs p53's repressive side. *Cell Cycle* **2016**, *15*, 2852–2853. [CrossRef] [PubMed]
152. Filippova, M.; Song, H.; Connolly, J.L.; Dermody, T.S.; Duerksen-Hughes, P.J. The human papillomavirus 16 E6 protein binds to tumor necrosis factor (TNF) R1 and protects cells from TNF-induced apoptosis. *J. Biol. Chem.* **2002**, *277*, 21730–21739. [CrossRef] [PubMed]
153. Filippova, M.; Parkhurst, L.; Duerksen-Hughes, P.J. The human papillomavirus 16 E6 protein binds to Fas-associated death domain and protects cells from Fas-triggered apoptosis. *J. Biol. Chem.* **2004**, *279*, 25729–25744. [CrossRef] [PubMed]
154. Garnett, T.O.; Filippova, M.; Duerksen-Hughes, P.J. Accelerated degradation of FADD and procaspase 8 in cells expressing human papilloma virus 16 E6 impairs TRAIL-mediated apoptosis. *Cell Death Differ.* **2006**, *13*, 1915–1926. [CrossRef] [PubMed]
155. Hemmi, H.; Takeuchi, O.; Kawai, T.; Kaisho, T.; Sato, S.; Sanjo, H.; Matsumoto, M.; Hoshino, K.; Wagner, H.; Takeda, K.; et al. A Toll-like receptor recognizes bacterial DNA. *Nature* **2000**, *408*, 740–745. [CrossRef] [PubMed]
156. Hasan, U.A.; Bates, E.; Takeshita, F.; Biliato, A.; Accardi, R.; Bouvard, V.; Mansour, M.; Vincent, I.; Gissmann, L.; Iftner, T.; et al. TLR9 expression and function is abolished by the cervical cancer-associated human papillomavirus type 16. *J. Immunol.* **2007**, *178*, 3186–3197. [CrossRef] [PubMed]
157. Fiola, S.; Gosselin, D.; Takada, K.; Gosselin, J. TLR9 contributes to the recognition of EBV by primary monocytes and plasmacytoid dendritic cells. *J. Immunol.* **2010**, *185*, 3620–3631. [CrossRef] [PubMed]
158. Lund, J.; Sato, A.; Akira, S.; Medzhitov, R.; Iwasaki, A. Toll-like receptor 9-mediated recognition of Herpes simplex virus-2 by plasmacytoid dendritic cells. *J. Exp. Med.* **2003**, *198*, 513–520. [CrossRef] [PubMed]
159. Ewald, S.E.; Engel, A.; Lee, J.; Wang, M.; Bogyo, M.; Barton, G.M. Nucleic acid recognition by Toll-like receptors is coupled to stepwise processing by cathepsins and asparagine endopeptidase. *J. Exp. Med.* **2011**, *208*, 643–651. [CrossRef] [PubMed]
160. Hasan, U.A.; Zannetti, C.; Parroche, P.; Goutagny, N.; Malfroy, M.; Roblot, G.; Carreira, C.; Hussain, I.; Muller, M.; Taylor-Papadimitriou, J.; et al. The human papillomavirus type 16 E7 oncoprotein induces a transcriptional repressor complex on the Toll-like receptor 9 promoter. *J. Exp. Med.* **2013**, *210*, 1369–1387. [CrossRef] [PubMed]
161. Lau, L.; Gray, E.E.; Brunette, R.L.; Stetson, D.B. DNA tumor virus oncogenes antagonize the cGAS-STING DNA-sensing pathway. *Science* **2015**, *350*, 568–571. [CrossRef] [PubMed]
162. Karim, R.; Meyers, C.; Backendorf, C.; Ludigs, K.; Offringa, R.; van Ommen, G.J.; Melief, C.J.; van der Burg, S.H.; Boer, J.M. Human papillomavirus deregulates the response of a cellular network comprising of chemotactic and proinflammatory genes. *PLoS ONE* **2011**, *6*, e17848. [CrossRef] [PubMed]
163. Reiser, J.; Hurst, J.; Voges, M.; Krauss, P.; Munch, P.; Iftner, T.; Stubenrauch, F. High-risk human papillomaviruses repress constitutive kappa interferon transcription via E6 to prevent pathogen recognition receptor and antiviral-gene expression. *J. Virol.* **2011**, *85*, 11372–11380. [CrossRef] [PubMed]

164. Rincon-Orozco, B.; Halec, G.; Rosenberger, S.; Muschik, D.; Nindl, I.; Bachmann, A.; Ritter, T.M.; Dondog, B.; Ly, R.; Bosch, F.X.; et al. Epigenetic silencing of interferon-kappa in human papillomavirus type 16-positive cells. *Cancer Res.* **2009**, *69*, 8718–8725. [CrossRef] [PubMed]
165. Cortes Gutierrez, E.I.; Garcia-Vielma, C.; Aguilar-Lemarroy, A.; Vallejo-Ruiz, V.; Pina-Sanchez, P.; Zapata-Benavides, P.; Gosalvez, J. Expression of the HPV18/E6 oncoprotein induces DNA damage. *Eur. J. Histochem.* **2017**, *61*, 2773. [CrossRef] [PubMed]
166. Bester, A.C.; Roniger, M.; Oren, Y.S.; Im, M.M.; Sarni, D.; Chaoat, M.; Bensimon, A.; Zamir, G.; Shewach, D.S.; Kerem, B. Nucleotide deficiency promotes genomic instability in early stages of cancer development. *Cell* **2011**, *145*, 435–446. [CrossRef] [PubMed]
167. Pihan, G.A.; Purohit, A.; Wallace, J.; Knecht, H.; Woda, B.; Quesenberry, P.; Doxsey, S.J. Centrosome defects and genetic instability in malignant tumors. *Cancer Res.* **1998**, *58*, 3974–3985. [PubMed]
168. Duensing, S.; Duensing, A.; Crum, C.P.; Munger, K. Human papillomavirus type 16 E7 oncoprotein-induced abnormal centrosome synthesis is an early event in the evolving malignant phenotype. *Cancer Res.* **2001**, *61*, 2356–2360. [PubMed]
169. Duensing, S.; Lee, L.Y.; Duensing, A.; Basile, J.; Piboonniyom, S.; Gonzalez, S.; Crum, C.P.; Munger, K. The human papillomavirus type 16 E6 and E7 oncoproteins cooperate to induce mitotic defects and genomic instability by uncoupling centrosome duplication from the cell division cycle. *Proc. Natl. Acad. Sci. USA* **2000**, *97*, 10002–10007. [CrossRef] [PubMed]
170. Duensing, S.; Munger, K. The human papillomavirus type 16 E6 and E7 oncoproteins independently induce numerical and structural chromosome instability. *Cancer Res.* **2002**, *62*, 7075–7082. [PubMed]
171. Warren, C.J.; Xu, T.; Guo, K.; Griffin, L.M.; Westrich, J.A.; Lee, D.; Lambert, P.F.; Santiago, M.L.; Pyeon, D. APOBEC3A functions as a restriction factor of human papillomavirus. *J. Virol.* **2015**, *89*, 688–702. [CrossRef] [PubMed]
172. Vieira, V.C.; Leonard, B.; White, E.A.; Starrett, G.J.; Temiz, N.A.; Lorenz, L.D.; Lee, D.; Soares, M.A.; Lambert, P.F.; Howley, P.M.; et al. Human papillomavirus E6 triggers upregulation of the antiviral and cancer genomic DNA deaminase APOBEC3B. *MBio* **2014**, *5*. [CrossRef] [PubMed]
173. Warren, C.J.; Westrich, J.A.; Doorslaer, K.V.; Pyeon, D. Roles of APOBEC3A and APOBEC3B in Human Papillomavirus Infection and Disease Progression. *Viruses* **2017**, *9*. [CrossRef] [PubMed]
174. Sheehy, A.M.; Gaddis, N.C.; Choi, J.D.; Malim, M.H. Isolation of a human gene that inhibits HIV-1 infection and is suppressed by the viral Vif protein. *Nature* **2002**, *418*, 646–650. [CrossRef] [PubMed]
175. Zheng, Y.H.; Irwin, D.; Kurosu, T.; Tokunaga, K.; Sata, T.; Peterlin, B.M. Human APOBEC3F is another host factor that blocks human immunodeficiency virus type 1 replication. *J. Virol.* **2004**, *78*, 6073–6076. [CrossRef] [PubMed]
176. Hultquist, J.F.; Lengyel, J.A.; Refsland, E.W.; LaRue, R.S.; Lackey, L.; Brown, W.L.; Harris, R.S. Human and rhesus APOBEC3D, APOBEC3F, APOBEC3G, and APOBEC3H demonstrate a conserved capacity to restrict Vif-deficient HIV-1. *J. Virol.* **2011**, *85*, 11220–11234. [CrossRef] [PubMed]
177. Mori, S.; Takeuchi, T.; Ishii, Y.; Yugawa, T.; Kiyono, T.; Nishina, H.; Kukimoto, I. Human Papillomavirus 16 E6 Upregulates APOBEC3B via the TEAD Transcription Factor. *J. Virol.* **2017**, *91*. [CrossRef] [PubMed]
178. Westrich, J.A.; Warren, C.J.; Klausner, M.J.; Guo, K.; Liu, C.W.; Santiago, M.L.; Pyeon, D. Human Papillomavirus 16 E7 Stabilizes APOBEC3A Protein by Inhibiting Cullin 2-Dependent Protein Degradation. *J. Virol.* **2018**, *92*. [CrossRef] [PubMed]
179. Warren, C.J.; Van Doorslaer, K.; Pandey, A.; Espinosa, J.M.; Pyeon, D. Role of the host restriction factor APOBEC3 on papillomavirus evolution. *Virus Evol.* **2015**, *1*. [CrossRef] [PubMed]
180. Zhang, Y.; Koneva, L.A.; Virani, S.; Arthur, A.E.; Virani, A.; Hall, P.B.; Warden, C.D.; Carey, T.E.; Chepeha, D.B.; Prince, M.E.; et al. Subtypes of HPV-Positive Head and Neck Cancers Are Associated with HPV Characteristics, Copy Number Alterations, PIK3CA Mutation, and Pathway Signatures. *Clin. Cancer Res.* **2016**, *22*, 4735–4745. [CrossRef] [PubMed]
181. Nichols, A.C.; Palma, D.A.; Chow, W.; Tan, S.; Rajakumar, C.; Rizzo, G.; Fung, K.; Kwan, K.; Wehrli, B.; Winquist, E.; et al. High frequency of activating PIK3CA mutations in human papillomavirus-positive oropharyngeal cancer. *JAMA Otolaryngol. Head Neck Surg.* **2013**, *139*, 617–622. [CrossRef] [PubMed]
182. Siddiqui, M.A.; Perry, C.M. Human papillomavirus quadrivalent (types 6, 11, 16, 18) recombinant vaccine (Gardasil). *Drugs* **2006**, *66*, 1263–1271. [CrossRef] [PubMed]

183. Donovan, B.; Franklin, N.; Guy, R.; Grulich, A.E.; Regan, D.G.; Ali, H.; Wand, H.; Fairley, C.K. Quadrivalent human papillomavirus vaccination and trends in genital warts in Australia: Analysis of national sentinel surveillance data. *Lancet Infect. Dis.* **2011**, *11*, 39–44. [CrossRef]
184. Korostil, I.A.; Ali, H.; Guy, R.J.; Donovan, B.; Law, M.G.; Regan, D.G. Near elimination of genital warts in Australia predicted with extension of human papillomavirus vaccination to males. *Sex Trans. Dis.* **2013**, *40*, 833–835. [CrossRef] [PubMed]
185. Schiller, J.T.; Castellsague, X.; Garland, S.M. A review of clinical trials of human papillomavirus prophylactic vaccines. *Vaccine* **2012**, *30* (Suppl. 5), F123–F138. [CrossRef] [PubMed]
186. Joura, E.A.; Giuliano, A.R.; Iversen, O.E.; Bouchard, C.; Mao, C.; Mehlsen, J.; Moreira, E.D., Jr.; Ngan, Y.; Petersen, L.K.; Lazcano-Ponce, E.; et al. A 9-valent HPV vaccine against infection and intraepithelial neoplasia in women. *N. Engl. J. Med.* **2015**, *372*, 711–723. [CrossRef] [PubMed]
187. Jung, H.S.; Rajasekaran, N.; Ju, W.; Shin, Y.K. Human Papillomavirus: Current and Future RNAi Therapeutic Strategies for Cervical Cancer. *J. Clin. Med.* **2015**, *4*, 1126–1155. [CrossRef] [PubMed]
188. Shukla, S.; Bharti, A.C.; Hussain, S.; Mahata, S.; Hedau, S.; Kailash, U.; Kashyap, V.; Bhambhani, S.; Roy, M.; Batra, S.; et al. Elimination of high-risk human papillomavirus type HPV16 infection by 'Praneem' polyherbal tablet in women with early cervical intraepithelial lesions. *J. Cancer Res. Clin. Oncol.* **2009**, *135*, 1701–1709. [CrossRef] [PubMed]
189. Ferrara, N.; Hillan, K.J.; Novotny, W. Bevacizumab (Avastin), a humanized anti-VEGF monoclonal antibody for cancer therapy. *Biochem. Biophys. Res. Commun.* **2005**, *333*, 328–335. [CrossRef] [PubMed]
190. Tewari, K.S.; Sill, M.W.; Penson, R.T.; Huang, H.; Ramondetta, L.M.; Landrum, L.M.; Oaknin, A.; Reid, T.J.; Leitao, M.M.; Michael, H.E.; et al. Bevacizumab for advanced cervical cancer: Final overall survival and adverse event analysis of a randomised, controlled, open-label, phase 3 trial (Gynecologic Oncology Group 240). *Lancet* **2017**, *390*, 1654–1663. [CrossRef]

Permissions

All chapters in this book were first published by MDPI; hereby published with permission under the Creative Commons Attribution License or equivalent. Every chapter published in this book has been scrutinized by our experts. Their significance has been extensively debated. The topics covered herein carry significant findings which will fuel the growth of the discipline. They may even be implemented as practical applications or may be referred to as a beginning point for another development.

The contributors of this book come from diverse backgrounds, making this book a truly international effort. This book will bring forth new frontiers with its revolutionizing research information and detailed analysis of the nascent developments around the world.

We would like to thank all the contributing authors for lending their expertise to make the book truly unique. They have played a crucial role in the development of this book. Without their invaluable contributions this book wouldn't have been possible. They have made vital efforts to compile up to date information on the varied aspects of this subject to make this book a valuable addition to the collection of many professionals and students.

This book was conceptualized with the vision of imparting up-to-date information and advanced data in this field. To ensure the same, a matchless editorial board was set up. Every individual on the board went through rigorous rounds of assessment to prove their worth. After which they invested a large part of their time researching and compiling the most relevant data for our readers.

The editorial board has been involved in producing this book since its inception. They have spent rigorous hours researching and exploring the diverse topics which have resulted in the successful publishing of this book. They have passed on their knowledge of decades through this book. To expedite this challenging task, the publisher supported the team at every step. A small team of assistant editors was also appointed to further simplify the editing procedure and attain best results for the readers.

Apart from the editorial board, the designing team has also invested a significant amount of their time in understanding the subject and creating the most relevant covers. They scrutinized every image to scout for the most suitable representation of the subject and create an appropriate cover for the book.

The publishing team has been an ardent support to the editorial, designing and production team. Their endless efforts to recruit the best for this project, has resulted in the accomplishment of this book. They are a veteran in the field of academics and their pool of knowledge is as vast as their experience in printing. Their expertise and guidance has proved useful at every step. Their uncompromising quality standards have made this book an exceptional effort. Their encouragement from time to time has been an inspiration for everyone.

The publisher and the editorial board hope that this book will prove to be a valuable piece of knowledge for researchers, students, practitioners and scholars across the globe.

List of Contributors

Alfredo Cruz-Gregorio
Programa de Maestría y Doctorado en Ciencias Bioquímicas, Facultad de Química, Universidad Nacional Autónoma de México, Ciudad Universitaria, Mexico City 04510, Mexico

Adela Carrillo-García, Adriana Contreras-Paredes and Elizabeth Ortiz-Sánchez
Unidad de Investigación Biomédica en Cáncer, Instituto Nacional de Cancerología (INCan)/Instituto de Investigaciones Biomédicas, Universidad Nacional Autónoma de México (UNAM), Mexico City 14080, Mexico

Imelda Martínez-Ramírez
Programa de Maestría y Doctorado en Ciencias Bioquímicas, Facultad de Química, Universidad Nacional Autónoma de México, Ciudad Universitaria, Mexico City 04510, Mexico
Unidad de Investigación Biomédica en Cáncer, Instituto Nacional de Cancerología (INCan)/Instituto de Investigaciones Biomédicas, Universidad Nacional Autónoma de México (UNAM), Mexico City 14080, Mexico

Marcela Lizano
Unidad de Investigación Biomédica en Cáncer, Instituto Nacional de Cancerología (INCan)/Instituto de Investigaciones Biomédicas, Universidad Nacional Autónoma de México (UNAM), Mexico City 14080, Mexico
Departamento de Medicina Genómica y Toxicología Ambiental, Instituto de Investigaciones Biomédicas, Universidad Nacional Autónoma de México (UNAM), Mexico City 04510, Mexico

Ugo Moens
Department of Medical Biology, Faculty of Health Sciences, University of Tromsø—The Arctic University of Norway, 9037 Tromsø, Norway
Molecular Inflammation Research Group, Department of Medical Biology, Faculty of Health Sciences, University of Tromsø, 9037 Tromsø, Norway

Valeria Pietropaolo
Department of Public Health and Infectious Diseases, "Sapienza" University of Rome, 00185 Rome, Italy

Carla Prezioso
Department of Public Health and Infectious Diseases, "Sapienza" University of Rome, 00185 Rome, Italy
IRCSS San Raffaele Pisana, Microbiology of Chronic Neuro-degenerative Pathologies, 00160 Rome, Italy

Lucia Signorini, Maria Dolci, Rosalia Ticozzi, Pasquale Ferrante and Serena Delbue
Biomedical, Surgical and Dental Sciences, University of Milano, 20133 Milano, Italy

Evaldo Favi and Mariano Ferraresso
Department of Clinical Sciences and Community Health, University of Milano, 20122 Milano, Italy
Kidney Transplantation, Fondazione IRCCS Ca' Granda, Ospedale Maggiore Policlinico, 20122 Milano, Italy

Caterina Colico
Kidney Transplantation, Fondazione IRCCS Ca' Granda, Ospedale Maggiore Policlinico, 20122 Milano, Italy

Giuseppe Basile
Service of Legal Medicine, San Siro Clinical Institute, 20148 Milano, Italy

Kersti Nilsson, Chengjun Wu and Stefan Schwartz
Department of Laboratory Medicine, Lund University, 221 84 Lund, Sweden

Bartłomiej Drop
Department of Information Technology and Medical Statistics, Medical University of Lublin, 20-059 Lublin, Poland

Małgorzata Strycharz-Dudziak
Chair and Department of Conservative Dentistry with Endodontics, Medical University of Lublin, 20-059 Lublin, Poland

Ewa Kliszczewska and Małgorzata Polz-Dacewicz
Department of Virology, Medical University of Lublin, 20-059 Lublin, Poland

Katrin Carow, Mandy Gölitz, Maria Wolf, Norman Häfner, Lars Jansen, Ingo B. Runnebaum and Matthias Dürst
Department of Gynecology, Jena University Hospital—Friedrich Schiller University Jena, 07747 Jena, Germany

Heike Hoyer
Institute of Medical Statistics, Information Sciences and Documentation, Jena University Hospital—Friedrich Schiller University Jena, 07743 Jena, Germany

Elisabeth Schwarz
Research Program Infection and Cancer, DKFZ, 69120 Heidelberg, Germany

Domenico Mattoscio
Department of Medical, Oral and Biotechnology Science, University of Chieti-Pescara, 66100 Chieti, Italy
Center on Aging Science and Translational Medicine (CeSI-MeT), University of Chieti-Pescara, 66100 Chieti, Italy

Alessandro Medda and Susanna Chiocca
Department of Experimental Oncology, European Institute of Oncology, 20139 Milan, Italy

Kimon Chatzistamatiou
2nd Department of Obstetrics and Gynecology, Aristotle University of Thessaloniki, Hippokratio General Hospital, 54642 Thessaloniki, Greece

Theodoros Moysiadis
Institute of Applied Biosciences, Centre for Research & Technology-Hellas, 57001 Thessaloniki, Greece

Dimos Vryzas
Medical School, Democritus University of Thrace, 68100 Alexandroupolis, Greece

Ekaterini Chatzaki
Laboratory of Pharmacology, Medical School, Democritus University of Thrace, 68100 Alexandroupolis, Greece

Andreas M. Kaufmann
Department of Gynecology, Charité-Universitaetsmedizin Berlin, Campus Benjamin Franklin, 10117 Berlin, Germany

Isabel Koch, Erwin Soutschek and Oliver Boecher
Mikrogen GmbH, 82061 Neuried, Germany

Athena Tsertanidou and Theodoros Agorastos
4th Department of Obstetrics and Gynecology, Aristotle University of Thessaloniki, Hippokratio General Hospital, 54642 Thessaloniki, Greece

Nikolaos Maglaveras
Lab of Computing and Medical Informatics, Department of Medicine, Aristotle University of Thessaloniki, 54124 Thessaloniki, Greece

Pidder Jansen-Duerr
Research Institute for Biomedical Aging Research, University of Innsbruck, A-6020 Innsbruck, Austria

Xiaobo Song
Host Microbe Interaction Research Group, Department of Medical Biology, Faculty of Health Sciences, University of Tromsø, 9037 Tromsø, Norway

Marijke Van Ghelue
Department of Medical Genetics, University Hospital Northern-Norway, 9038 Tromsø, Norway

John A. Lednicky
Department of Environmental and Global Health, College of Public Health and Health Professions, University of Florida, Gainesville, FL 32603, USA

Bernhard Ehlers
Division 12, Measles, Mumps, Rubella and Viruses Affecting Immunocompromised Patients, Robert Koch Institute, 13353 Berlin, Germany

Henning Mothes
Department of General, Visceral and Vascular Surgery, Jena University Hospital—Friedrich-Schiller-University; 07747 Jena, Germany

Katrin Beer
Department of Gynecology, Jena University Hospital—Friedrich-Schiller-University, 07747 Jena, Germany

Anja Lidwina Geßner
Department of General, Visceral and Vascular Surgery, Jena University Hospital—Friedrich-Schiller-University; 07747 Jena, Germany
Department of Gynecology, Jena University Hospital—Friedrich-Schiller-University, 07747 Jena, Germany

Angelika Borkowetz
Department of Urology, Technische Universität Dresden; 01307 Dresden, Germany

Michael Baier
Institute for Medical Microbiology, Jena University Hospital—Friedrich-Schiller-University, 07747 Jena, Germany

Angela Göhlert
Institute for Pathology, Jena University Hospital—Friedrich-Schiller-University, 07743 Jena, Germany

Torsten J. Wilhelm
Department of Surgery, University Medical Centre Mannheim, 68167 Mannheim, Germany

Alexander Thumbs and Eric Borgstein
Department of Surgery, Queen Elizabeth Central Hospital—College of Medicine, Blantyre 3, Malawi

Torbjörn Ramqvist, Anders Näsman, Bo Franzén, Cinzia Bersani, Susanne Becker, Linnea Haeggblom and Tina Dalianis
Department of Oncology-Pathology, Karolinska Institutet, Karolinska University Hospital, 171 76 Stockholm, Sweden

Andrey Alexeyenko
Department of Microbiology, Tumor and Cell Biology (MTC), Karolinska Institutet, 171 77 Stockholm, Sweden
National Bioinformatics Infrastructure Sweden, Science for Life Laboratory, 17121 Solna, Sweden

Aeneas Kolev and Eva Munck-Wikland
Department of Clinical Science and Technology (CLINTEC), Karolinska Institutet, Karolinska University Hospital, 171 76 Stockholm, Sweden

Giovanni Barillari
Department of Clinical Sciences and Translational Medicine, University of Rome Tor Vergata, 1 via Montpellier, 00133 Rome, Italy

Paolo Monini, Cecilia Sgadari and Barbara Ensoli
National HIV/AIDS Research Center, Istituto Superiore di Sanità, 299 viale Regina Elena, 00161 Rome, Italy

Margarita-Maria Panou, Emma L. Prescott, Daniel L. Hurdiss, Gemma Swinscoe, Ethan L. Morgan, Marietta Müller, Michelle Antoni, David Kealy, Neil A. Ranson and Andrew Macdonald
Faculty of Biological Sciences and Astbury Centre for Structural and Molecular Biology, University of Leeds, Leeds LS2 9JT, UK

Michael Hollinshead, Laura G. Caller, Louisa Carlisle and Colin M. Crump
Department of Pathology, University of Cambridge, Tennis Court Road, Cambridge CB2 1QP, UK

Nicole S. L. Yeo-The and Yoshiaki Ito
Cancer Science Institute of Singapore, National University of Singapore, Singapore 117599, Singapore
NUS Graduate School for Integrative Sciences and Engineering, National University of Singapore, Singapore 117456, Singapore

Sudhakar Jha
Cancer Science Institute of Singapore, National University of Singapore, Singapore 117599, Singapore
Department of Biochemistry, Yong Loo Lin School of Medicine, National University of Singapore, Singapore 117596, Singapore

Index

www.ingramcontent.com/pod-product-compliance
Lightning Source LLC
LaVergne TN
LVHW080956210926
848485LV00004B/839